AAPG TREATISE OF PETROLEUM GEOLOGY

The American Association of Petroleum Geologists
gratefully acknowledges and appreciates the leadership and support
of the AAPG Foundation in the development of the
Treatise of Petroleum Geology

STRUCTURAL TRAPS VI

COMPILED BY
NORMAN H. FOSTER
AND
EDWARD A. BEAUMONT

TREATISE OF PETROLEUM GEOLOGY
ATLAS OF OIL AND GAS FIELDS

PUBLISHED BY
THE AMERICAN ASSOCIATION OF PETROLEUM GEOLOGISTS
TULSA, OKLAHOMA 74101, U.S.A.

ISBN: 0-89181-588-0
ISSN: 1043-6103

Available from:
The AAPG Bookstore
P.O. Box 979
Tulsa, OK 74101-0979

Phone: (918) 584-2555
Telex: 49-9432
FAX: (918) 584-0469

Association Editor: Susan Longacre
Science Director: Gary D. Howell
Publications Manager: Cathleen P. Williams
Special Projects Editor: Anne H. Thomas
Science Staff: William G. Brownfield
Project Production: Custom Editorial Productions, Inc.

TABLE OF CONTENTS

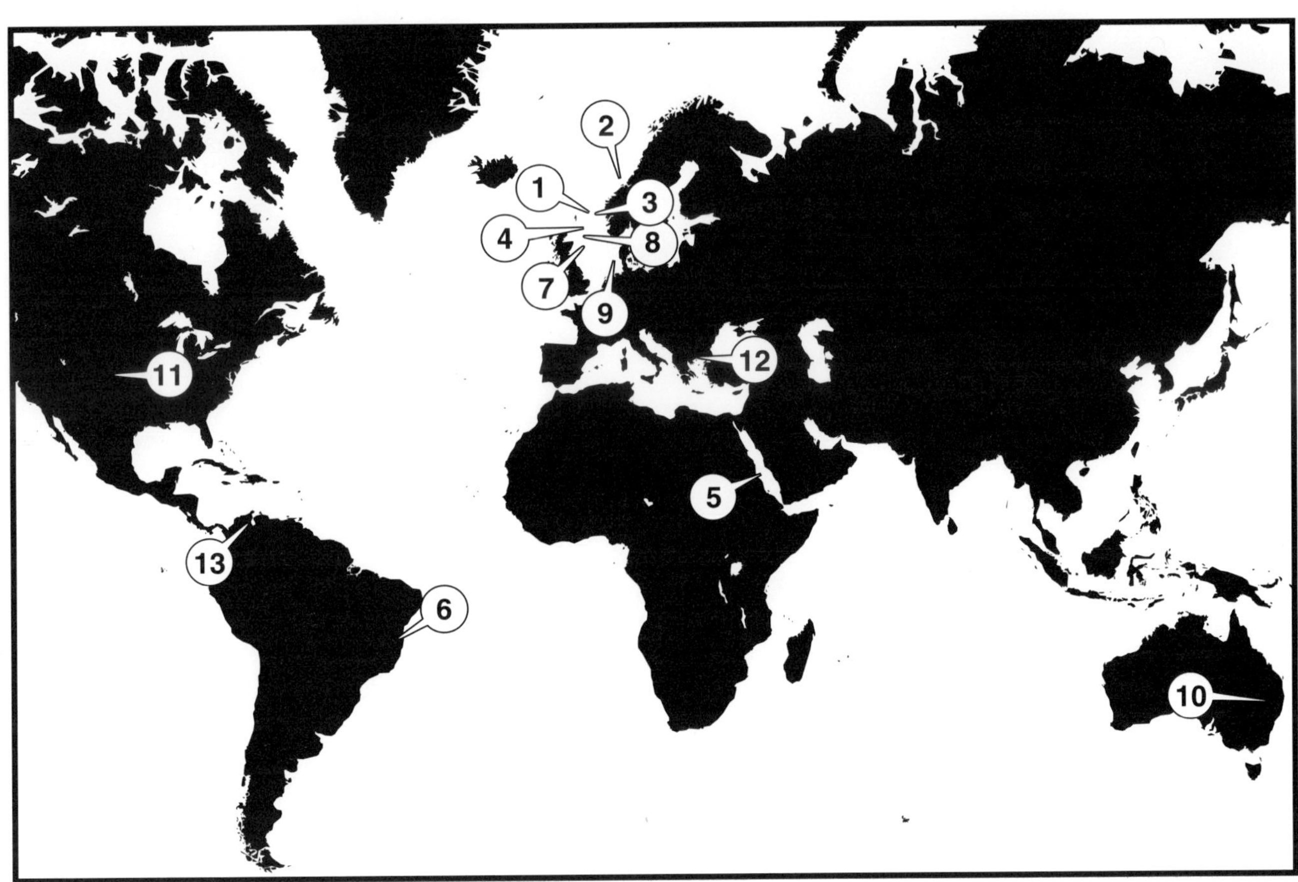

Treatise of Petroleum Geology
Advisory Board

Susan E. Palmer
Arthur J. Pansze
John M. Parker
Stephen J. Patmore
Dallas L. Peck
William H. Pelton
Alain Perrodon
James A. Peterson
R. Michael Peterson
Edward B. Picou, Jr.
Max Grow Pitcher
David E. Powley
William F. Precht
A. Pulunggono
Bailey Rascoe, Jr.
R. Randy Ray
Dudley D. Rice
Edward P. Riker
Edward C. Roy, Jr.
Eric A. Rudd
Floyd F. Sabins, Jr.
Nahum Schneidermann
Peter A. Scholle
George L. Scott, Jr.
Robert T. Sellars, Jr.
Faroog A. Sharief
John W. Shelton
Phillip W. Shoemaker
Synthia E. Smith
Robert M. Sneider
Frank P. Sonnenberg
Stephen A. Sonnenberg
William E. Speer
Ernest J. Spradlin
Bill St. John
Philip H. Stark
Richard Steinmetz
Per R. Stokke
Denise M. Stone
Donald S. Stone
Douglas K. Strickland
James V. Taranik
Harry Ter Best, Jr.
Bruce K. Thatcher, Jr.
M. Ray Thomasson
Jack C. Threet
Bernard Tissot
Don F. Tobin
Don G. Tobin
Donald F. Todd
Harrison L. Townes
M. O. Turner
Peter R. Vail
B. van Hoorn
Arthur M. Van Tyne
Kent Lee Van Zant
Ian R. Vann
Harry K. Veal*
Steven L. Veal
Richard R. Vincelette
Fred J. Wagner, Jr.
William A. Walker, Jr.
Carol A. Walsh
Anthony Walton
Douglas W. Waples
Harry W. Wassall, III
W. Lynn Watney
N. L. Watts
Koenradd J. Weber
Robert J. Weimer
Dietrich H. Welte
Alun H. Whittaker
James E. Wilson, Jr.
Thomas Wilson
John R. Wingert
Martha O. Withjack
P. W. J. Wood
Homer O. Woodbury
Walter W. Wornardt
Marcelo R. Yrigoyen
Mehmet A. Yukler
Zhai Guangming
Robert Zinke

* Deceased

American Association of Petroleum Geologists Foundation
Treatise of Petroleum Geology Fund*

Major Corporate Contributors
($25,000 or more)

Amoco Production Company
BP Exploration Company Limited
Chevron Corporation
Exxon Company, U.S.A.
Mobil Oil Corporation
Oryx Energy Company
Pennzoil Exploration and Production Company
Shell Oil Company
Texaco Foundation
Union Pacific Foundation
Unocal Corporation

Other Corporate Contributors
($5,000 to $25,000)

ARCO Oil & Gas Company
Ashland Oil, Inc.
Cabot Oil & Gas Corporation
Canadian Hunter Exploration Ltd.
Conoco Inc.
Marathon Oil Company
The McGee Foundation, Inc.
Phillips Petroleum Company
Transco Energy Company
Union Texas Petroleum Corporation

Major Individual Contributors
($1,000 or more)

John J. Amoruso
Thornton E. Anderson
C. Hayden Atchison
Richard A. Baile
Richard R. Bloomer
A. S. Bonner, Jr.
David G. Campbell
Herbert G. Davis
George A. Donnelly, Jr.
Paul H. Dudley, Jr.
Lewis G. Fearing
Lawrence W. Funkhouser
James A. Gibbs
George R. Gibson
William E. Gipson
Mrs. Vito A. (Mary Jane) Gotautas
Robert D. Gunn
Merrill W. Haas
Cecil V. Hagen
Frank W. Harrison
William A. Heck
Roy M. Huffington
J. R. Jackson, Jr.
Harrison C. Jamison
Thomas N. Jordan, Jr.
Hugh M. Looney
Jack P. Martin
John W. Mason
George B. McBride
Dean A. McGee
John R. McMillan
Lee Wayne Moore
Grover E. Murray
Rudolf B. Siegert
Robert M. Sneider
Estate of Mrs. John (Elizabeth) Teagle
Jack C. Threet
Charles Weiner
Harry Westmoreland
James E. Wilson, Jr.
P. W. J. Wood

The Foundation also gratefully acknowledges the many who have supported this endeavor with additional contributions.

*Based on contributions received as of September 30, 1992.

PREFACE

The Atlas of Oil and Gas Fields and the Treatise of Petroleum Geology

The *Treatise of Petroleum Geology* was conceived during a discussion held at the annual AAPG meeting in 1984 in San Antonio, Texas. This discussion led to the conviction that AAPG should publish a state-of-the-art textbook in petroleum geology, aimed not at the student, but at the practicing petroleum geologist. The textbook gradually evolved into a series of three different publications: the Reprint Series, the Atlas of Oil and Gas Fields, and the Handbook of Petroleum Geology. Collectively these publications are known as the *Treatise of Petroleum Geology*, AAPG's Diamond Jubilee project commemorating the Association's 75th anniversary in 1991.

With input from the Advisory Board of the Treatise of Petroleum Geology, we designed this set of publications to represent, to the degree possible, the cutting edge in petroleum exploration knowledge and application: the Reprint Series to provide useful and important published literature; the Atlas to comprise a collection of detailed field studies that illustrate the many ways oil and gas are trapped and to serve as a guide to the petroleum geology of basins where these fields are found; and the Handbook as a professional explorationist's guide to the latest knowledge in the various areas of petroleum geology and related disciplines.

The Treatise Atlas is part of AAPG's long tradition of publishing field studies. Notable AAPG field study compilations include *Structure of Typical American Fields*, published in 1929 and edited by Sidney Powers; and Memoir 30, *Giant Fields of 1968-1978*, published in 1981 and edited by Michel T. Halbouty. The Treatise Atlas continues that tradition but introduces a format designed for easier access to data.

Hundreds of geologists participated in this first compilation of the Atlas. Authors are from all parts of the industry and numerous countries. We gratefully acknowledge the generous contribution of their knowledge, resources, and time.

Purpose of the Atlas

The purpose of the Atlas is twofold: (1) to help exploration and development geologists become more efficient by increasing their awareness of the ways oil and gas are trapped, and (2) to serve as a reference for both the petroleum geology of the fields described and the basins in which they occur.

Imagination is the primary tool of the explorationist. Wallace E. Pratt once said that the unfound field must first be sought in the mind. In part, what is imagined is based on what is remembered; memory is the direct link to what is created in the mind. To create ideas that lead to the discovery of new fields, the mind of the geologist builds from its knowledge of petroleum geology. To that end, the Atlas of field studies will be a primary source for locating much of the information necessary for creating prospects and will provide a connection to the phenomenon of oil and gas traps.

Next to the firsthand experience of having prospects tested with the drill bit, studying the many facets and concepts of developed fields is perhaps the best way for the geologist to develop the ability to create plays and prospects. Also, familiarity with the many ways oil and gas are trapped allows the geologist to see through the noise inherent to exploration data and to close gaps in that data.

Format of the Atlas

To facilitate data access, all field studies in the Atlas follow the same format. Once users become familiar with this format, they will know where to look for the information they seek. Different fields from different parts of the world can be easily compared and contrasted.

The following is a generalized format outline for field studies in the Atlas:

- Location
- History
  - Pre-Discovery
  - Discovery
  - Post-Discovery
- Discovery Method
- Structure
  - Tectonic History
  - Regional Structure
  - Local Structure
- Stratigraphy
- Trap
  - General Description
  - Reservoir(s)
  - Source(s)
- Exploration Concepts

Criteria for Inclusion of a Field

Fields described in the Atlas are selected using two main criteria: (1) trap type, and (2) geographic distribution. Our ultimate goal for the Atlas is to include a field study from each major petroleum-producing province and to include an example of each known trap type. Size or economic importance are not, of themselves, criteria. Many fields that are not giants are included because they are geologically unique, because they are significant examples of geological investigation and original thinking, or because they are historically important, having led to the discovery of many other fields.

Grouping of Fields into Separate Volumes

We considered several ways to group fields in these volumes. We chose trap type because the purpose of the Atlas is to make exploration geologists more effective oil and gas trap finders, regardless of where they search for traps.

Grouping oil and gas field studies into separate volumes by trap type is a difficult exercise. We decided to group the fields into volumes by designating them as structural or stratigraphic traps. Most traps are a combination of both structure and stratigraphy. Some traps are obviously more a consequence of one than the other, but many are not. The continuum that exists between purely stratigraphic and purely structural traps is what makes grouping difficult. A further complication is that many fields contain more than one trap type.

Papers selected for *Structural Traps VI*

This volume in the Atlas of Oil and Gas Fields series contains studies of fields with traps that are mainly structural in nature. Like many structural traps, there is a strong component of stratigraphic control in most of the traps of these fields.

Twelve fields described in this volume are outside the United States; only one, Cahoj, is in the United States (Kansas). Seven fields are in the North Sea.

Cormorant, Draugen, Gullfaks, Piper, Ramadan, and Rio Itaúnas all contain traps associated with block faulting. Cormorant, Gullfaks, and Piper have traps in tilted fault blocks. Rio Itaúnas's trap is in a horst and Draugen's trap is an anticline at the top of a horst that has been modified by erosion.

Cod, Maureen, and Dan fields have anticlinal traps that formed as a result of salt movement. Silver Springs/Renlim is a field whose trap could easily be classified as stratigraphic. It is in an onlap of sandstones eroded from a group of granite outliers. After sand was deposited around the granite outliers, the trap's structure was enhanced by differential compaction. Cahoj's trap is in an anticline. Prinos is a faulted rollover anticline associated with a large listric normal fault. Rosario's trap is a faulted anticline.

Drilling and discovery show how closely concept matches reality. Knowing the history of discovery may help explorationists realize that problems, seemingly insoluble at one time, were eventually solved. It is also instructive to learn of the sequence of thinking that solved these problems, as well as the role of serendipity in discovery. If luck has a role in successful exploration, we still come away with the sense that creativity, logic, and knowledge guide the explorationist to those circumstances where the serendipity may occur.

Careful study of these fields will enhance the prospect generator's knowledge base and, consequently, his or her ability to apply that knowledge toward future prospecting.

Norman H. Foster
Edward A. Beaumont, Editors

Cormorant Field—U.K.
East Shetland Basin, North Sea

B. K. HOWE
Shell U.K. Exploration and Production*
Aberdeen, Scotland

FIELD CLASSIFICATION

BASIN: North Sea
BASIN TYPE: Rift
RESERVOIR ROCK TYPE: Sandstone
RESERVOIR AGE: Jurassic
PETROLEUM TYPE: Oil
TRAP TYPE: Four Tilted Fault Blocks
RESERVOIR ENVIRONMENT OF DEPOSITION: Wave-Dominated Coastal Deltaic Complex and Submarine Fan
TRAP DESCRIPTION: Complex of four tilted fault blocks where trap is in truncated reservoirs away from main bounding faults

LOCATION

The Cormorant oil field is one of the larger accumulations in the Brent Province of the U.K. northern North Sea. It is situated approximately 90 mi (150 km) northeast of the Shetland Islands in 500 ft (150 m) of water and straddles the Shell/Esso license blocks 211/21 and 211/26 (Figures 1 and 3).

The structural configuration of the northern North Sea is dominated by the north-south-trending Viking graben system. To the west of the graben lies the East Shetland basin. The Cormorant field is situated on the western side of this basin. It is bound by the Tern/Eider horst block and East Shetland platform to the west and northwest (Figure 2). Twenty-five miles (40 km) east (toward the graben center) lie the major Brent and Statfjord fields.

Hydrocarbons are produced from the Middle Jurassic Brent Group. The estimated ultimate recovery of the Cormorant field as of 1 January 1988 is 586 million bbl of oil. Thus, the Cormorant field qualifies as a giant field although it has not been ranked by Carmalt and St. John (1986).

*Current address: Brunei Shell Petroleum, Brunei

HISTORY

Pre-Discovery

After the discovery of the Brent (Jurassic), Ekofisk (Cretaceous), and Forties (Paleocene) fields, considerable attention focused on the area.

In June 1971, the British Government announced that in the fourth round allocation, 426 blocks in British waters would be offered to the petroleum industry. As an experiment, 15 of these license blocks would be awarded by competitive bidding, the first auction of blocks ever held for the North Sea.

Block 211/21 was one of the 15 auctioned and had been mapped by Shell/Esso using a regional seismic survey. A very large Paleocene structure, extending into Block 211/26 to the south, had been identified. As in the Brent area, the principal play was thought to be the large Paleocene drape structure although sub-Cretaceous tilted blocks were recognized. The Tertiary section was much thicker in the 211/21 area than that penetrated in the Brent field discovery well 211/29-1. This was interpreted to be the result of deposition of Paleocene sands, possibly analogous to those of the Forties fields (turbidites), some 125 mi (200 km) south. The combination of two possible hydrocarbon plays (Paleocene and sub-Cretaceous) and the knowledge of the Brent field discovery approximately 20 mi (35 km) to the east encouraged Shell/Esso to make a high bid of £21,050,001.

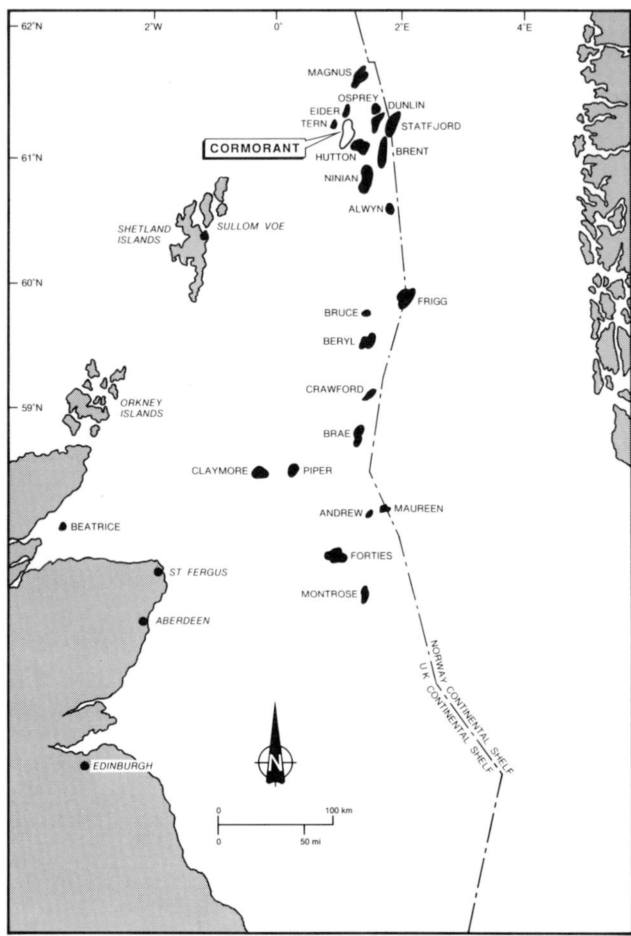

Figure 1. Location map of the oil fields in the northern North Sea, showing the position of the Cormorant field.

Bids ranged from £3,200 to the high Shell/Esso bid which exceeded all others. The total cash exposure of the industry was £135 million. Block 211/21 was awarded to Shell/Esso and became known as the "Golden Block." The block became part of production license P232, which was granted in the fourth round allocation on 15 March 1972 for an initial period of six years. Eight additional blocks were also included in the license. This gave Shell/Esso an excellent coverage of the new hydrocarbon province. The license (excluding Block 211/21) carried an obligation to drill six wells.

Regional correlations were made between 211/26 and the Brent well 211/29-1, using a seismic grid (with an average spacing of 1.2 mi or 2 km between east-west lines and 1.9 mi or 3 km between north-south lines) and a magnetic survey. Structural maps were prepared at five levels: near the top of the Paleocene, near the base of the Tertiary, "X" or Kimmerian unconformity, the first reflector beneath the unconformity (now the top of the Brent, Figure 3), and the deepest mappable reflector (now basement). Additionally, an isopach map of the Paleocene interval was constructed. This mapping supported both the Paleocene and the Jurassic plays.

The Paleocene play concept relied on two primary pieces of evidence. The first was an abrupt change in thickness of the Paleocene interval compared to that in the Brent area. This was interpreted to be possibly analogous to conditions farther south where sandstones were known to exist west of a Paleocene shelf edge having similar seismic expression. The second piece of evidence was that the stacking velocities to the X-unconformity level were somewhat higher than those observed in the 211/29 area, and it was assumed that this was due to a thicker Paleocene section as a result of change from shale (observed in 211/29-1) to sandstone.

Discovery

Well 211/26-1 (Figure 3) was spudded on 18 June 1972 (target location 61°07′43″N, 01°06′24″E). The primary objective of the well was to evaluate the large Paleocene anticline. The sub-Kimmerian structure was the second objective; it was to be tested at the highest reliably picked location given the seismic evidence of erosion on the crest (Figure 3). The well drilled through a Tertiary claystone section before encountering the Paleocene sandstones as expected at 4400 ft (1341 m) (all depths are subsea). These sandstones were of limited development, being less than 100 ft (30 m) thick. Individual sandstone porosities ranged up to 30%. No oil shows were recorded, and logs indicated no hydrocarbons. Below these sandstones, down to 8600 ft (2621 m), the Cretaceous section was predominantly shale with marls and some limestones. The X-unconformity seismic event predicted at 8860 ft (2621 m) appeared to correlate with the contact between the Lower Cretaceous and the lower Kimmeridgian organic shale observed in the Brent well 211/29-1. The Kimmeridgian organic shale overlies the upper Bathonian Heather shales at 8820 ft (2688 m), which in turn overlies the Bajocian Brent Group sandstones at 8864 ft (2702 m). A 236 ft (72 m) oil-stained sandstone section of the Brent Group was penetrated and cored. The Brent Group overlies 225 ft (69 m) of bioturbated, glauconitic marine shales of the Dunlin Group at 9100 ft (2774 m).

The only significant paleontological evidence of age in the Jurassic was the presence of *Lingulina tenera* and *Astacolus stilla* at 9180 ft (2798 m) in the Dunlin Group indicating a Sinemurian/Pliensbachian age.

A predominantly red bed sequence containing coarse-grained, poorly sorted, rounded sandstones was penetrated below the Dunlin Group between 9325 ft (2842 m) and 10,100 ft (3078 m). The sandstones had no hydrocarbon shows. This interval was underlain by a 500 ft (152 m) thick Upper Triassic (Norian age) red-brown claystone sequence. Below this a conglomerate containing angular pebbles and

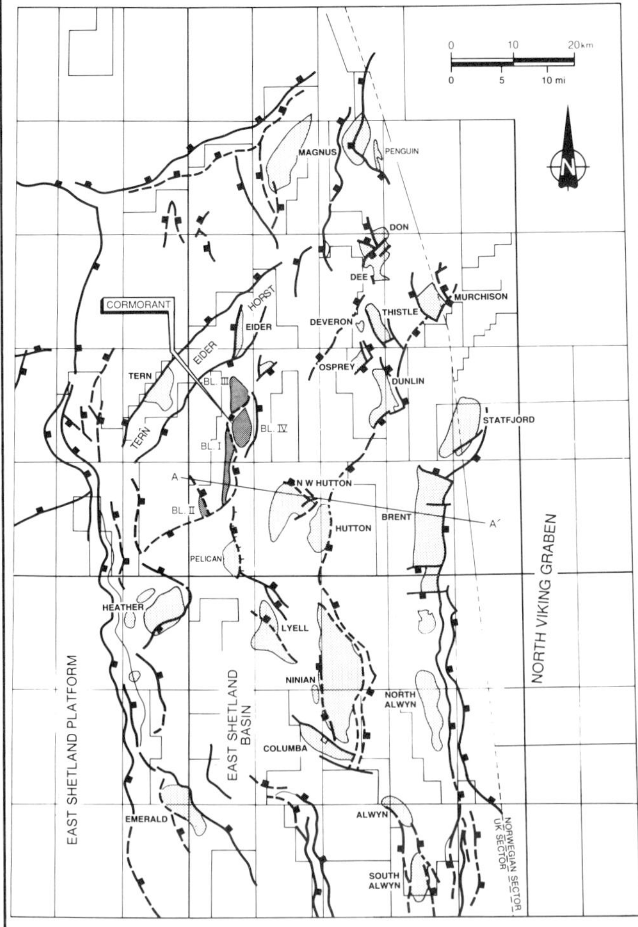

Figure 2. Regional map of the northern North Sea emphasizing the predominant fault trends (north-south, northeast-southwest, and northwest-southeast). Major oil fields in the East Shetland basin are also depicted. Cross section A-A′ is shown in Figure 8.

granules of basement in a sandy matrix was followed by sub-Devonian metamorphic basement at 10,680 ft (3255 m), penetrated 100 ft (30 m) to the total depth of the well (10,780 ft, 3286 m). Hydrocarbon indications were noted in the conglomerate and in fractures in the basement, but no flow was obtained upon testing.

The Jurassic Brent sandstone interval was production tested in four intervals. The two bottom intervals produced 4500 bbl/d of 36° API oil with a gas/oil ratio (GOR) of 300 scf/bbl on a 48/64-in. choke with a drawdown, at the formation, of 1300 psi (8980 kPa).

The two top intervals produced 7728 bbl/d of 36° API oil with a GOR of 340 scf/bbl on a 48/64-in. choke with a drawdown, at the formation, of 400 psi (2760 kPa).

Thus, the tests proved *a net* 123 ft (37 m) of Jurassic oil sandstone capable of flows up to 7800 BOPD. Since the tests produced no water and deeper water-bearing sandstones appeared (from wireline tests) to belong to a different pressure regime, the position of the oil-water contact (OWC) could not be deduced. Only 90 million bbl of recoverable oil were estimated on the basis of the observed hydrocarbon column.

The well was abandoned on 12 September 1972 at a total cost of £1,302,000.

Post-Discovery

Appraisal Activity

The appraisal of the Cormorant field was a complex process. From the time of discovery (1972) to the development of South Cormorant (Block 211/26 area 1979), eight appraisal wells were drilled to delineate the field. Another eight appraisal wells were drilled to define the central and northern areas (Block 211/21 area).

The accumulation penetrated by the discovery well (211/26-1) was thought to extend northward into Block 211/21 and to have a similar structure (a large gently dipping fault block) and thickness (approximately 800 ft; 244 m) to those observed in the Brent field to the east. Early appraisal activity (well 211/21-1A, Figure 3) quickly disproved this concept by showing a condensed reservoir sequence. A clearer picture of the fault pattern emerged from new seismic (acquired from 1975 to 1978) showing five separate structural areas. These were named Blocks I, II, III, IV, and V (Block V was later proven to be part of Block I) (Figure 3). The estimated total recoverable reserves of 220 million bbl oil were unevenly distributed among the blocks. Economic evaluations showed the field to be unattractive for development. However, using a platform sited in the South Cormorant area as a major oil trunkline terminal serving other future fields in the area introduced attractive incentives, and, consequently, the appraisal of the field continued.

Four of the five structural traps (Blocks I, II, III, and IV) were proven to contain oil in the Jurassic Brent Group. Wells 211/26-2 and 3 (Figure 3), drilled in Block II in the southwest of the field in 1974, found hydrocarbons in the Triassic red beds (Cormorant Formation). This was the first oil-bearing Triassic interval to be discovered in the Brent Province.

Historical Overview of Development Activity

The geographical spread of the reservoir blocks (16 mi or 25 km north to south) was the main complication in the development planning of the Cormorant field. The discovery of small oil reservoirs in the Triassic, in Block II, added further complication. To help in the planning, Block I was divided into three parts, Ia, Ib, and Ic (Figures 6A and 6B).

Adequate drainage of the area required a number of platforms in combination with subsea completions.

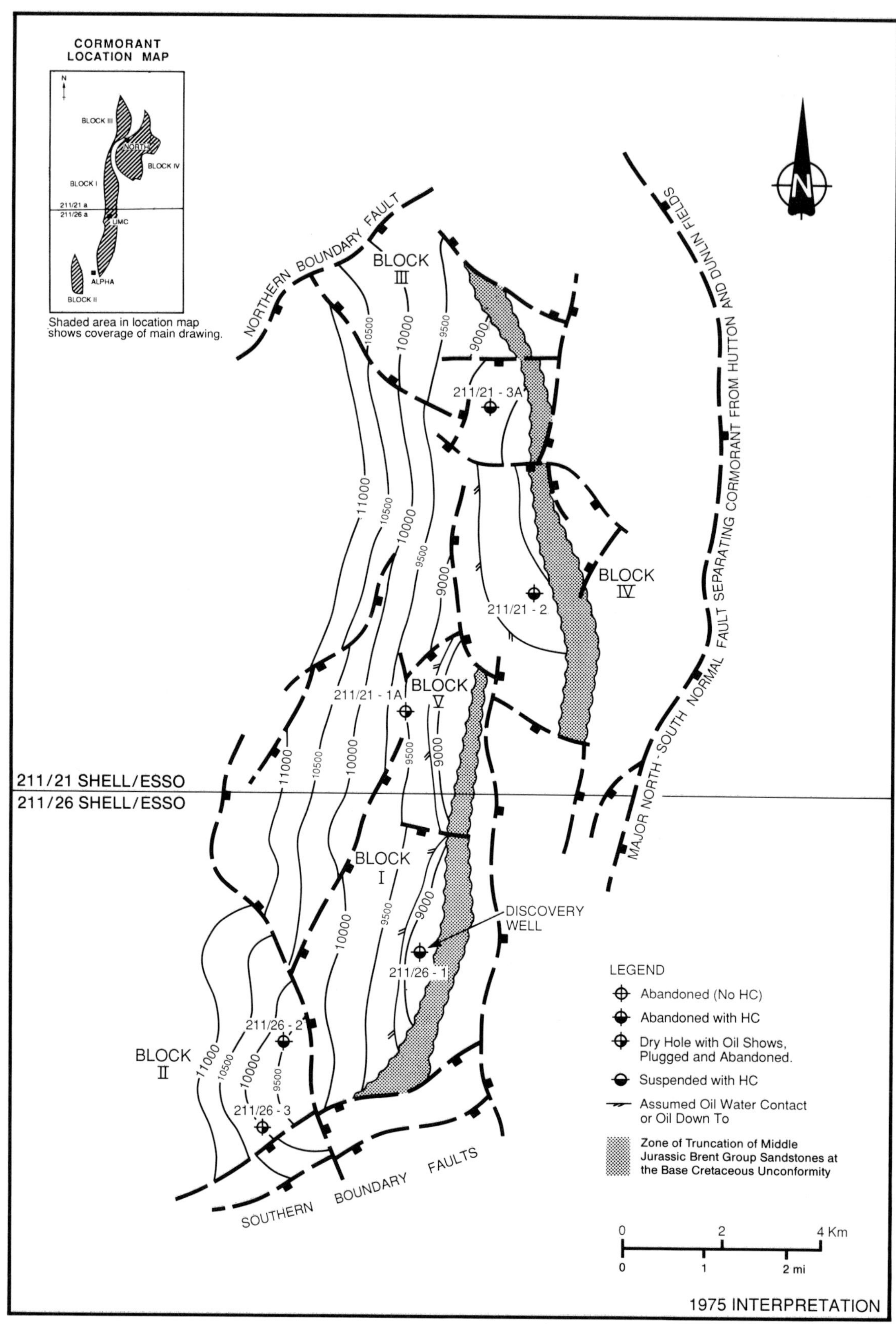

Figure 3. Top of Brent Group structure map (1975) showing the location of the discovery and first five appraisal wells. Blocks I, II, III, and IV were proven hydrocarbon bearing. Block V, as outlined here, was later proven to be part of Block I. Contours are in feet subsea. Contour interval, 500 ft (152 m).

Of the development schemes that were considered for the southern area (Blocks Ia and II, license block 211/26), the most favored involved the development of the Jurassic from a gravity platform with crestal producers and downflank injectors. The northern half of Block I, which was outside the platform reach, was to be drained by subsea completions tied into an underwater manifold center (UMC). This, in turn, was to be linked to the Cormorant Alpha platform (Figure 4).

Development plans for the northern Cormorant area (Blocks Ic, III, IV, and V) were evaluated using several different schemes because the size and shape of the accumulations were not clear from appraisal drilling. These schemes included multiple platforms, a platform plus underwater development, full development by a floating production system, and a combination of platform and floating systems. It was decided to use a conventional platform plus underwater development for areas beyond the platform reach, even though the latter concept was still new technology (Figure 5). (Block V was later found to be part of Block I.)

Current Development

The current Cormorant field development plan splits the field into three main areas. Blocks II and Ia are drained by the Cormorant Alpha platform (CA), Blocks III, IV, and Ic are developed via the North Cormorant platform (NC), and the area in the middle, Block Ib, is drained using subsea technology in the form of a UMC that is tied into the Cormorant Alpha platform (Figures 6A and B) (Watters et al., 1985). The base development plans were submitted to the British Government in 1976 (South Cormorant), 1978 (North Cormorant), and 1981 (UMC).

In Blocks I and III, general field development has gone according to the plan. In Blocks II and IV, however, diagenesis and fault-related transmissibility problems have caused changes to the base plan.

Block I primary development was concluded in 1988. Thereafter, the block has been drained by a row of ten updip producers supported by a row of seven downflank injectors. During the secondary development phase, oil recovery will be enhanced by sidetracking the producers into the truncated crestal area to recover attic oil by recompletions and by infill drilling.

Production from Block II began on 11 December 1979. Early production behavior indicated that the block consisted of two producible areas: a crestal and a downflank area, separated by a north–south-trending transmissibility barrier. Seismic data partly revealed this barrier as a series of faults with limited offsets (50 ft; 15 m). The original development plan was modified to allow these two areas to be developed separately.

Primary development of the crestal area was completed in 1985. Primary development of the downflank area was completed in 1988 following the drilling of two injection wells. Subsequently Block II will be drained by three producers supported by six injectors.

The independent development of crestal and flank areas coupled with a high bubble-point pressure means that stringent controls are placed on the allowable operating pressures for the block. The secondary development phase will consist of drilling new infill wells and sidetracks as well as recompleting wells.

Development of the Triassic Cormorant Formation has not yet commenced; the project is still in the appraisal stage. Production tests on the Cormorant Formation show rapid declines implying a rather limited reservoir connectivity. The remaining two wells to be drilled in Block II will have Triassic appraisal as a secondary objective. The STOIIP currently attributable to the Triassic sandstones is 100 million bbl of oil. In view of the uncertainties in the production test results, no reserves are presently assigned to the Cormorant Formation.

The development plan for Block III allowed for flexibility owing to the complex geology and its effects on natural water influx, the nature of faulting in the downdip area, and the extent of producible hydrocarbons beyond the drainage area of wells drilled from the platform. Production started from the block on 1 April 1982.

Evidence of good pressure communication within the reservoir led to a modification of the initial development plan to one rather than two rows of producer wells. The primary development phase was concluded in 1985 and resulted in four producers supported by four injectors.

The need for subsea development of the northern area beyond the platform reach has been proven unnecessary by a reservoir simulation study of the block. Secondary development therefore will concentrate on infill locations and sidetracks to recover attic oil.

The development plans for Cormorant Block IV changed considerably as the complex structural picture of the block became clearer. The initial seismic interpretation showed a relatively unfaulted structure, and therefore, the development plan called for a row of five downdip injectors and two rows of five crestal producers plus a further three to five satellite injectors in the south.

A 3D seismic survey was shot in 1979, and Block IV production began on 14 February 1982. The production history of the first wells drilled confirmed that the reservoir was compartmentalized into many small fault blocks, each requiring separate development. A revision of the development plan became necessary.

The revised strategy involved a development pattern with dedicated producer/injector pairs for each subblock. Uncertainties regarding the structural interpretation, OWC position, and reservoir and fluid properties necessitated a flexible drilling

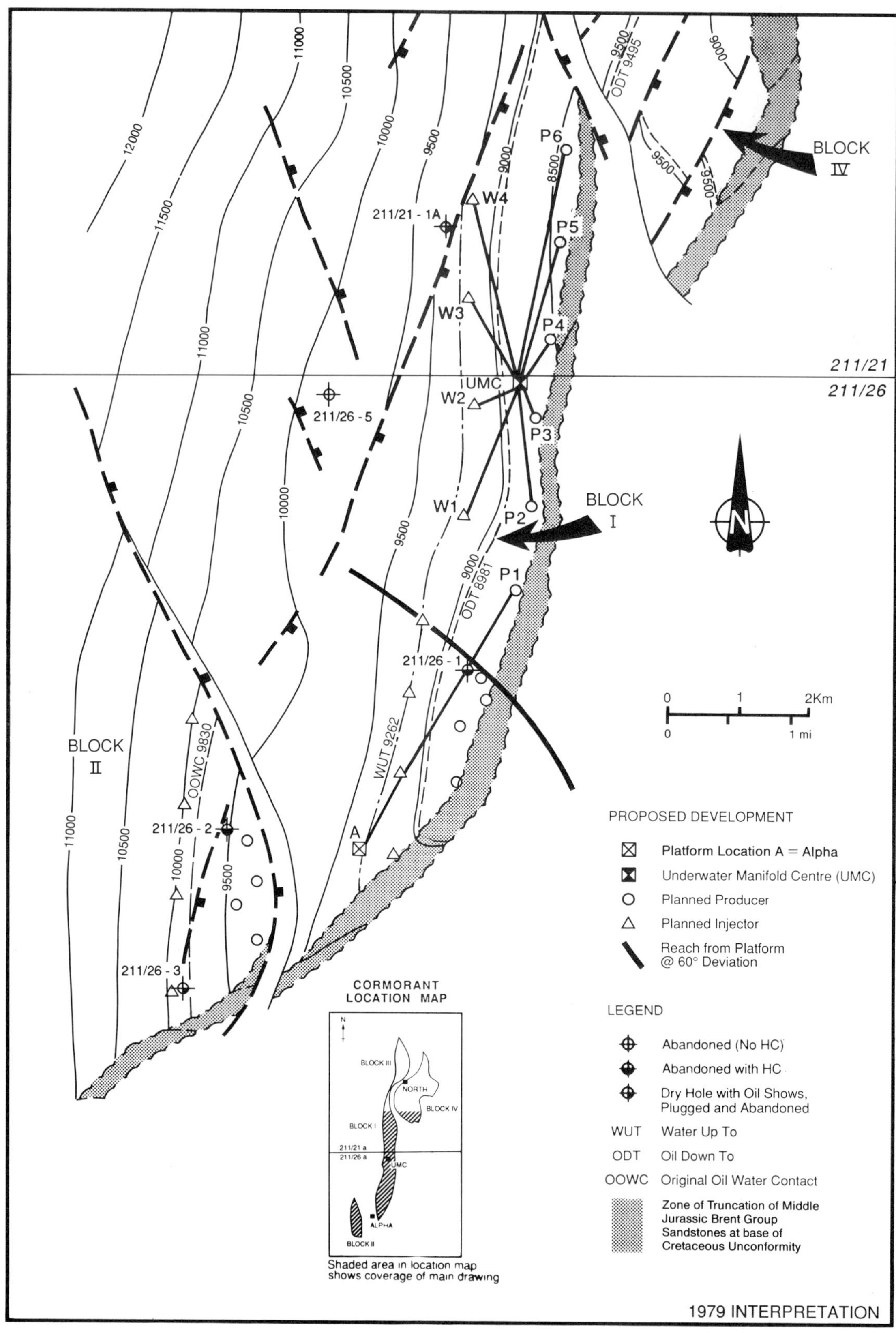

Figure 4. South Cormorant proposed development plan, showing the positions of the gravity platform and underwater manifold center (UMC). Well P1 was planned and later drilled as a satellite linked directly to the Cormorant Alpha platform. Contours are at the top of the Brent Group, in feet subsea. Contour interval, 500 ft (152 m).

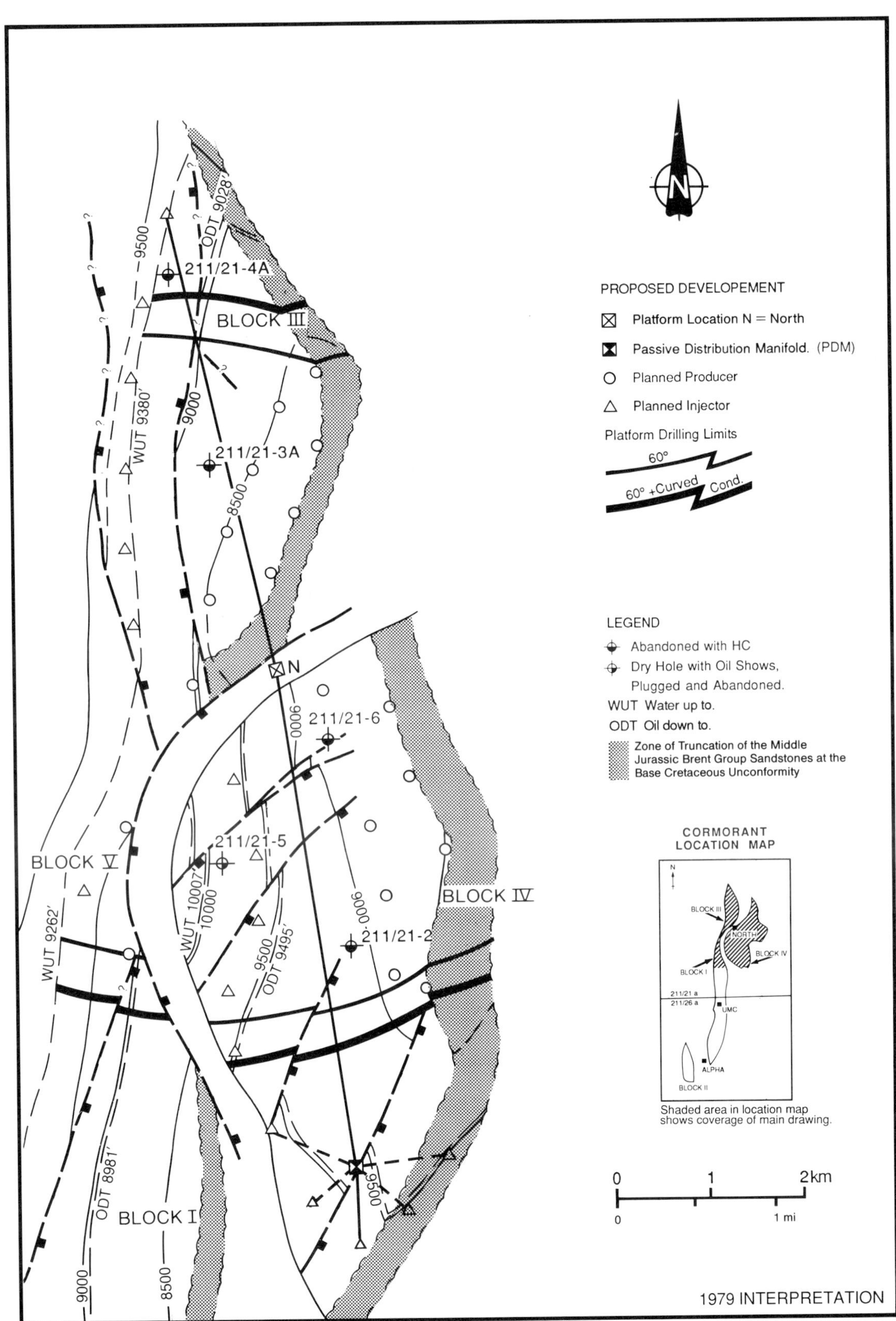

Figure 5. North Cormorant proposed development plan. This plan was very flexible, allowing for uncertainties in reservoir size, shape, and continuity. Planned positions of satellite wells and the passive distribution manifold (PDM) are shown. Contours are at the top of the Brent Group, in feet subsea. Contour interval, 500 ft (152 m).

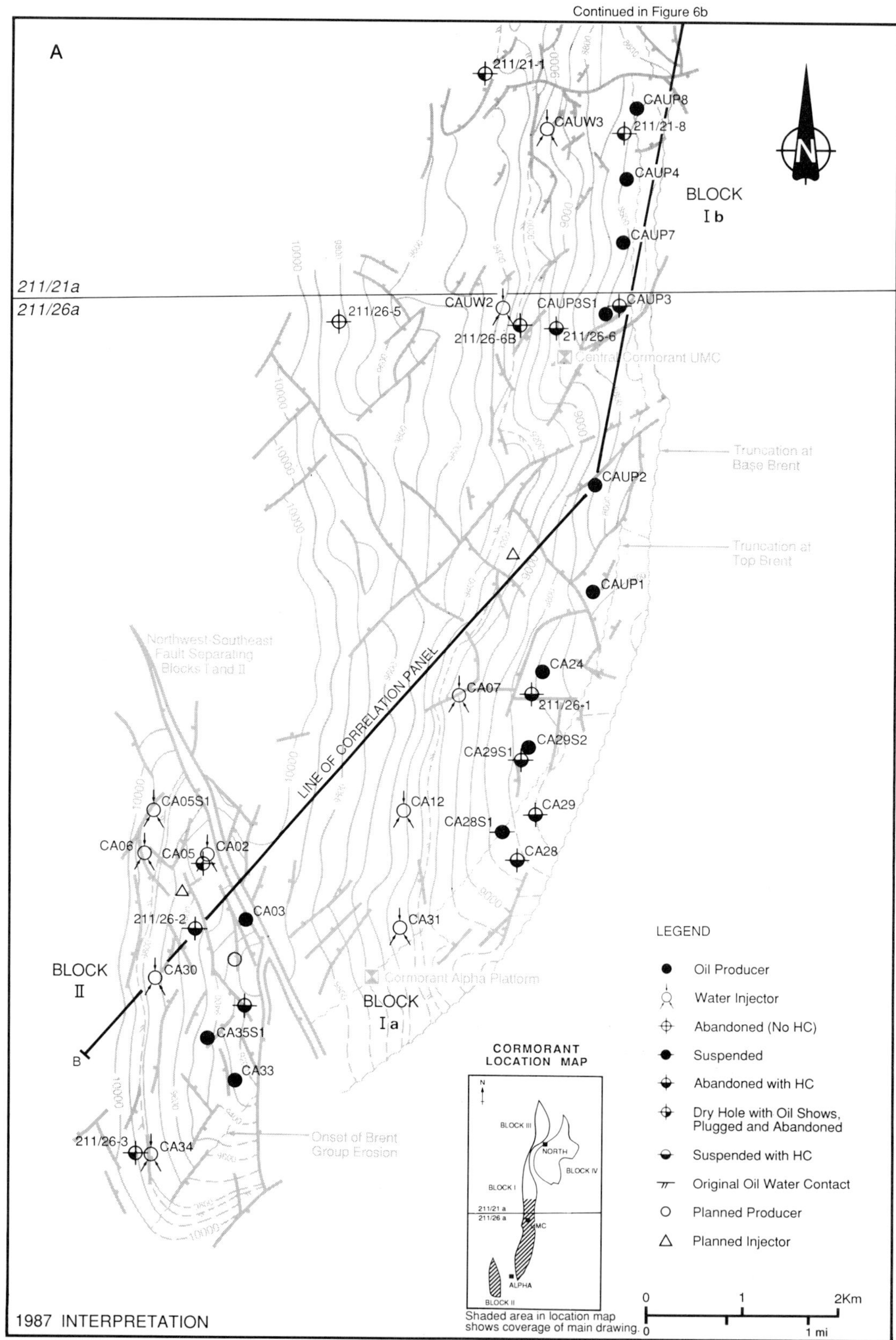

Figure 6 (A and B). The current (1989) development plan. Development of Blocks I and III is almost complete. Infill drilling and sidetracking to recover attic oil will be part of the secondary development phase. Blocks II and IV are still in the first phase. For these two blocks, the locations of planned producers and injectors is shown. Contours are at the top of the Brent Group, in feet subsea. Contour interval, 100 ft (30 m). Evidence for the structural complexity comes from 3D seismic data. The structural cross section C-C′ in Figure 12, stratigraphic correlation panel B-B′ in Figure 17, and line of seismic comparison in Figure 7 are also located.

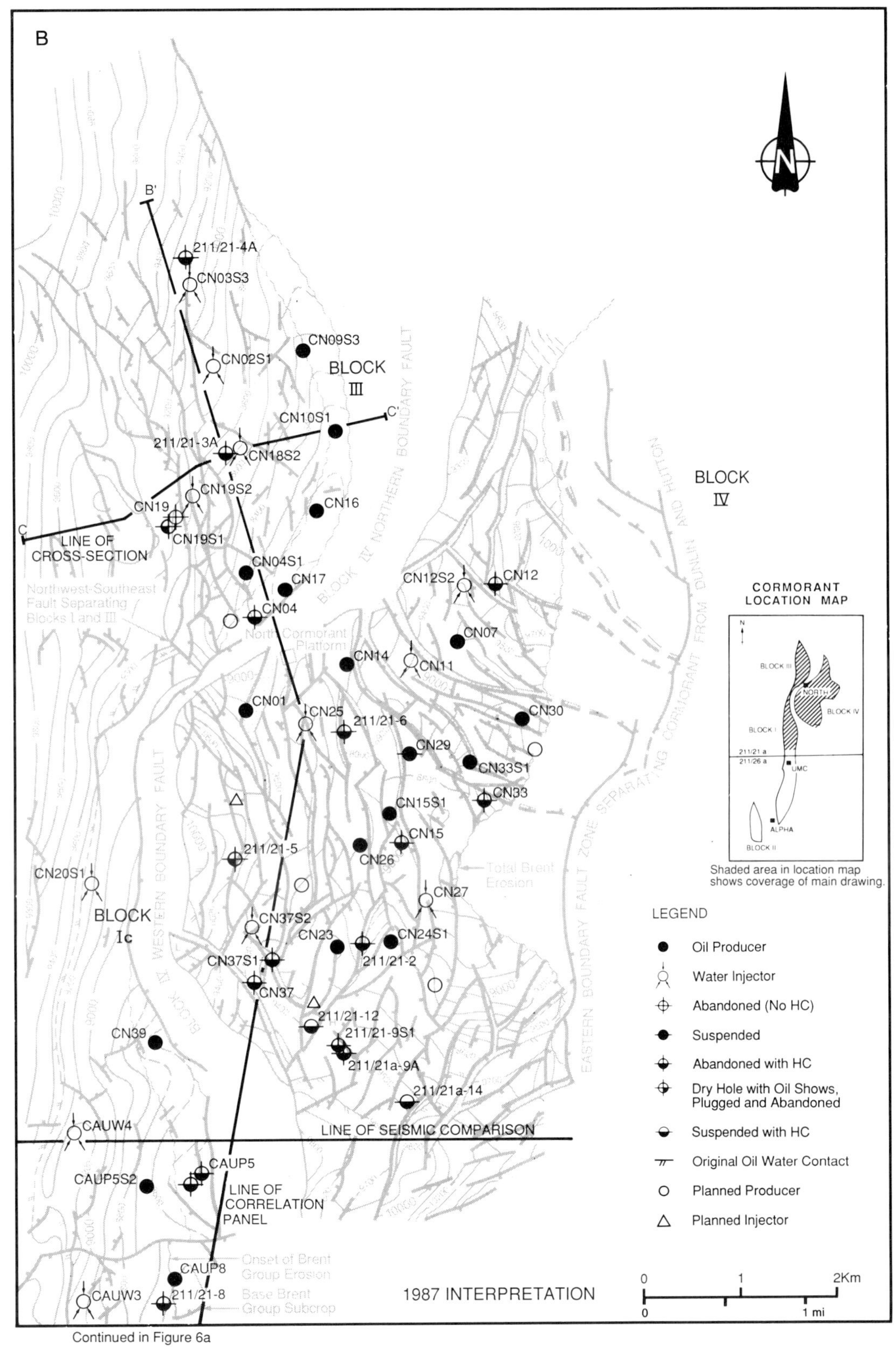

Figure 6. Continued

sequence and also led to a new 3D seismic survey being acquired in 1984 (see Figure 7).

As part of the revised strategy, the originally planned underwater passive distribution manifold to provide injection support in the south of the block was no longer considered to be a feasible project. An alternative plan was adopted that involved drilling highly deviated injection wells from the platform.

Secondary development will involve sidetracks and recompletions to recover some of the oil bypassed because of the heavy faulting of the block.

DISCOVERY METHOD

After acquiring Blocks 211/21 and 211/26 in the fourth-round allocation (1972), Shell/Esso shot six east-west seismic lines and obtained two north-south lines by data trade. The addition of these to the existing data formed a grid with an average spacing of 1.2 mi (2 km) between lines in the east-west direction and 1.9 mi (3 km) between lines in a north-south direction. Additionally, a magnetic survey of the area was purchased. Based on this information, regional correlations and comparisons were made between Block 211/26 and the Brent discovery well 211/29-1. From these studies, two possible plays were identified and evaluated by drilling. Well 211/26-1 found hydrocarbon-bearing Middle Jurassic sandstones.

STRUCTURE

Tectonic History

The structural history of the northern North Sea is dominated by two distinct tectonic phases. The first, the pre-Devonian Caledonian tectonic phase, imparted a predominantly northeast-southwest grain. The effects of this phase were largely overprinted by the second, late Kimmerian extensional tectonic phase that began in the Early Jurassic and culminated in the Early Cretaceous. The latter extensional phase caused the rifting that resulted in the development of the Viking graben to the east. Movement took place along north-south and basement reactivated northeast-southwest faults.

In the Bathonian, the fault blocks produced by rifting underwent rotation about a north-south axis causing many of them to dip gently westward (Figure 8). The normal faults associated with this rotation are characterized by their en echelon listric form as seen on seismic maps and cross sections. Toward the end of the Kimmerian tectonic phase (from Bathonian to Early Cretaceous), repeated changes in relative sea level resulted in a series of unconformities. The two major unconformities in the Cormorant area are:

1. The middle Callovian Unconformity is defined as the base of the upper Heather Formation. There is no evidence in the Cormorant field area for strong erosion at this unconformity. The underlying middle and lower Heather members are consistently present although in some areas they are thinly developed. The unconformity marks a depositional hiatus rather than a strong phase of erosion.
2. The "X" or basal Cretaceous unconformity at the base of the Cromer Knoll Group is a major regional unconformity and a good seismic reflector. Lower Cretaceous marls of the Cromer Knoll Group were deposited as an onlap sequence. Over the structural highs, the unit is represented by 100 to 200 ft (30 to 60 m) of marls and shales of Albian/Aptian age. In the deeper parts of the basin, sedimentation was continuous and a thickness of several hundred feet was attained.

By the Turonian, fault-related subsidence was replaced by a "sag" period. Superimposed upon this are the effects of isostatic subsidence of the crust in response to loading by the chalky mudstone of the Shetland Group (Tertiary).

Regional Structure

The Cormorant field consists of four separate oil accumulations along the crest of one of the elongated north–south-striking, westward-tilted large fault blocks that characterize the East Shetland basin (as intepreted from 2D seismic data). To the east a major east-dipping normal fault with a throw of approximately 3000 ft (900 m) separates the complex Cormorant fault block from the Hutton and Dunlin fields. To the west, the block dips gradually to the Tern/Eider horst and East Shetland platform (Figures 2 and 8). The northerly and southerly boundaries of the block are defined by large, south-dipping, northeast–southwest-trending transfer faults (cf. Gibbs, 1984).

Three predominant structural trends may be identified in the Cormorant region of the East Shetland basin (Figure 9):

1. North–south-striking faults of variable length and throw are present throughout the area.
2. Northeast–southwest-trending faults occur predominantly to the north of Cormorant. (The faults bounding the Tern/Eider horst block and the Block IV northern boundary fault are examples.)
3. Northwest-southeast faults are generally developed to the south of Cormorant. (The major fault offsetting Block II from Block I and the parallel fault systems bounding the Ninian field are examples.)

The north-south fault set comprises major east-dipping normal faults that are related to the Mesozoic extension of the Viking graben system. They define the major elongated westward-tilted fault blocks that

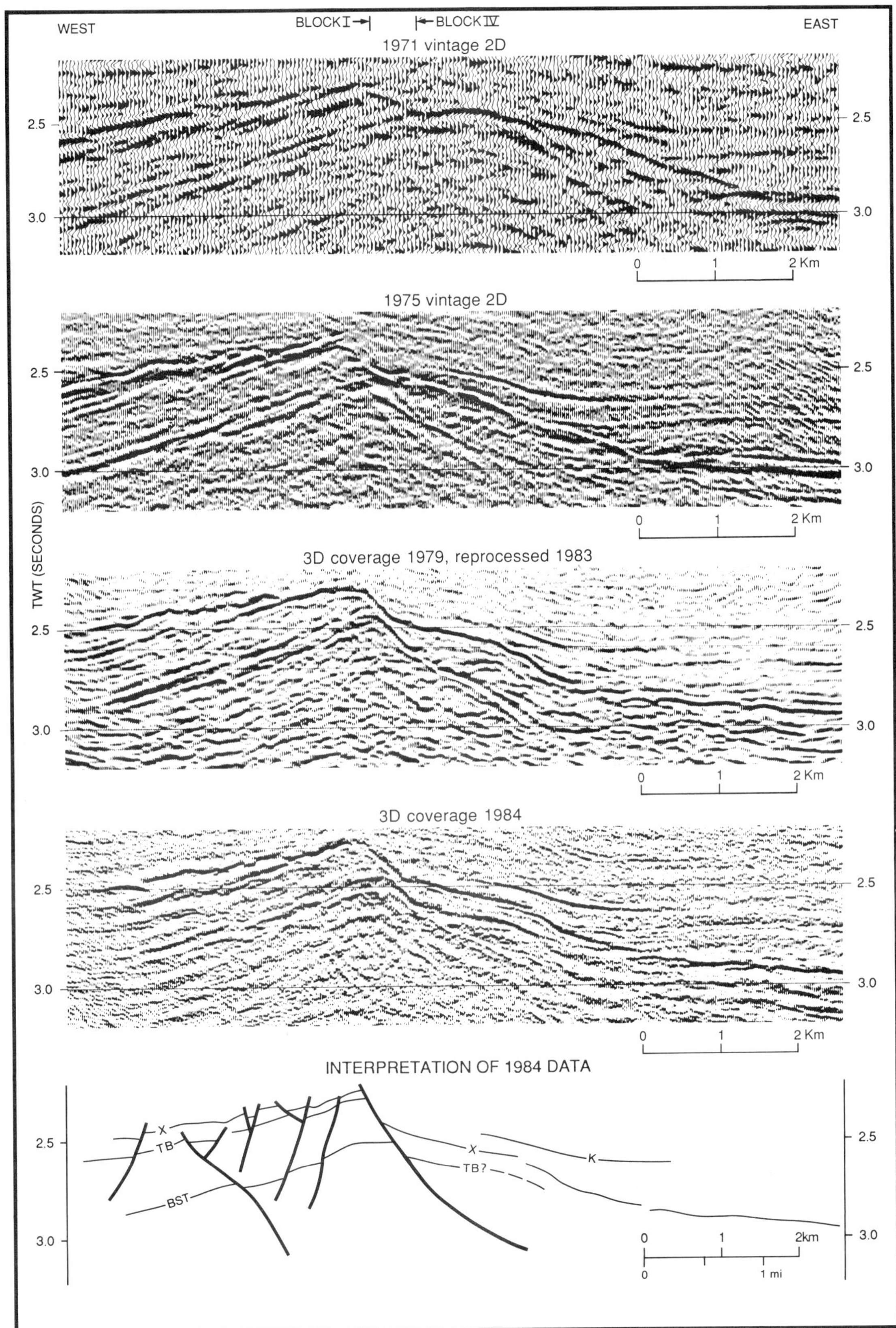

Figure 7. West-east seismic line illustrating the comparison of 2D and 3D seismic data from 1971 to 1984 across Cormorant Blocks I and IV. The base of the Brent Group is poorly defined even on the 1984 survey line, notwithstanding its much higher data sampling density. The horizontal scale for all seismic lines and the intepretation are the same. The line of section is shown in Figure 6B. K, base of intra-Cretaceous limestone reflector. X, basal Cretaceous unconformity reflector. TB, top of Brent (Middle Jurassic) reflector. BST, top of basement (sub-Devonian) reflector.

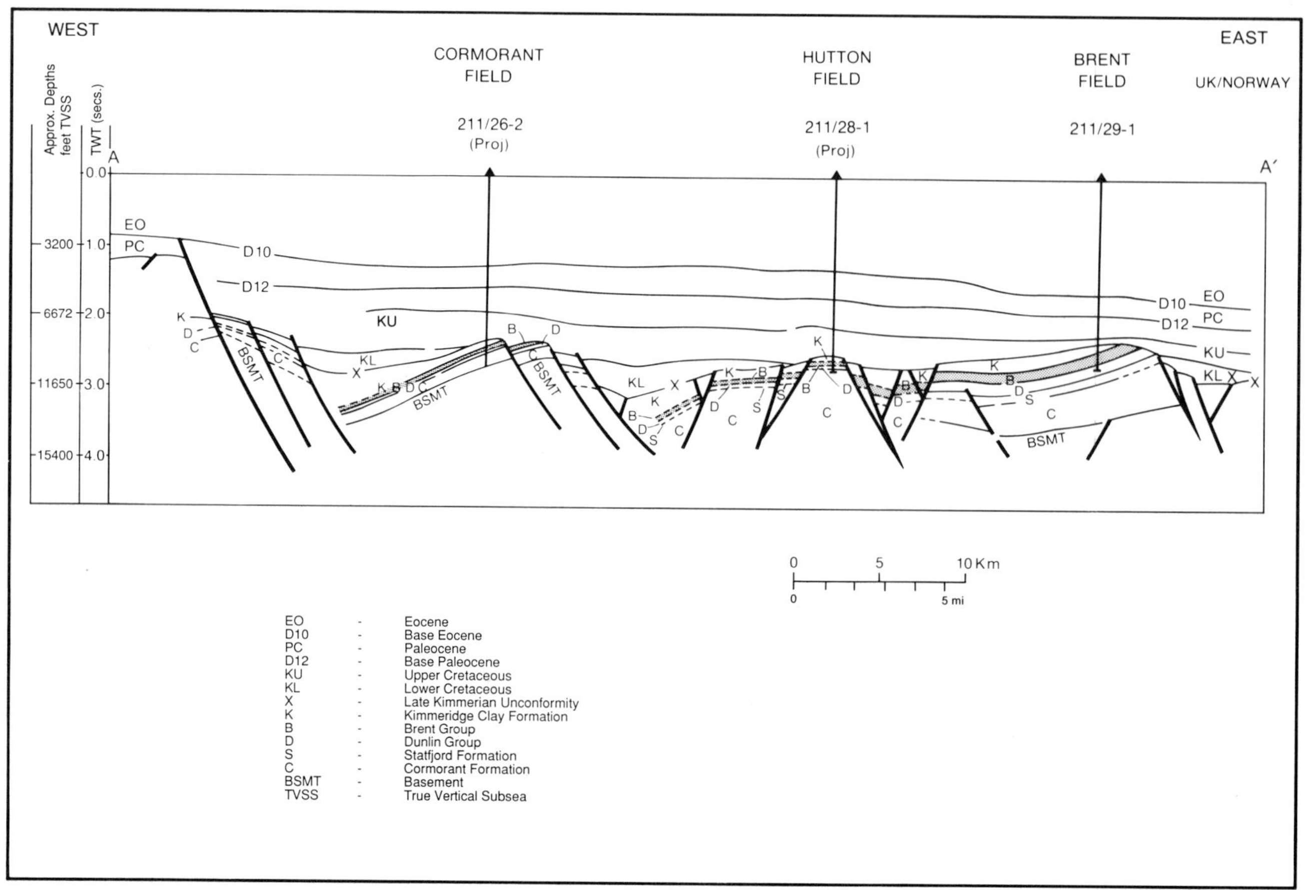

Figure 8. East Shetland basin regional schematic west-east cross section A–A′, showing the listric normal faulting and westward rotation of the major fault blocks. The approximate depths are based on the average velocity function that is most representative for the area of the section. (TWT, two-way travel time.) The location of section A–A′ is shown in Figure 2.

characterize this region. The northeast-southwest and northwest-southeast sets are interpreted as predominantly oblique-slip faults with left-lateral and right-lateral offsets, respectively (Speksnijder, 1987) (Figure 10). The orientation and displacement of these faults is thought to have been governed by reactivation along pre-existing trends in the sub-Devonian basement; consequently, the northeast-southwest set is commonly assigned to the "Caledonian trend." Although the northwest-southeast and northeast-southwest sets are therefore influenced by the basement reactivation, the two sets effectively act as transfer faults associated with regional east-west extension of the Viking graben system (cf. Gibbs, 1984).

Local Structure

The Cormorant field is located in a large block bounded to the north and south by two major northeast-southwest transfer faults and to the east by a major north-south normal fault that is downthrown approximately 3000 ft (914 m) to the east (Figure 3).

The four accumulations (Blocks I, II, III, and IV) are trapped in the crestal areas of three smaller fault blocks within the large block (Figure 11). The largest, a westward-tilted block, dips at 8 to 12° and measures 8 mi (25 km) from north to south. Blocks I and III occur along the axial part of this large structure, and the highest point of the reservoir is at 8200 ft subsea (2499 m). (Figure 12 shows a west-east cross section across Block III.) Blocks I and III are separated by a northwest-southeast-trending fault with a vertical throw of 400 ft (122 m) (Figure 6B).

Block II is a small fault trap that lies to the southwest of the main fault block (Figure 6A). It is a gently plunging (12°) north-northeast-south-southwest oriented anticline that is fault bounded to the east. The Brent reservoir culminates at a depth of 9600 ft (2926 m) and is the deepest culmination of the four fault blocks. Thus it has the most severe depth-related reservoir deterioration.

Block IV represents the down-faulted crestal portion of the major fault block. It was detached from the main Cormorant (Block I/III) structure by a combination of north-south normal faults and northeast-southwest–northwest-southeast oblique-slip faults.

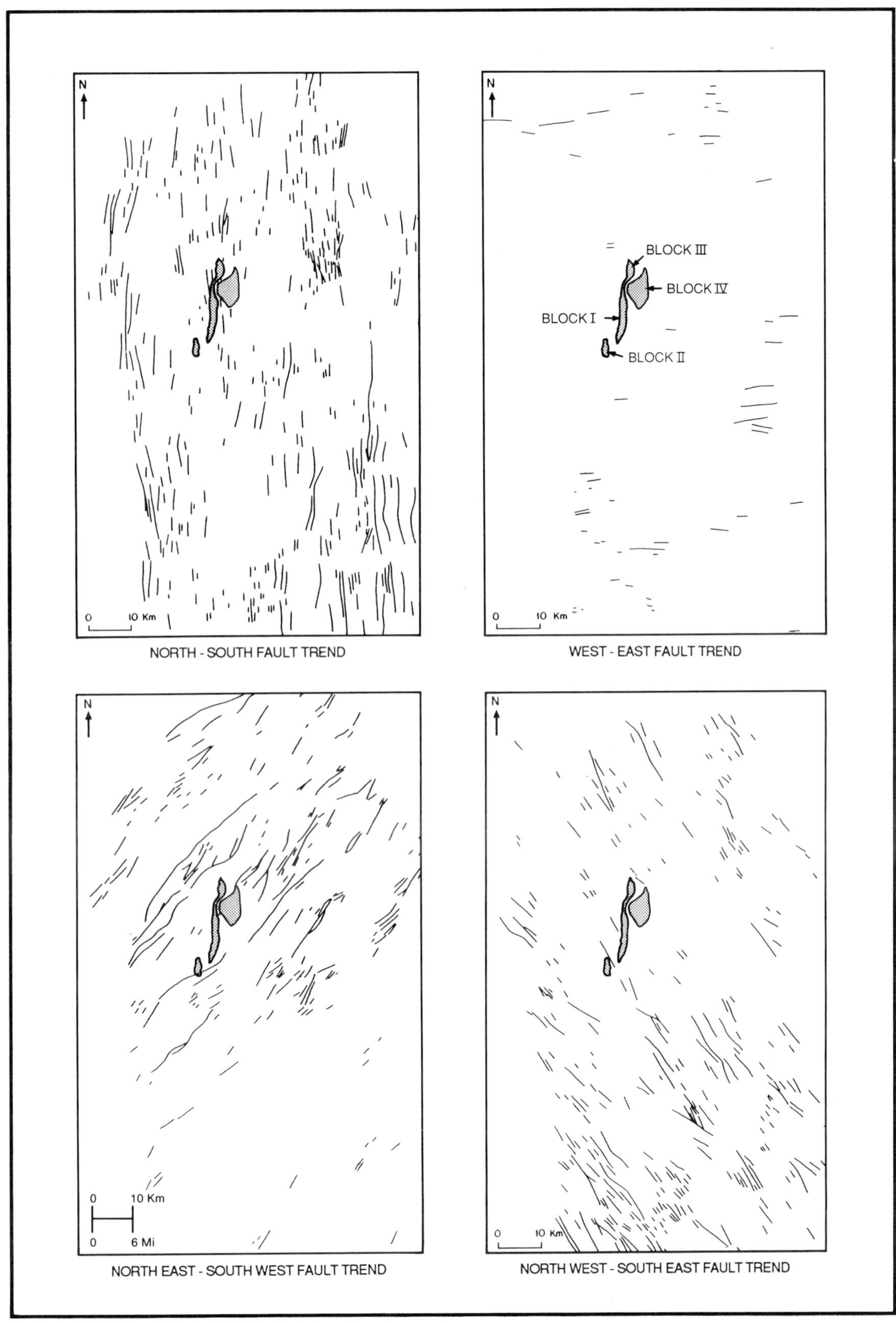

Figure 9. The predominant structural trends of the East Shetland basin in relation to the position of the Cormorant field (after Speksnijder, 1987). Note that the northeast-southwest-striking faults are longer and occur more densely in the northern part of the map, while the northwest-southeast family are more abundant south of Cormorant. All faulting is interpreted from seismic data.

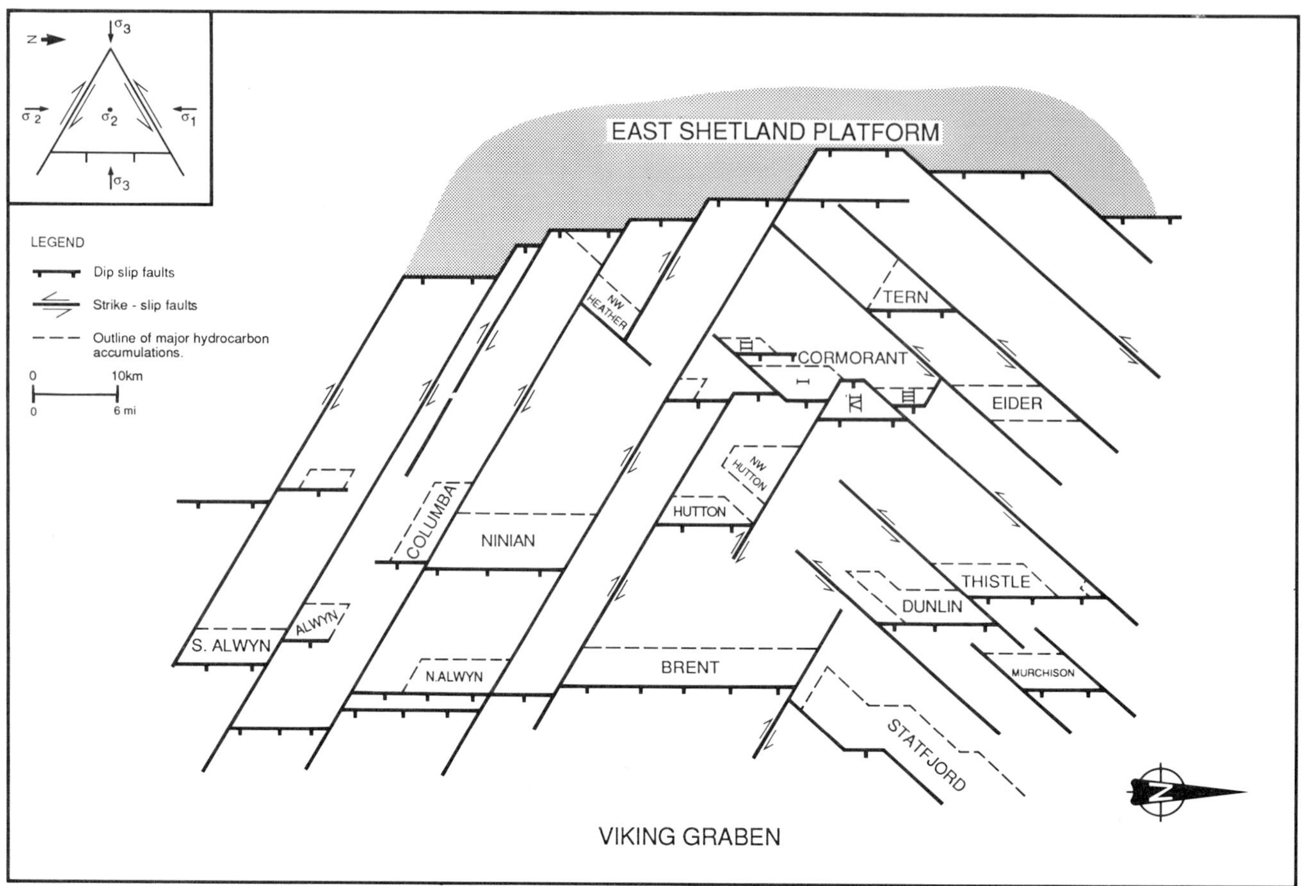

Figure 10. Schematic structural framework of the East Shetland basin. Only major faults of the Caledonian in the pre-Devonian basement level are indicated. The labeling of faults as dip-slip or strike-slip refers to the predominant component of the fault. Their relation to major hydrocarbon accumulations at higher structural levels is expressed by the schematic outline of the fields (after Speksnijder, 1987). The diagram in the corner shows the interpretation of the general stress regime around Block IV.

These faults resulted in an easterly displacement of Block IV relative to Blocks I and III. This movement occurred during the major phase of regional east-west extension (Speksnijder, 1987). In the crestal region of Block IV, the Brent reservoir culminates at a depth of approximately 9000 ft (2743 m).

Faulting in the Cormorant area is primarily recognized on 3D migrated seismic data and is mapped at three horizons, the X-unconformity, the top of the Brent, and the basement. Production behavior has been used to recognize pressure barriers within the field. These may or may not have a clear seismic expression as faults or fault zones.

Synthetic seismograms are used to tie formation tops to seismic markers and are the basis for establishing correlations and estimating fault throws. Blocks I and III are relatively unfaulted, and seismic correlation of horizons in these blocks is generally good. Some north-south-striking small-scale normal faults have been mapped. These faults locally may act as partial transmissibility barriers.

Fault trends in Blocks II and IV are more difficult to define than in Blocks I and III. This has led to the acquisition of new 3D seismic surveys over Block IV in 1984 and over Block II in 1988 to improve the horizon and fault definitions (Gaarenstroom, 1984).

Block IV is dissected by a complex set of interconnected faults (Figure 11) that can broadly be divided into synsedimentary north–south-trending normal faults with large vertical offsets, present in the western area of Block IV, and northwest–southeast- and northeast–southwest-trending faults occurring in the northern and southern areas. Although there is no direct evidence for lateral fault movements within the block, some lateral offset could be apportioned to the northwest–southeast- and northeast–southwest-trending sets in light of the regional fault model. This would imply that the small-scale northeast-southwest and northwest-southeast faults could be acting as local conjugates or antithetic partners to the regional transfer faults.

The local structural framework was formed during the regional Kimmerian tectonic phase that started in the Early Jurassic and culminated in the Early Cretaceous. Although rotation of the main fault block occurred in the Early Cretaceous, thickness varia-

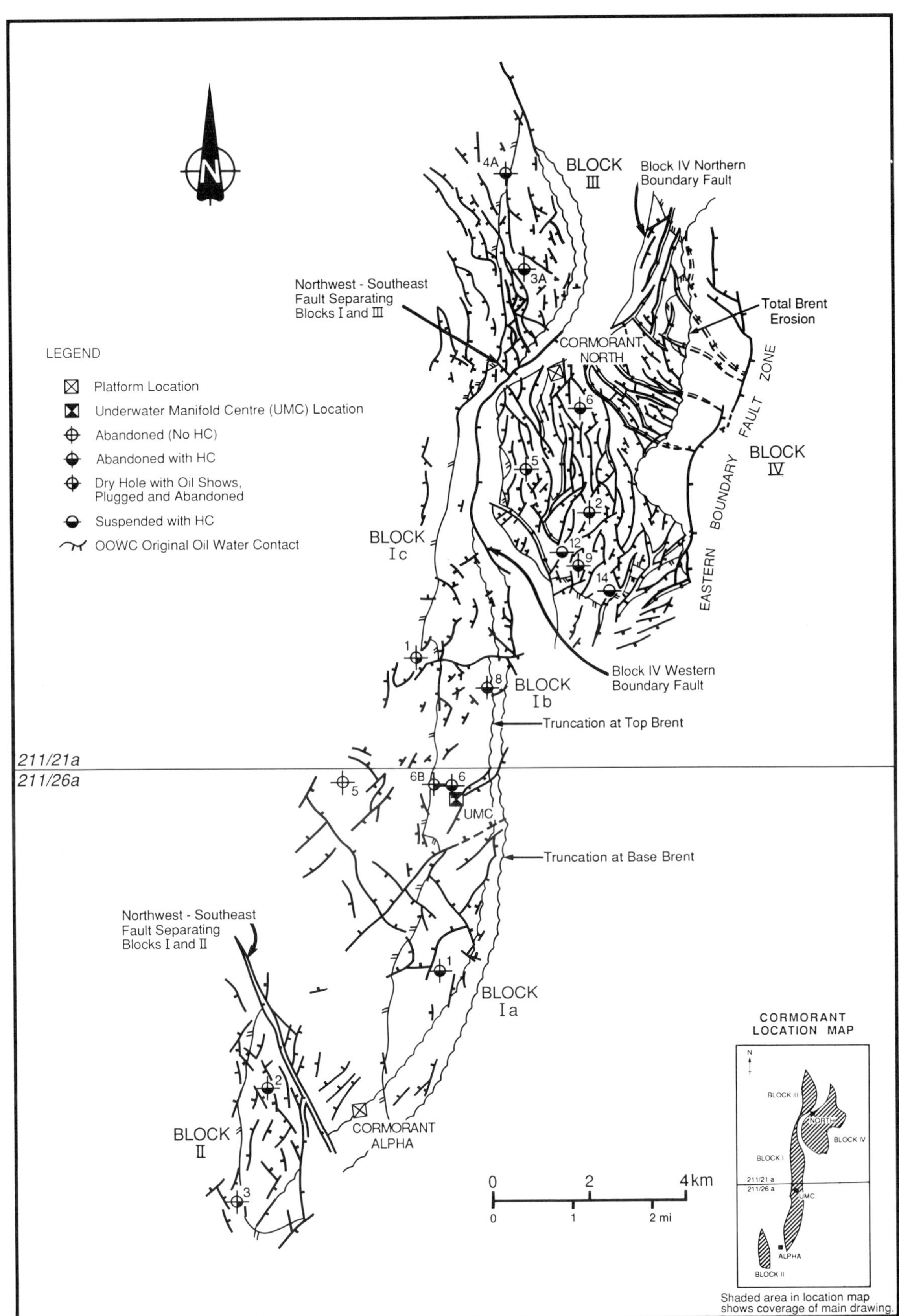

Figure 11. The local structural framework of the Cormorant field at the top of the Brent Group, outlining the four oil accumulations. Note the structural complexity of Block IV in relation to the other fault blocks. Evidence for the structural model comes from 3D seismic data.

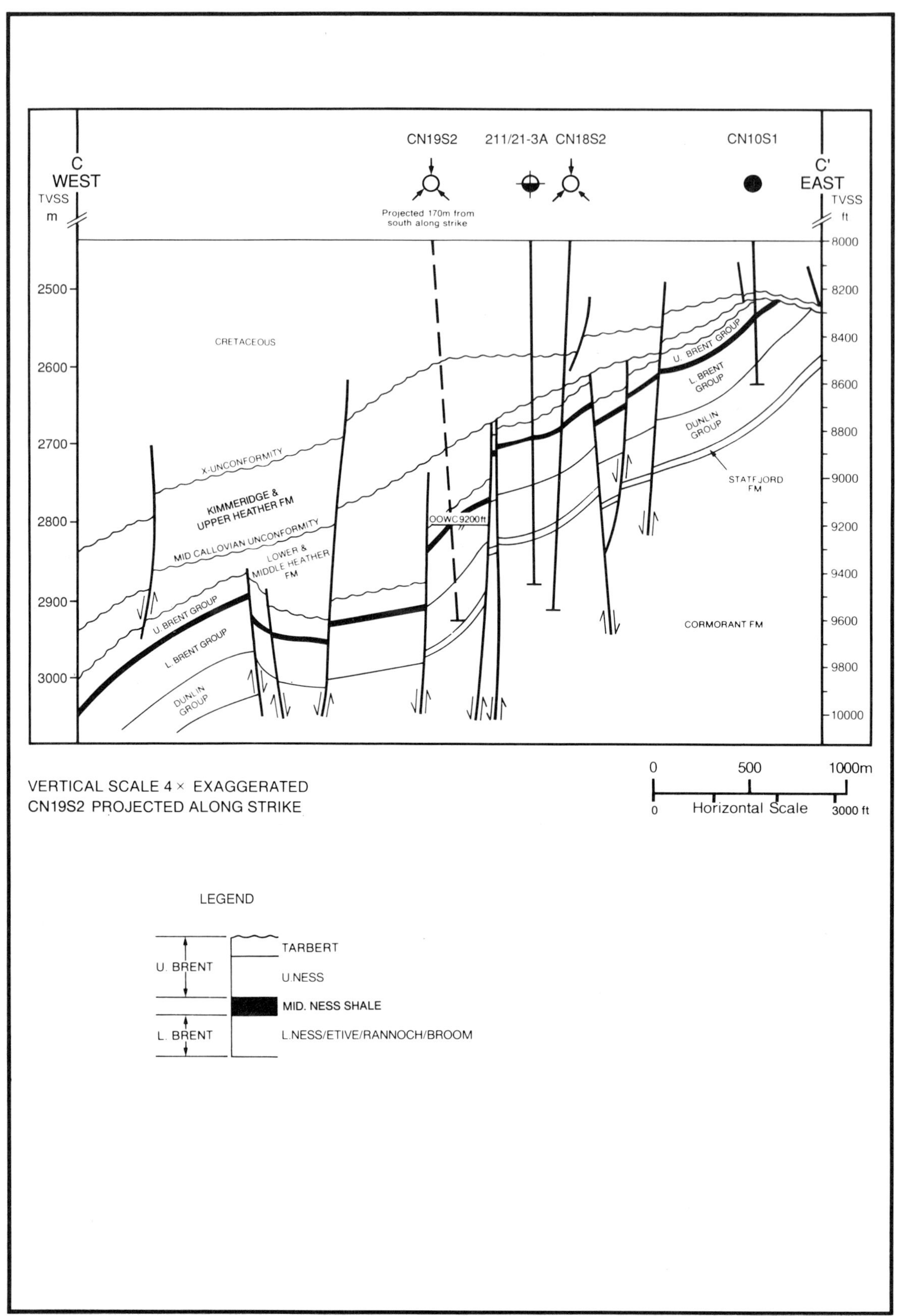

Figure 12. West-east structural section C-C′ across Cormorant Block III showing total erosion of the Brent Group at the crest of the structure. The uneroded sequence in this fault block is about 350 ft thick. The location of section C-C′ is shown in Figure 6B.

tions suggest that the fault separating Block IV from its neighbors was active at least from the Early Jurassic. In contrast, similar Brent thicknesses in Block II and south of Block I suggest that the fault separating them did not start moving until the Early Cretaceous.

STRATIGRAPHY

A schematic stratigraphic table and generalized lithology of the overburden is given in Figure 13. The basement consists of garnet-bearing mica schists and severely weathered pelitic gneisses from 10,500 ft (3200 m). Radiometric age dating in a basement sample from 211/21-1 confirmed a pre-Devonian age.

With the onset of the Triassic, the Viking graben rift system developed as a major feature. Sediments eroded from the Shetland platform were deposited unconformably on the basement in a distal flood plain environment characterized by intermittent flash floods. The Cormorant Formation consists of 1000 to 1500 ft (304 to 457 m) of red-brown silty claystones interbedded with fine- to medium- and occasionally coarse-grained sandstones. These sandstones are commonly argillaceous or cemented by calcite and generally form poor reservoir rocks.

The Lower Jurassic Statfjord Formation conformably overlies the Cormorant Formation. The Statfjord is represented by a section of 10 to 30 ft (3 to 10 m) thick arenaceous, calcareous sandstones. These were deposited initially in a fluvial and later in a shallow marine environment. These sediments are equivalent to Unit I (850 ft; 259 m) of the Statfjord reservoir occurring in the Brent field to the east (Livera and Gdula, personal communication). As rifting progressed, the sea invaded the entire graben system and the Dunlin Group was deposited. It consists of 150 to 200 ft (45 to 61 m) of marine shales and siltstones. These shales constitute the basal seal for the Brent Group reservoirs. The Dunlin Group is subdivided into four formations as proposed by the Institute of Geological Sciences (IGS) (Deegan and Scull, 1977), from oldest to youngest: Amundsen, Burton, Cook, and Drake. The top of the interval is clearly marked by the base of the arenaceous Brent Group.

During the Middle Jurassic, a widespread lowering of sea level led to emergence and erosion of the Shetland platform and the Central North Sea high to the west and south, respectively, of the Cormorant area. Erosional products from these structural highs were deposited as a north-northeasterly accreting coastal-deltaic system, the Brent Group. This system gradually encroached across and filled the differentially subsiding shallow marine basin with sediment. In the Cormorant field area, these deposits are up to 450 ft (137 m) thick.

Marine sedimentation continued in the Late Jurassic (Bathonian–Callovian) with deposition of the Humber Group, subdivided into 640 ft (195 m) of the Heather Shale Formation and 100 ft (30 m) of the Kimmeridge Clay Formation. This brought the deltaic conditions of Brent Group sedimentation to a close.

The Heather Formation has been subdivided into lower, middle, and upper members, with the middle and upper members separated by the middle Callovian unconformity. Over most of the field all three units can usually be recognized. However, the formation shows marked thickness variations from several hundred feet on the flanks to less than 100 ft (30 m) over the crests of Blocks I, II, III, and IV. Rotation of the fault blocks was active at the time of deposition, and consequently, the Brent and Dunlin units were truncated. The Heather Formation forms the upper seal to the Brent Group reservoirs.

Organic-rich Kimmeridge Clay overlies the Heather Formation. It is a dark gray-brown, generally not calcareous, carbonaceous claystone. The Kimmeridge Clay Formation averages 100 ft (30 m) thick and forms the source rock for the Brent Group reservoirs.

The lower part of the Cretaceous is formed by the Cromer Knoll Group. Its lithology consists mainly of limestones, marls, and red-brown claystone and is some 125 ft (38 m) thick. It unconformably overlies the Kimmeridge Clay Formation. Where the Humber Group has been totally eroded below the X-unconformity, the Cromer Knoll marl and limestone formations locally form the upper seal to the Brent reservoir.

The Shetland Group unconformably overlies the Cromer Knoll Group. In the Cormorant area, the lower part of the Shetland Group forms a distinct lithological unit, consisting of limestones overlain by claystones. The upper part is, however, predominantly gray marl and calcareous claystones. The thickness of the Shetland Group averages 3500 ft (1067 m) in the Cormorant area. Toward the end of the Late Cretaceous, the Laramide tectonic phase produced a regional unconformity that marks the Cretaceous–Tertiary boundary.

Tertiary sedimentation is represented by some 4500 ft (1372 m) of fairly unconsolidated marine clays and sands. The lower zone, the Lista Formation of the Montrose Group, is overlain by a regionally correlatable volcanic ash deposit known as the Balder Tuff Formation of the Rogaland Group. Above this tuff is a sequence of claystones and sands known as the Hutton Clay and Hutton Sand formations of the Nordland/Hordaland Group. Neither is hydrocarbon prospective in the Cormorant area.

TRAP

All four Cormorant oil accumulations have similar trapping conditions. Blocks I, II, and III are dip-closed to the west and closed to the east by faulting and erosion that juxtaposes the reservoir against Cretaceous mudstones (Figures 6A and 6B). Block

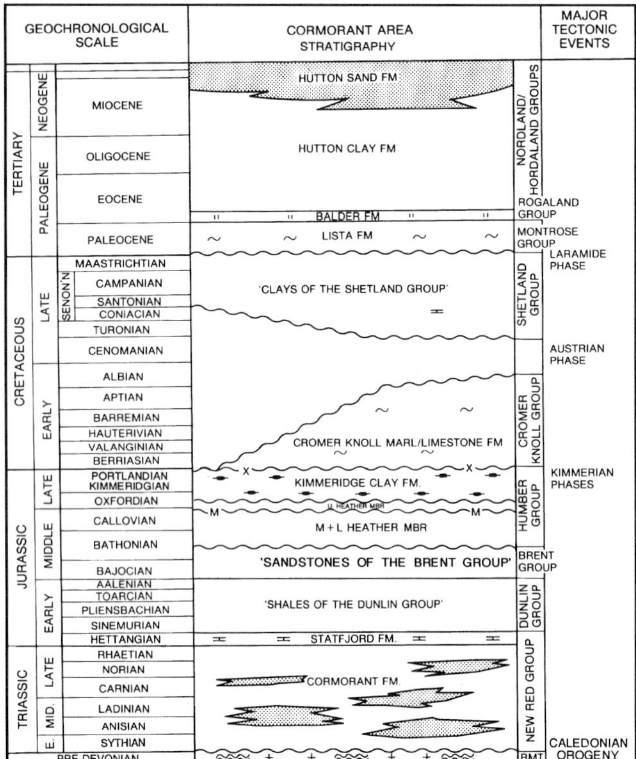

Figure 13. General stratigraphic table for the Cormorant field. Cormorant produces from the Bajocian-Bathonian sandstones of the Brent Group. The Kimmeridge Clay is the hydrocarbon source rock for the Jurassic reservoirs of the northern North Sea. On the crest of the tilted fault blocks the Brent Group reservoirs are unconformably overlain by the Heather, Kimmeridge Clay, and Cromer Knoll groups. M, middle Callovian unconformity. X, unconformity beneath the Cretaceous.

IV is trapped by major east-dipping faults on the west and east and closed by dip to the north and south. The top and lateral seals of all the blocks are marine claystones of the Jurassic Humber Group. These unconformably overlie the Brent Group reservoir on the crest of the structures.

Existing log and pressure data indicate that each block has a different original oil-water contact (OOWC): in Block I, 9250 ft (2819 m); in Block II, 9830 ft (2996 m); and in Block III, 9200 ft (2804 m). The OOWC in Block IV was assumed to be 9600 ft (2926 m) from early appraisal drilling. However, a comprehensive review of log, pressure, geological, and seismic data obtained from ongoing development drilling has shown a variation of fluid levels from 9600 to 10,000 ft (2926 to 3048 m) (Figure 14). Some uncertainty still exists as to the exact positions of the fault boundaries between the subblocks. The majority of these fault boundaries have been partially broken down because of large pressure differentials (on the order of 600 to 1000 psi; 4100 to 6900 kPa) created across them as a result of depletion.

The spill points for the Cormorant blocks are unknown, as the original oil-water contacts (OOWCs) in each block ultimately terminate against a major boundary fault. The exact juxtaposition geometries along these faults are unknown.

The field is characterized in all four blocks by an eroded crestal area. In Blocks I, III, and IV, truncation extends into the Dunlin Group. In addition, the transgressive sand of the Brent Group Tarbert Formation is not present anywhere in Block I, either as a result of erosion or nondeposition (see Figures 16, 17, and 19). The exact areas and extent of erosion are more difficult to define in Block II owing to the poorer quality 2D seismic data over this block and a lack of well data. However, it appears that the upper reservoir is eroded on the crest of the structure. The pattern of erosion in the crestal regions (eastern edge) of Block IV is complex. The upper Ness Formation appears to have completely eroded at the highest point on the structure, and the erosion products redeposited down structure to the west as part of the Tarbert Formation. Detailed stratigraphic correlation shows that the Tarbert Formation onlaps the upper Ness Formation from west to east.

The traps were formed by the Middle Cretaceous, well before oil migration charged the structures during the Eocene. Overpressures of 1110, 1010, and 1005 psia (7.65, 6.96, and 6.93 MPa) for Blocks I to III, respectively, and between 1150 and 1270 psia (7.93 and 8.76 MPa) for Block IV were probably developed during the Tertiary. Figure 15 shows the burial history curve for well 211/21-1.

RESERVOIR

Stratigraphy and Depositional Setting of the Brent Group

The Brent Group comprises marine and nonmarine deltaic sedimentary rocks of Middle Jurassic, Bathonian age. It is divided into five formations which are, from top to base: Tarbert, Ness, Etive, Rannoch, and Broom. They are the result of two distinct depositional systems. The older one is represented by the Broom Formation, consisting of basal coarse sandstones of submarine fan delta origin. The remainder of the Brent Group (Rannoch, Etive, Ness, and Tarbert formations) comprises the younger second system (Brown et al., 1987), which is interpreted as a diachronous, wave-dominated, coastal deltaic complex that prograded in a north-northeast direction (regionally) within the Viking graben system (Figure 16) (Budding and Inglin, 1981). Differential subsidence and faulting during this progradational phase expressed itself in thickness variations of the system and rarely affected the expression of lithofacies types (Johnson and Stewart, 1985). Figure 17 shows a north-south stratigraphic correlation across the Cormorant field, from Block III to Block II.

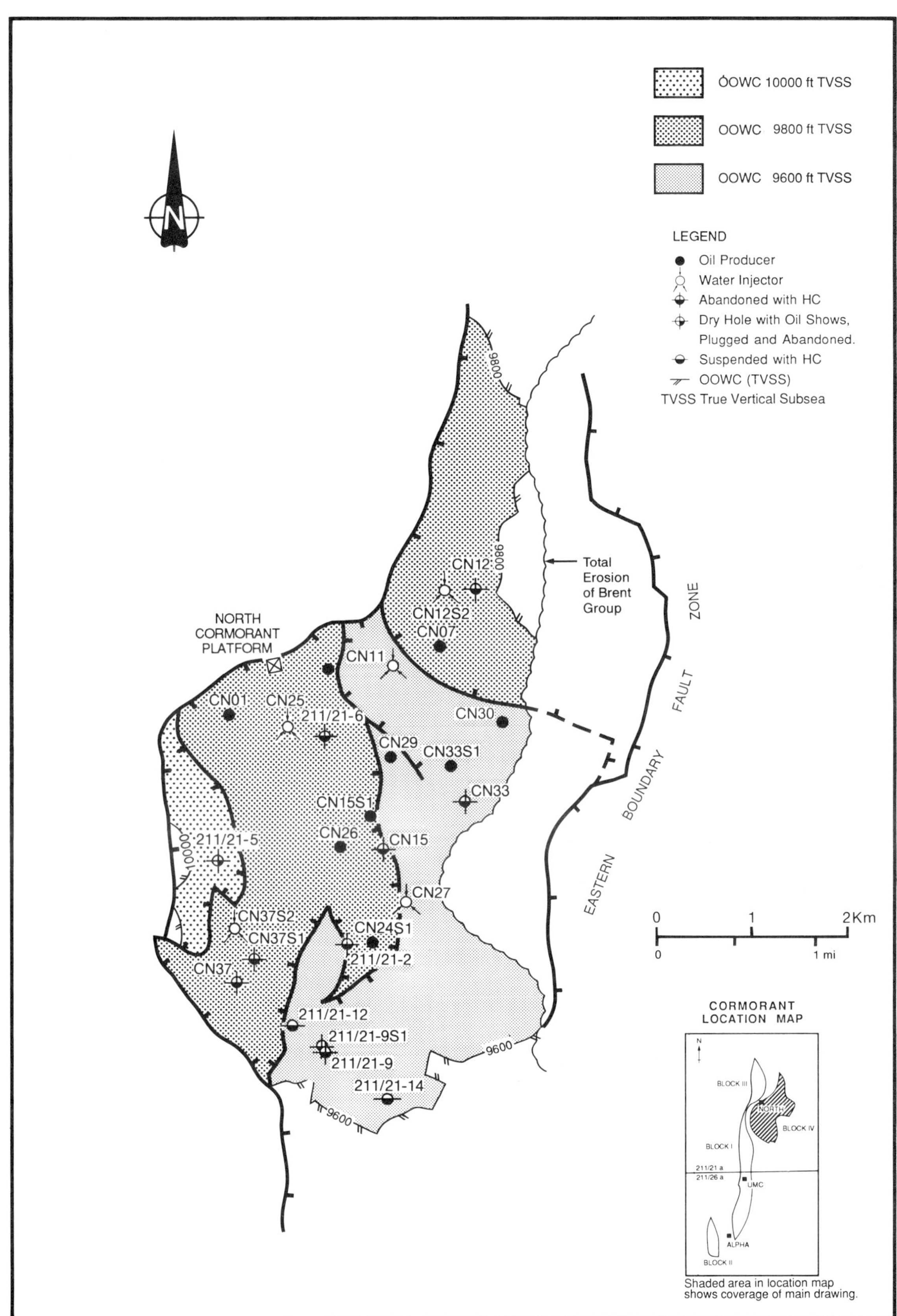

Figure 14. The complex hydrocarbon distribution in Block IV. Uncertainty exists as to the exact boundaries separating the different original oil-water contact (OOWC) areas. The OOWC in the northern and eastern areas has not yet been established owing to the lack of well control in these areas.

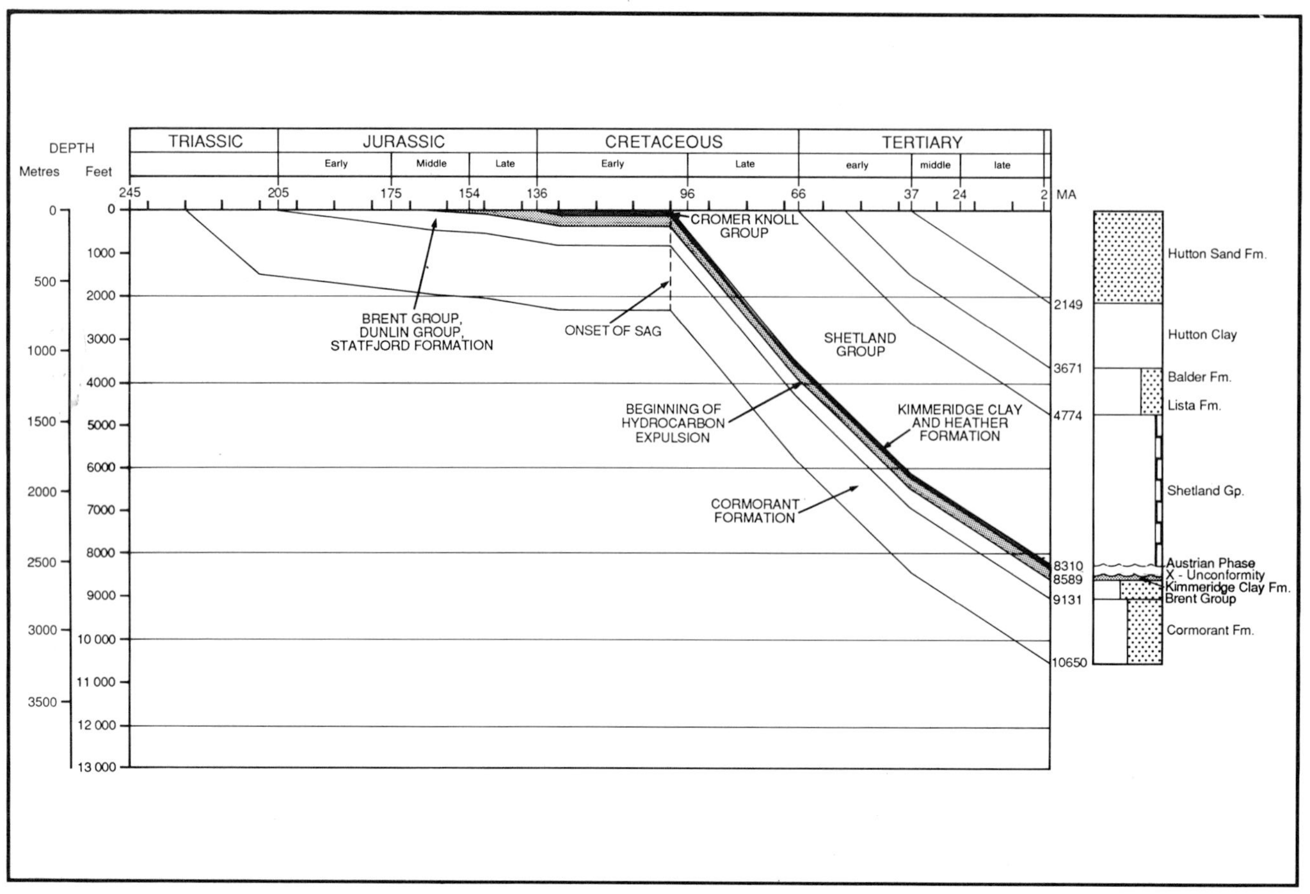

Figure 15. The burial history curve for 211/21-1, highlighting the onset of sag and beginning of hydrocarbon expulsion for the area. Note that erosion by the X-unconformity is minimal at this downflank location.

These five formations are subdivided into subunits for effective reservoir management on a block by block basis.

The main stratigraphic characteristics of the formations in the Brent Group and their variation on a regional scale over the four fault blocks are reviewed in Table 1.

Broom Formation: The Broom thickness varies from 10 to 55 ft (2 to 15 m), pinching out toward the east and graben center (Figure 18). The thickness trends suggest that the formation acted as an important early phase of lateral infill. Reservoir quality is variable but generally poor. Porosities average 20% and core air permeabilities are less than 200 md (Figure 19).

Rannoch Formation: Thickness variations in the Rannoch range from 50 to 102 ft (15 to 30 m) and exhibit a north-south trend, roughly in line with the delta progradation direction (Figure 18). This trend is considered to be related to fluctuations in the rates of sediment input and basin subsidence (Budding and Inglin, 1981). The reservoir quality is poor, with porosities on the order of 17% and permeabilities 100 to 300 md.

Etive Formation: In the Cormorant field, the Etive sandstones are 30 to 98 ft (10 to 30 m) thick (Figure 18). These marked variations are explained by facies changes within the Etive between foreshore, barrier, and channel sandstones. Individually these sandstones commonly have average porosities of 25%. Permeabilities are generally in the range of 500 to 6000 md. These excellent reservoir properties ensure good vertical and lateral communication.

Ness Formation: The Ness Formation in the Cormorant field can be divided into three members: upper and lower Ness sandstone members separated by the prominent 15 to 20 ft (4 to 6 m) thick middle Ness shale member (Figure 20).

The lower Ness member is intimately associated with the Etive Formation, and lateral transition between the two formations can be demonstrated across Block IV. The unit thins from south to north, from 70 ft (21 m) in Blocks I and II to 15 ft (4 m) in Block III. In the Eider field 15 km (9 mi) to the north it is absent. The same northward thinning occurs in Block IV where thicknesses decrease from 70 to 20 ft (21 to 6 m). Reservoir quality is variable, and the overall degree of vertical continuity is poor.

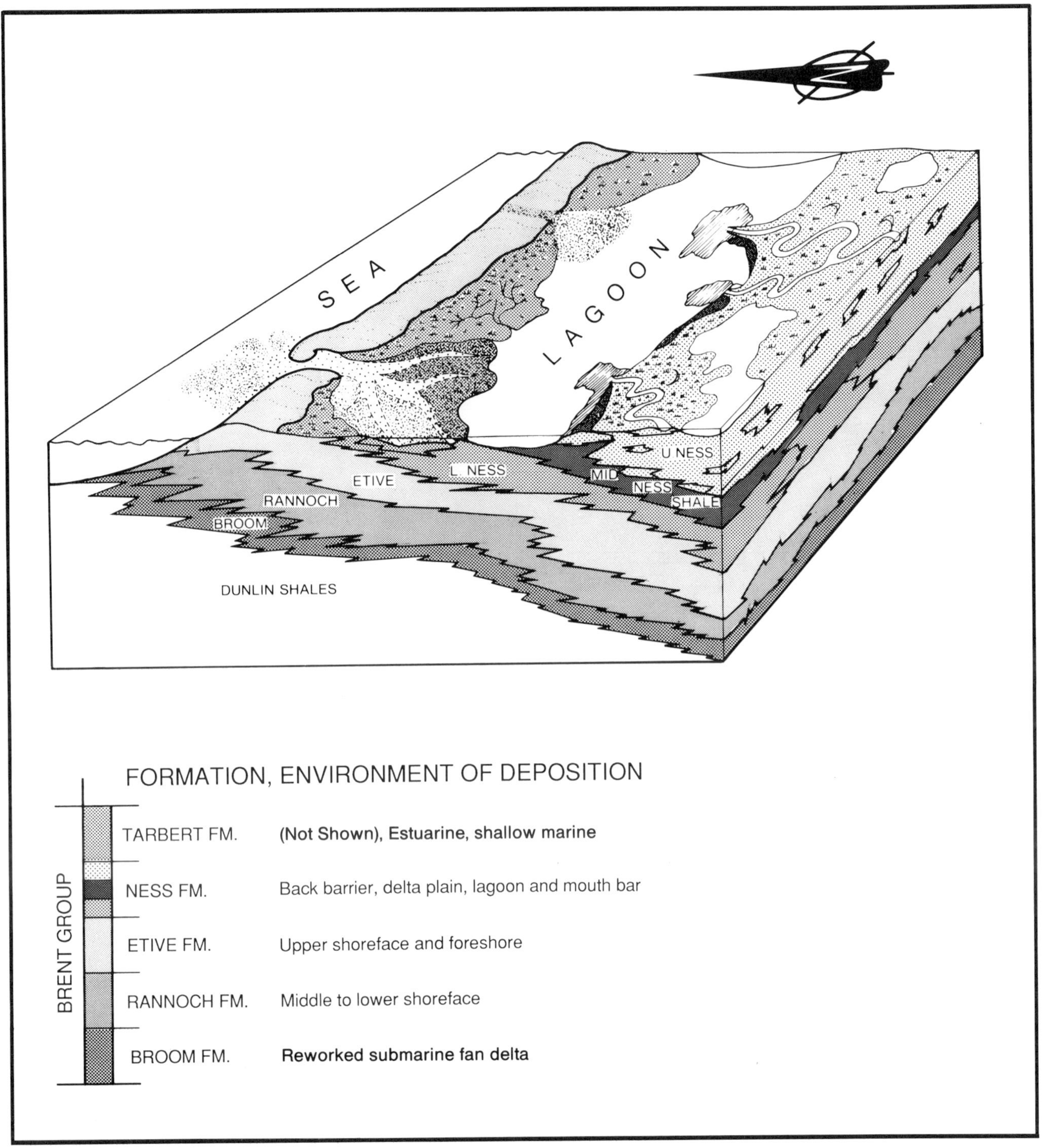

Figure 16. Schematic depositional model for the Brent Group in the north Viking graben system (after Budding and Inglin 1981). This wave-dominated coastal deltaic complex prograded in a north-northeast direction.

The middle Ness shale is a nonreservoir unit 15 to 20 ft (4 to 6 m) thick. This forms a transmissibility barrier dividing the Brent Group into an upper (Tarbert and upper Ness) and a lower reservoir (lower Ness, Etive, Rannoch, and Broom).

The upper Ness sandstones are similar to those of the lower Ness. Thickness averages some 50 ft (15 m) but shows marked variation resulting from erosional thinning or removal in some crestal areas. The unit achieves maximum thickness (190 ft; 58 m) in the western part of Block IV, probably as a result of rapid subsidence along the western boundary fault system. The sandstones commonly consist of good reservoir material interbedded with shales and occasionally coals. Porosities average 21% and permeabilities are on the order of 500 to 1000 md.

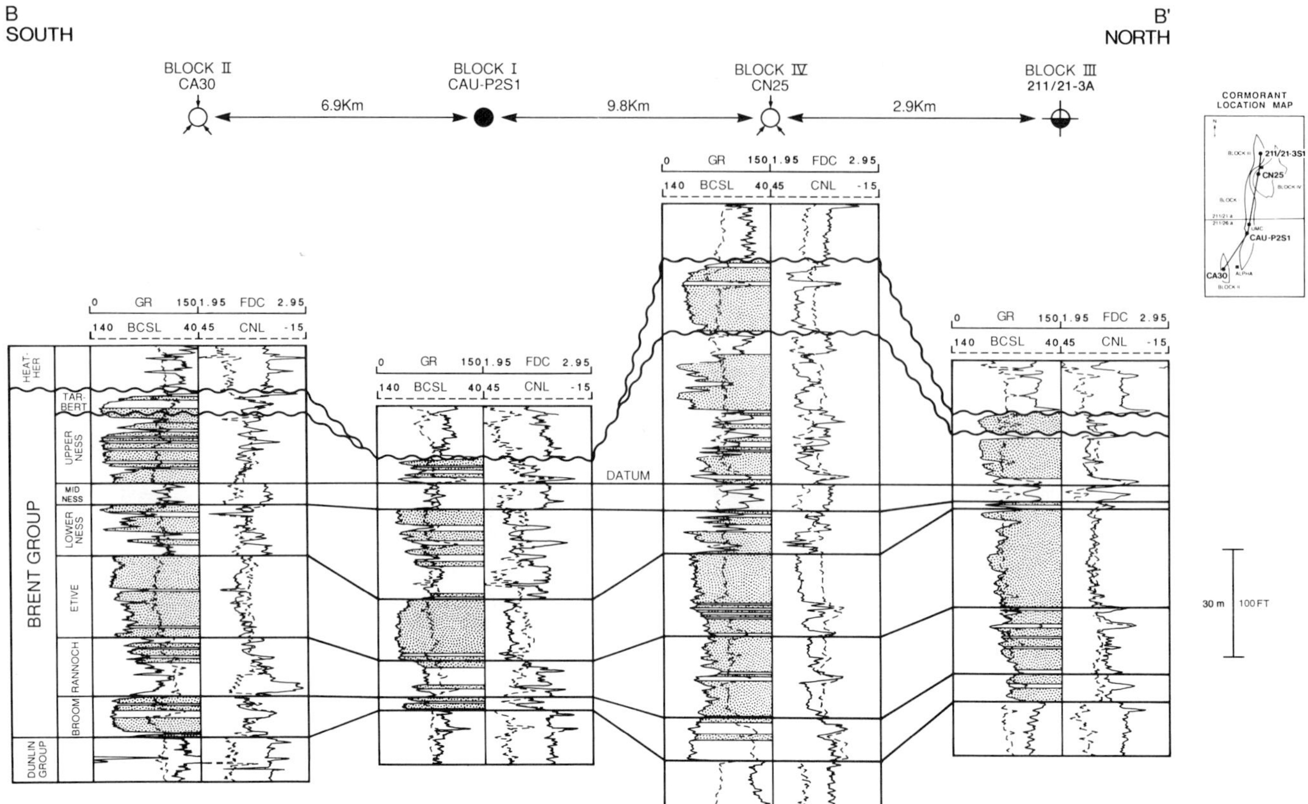

Figure 17. Stratigraphic correlation showing the primary formations of the Brent Group. The thickness and reservoir quality of each formation varies from block to block. Subdivision for field development purposes therefore differs per block. The line of section B–B′ is shown in Figures 6A and extending across 6B. The log traces displayed are gamma ray (GRS), sonic (BCSL), density (FDC), and neutron (CNL).

Tarbert Formation: The thickness of the Tarbert varies from block to block (Figure 20). In Block I, it is totally absent either as a result of nondeposition or erosion. In Blocks III, II, and IV, it is eroded on the crest of the structures and thickens downflank (maximum thickness 104 ft or 30 m in Block IV). In general the Tarbert Formation is a relatively good reservoir rock with an average porosity of 22%. Permeabilities are variable but typically 100 to 500 md.

Each formation constitutes a particular association of facies that was deposited in a unique part of the coastal-deltaic complex that together can be over 500 ft (152 m) thick.

The Broom Formation is intepreted to represent prograding fan deltas in a shallow marine setting, which underwent some later marine reworking prior to the advancement of the Brent delta (Graue et al., 1987). The Rannoch, Etive, and Ness formations form the transition from shoreface to emergent barrier and finally lagoonal environments of the wave-dominated Brent delta system. The Rannoch Formation forms the storm-dominated, offshore to shoreface regressive sequence that grades into the upper shoreface to foreshore, and then into the emergent barrier environment of the Etive. The Ness Formation represents the back barrier deposition in this system. It includes washover fans, lower delta plain deposits, and channel fills. The Tarbert Formation is interpreted as a transgressive formation formed as the Brent delta was submerged owing to a relative rise in sea level.

Diagenesis and Its Effect on Reservoir Quality

The Brent Group sandstones range from very fine to coarse-grained variably sorted arenites. Texturally the deposits coarsen upward from Rannoch to Etive formations and have an improvement in sorting from Rannoch to lower Ness formations.

A trend of increasing quartz, decreasing feldspar, mica, and clay content is seen within the Rannoch to the lower Ness sequence. The decrease in feldspar continues into the upper Ness Formation, whereas mica, kaolinite, and authigenic silica become slightly more abundant in the upper Ness than in the lower Ness Formation.

Petrographic analysis indicated that all the reservoir strata underwent a broadly similar pattern of postdepositional alteration. The only control upon diagenesis exerted by the depositional environment and associated pore waters would appear to be the presence of early authigenic illite and clay rim

Table 1. The main lithological and depositional characteristics of the Brent formations.

FORMATION	RESERVOIR UNIT	GENERAL CHARACTERISTICS	SEQ	ENVIRONMENT	GROSS THICKNESS (FEET)	SAND BODY DISTRIBUTION
		GRADATIONAL BOUNDARY				
TARBERT	BLOCK I	Absent, due to erosion or non-deposition		ESTUARINE/ SHALLOW MARINE	0-100 FT	LATERALLY CONTINUOUS IN BLOCK II and III DISCONTINUOUS IN BLOCK IV ABSENT IN BLOCK I
	BLOCK II, BLOCK III AND BLOCK IV	In Block II 2-30 ft thick. In Block III 5-18 ft thick, in Block IV up to 100 ft thick medium-coarse grained homogeneous sandstones with occasional gravel layers, mudclasts and carbonaceous debris; increased shale content where reworked Upper Ness Shales				
		SHARP EROSIONAL BOUNDARY				
UPPER NESS		Interbedded sandstones, shales and coals (in Block III coarsening upwards to a blocky sandstone package with a shale break in the middle and at the top). In Block IV Upper Ness is thickened due to activation of main boundary fault displacing Block IV relative to main Cormorant structure		DELTA PLAIN AND BAY FILLS	35-120 FT	VARIABLE FAN SHAPED (MOUTH BAR) AND ELONGATE CHANNELS E-W EXTENT 1-2 Km N-S EXTENT 2-4 Km
		GRADATIONAL BOUNDARY				
MID NESS SHALE		Shale + minor ripple (wave + current) laminated sandstones. Bioturbated extensive back barrier lagoon		LAGOON	15-20 FT	—
		SHARP BOUNDARY				
LOWER NESS		Interbedded sandstones, shales and coals - shales: grey laminated muds, wave rippled, slumped bioturbated - sandstones: medium grained, cross bedded to fine grained micaceous, bioturbated, rippled. Main reservoir sandstones are in channels: crossbedded at the base passing into rippled sandstones capped by deformed silts, muds and coals. Thickness 70 ft in Block IV, Block I and Block II, decreasing to 20 ft in Block III		DELTA PLAIN and BACK BARRIER	20-70 FT	ELONGATE CHANNELS FAN SHAPED MINOR MOUTH BARS SPLAYS
		SHARP BOUNDARY				
ETIVE		Fine to medium grained sands generally cross bedded and horizontally stratified at base, partly stratified mottled or rippled sandstone above, mud, coal clasts and clay drapes occasionally present		UPPER SHORE FACE AND FORESHORE AREAS (BEACH, DUNE, CHANNEL/ TIDAL INLET	30-100 FT	LATERALLY EXTENSIVE
		GRADATIONAL BOUNDARY				
RANNOCH		Bioturbated argillaceous and micaceous silts passing upward into horizontal /low angled stratified micaceous silts and fine sandstones with common calcite cemented horizons		MIDDLE-LOWER SHORE FACE	60-100 FT	LATERALLY EXTENSIVE
		SHARP BOUNDARY				
BROOM		Clay rich and burrowed, faintly laminated or cross bedded coarse grained sandstone.		REWORKED FAN DELTA	10-35 FT (BLOCKS I, II and III) 15-55 FT (BLOCK IV)	LATERALLY EXTENSIVE INHOMO – GENEOUS DISTRIBUTION OF GRAIN SIZE.
		SHARP BOUNDARY				

LEGEND SEQUENCE NOTATION ▼ COARSENING UPWARDS ▲ FINING UPWARDS MAIN SHALE LAYERS (schematic)

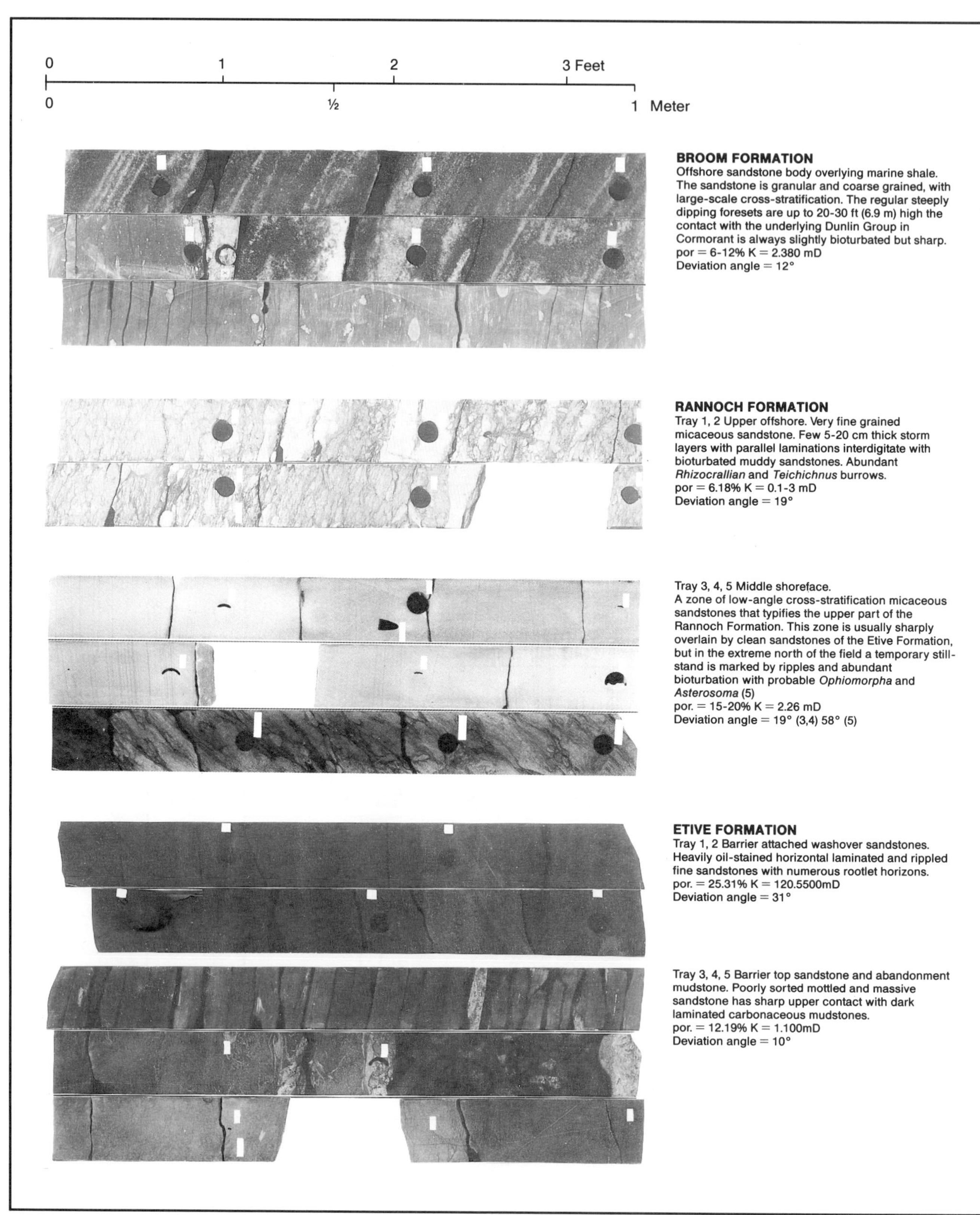

Figure 18. Broom, Rannoch, and Etive core photographs.

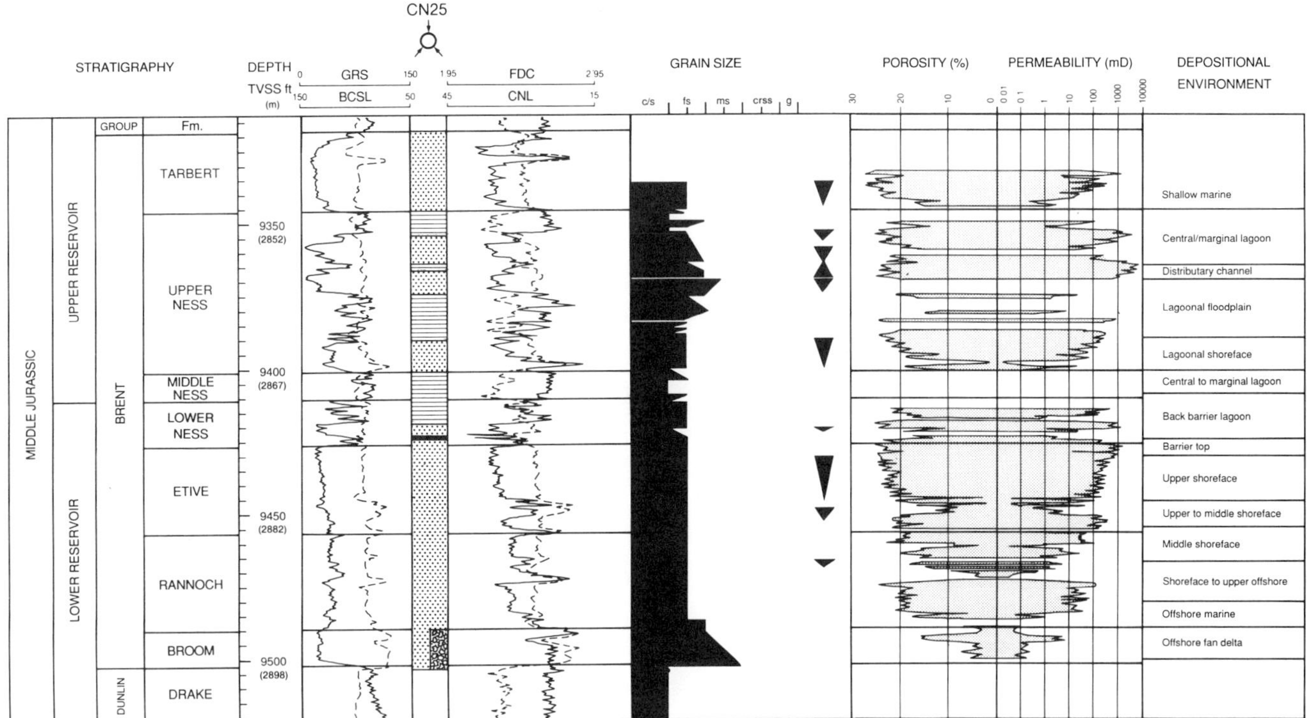

Figure 19. Type log of the Brent Group for the Cormorant field showing a typical porosity/permeability profile and environmental interpretation. The depth scale is in feet subsea. (TVSS, true vertical subsea.)

chlorite cements. These are generally present in sandstones of marine origin. Further diagenetic modifications involving the precipitation of a range of authigenic cement phases are common to the entire reservoir sequence.

For the Brent Group sandstones, in general, kaolinite, detrital feldspar, and mica are particularly important in the control of reservoir quality. Mica flakes and laminae reduce primary porosity and permeability and are also sites of kaolinite formation that further reduces primary pore space. The leaching of feldspars appears to be the main source of alumina and silica for the formation of authigenic quartz, which also causes the deterioration of the reservoir quality by decreasing pore size. The two important controls on permeability are kaolinite and grain size that influence pore-throat size.

Chlorite comprises 5 to 55% of the less than 2μ clay fraction in the reservoir. Authigenic chlorite generally occurs in the form of well-developed pore-lining crystal plates 1μ to 5μ in diameter. It has only been recorded in the Broom Formation (Budding and Inglin, 1981; Kantorowicz, 1984). Recent analysis from Cormorant Block IV, however, has identified authigenic chlorite in the Rannoch, Etive, and lower Ness formations, in certain intervals and in certain wells. This suggests that chlorite is not just restricted to freshwater environments (Kantorowicz, 1984) in the Cormorant area but that marine conditions were also favorable to the growth of chlorite cements.

The Effects of Illite on Permeability

In South Cormorant, mainly Block II, the effects of illite clays on reservoir quality have been quite marked. Illite occurs during early and late stage diagenesis. The pore-lining early illite is ubiquitous as a coating on detrital clay, quartz, and feldspar grains in the reservoir sandstones. The late illite occurs as platy and fibrous webs "blooming" out between quartz overgrowths, forming fibrous bridges across pore throats and also replacing kaolinite grains. This pore-filling illite is only formed below the oil-water contact (Figures 21A and B) and is present in extremely small quantities in the sandstones, comprising about 1% of the bulk-rock volume. The morphology and distribution rather than quantity of illite has the greatest influence on permeability. The fibrous, platy pore-filling types have large surface areas and block pore throats, drastically reducing permeability.

Porosity/Permeability Relationships

Historically porosity/permeability relationships within the Brent Group reservoir of the Cormorant field have been defined by a log permeability versus in situ porosity for each formation per block (I, II, III, IV). However, in the majority of instances, as permeability is largely affected by grain size, sorting, and diagenetic minerals (illite and kaolinite),

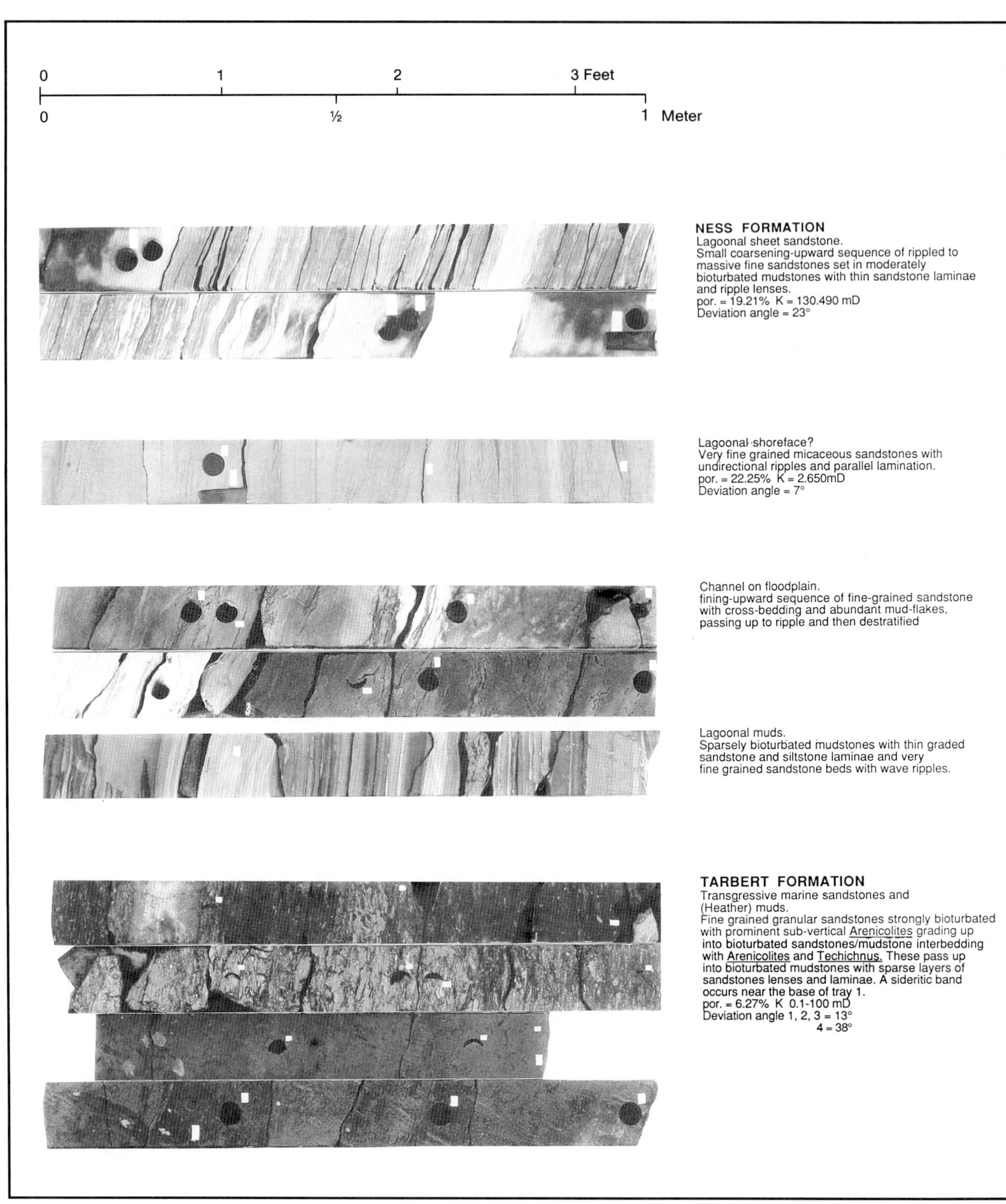

Figure 20. Ness and Tarbert core photographs.

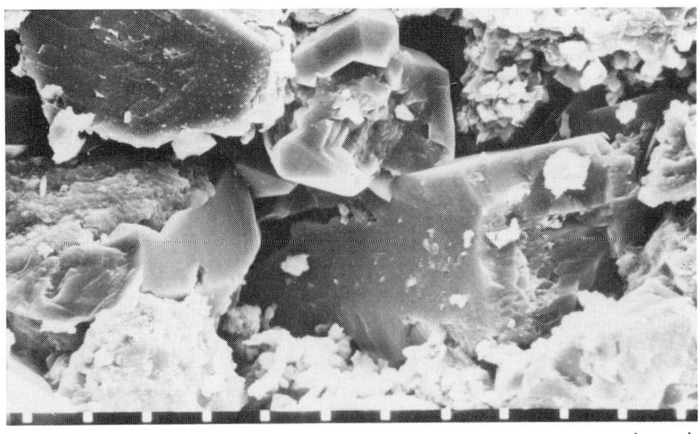

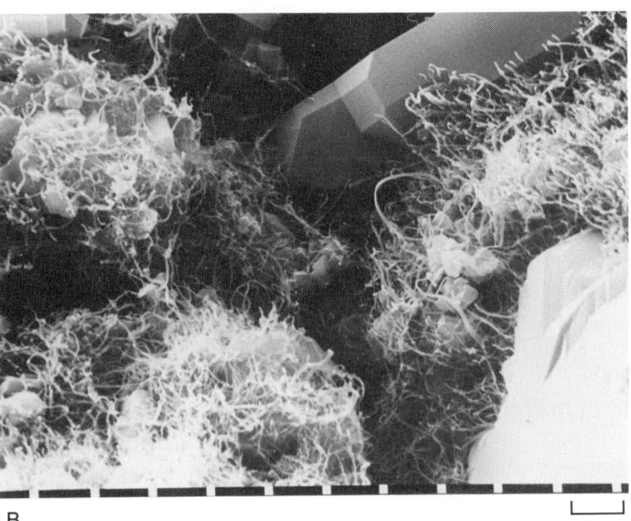

Figure 21. SEM micrographs showing development of illite in the water leg. (A) Oil leg. The grains show overgrowth of quartz; illite is not extensively developed. (B) Water leg. Shows blooming illite. Thin elongate fibers occur projecting between and beneath areas of quartz overgrowth.

refinement of this relationship is ongoing to achieve a better permeability measurement.

Porosity for each individual formation is depth related; porosity decreases with depth at an average rate of 4% per 1000 ft (305 m). This is thought to be mainly caused by compaction of diagenetic minerals (authigenic quartz, feldspars). No jump in porosity values across the OWC is observed.

Hydrocarbon Characteristics

Reservoir fluid properties differ from block to block. Although the oil composition between blocks varies, it stays within the paraffinic-naphthenic class of Tissot and Welte (1984). Initially the crude oil in all four accumulations contained no sulfur. The oil in all blocks is undersaturated and therefore there are no gas caps. The viscosity at the operating pressures (3500 to 4500 psia; 24 to 31 MPa) varies between 0.5 and 1.1 cp. No oil/water interfacial tension measurements have been performed on Cormorant fluid samples. A value of 15 dyne/cm is assumed. Hydrocarbon characteristics for each block are shown in Table 2.

Table 2. The oil characteristics in the Cormorant field.

	Block			
Characteristic	I	II	III	IV
Gravity ° API	33.8	34.3	36.6	34.4
Bubble point Pressure, psia (MPa)	1770 (12.20)	2970 (20.48)	1040 (7.170)	1390 (9.584)
Formation volume Factor B (rb/stb)*	1.3	1.5	1.2	1.2
Solution gas/oil Ratio (scf/stb)	470	770	200	310

*rb/stb, reservoir bbl/stock tank bbl; scf/stb, standard ft^3/stock tank bbl.

Fluid Flow Characteristics

In the Brent Group reservoir sequence, the middle Ness shale is the single major fieldwide barrier to flow. It is up to 20 ft (6 m) thick and forms a perfect barrier to vertical flow. Pressure measurements indicate that the sandstones above and below are in different pressure regimes. Although other shale intervals restrict vertical permeability, they are "block specific." In Block I the lower Ness is characterized by a shale and coal layer near the base. This can be up to 40 ft (12 m) thick. In Block IV the well-developed upper Ness Formation contains two shaly strata that can equal the middle Ness shale in thickness. They both form total or partial barriers to flow.

On a more local scale, especially in the lower Ness and Rannoch formations, calcite stringers called "doggers" are baffles to reservoir flow. These doggers can be seen in cores and well logs to be only a few feet thick. They cannot be correlated across the field or even from well to well. Thus their effect on vertical permeability is considered minimal. The heterogeneity of the lower part of the Rannoch Formation generally acts as a barrier between the Etive and Broom formations because of the abundance of shales and of the highly cemented nature of the very fine grained sandstones. Further effects of diagenesis can be seen in Block II where small quantities of fibrous illites in the water leg have severely reduced the Etive and Ness permeabilities.

Faults also form major flow restrictions to differing degrees and will be discussed in a later section.

The range of inflow performance can vary considerably; therefore, the values quoted in Table 3 should be regarded as indicators only.

Table 3. The fluid flow characteristics of the four Cormorant reservoir blocks.

	Permeability-feet kh (md-ft)	Productivity Index (PI) (bbl/d/psi)	Injectivity Index (II) (bbl/d/psi)
Block I	2900	28	10
Block II	3800	9	3
Block III	75000	49	11
Block IV	72500	60	12

The current average well rates for Blocks I and II are 5000 BOPD. Block I reached peak production in 1988 and Block II in 1989. The Block II peak would have been reached earlier but handling constraints on the platform caused a delay. Block III shows average well rates of 5000 BOPD and reached a peak of 15,000 BOPD in 1985. In Block IV, individual well rates vary from 500 to 20,000 BOPD, and the peak level has not yet been attained.

Development Considerations

In Blocks I, II, and III where good intra-reservoir continuity was expected, the producers are situated on the crest of the blocks with a well spacing of 2300 to 3950 ft (700 to 1200 m). The producers are supported by injectors that are, in most cases, located 3280 to 4220 ft (1000 to 1500 m) downdip, near the oil-water contact or just within the oil leg. Some infill wells have been drilled to improve ultimate recovery.

The situation in Block IV is complicated by dense faulting that heavily compartmentalizes the reservoir. Since some of the faults are sealing, major subblocks are developed by dedicated producer/injector pairs. Thus, the well spacing is a function of the subblock size and the expected tortuosity of the flow path from injector to producer. As field development progresses and differential pressures emerge between subblocks, the fault seals break locally, resulting in partial communication.

The widely varying permeabilities and reservoir qualities of the Brent Group formations cause production and injection to concentrate preferentially in the better quality Etive and Ness sandstones, which results in an uneven flood front that impairs recovery. To minimize bypassed oil and improve sweep efficiency, high-permeability zones are only perforated over a few feet, whereas the lower permeable zones are perforated over the full interval (Stiles and McKee, 1986; Stiles and Valenti, 1987).

Channel sandstones, with large average porosities, are particularly common in the Etive and Ness of Block I and serve as preferential fluid conduits causing early water breakthrough in the producers. Channel identification in wells is possible from logs and cores, but predictions of their lateral extent and direction is difficult because of their meandering nature.

The recovery factor is expected to be 40% for the Brent Group reservoir in Blocks II and III where water displacement by injection is relatively unimpeded by reservoir heterogeneities. However, the recovery factor for Block IV is expected to be lower since hydrocarbon displacement is impeded by flow barriers and severe compartmentalization. Therefore, for each subblock the uncertainty in recovery factor has been assessed probabilistically. This assessment indicates a 33% recovery factor. In Block I the potential of the UMC producers tends to be limited by the pressure drop in the dual 3.5-in. tubing strings, by the 3-in. flowlines in the case of satellite wells, and by the 8-in. lines between the UMC and the Cormorant Alpha platform. Flow rates are consequently only about half of what could be achieved by similar single-string platform wells, without, however, affecting the estimated ultimate recovery of 40%.

Faulting and Its Effects on Transmissibility

Performance data in Block I have shown a barrier between the northern producers and southern injectors. This apparent barrier could be due to a combination of thin Etive development and small faults (50 ft or 15 m; i.e., too small to be seen on seismic data) juxtaposing sandstone and shaly strata. In Block II, a similar limited pressure communication is seen between the crestal producers and mid-flank wells in the upper Ness and Broom formations, suggesting a transmissibility barrier. It therefore appears that the crestal and flank areas are separated by a series of small faults that could cause seals in the Ness as a result of the shaly nature of the interval.

The largest effect of fault seals on field development is in Block IV where extensive fault-seal studies had to be made. Sealing could result either from total reservoir offset by a fault juxtaposing reservoir to nonreservoir sections or from a reduction in permeability along fault planes caused by shale smearing and/or preferential cementation. In Block IV, although no total reservoir offset occurs, shaly intervals within the Brent sequence (notably the middle Ness) do cause partial transmissibility barriers. Thus, clay smearing along the fault planes is thought to be an important factor.

In Block IV fault throws are not thought to be large enough to produce a total seal by clay smearing. However, some sealing has been observed, and in general a better communication exists in the "cleaner" lower reservoir than in the "shaly" upper reservoir. Pressure and production data suggest that where seals occur they cannot withstand large pressure differentials. Owing to the high production/injection ratios in the block, most of these seals have been broken. However, whereas "isolated" subblocks may exhibit depleted pressures, producer/injector pairs are still required in each subblock for effective drainage.

SOURCE

The organic-rich black shales of the Upper Jurassic Kimmeridge Clay Formation (Humber Group) are the most prominent oil source rocks of the northern North Sea (Cornford, 1986). The thickness of this formation varies greatly over the area, reaching a maximum of 535 ft (163 m) around the Cormorant field. Although the Middle and Lower Jurassic sequences could be alternative source rocks, studies show that no correlation exists between the composition of their organic constituents and the oil. It is therefore unlikely that these mudstones contributed to the hydrocarbons in the Cormorant area.

The Kimmeridge Clay Formation is a very rich source rock with an average total organic carbon content of 5.6% and a maximum of 12.5%. The main maceral of the source rock is the hydrogen-rich, structureless organic matter associated with framboidal pyrite. Characteristic of the mudstone is the common occurrence of dinoflagellates. The clay shows a minor land plant contribution in the alkane distribution of immature samples, although after maturation a significant landplant contribution is hardly recognizable. The Kimmeridge Clay also has a high phytane/nC_{18} ratio, a characteristic that is also observed in the Cormorant hydrocarbons. Thus, oil/source rock correlations are generally good in the North Sea. The kerogen type is II, indicating source organisms were spores and cuticles. The estimated vitrinite reflectance (R_0) is 0.6%.

Hydrocarbon expulsion began 65 Ma and peaked between 40 and 50 Ma (Figure 15). The migration path was probably from the Ninian/Alwyn area in the south, northward to Cormorant, Tern, and Eider (see Figure 22).

In terms of yield, Goff (1983) has calculated an overall efficiency of generation, migration, and entrapment for the East Shetland basin to be between 20 and 30%.

EXPLORATION AND DEVELOPMENT CONCEPTS

The exploration for and discovery of the Cormorant field was largely made with the knowledge of the Brent field some 25 mi (40 km) to the east. The discovery well tested two exploration plays (Paleocene and Jurassic) and found only the Middle Jurassic Brent Group hydrocarbon bearing. Subsequent appraisal and development of the field experienced major problems in terms of structural control and reservoir communication. Three-dimensional seismic has been an essential tool in field development and has been used extensively in defining the numerous faults that characterize the field, in particular Cormorant Block IV. Had this technique been available at the time of discovery, delineation and recognition of four separate structural accumulations may have been possible. Some of the uncertainties in appraisal and development planning could have been removed.

Early development plans for the Cormorant field have all been modified based on the results of early development activity and reservoir geology/engineering simulation studies. Detailed core description has led to two main changes in development. The first has involved the recognition and definition of the stratification of reservoir sandstones and the permeability contrasts among them. Consequently, the completion policy was modified from one of waterflooding all layers concurrently to partially perforating the higher permeability sandstones to achieve an equal floodfront movement rate through all reservoir layers (Stiles and Valenti, 1987). The second has highlighted the importance of diagenetic effects on the reservoir and resulted in repositioning of the injectors and ruling out the original plan of a passive distribution manifold scheme to provide injection support in Block IV.

Finally, much has been learned concerning field development using the underwater manifold center. Although cheaper than a conventional platform, the high cost of drilling satellite wells, expensive workovers, and severe operating restrictions make the costs extremely high and reservoir monitoring activity difficult. Wherever possible, locations that were originally planned as UMC wells are drilled as highly deviated (70°) wells from the platforms.

ACKNOWLEDGMENTS

The author expresses her appreciation to Shell and Esso Exploration and Production for giving permission to publish this paper. Appreciation is also given to J. Marshall and M. Bentley for fruitful discussions and advice. I also wish to acknowledge unpublished Shell Expro studies by many former and present colleagues. Thanks must also go to the drafting team and to the northern word processing section.

REFERENCES CITED

Brown, S., P. C. Richards, and A. R. Thompson, 1987, Patterns in the deposition of the Brent Group (Middle Jurassic) U.K.

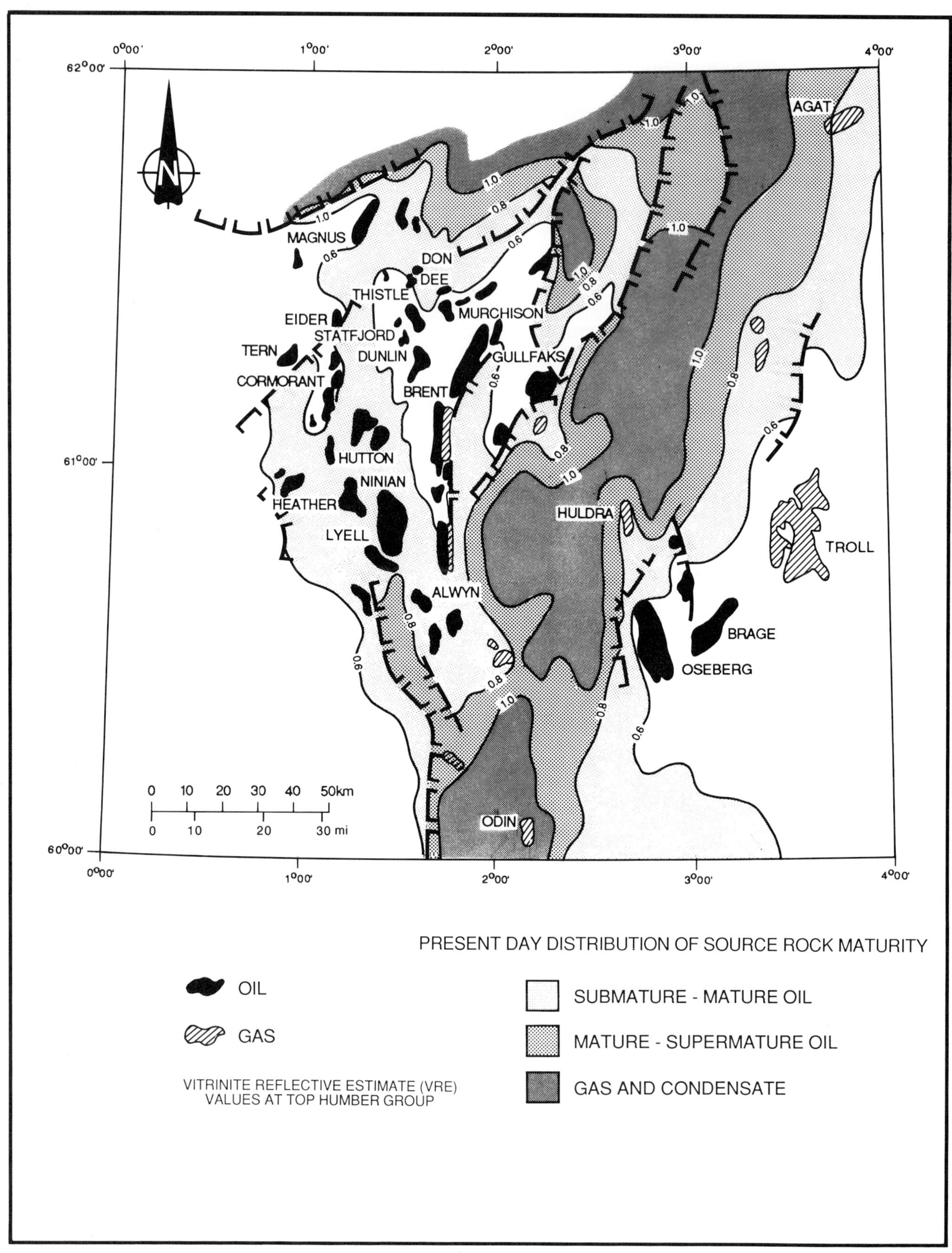

Figure 22. The distribution of oil and gas source rock maturities in the northern North Sea. Vitrinite reflectance estimate values refer to the top of the Humber Group (Kimmeridge Clay Formation).

North Sea, *in* J. Brooks and K. W. Glennie, eds., Petroleum geology of North West Europe: London, Graham and Trotman, v. 2, p. 899-915.

Budding, M. C., and H. F. Inglin, 1981, A reservoir geological model of the Brent sands in South Cormorant, *in* L. V. Illing and G. D. Hobson, eds., Petroleum geology of the continental shelf of north-west Europe: London, Heyden and Sons, p. 326-335.

Carmalt, S.W., and B. St. John, 1986, Giant oil and gas fields, *in* M. T. Halbouty, ed., Future petroleum provinces of the world: AAPG Memoir 40, p. 11-54.

Cornford, C., 1986, Source rocks and hydrocarbons of the North Sea, *in* K. W. Glennie, ed., Introduction to the petroleum geology of the North Sea: Oxford, Blackwell Scientific Publications, p. 133-160.

Deegan, C. E., and B. J. Scull, 1977, A proposed standard lithostratigraphic nomenclature for the central and northern North Sea: Report of the Institute of Geological Sciences 77/25.

Gaarenstroom, L., 1984, The value of 3d seismic in field development: Society of Petroleum Engineers, 13049, presented at 59th Annual Technical Conference and Exhibition held in Houston, Texas, September 16-19, 1984.

Gibbs, A. D., 1984, Structural evolution of extensional basin margins: Journal of Geological Society of London, v. 141, p. 609-620.

Goff, J. C., 1983, Hydrocarbon generation and migration from Jurassic source rocks in the E. Shetland basin and Viking graben of the northern North Sea: Journal of the Geological Society of London, v. 140, p. 445-474.

Graue, E., W. Helland-Hansen, J. Johnson, L. Lomo, A. Nottvedt, K. Ronning, A. Ryzeth, and R. Steel, 1987, Advance and retreat of Brent delta system, Norwegian North Sea, *in* J. Brooks and K. W. Glennie, eds., Petroleum geology of North West Europe, v. 2: London, Graham and Trotman, p. 915-939.

Johnson, H. D., and D. J. Stewart, 1985, Role of clastic sedimentology in the exploration and production of oil and gas in the North Sea, *in* P. J. Brenchley and B. P. J. Williams, eds., Sedimentology: recent developments and applied aspects: Special Publication of Geological Society of London, v. 18, Oxford, Blackwell Scientific Publications, p. 248-310.

Kantorowicz, J., 1984, The nature, origin and distribution of authigenic clay minerals from Middle Jurassic Ravenscar and Brent Group sandstones: Clay Minerals, v. 19, p. 359-375.

Speksnijder, A., 1987, The structural configuration of Cormorant Block IV in context of the northern Viking graben structural framework: Geologie en Mijnbouw, v. 65, p. 357-379.

Stiles, J. H., and J. W. McKee, 1986, Cormorant: development of a complex field: Society of Petroleum Engineers, 15504, presented at 61st Annual Technical Conference and Exhibition held in New Orleans, Louisiana, October 5-8, 1986.

Stiles, J. H., and N. P. Valenti, 1987, The use of detailed reservoir description and simulation studies in investigating completion strategies, Cormorant—UK North Sea: Society of Petroleum Engineers, 6553/1, presented at Offshore Europe 87, Aberdeen, September 8-11, 1987.

Tissot, B. P., and D. H. Welte, 1984, Petroleum formation and occurrence: Berlin, Springer-Verlag, p. 678.

Watters, D. G., M. C. T. Pagan, and D. M. J. Provan, 1985, Cormorant field, United Kingdom North Sea. Synergistic approach to optimum development of a complex reservoir (abs.): AAPG Bulletin, v. 69, p. 315.

SUGGESTED READING

The volumes produced after the three U.K. Organised Petroleum and Continental Shelf of North-West Europe Conferences in 1975, 1981, and 1988 provide background information.

Petroleum and the continental shelf of north-west Europe, 1975: London, Applied Science Publishers.

Petroleum geology of the continental shelf of north-west Europe, 1981: London, Heyden and Sons.

Petroleum geology of north west Europe, 1987: London, Graham and Trotman.

Introduction to the petroleum geology of the North Sea, 1986: Oxford, Blackwell Scientific Publications.

Appendix 1. Field Description

Field name .... *Cormorant field*
Ultimate recoverable reserves .... *586 MMstb (million stock tank barrels) of oil*

Field location:

- **Country** .... *United Kingdom*
- **State** .... *Offshore Block 211/21, 211/26*
- **Basin/Province** .... *North Sea East Shetland basin*

Field discovery:

- **Year first pay discovered** .... *Middle Jurassic Brent Group 1972*

Discovery well name and general location:

- **First pay** .... *211/26-1 (60°07′43″N, 01°06′24″E)*

Discovery well operator .... *Shell U.K. Exploration and Production Ltd.*

IP .... *7728 stb/d, 48/64-in. choke with flowing tubing head pressure of 1050 psi (7.240 MPa) (well 211/26-1)*

All other zones with shows of oil and gas in the field:

Age	Formation	Type of Show
Triassic	*Cormorant*	*Oil*

Geologic concept leading to discovery and methods used to delineate prospect

A regional seismic survey indicated the presence of a large (Forties-like) Paleocene drape structure underlying Blocks 211/21 and 211/26 along with poorly defined sub-Cretaceous tilted blocks, potentially comparable to the known Brent field. The first well (211/26-1) showed the Paleocene prospect to be thin and water-bearing but the sub-Cretaceous to be oil-bearing.

Structure:

Province/basin type .... *Bally 1211; Klemme IIIA*

Tectonic history

The field lies in the East Shetland basin, on the western flank of the Viking graben rift system. Rifting began in the Early Triassic in response to crustal extension in the Arctic-North Atlantic domain. The Caledonian basement was strongly downfaulted, and thick graben-fill sequences comprising the reservoir intervals were deposited during the Triassic and Jurassic. Continual subsidence along large normal faults gradually tilted the intervening fault blocks to produce the general structural configuration seen today. The upper Kimmerian "X" unconformity was produced by local erosion of the updip edges of the fault blocks, after which post-rifting thermal sag allowed burial of the succession to present depths.

Regional Structure

The field is located in a major rotated fault block within which three structural trends intersect. The first represents a north-south-striking fault system developed as an accommodation to graben extension; the second is defined by northeast-southwest-striking faults; and the third is made of northwest-southeast-striking faults. The two latter trends are assumed to result from pre-existing trends in the Caledonian basement to the north and south of the field, respectively.

Local Structure

The field extends over four fault blocks: Blocks I, II, III, and IV. Blocks I and III occur along the axial part of a westward-tilted fault block bounded to the east by faulting and erosion and to the west by dip closure. Block II is a separate accumulation which lies to the southwest of Block I. It is a north-northeast-to south-southwest-trending anticline that is fault bounded to the east and east-northeast. Block IV is a large downthrown fault block situated to the east of Blocks I and III, formed as the result of crestal collapse of the main fault block. It is associated with a major east-dipping listric normal fault.

Trap:

Trap Types

The reservoirs in Blocks I, II, and III have dip closure to the west and are partly fault-bounded, partly truncated by erosion to the east. Block IV is mostly fault-bounded and partly dip-closed.

Basin stratigraphy (major stratigraphic intervals from surface to deepest penetration in field):

Chronostratigraphy	Formation	Depth to Top in ft (m)
Upper Cretaceous	*Shetland Group*	*5400 (1650)*
Lower Cretaceous	*Cromer Knoll Group*	*8400 (2550)*
Bathonian-Portlandian	*Humber Group*	*8500 (2600)*
Bajocian-Bathonian	*Brent Group*	*8700 (2650)*
Sinemurian-Aalenian	*Dunlin Group*	*9000 (2750)*
Rhaetic-Hettangian	*Statfjord Formation*	*9300 (2840)*
Triassic	*Cormorant Formation*	*9350 (2850)*

Reservoir characteristics:

Number of reservoirs *1*
Formations .... *Brent Group (Broom, Rannoch, Etive, Ness, and Tarbert formations)*
Ages .... *Bajocian-Bathonian*
Depths to top of reservoirs .... *8700 ft (2650 m) ss (8000 ft, 2440 m, on crest)*
Gross thickness (top to bottom of producing interval) .... *300 (200 to 500) ft; 90 (60 to 150) m*
Net thickness—total thickness of producing zones .... *210 (140 to 350) ft; 65 (40 to 110) m*
Lithology .... *Fine to medium-grained, moderately well-sorted subarkosic and micaceous sandstones*
Porosity type .... *Primary intergranular with some secondary porosity*
Average porosity .... *12 to 28%*
Average permeability .... *Oil, 10 md to 10 d; water, 3 md to 3 d*

Seals:

Upper

Formation, fault, or other feature .... *Sealed by upper Jurassic mudstones*

Lateral

Formation, fault, or other feature .... *Sealed by faults and juxtaposition with mudstones*

Source:

Formation and age .... *Kimmeridge Clay Formation (Oxfordian-Rayazanian)*
Lithology .... *Shale/mudstone*
Average total organic carbon (TOC) *5.6%*
Maximum TOC *12.5%*
Kerogen type (I, II, or III) .... *II*
Vitrinite reflectance (maturation) .... $R_o = 0.6$ *(maximum generation 0.7-1.0% at 10,700 ft, 3250 m, ss)*
Time of hydrocarbon expulsion .... *65 Ma to present, peak from 40 to 50 Ma*
Present depth to top of source .... *8500 ft (2600 m) ss*
Thickness .... *Up to 535 ft (163 m) in the Cormorant area*
Potential yield .... *80 L/m³*

Appendix 2. Production Data

Field name ... *Cormorant field*

Field size:

- **Proved hectares** ... *2774*
- **Number of wells all years** ... *72 (including injectors)*
- **Current number of wells (1/1/88)** ... *29 oil producers; 21 water injectors*
- **Well spacing** ... *2300 to 3940 ft (700 to 1200 m)*
- **Ultimate recoverable** ... *586 MMstb of oil*
- **Cumulative production(1/1/1987)** ... *254 MMstb of oil*
- **Annual production (1987)** ... *44 MMstb of oil*
- **Present decline rate (1987)** ... *13% (reserves per annum, pa)*
  - **Initial decline rate** ... *5% (1983 both Alpha and North platforms producing)*
  - **Overall decline rate** ... *10% (reserves pa to end 1987)*
- **Annual water production (1987)** ... *21 MM bbl*
- **Remaining reserves (1/1/88)** ... *332 MMstb of oil*
- **In place, per acre foot** ... *138 stb/ac-ft*
- **Primary recovery** ... *N/A; water injection from onset*
- **Secondary recovery** ... *586 MMstb of oil*
- **Enhanced recovery, estimate potential** ... *135 MMstb of oil*
- **Cumulative water production (1/1/1988)** ... *68 MM bbl*

Drilling and casing practices:

Casing program

Conductor, 30-in. at 650 ft (200 m) ss; 20-in. at 1850 ft (560 m) ss; 13⅜-in. at 5400 ft (1650 m) ss (100 ft [30 m] into the Shetland Marl); 9⅝-in. at 8200 ft (2500 m) ss (300 ft [100 m] TVD above the Kimmeridge Clay); liner, 7-in. from 8650 ft (2640 m) ss to TD (through the reservoir section)

Drilling Mud ... *Gyp-ligno mud from surface to 12¼-in. hole section; invert oil emulsion mud from 12¼-in. hole section to TD*

Bit Program ... *17½-in. to 5400 ft (1650 m); rockbit; 12¼-in. and 8½-in. to 8200 ft (2500 m) and TD respectively: PDC*

Completion practices:

- **Interval(s) perforated** ... *Initially commingled; currently selective perforations to isolate high permeability layers*
- **Well treatment** ... *Occasionally acidize injectors Blocks II and IV (not common)*

Formation evaluation:

- **Logging suites** ... *LDT (or FDC)/CNL/GR/Sonic/ISF from 200 ft (60 m) above; X-unconformity to TD (including the reservoir interval)*
- **Testing practices** ... *Each well tested bimonthly through the test separator for 4-8 hrs; BS+W, flow rates, and reservoir pressure are monitored*
- **Mud logging techniques** ... *Reservoir zones only*

Oil characteristics:

- **Type** ... *Paraffinic-naphthenic*
- **API gravity** ... *33.8° to 36.0° API*
- **Initial GOR** ... *200 to 770 Mscf/stb*
- **Sulfur, wt%** ... *Nil*
- **Viscosity, SUS** ... *0.5 to 1.1 cp*
- **Pour point** ... *ASTM max. 9°F, min -23°F*
- **Gas-oil distillate** ... *NA*

Field characteristics:

Average elevation .... *160 to 170 ft (49-52 m) (platform above sea level)*
Initial pressure .... *Block I, 4960 psi (34.2 MPa) at 8750 ft ss*
Block II, 5165 psi (35.6 MPa) at 9500 ft ss
Block III, 4825 psi (33.3 MPa) at 8750 ft ss
Block IV, 5265 psi (36.3 MPa) at 9100 ft ss
Operating pressure .... *3500 to 4500 psi (24.1 to 31.0 MPa) at datum (1/1/88)*
Pressure gradient .... *0.33 psi/ft (7.46 kPa/m)*
Temperature .... *Blocks I and III, 195°F at 8750 ft ss (90.5°C at 2669 m ss)*
Block II, 225°F at 9500 ft ss (107.2°C at 2898 m ss)
Block IV, 210°F at 9100 ft ss (98.9°C at 2776 m ss)
Geothermal gradient .... *34 to 35°C/km (surface = 4.4°C); 2.89 to 2.90°F/100 ft (surface = 40°F)*
Drive .... *Downflank water injection*
Oil column thickness .... *1000 ft or 305 m (1400 ft or 427 m in Block IV)*
Oil-water contact .... *Block I, 9250 ft (2821 m) ss*
Block II, 9830 ft (2998 m) ss
Block III, 9200 ft (2806 m) ss
Block IV, 9600 ft (2928 m) ss (9800 and 10,000 ft or 2989 and 3050 m in some subblocks)
Connate water .... *10% (from capillary pressure curves)*
Water salinity, TDS .... *17 g/L (14 to 18 g/L)*
Resistivity of water .... *0.14 ohm-m (0.12 to 0.17 ohm-m)*
Bulk volume water (%) .... *NA*

Transportation method and market for oil and gas:

Oil is transported through a 36-in. pipeline to Sullom Voe in the Shetland Islands via the Cormorant Alpha platform. Gas is transported through a 36-in. pipeline to St. Fergus, Scotland. Natural gas liquids recovered on the platform are added into the oilstream. Gas is sold under contract to British Gas.

Draugen Field—Norway
North Sea Basin, Haltenbanken Area

M. J. B. G. GOESTEN
Thai Shell Exploration and Production Ltd.
Bangkok, Thailand

P. H. NELSON
Shell International Petroleum Maatschappij B. V.
The Hague, Netherlands

FIELD CLASSIFICATION

BASIN: North Sea
BASIN TYPE: Rift
RESERVOIR ROCK TYPE: Sandstone
TRAP TYPE: Anticline and Truncation Associated with Tilted Fault Block
RESERVOIR ENVIRONMENT OF DEPOSITION: Shoreline to Offshore Marine
RESERVOIR AGE: Jurassic
PETROLEUM TYPE: Oil

LOCATION

The Draugen oil field is located in Block 6407/9, offshore Mid-Norway, approximately 160 km northwest of Trondheim (Figure 1). The average water depth in Block 6407/9 is 260 m. The Draugen oil field is situated near the southern end of a cluster of oil, gas, and condensate accumulations in the Haltenbanken area (Figure 2). It is estimated that 422×10^6 STBO (stocktank bbl of oil) and 95 bcf of gas will be recovered from the Draugen oil field.

HISTORY

In the 8th Concession Round (1983/84), Block 6407/9 was awarded as production license 093 to a group consisting of the following companies:

Den Norske Stats Oljeselskap (Statoil)	50%
A/S Norske Shell (Operator)	30%
BP Petroleum Development of Norway A/S	20%

Previous exploration in the area in 1981 and 1983 had resulted in the discovery of the Midgard gas/condensate and Tyrihans gas and oil fields 60 to 70 km to the north-northwest of the Draugen field. These discoveries and other exploration drilling that had not been successful established the existence of the same essential geological features in the Haltenbanken area as had proved to be the environment for prolific oil and gas fields in the Brent/Statfjord province in the Viking graben at the northern end of the North Sea, i.e., tilted fault blocks containing Middle and Lower Jurassic sandstone reservoirs capped and charged by the Upper Jurassic Kimmeridgian "hot" shale source rocks.

The field was discovered in June 1984 by exploration well 6407/9-1 (Figure 3), which encountered undersaturated oil in a previously unknown 49 m thick Upper Jurassic sandstone sequence at a depth of 1596 m SS (subsea). The sandstone was found to have excellent reservoir properties and a net oil column of 39 m. The well tested 41° API oil at a rate of 8500 BOPD. The field was named after *Draugen*, a fearsome sea figure of Norse fairy tales.

Five more appraisal wells drilled between November 1984 and February 1986 were logged, cored through the reservoir interval, and tested. The main objectives of the appraisal wells were to define (1) the structural configuration of the Upper Jurassic in the Draugen field, (2) the continuity of the hydrocarbon-bearing Upper Jurassic Rogn Formation sandstone (Figure 4), and (3) the feasibility of oil production and water injection.

STRUCTURE

Tectonic History

The structural and depositional history of the Haltenbanken area can be viewed in the context of the overall evolution of the Norwegian-Greenland rift system.

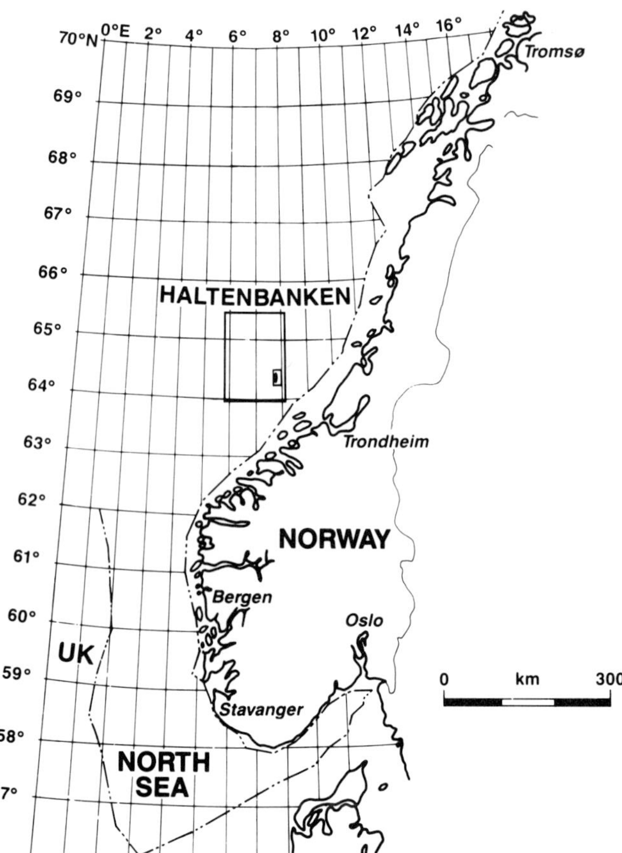

Figure 1. Location map showing the Haltenbanken area (large rectangle), Block 6407/9 (small rectangle), and Draugen field (dark area in small rectangle).

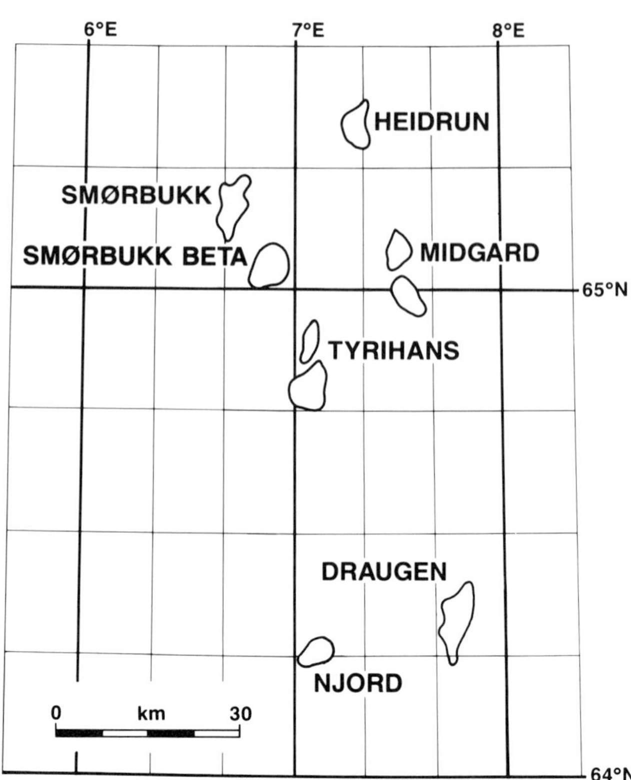

Figure 2. Map showing the various fields in the Haltenbanken area. The Draugen produces from the Middle Jurassic Garn and Upper Jurassic Rogn formations. Middle and Lower Jurassic Garn, Ile, and Tilje formations are the main reservoirs of the other fields.

The earliest post-Caledonian sedimentary record on the west coast of Norway dates from the Devonian. Intramontane grabens and pull-apart basins were created as a result of transcurrent fault movements during that time. These depressions were the sites of thick clastic deposition in both east Greenland and western Norway. During Carboniferous to Early Permian time, molasse-type clastic sequences were deposited in the newly developed rifts in response to a tensional regime that prevailed until the final crustal break-up in early Tertiary time. Similarities in the Mesozoic stratigraphy of east Greenland suggest that analogous upper Paleozoic sediments may be present offshore Mid-Norway. This is further substantiated by the seismic record which locally indicates that below the deepest identifiable horizon (Triassic or Upper Permian) a thick sedimentary section may be present. A major rifting pulse in the middle Permian in east Greenland, with faulting, tilting, dyke intrusion, and uplift, interrupted the continuity of sedimentation.

Toward the end of the Permian, the area between Greenland and Norway was sufficiently peneplained to allow marine waters to transgress from the north. In the central part of east Greenland, Upper Permian reefs are interbedded with black, organic-rich shales of proven oil source rock quality. The development of such facies in the Haltenbanken area remains speculative.

Clastic detritus was again introduced into the area during Triassic time. The Lower Triassic has not yet been recognized in Mid-Norway but is represented by marine shales in central east Greenland. Partial penetration of the Middle and Upper Triassic by wells on the Trøndelag platform of Mid-Norway (Figure 5) encountered a thick succession of red beds alternating with evaporites and shales. Fluviatile conditions continued with deposition of formations characterized by a 400 m thick coal-bearing sequence ranging in age from Rhaetian to early Pliensbachian.

A regional marine transgression occurred during Early Jurassic Pliensbachian time with deposition of some 200 m of tidally influenced deltaic sandstones. The overall Jurassic transgression continued in Toarcian–Aalenian time with deposition of shallow marine shales and sandstones. During the Middle Jurassic a massive sandstone, the Garn Formation, varying from 20 to 100 m in thickness, was deposited

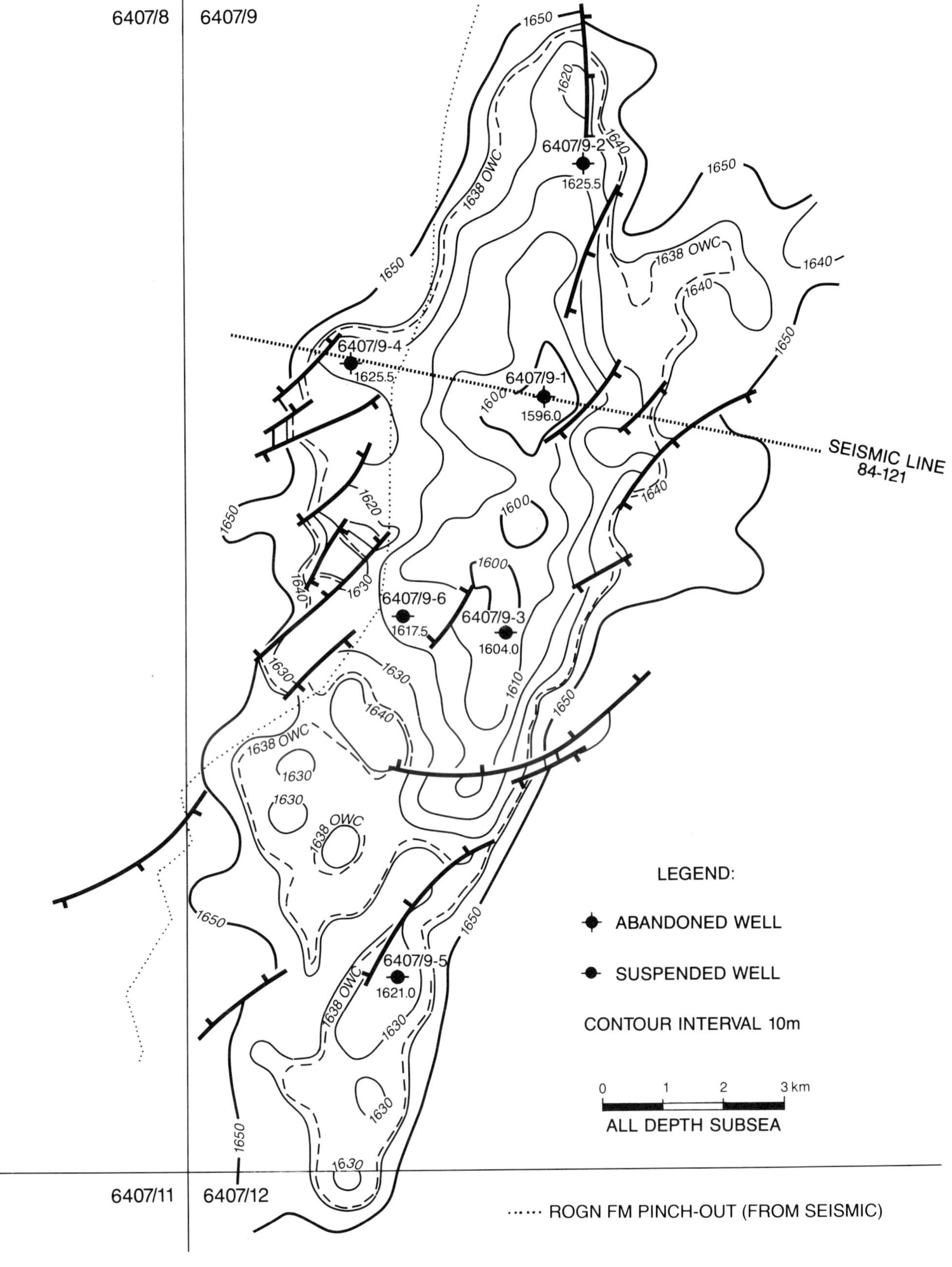

Figure 3. Structure map, top of Rogn Formation. Location of seismic line 84-121, Figure 13, is shown.

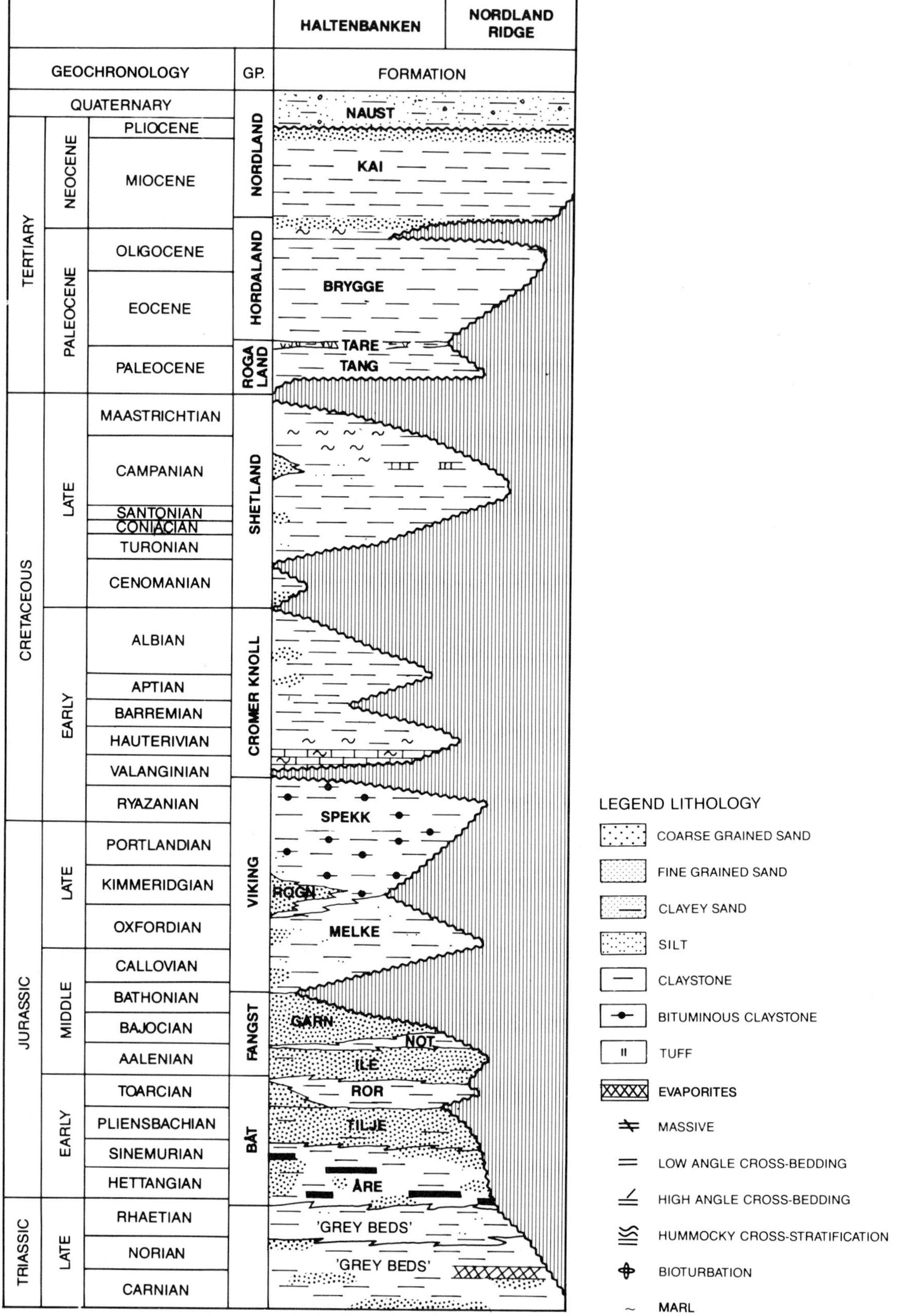

Figure 4. Stratigraphic section with lithologies, Draugen field (Dalland et al., 1988).

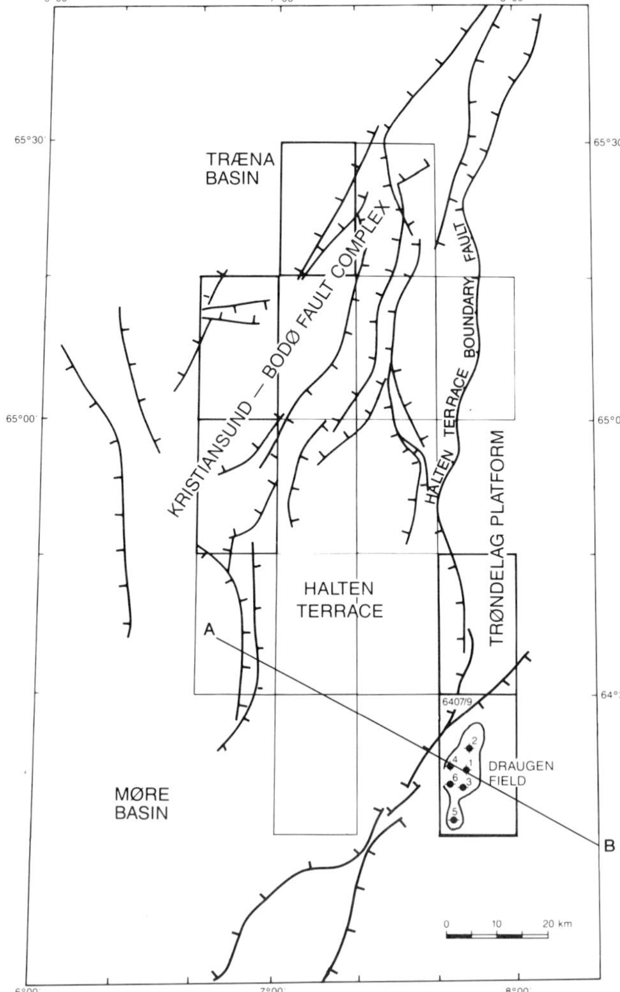

Figure 5. Tectonic map showing structural features and trends. License blocks, Draugen field, and the trace of cross section A–B (Figure 6) are also shown.

in a shoreface environment. Sedimentological investigations have indicated a westerly clastic source for these Jurassic sandstones.

This period was followed in the Middle to Late Jurassic by a rising sea level and block faulting related to the late Kimmerian rifting phase. Marine shales were deposited from Bathonian to Oxfordian time, and euxinic conditions resulted in organic-rich shales during the Kimmeridgian to Portlandian interval. In the Upper Jurassic Rogn Formation of the Draugen field, sands were derived from the erosion of structurally higher areas and deposited within the source rock interval.

The topography resulting from late Kimmerian tectonic activity controlled sedimentation patterns during Cretaceous time. The Tertiary of the Mid-Norwegian shelf comprises a predominantly coarsening-upward sequence. A seismic event within the lower part of this sequence is correlated with the base of the Eocene Tare Formation tuff marker. The tuff layers were deposited during a period of strong subaerial volcanic activity to the west of the area just prior to the initiation of oceanic crust formation and continental separation of Greenland and Norway.

Regional Structural Setting

The Draugen field is situated on the Trøndelag platform, an approximately 30 km wide, relatively undeformed monocline, separated by a regional southwest-northeast-trending fault zone from the highly deformed Halten terrace (Figure 5). Jurassic sediments, slightly increasing in thickness toward the axis of the platform, became progressively downfaulted toward the west on the Halten terrace (Figures 5 and 6). On the eastern margin, Middle and Upper Jurassic sediments outcrop on the sea floor. The Jurassic sediments are draped by a thin succession of Cretaceous sediments and a thick prograding wedge of Tertiary sediments. On the platform, synsedimentary faulting controlled local subsidence patterns and, thus, sedimentation. Over all structural highs, such as the Rogn high located south of Block 6407/9, thinning occurs as a result of non-deposition and erosion during the Middle Jurassic and, particularly, Late Jurassic and Early Cretaceous.

Draugen Structure

The Draugen field comprises an extremely flat, north-south-trending anticlinal structure, located close to the Halten terrace boundary faults, with an areal extent of approximately 20 by 6 km (Structural Map Figure 3 and seismic profile Figure 13). The highest structural relief is found in the north-central part of the field (Figures 3 and 8), with a maximum vertical closure of 45 m. This main northern anticline is separated from a lower relief southern lobe (vertical closure 17 m) by a structural saddle (Figures 3 and 7), approximately 4 km south of well 6407/9-3.

A number of relatively minor, northeast-southwest-trending faults with throws of generally less than 10 to 15 m dissect the structure. The northeast-southwest fault trend is also observed in other areas on the Trøndelag platform and represents the regional trend of faulting on the Halten terrace. Most of the faults are seen only at the base of the Upper Jurassic and at deeper levels, and not at the top of the Upper Jurassic. The increase in growth over the faults at deeper levels suggest synsedimentary activity from Triassic until Late Jurassic. A few faults also penetrate the Upper Jurassic.

STRATIGRAPHY

The stratigraphic sequence in the Haltenbanken area is given in Figure 4 and summarized below.

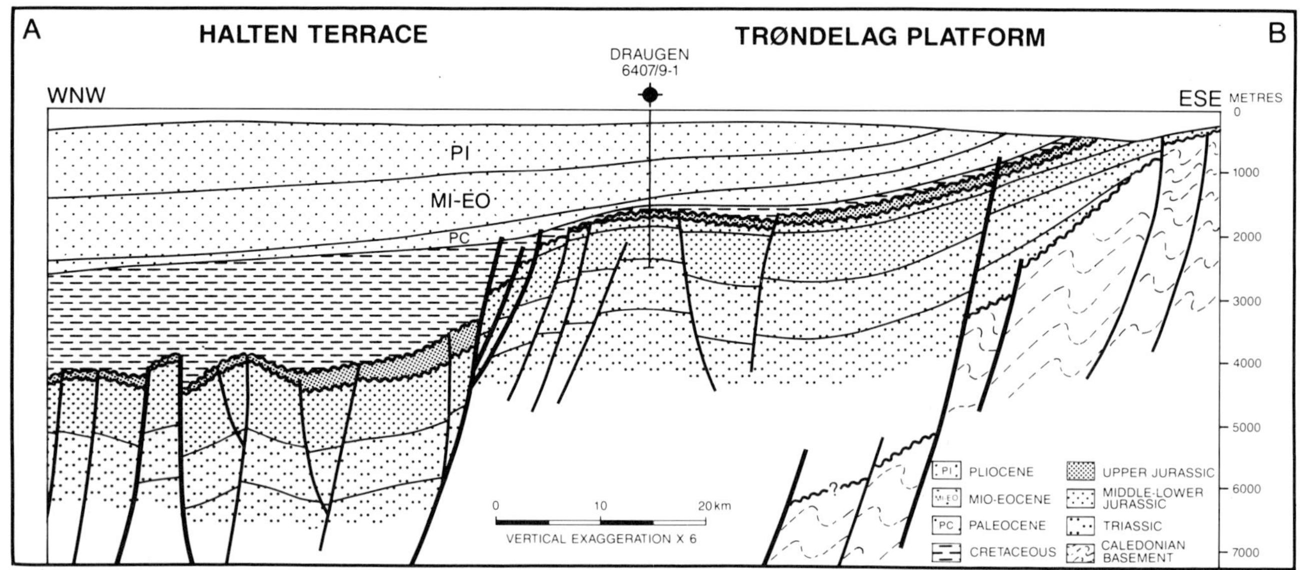

Figure 6. Regional cross section showing the faulted nature of the Haltenbanken area and the position of the Draugen field. Location of A–B is shown on Figure 5.

The Lower Jurassic consists of a transgressive succession of fluviatile sandstones and coaly shales of the Åre Formation, overlain by tidal heterolithic deposits of the Tilje Formation and open marine shales in the Ror Formation. The sedimentary sequence of the Ile Formation reflects a relative drop in sea level and is composed of 5 to 15 m thick kaolinitic sand beds interbedded with shales that grade upward into a 25 m thick silty shale of the Not Formation. Above the Not, the base of the Garn Formation consists of silty and sandy kaolinitic shales, grading into a coastal sequence of 10 to 15 m thick sandstones and thin, locally coaly claystones. The top of the formation is characterized by an unconformity representing a hiatus from early Bathonian to late Oxfordian.

Overlying this unconformity over most of the Haltenbanken area are the Upper Jurassic bituminous shales of the Spekk Formation. In the Draugen field, however, bituminous shales of late Oxfordian to early Kimmeridgian pass into a coarsening-upward sequence of the Rogn Formation. This is the main reservoir rock in the Draugen field.

The Cretaceous Cromer Knoll and Shetland groups consist of sequences of varicolored clays and gray claystone between 25 and 40 m in thickness. The lower Tertiary Rogaland and Hordaland groups are developed as gray claystones of 200 m and 600 m in thickness, respectively. The upper Tertiary Nordland group is comprised of a thick wedge of westward-prograding, coarsening-upward sequences grading from glauconitic and silty clays into clayey sands.

TRAP

The faulted anticlinal structure in which Rogn and Garn hydrocarbons are trapped has already been described above in *Draugen Structure* (Figure 3). This section will deal with the nature of the reservoir sandstones, seen in structural cross section in Figures 7 and 8. A seismic profile is shown by Figure 13.

The main oil-bearing reservoir is the sand sequence of the Rogn Formation. Approximately 96% of the oil is contained in the high-quality sands of Unit I, with a further 3% and 1% in the moderate quality Unit IIa and poor quality Unit IIb, respectively. The Garn Formation is waterbearing throughout most of the field, except in the central area of the western flank where the formation rises above the oil-water contact (Figure 8). All wells were cored and extensively logged, practices that allowed detailed lithological and mineralogical descriptions as well as sedimentological modeling of both formations.

Garn Formation Reservoir

The top part of the Garn Formation is oil-bearing in the west-central part of the structure (Figure 8), with a maximum vertical closure of approximately 18 m near well 6407/9-4.

The Garn Formation comprises two generally coarsening-upward sequences (Figures 9 and 10): Unit II is comprised of a 65–75 m thick coarsening-upward sequence of silty shales at the base, becoming

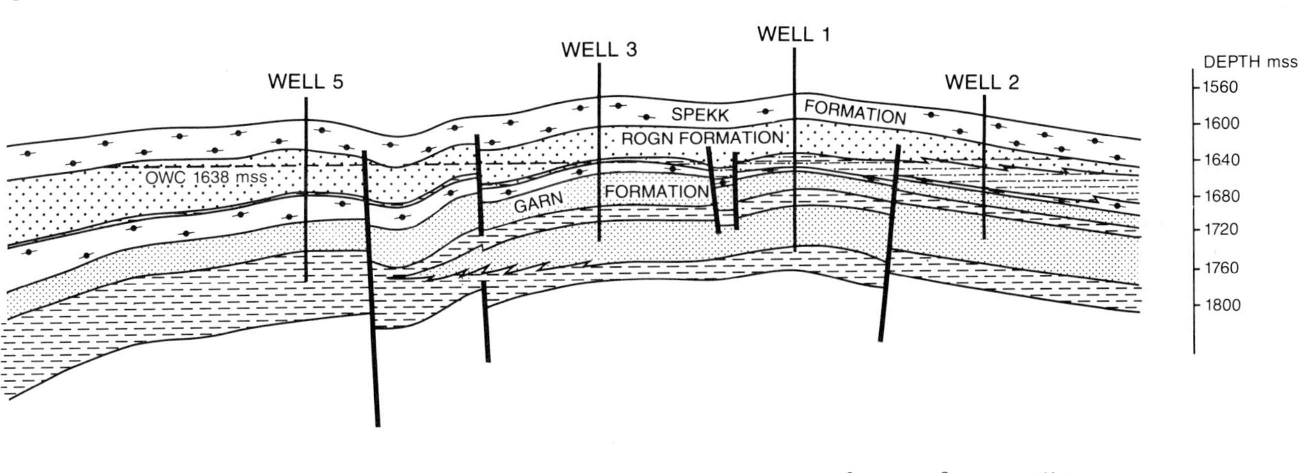

Figure 7. North-south structural cross section B-B'. See Figure 14 for location.

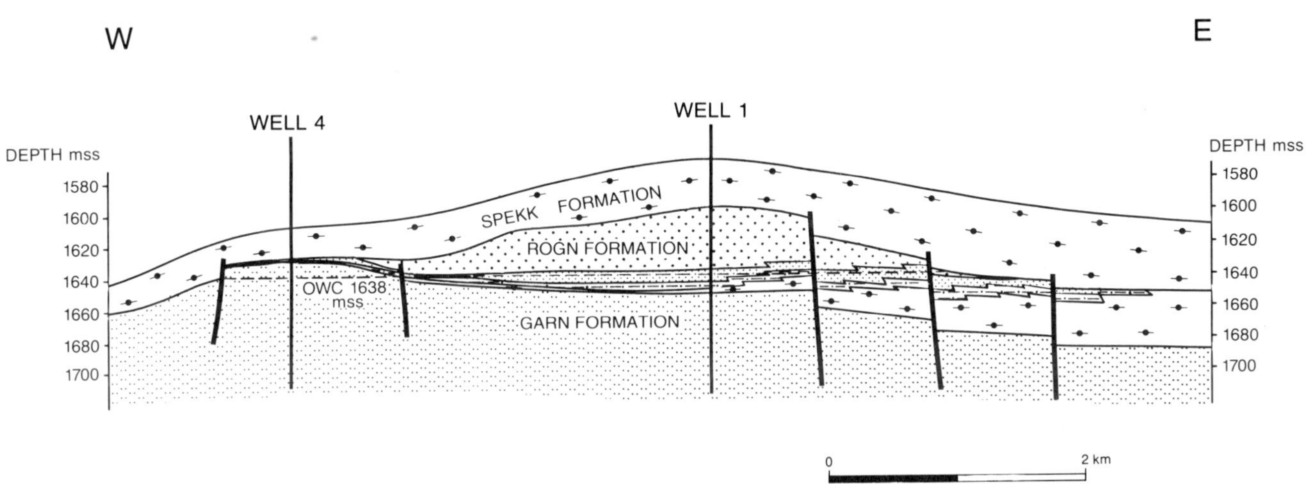

Figure 8. East-west structural cross section A-A'. See Figure 14 for location.

more sandy upward, and grading into coarse-grained, poorly sorted massive sands. Average porosity is 29%, and permeability ranges between 5 and 10 darcys. These sands are water bearing.

Unit I (Figures 9 and 10) consists of laminated, silty claystone (Unit Ib) at the base, grading upward into silty sandstone and well-sorted, low-angle cross-bedded and hummocky cross-stratified fine sandstones (Unit Ia). In the sandstones, the average porosity varies between 33 and 36% and average permeability ranges between 0.2 and 7 darcys.

The Garn Formation is interpreted to be a continuous progradational succession of shallow marine clays grading into coarse-grained nearshore bar sands that are overlain by back-bar shales and shoreface and possibly foreshore sands (Figures 10 and 15). The formation is extensive, with a uniform thickness across the field and subcropping on the seabed 40 km to the east. This is the main hydrocarbon-bearing formation in the greater Haltenbanken area.

Rogn Formation Reservoir

The Rogn Formation comprises an overall coarsening-upward unconsolidated sand sequence enclosed in bituminous shales. The formation thickness varies from 1 to 80 m and can be subdivided into three distinct lithological units (Figures 9, 11, and 12).

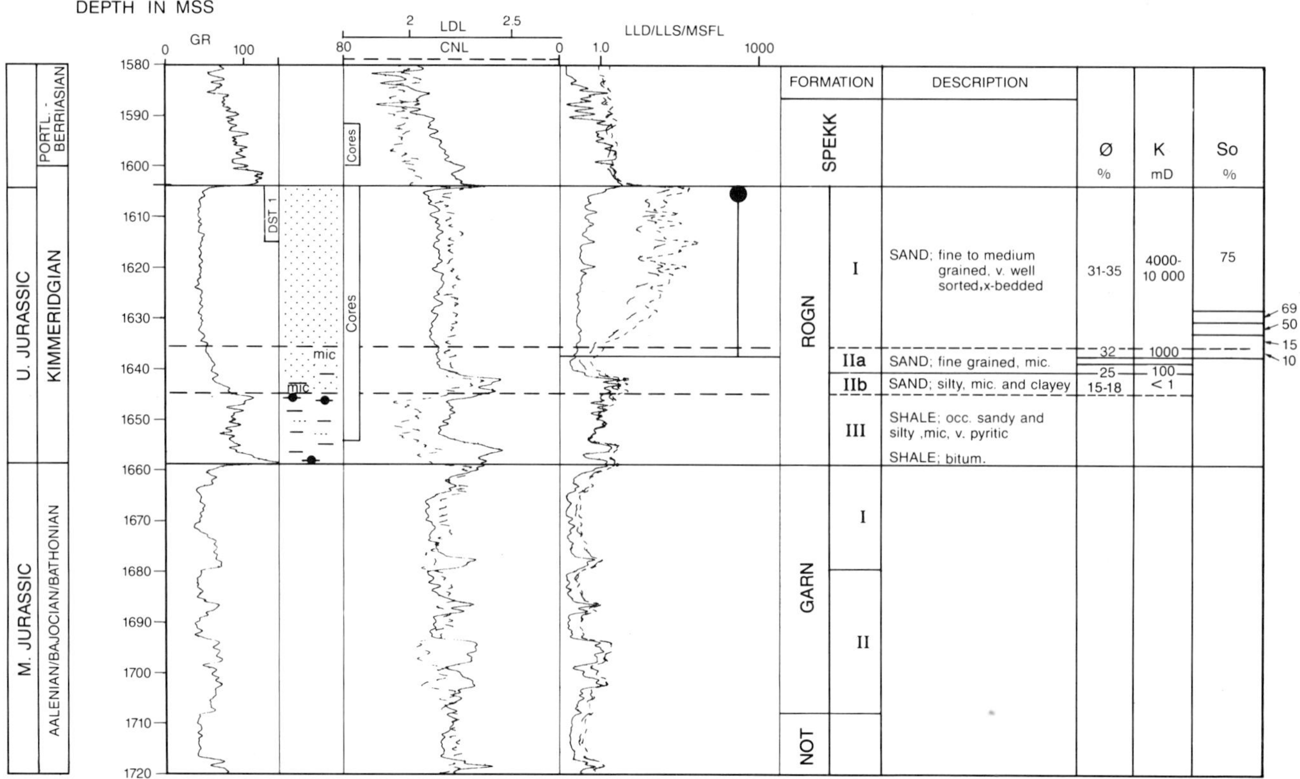

Figure 9. Type log for Draugen field showing zonation of Garn and Rogn formations with lithology and petrophysical data for the latter. See Figure 4 for symbols.

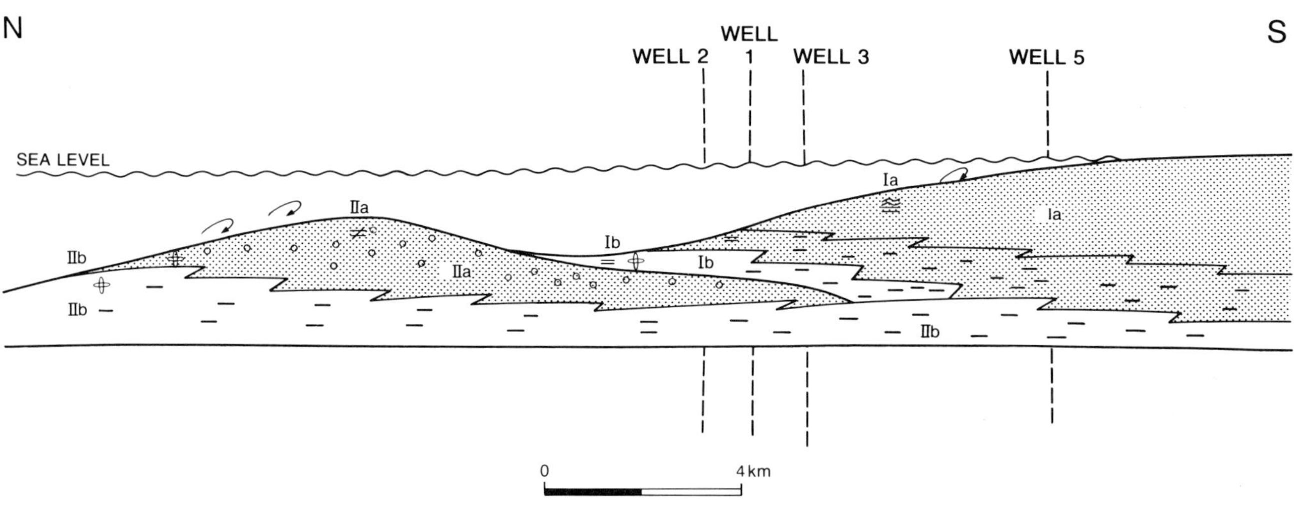

Figure 10. Diagrammatic north-south cross section across Draugen showing the environment of deposition of the Garn Formation. See Figure 4 for symbols. Locations of the wells are shown on Figure 3.

Unit I, the uppermost, consists of a moderately sorted, fine-grained sand containing a high percentage of dispersed coarse to very coarse grains. The sands are massive and bioturbated, and occasionally low-angle cross-bedding is observed. In the central part of the field, the coarse grains are less common, the sands are better sorted, and abundant cross-bedding is developed. The small size and degree of rounding of organic debris indicates moderate to high energy of deposition (Van der Zwan, 1990). Unit I is a continuous layer in a north-south direction, but it thins and pinches out within the oil zone on the western flank (Figures 7 and 8). In the central part of the field, average porosity is 32% and average

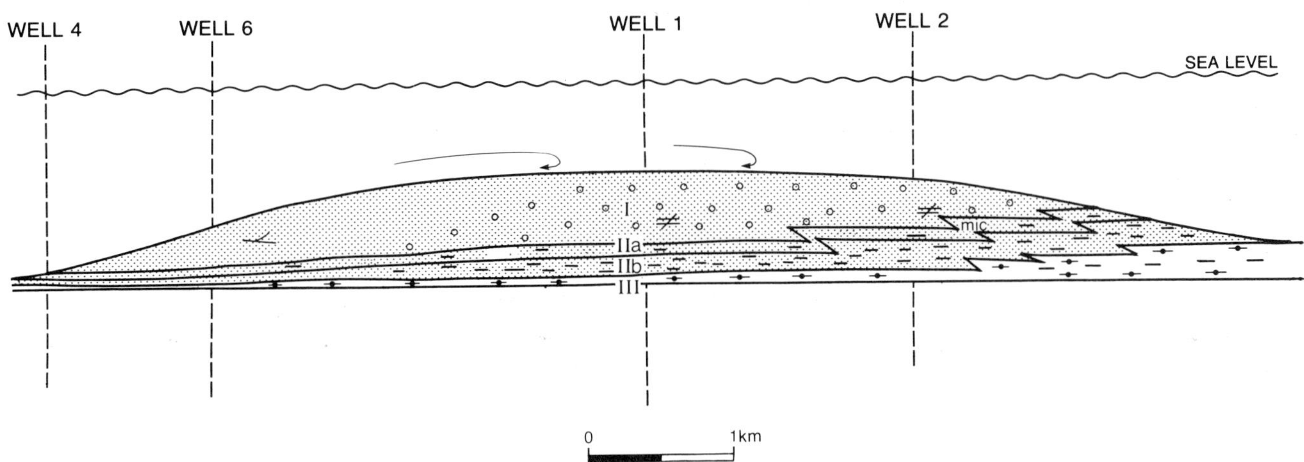

Figure 11. Diagrammatic east-west cross section across Draugen field showing the environment of deposition of the Rogn Formation. See Figure 4 for symbols.

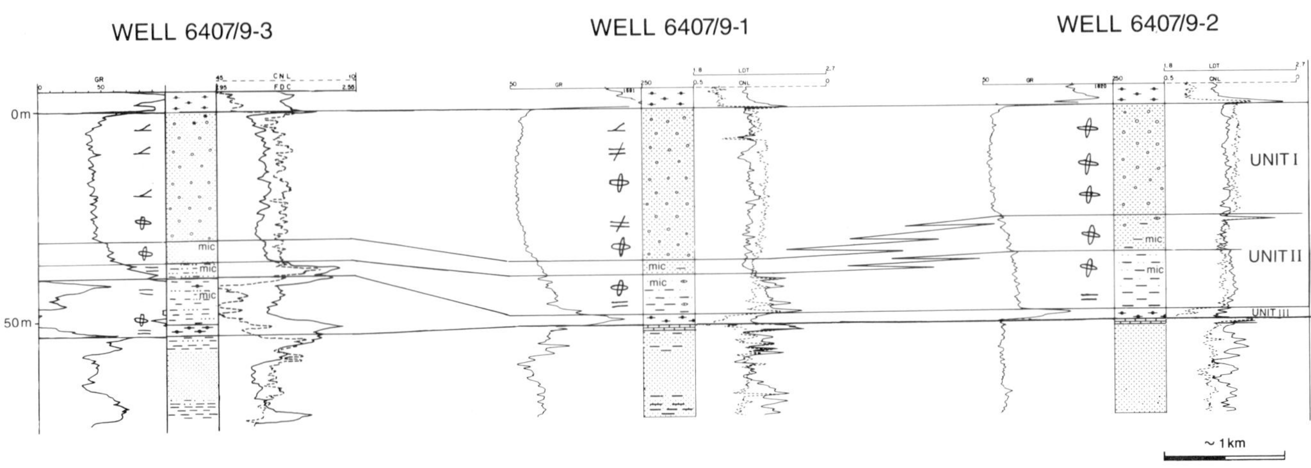

Figure 12. North-south correlation section showing the relationship of the Rogn subdivisions. See Figure 3 for locations of wells and Figure 4 for symbols.

permeability 6 to 7 darcys. Eastward and northward a gradual deterioration to 27% porosity and 2 darcy permeability is observed. Intraformational shales and other vertical transmissibility barriers are absent in Unit I.

Unit II is oil bearing in the central and west-central parts of the field. It consists of a transition from silty, sandy claystone and clayey micaceous sands at the base (Unit IIb) to fine-grained, micaceous sands at the top (Unit IIa). The sands are strongly bioturbated, and the original parallel lamination is only occasionally preserved. Calcite cemented intervals are common at the base of Unit IIb and an open marine fauna is found. Reservoir quality reflects the argillaceous and micaceous content of this unit; the upward decrease in mica and clay is related to a corresponding increase in permeability. In the top Unit IIa, average porosity ranges between 26 and 30% with low to moderately high permeabilities between 0.7 darcy in well 6407/9-3 and 0.2 darcy on the flanks. The poorest reservoir quality occurs at the base (Unit IIb) with porosities ranging from

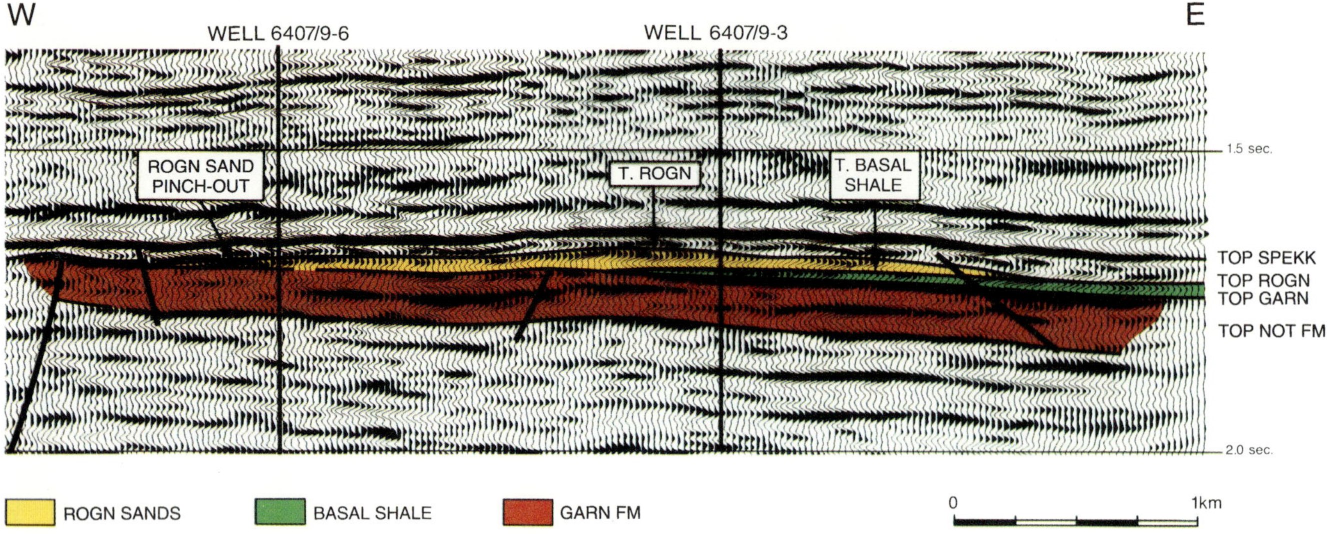

Figure 13. Seismic line 84-121, impedance version, across Draugen, showing the Rogn Formation sand bar. See Figures 3 and 14 for location.

10 to 25% and permeabilities from 1 to 50 md. Reservoir quality is expected to be deteriorated toward the shale-out in the east and north.

Unit III consists of nonreservoir shales with the thickest development (25–30 m) along the eastern flank. A rapid progressive thinning is observed toward the crestal areas and continues to the western flank where the shales are less than 1 m thick. Pyrite cement is common, and shales have an open-marine fauna. These bituminous, marine shales are encountered in all six wells and are believed to be fieldwide. Local erosion of the shales during deposition of the overlying low-energy silts and micaceous sands is considered unlikely.

From the standpoint of sedimentary environment, the Rogn Formation is a coarsening-upward sand wedge enclosed within open marine shales. The variations in grain size, mica/clay content, and palynofacies reflect an overall upward increase in sedimentation rate and energy conditions, from low-energy, slow deposition of the basal bituminous shales (Unit III) into low- to moderate-energy, slow deposition of the bioturbated micaceous sands and claystones (Unit II), and finally high- to moderate-energy, rapid deposition and reworking of the fine-grained bimodal sands (Unit I). The successive depositional processes are interpreted as the development of a shallow marine sand bar on top of offshore marine clays (Figure 11). The presence of offshore palynofacies supports an offshore marine environment.

Other examples of similar offshore bars are described in the literature of the formation of coarsening-upward lens-shaped sand bodies with limited lateral extent (Boyles and Scott, 1982; Brenner, 1987). However, each of these is commonly composed of a heterogeneous sequence with highly variable reservoir quality resulting from the stacking of individual sand bars and ridges.

Early in the appraisal phase of drilling it became clear that the Rogn sands were likely to be of very limited lateral extent. The 52 m thick coarsening-upward sand development in the central part of the field in well 6407/9-1 becomes more micaceous toward the north in well 6407/9-2, and the sands of unit I are thinner, indicating a gradual shale-out in a northerly direction (Figure 12). In the southeast, in well 6407/9-3, the basal shales are thicker, and finally in well 6407/9-4 the Rogn Formation was found to have thinned to 1.5 m, indicating a pinch out of the total Rogn Formation toward the west.

Delineation of the geometry of the sand bar was facilitated by the use of acoustic impedance seismic data. Interpretation and synthetic modeling revealed the following features (Figures 13 and 14):

- The Rogn sands are continuous in the north-south direction with minor thickness variations. The thickest sands are developed south of well 6407/9-5 and extend some 10 km south of Block 6407/9.
- The total Rogn Formation thins rapidly toward the west, leading to a north–south-trending pinch out. Well 6407/9-6 confirmed the presence of the pinch out derived from seismic interpretation.
- On the northeast and southeast flanks, the basal shale thickens at the expense of the sands, indicating a shale-out of the Rogn sands.

Depositional History

During the late Oxfordian and early Kimmeridgian, clays of the Melke and bituminous clays of the Spekk formations equivalents were deposited (Figure 15). Regional fault block tilting in the early Kimmeridgian resulted in the formation of small, partly enclosed basins where anoxic conditions prevailed. The

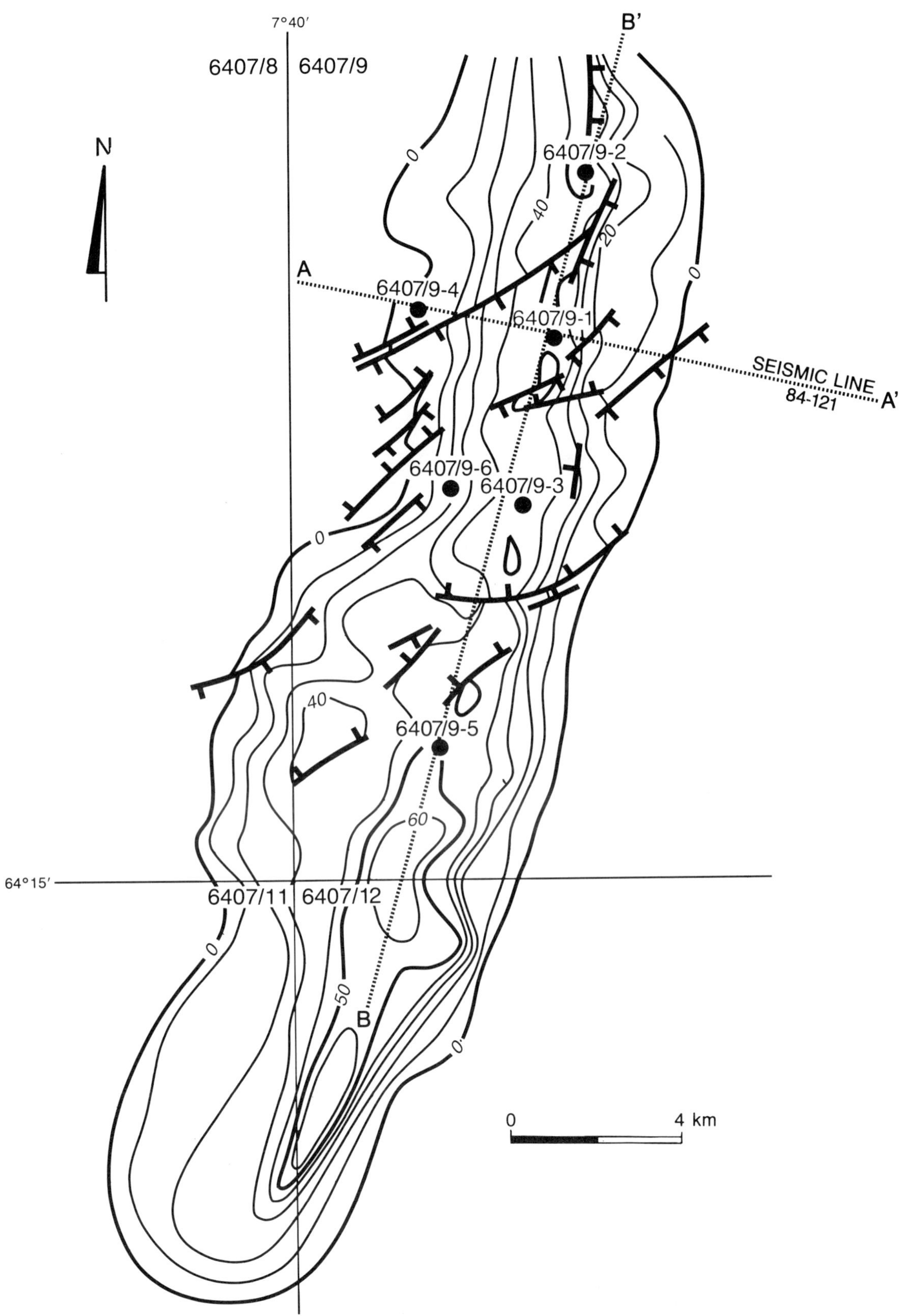

Figure 14. Isopach map of Rogn Formation sand bar as interpreted from seismic data. Locations of seismic line 84-121 (Figure 13), coincidental with cross section A–A′ (Figure 8), and cross section B–B′ (Figure 7) are shown. Contour interval, 10 m.

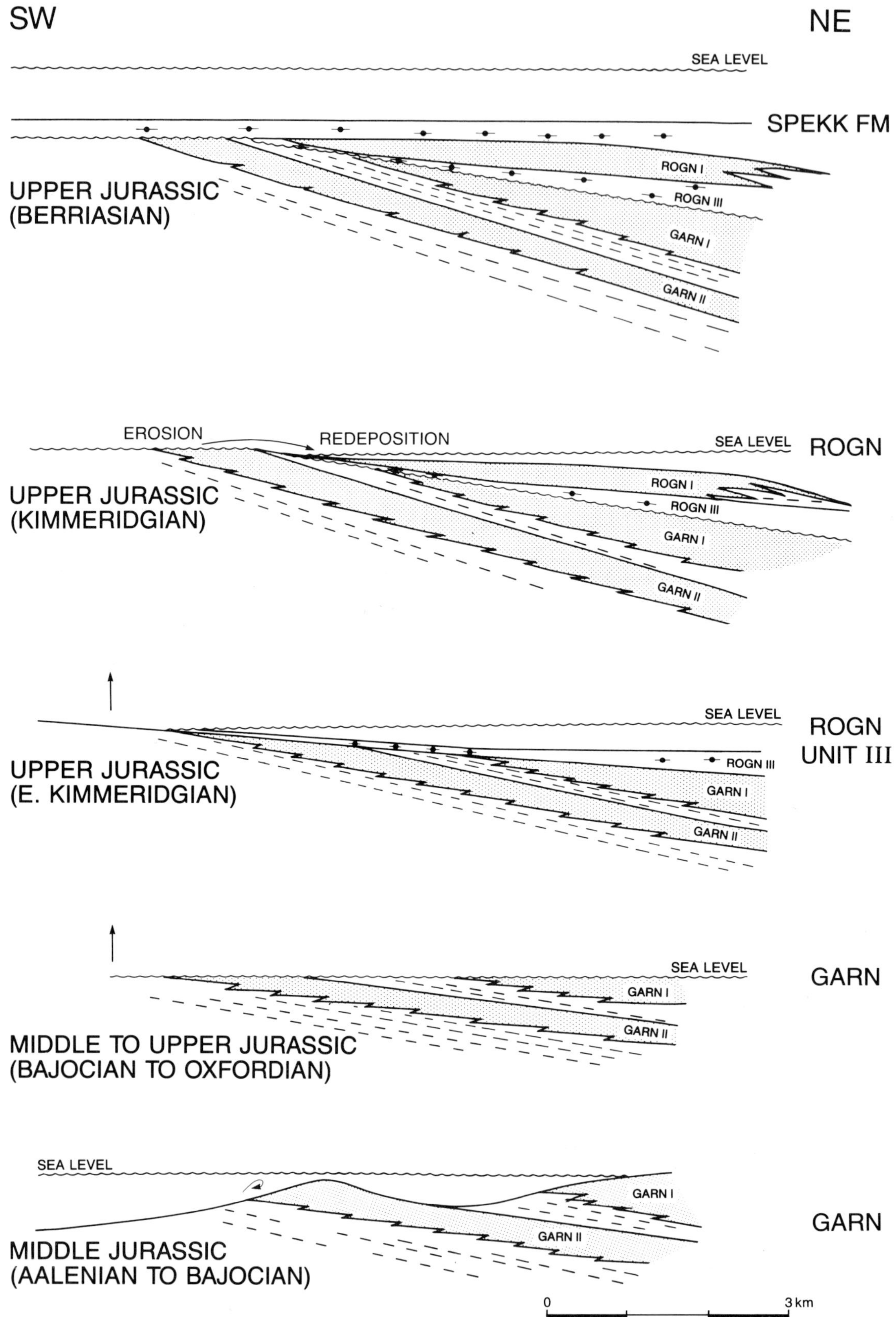

Figure 15. Diagrammatic cross sections across Draugen showing the depositional history of the Garn through Spekk formations. See Figure 4 for symbols.

southern part of the Trøndelag platform was slightly tilted toward the east and the Melke and Spekk formations onlap the Middle Jurassic unconformity.

A relative drop in sea level, caused by the tendency of rift margin upthrown fault blocks to rebound as the adjacent downthrown blocks continued to subside, led to the local emergence of the previously deposited Middle Jurassic succession in the west, i.e., the Garn Formation. The erosion products were transported and deposited probably by tidal and/or longshore currents, gradually building up one or more sand bars in the Draugen area as sedimentation rate kept pace with the rise in sea level. Once the bar had built up above storm-wave base, deposition was frequently interrupted by strong wave action, reworking and eroding the westward-facing shallow side of the bar and depositing coarse-grained, poorly sorted sediment as wash-over sands on the easterly facing side of the bar. This process created an asymmetric bar with initial current-related structures preserved in the west and poorly sorted, massive sands on the east side of the bar. Rapid drowning of the bar and the re-establishment of deeper marine conditions resulted in the deposition of offshore bituminous clays of the Spekk Formation, which then onlapped and draped over the Rogn sand bar.

Mineralogy

The Rogn and Garn formations are predominantly quartz arenites and occasionally arkosic arenites and lithic arenites. The detrital grains consist predominantly of monocrystalline quartz, polycrystalline quartz, K-feldspar with minor rock-fragments, plagioclase, and mica. Garnet is locally present as up to 8% of the bulk volume. Detrital clay minerals include kaolinite with minor illite, mixed layer clays, and traces of chlorite (glauconite).

The sands are unconsolidated and have been subjected to only minor compaction and diagenetic processes. Authigenic minerals are relatively rare and include feldspar overgrowths with localized pyrite, siderite, and calcite cements precipitated as discontinuous concretions. Minor amounts of authigenic kaolinite booklets occur in the pores. The quartz grains are subangular and often very irregular, with large embayments that are attributed to leaching of feldspar and garnet.

The sands in the Rogn Unit I and Garn Unit IIa have a bimodal grain-size distribution resulting in a pore throat size distribution as follows:

- Macropores: large pores with pore throat sizes between 150 and 26 microns.
- Minipores: pore throat sizes between 2.4 and 0.5 microns occurring as interparticle porosity in kaolinite booklets and matrix material, intragranular porosity in leached feldspars, and quartz embayments in quartz grains and composite grains.
- Micropores: pore throat sizes smaller than 0.08 micron, occurring as interparticle porosity in clay minerals.

Aquifer

Over large parts of the central and crestal areas of the field the oil-water contact (OWC) in the Rogn Formation is coincident with the base of the high-quality sands. In these areas, the relatively thin waterleg in the Rogn Formation is contained in poorer quality, micaceous sands of Unit II. On the western flank the formation is virtually absent at the level of the OWC, and on the eastern flank the high-quality sands have shaled out just below the OWC. Only on the northern and southern flanks do the high-permeability sands extend into the aquifer. Large aquifer volumes are not envisaged in the north. The Rogn sand bar extends approximately 10 km southward beyond the OWC and the total water-bearing volume is estimated to be six times the hydrocarbon-bearing volume.

The main aquifer is the regionally extensive Garn Formation that underlies the Rogn Formation. This sand, silt, and shale sequence is 110 m thick, has a large regional extent, and contains intervals of high porosity and permeability.

Communication between the Rogn and Garn formations is restricted by the widespread development of the basal Rogn shales. These bituminous shales, which vary in thickness from less than 1 m on the extremes of the western flank to more than 30 m on the southeast flank, are believed to be field wide. Locally the communication is further reduced by the tight calcite-cemented zones at the top of the Garn Formation and the base of the Rogn Unit IIb. Areas of localized communication may, however, exist because of faulting, where the fault offsets are greater than the thickness of the basal shales. In the central, northern, and southern flank areas, the shales are less than 5 m thick and fault throws are sufficient to result in juxtaposition of the Garn and Rogn sands.

The fault related communication may be restricted by: (1) clay smearing and realignment of mica, and (2) synsedimentary thickening of the basal shale in downthrown fault blocks.

Development Plan

The field is currently estimated to contain 1140 million stock tank bbl oil (STBO) of which 990 million STBO are in the Rogn Formation in the north-central accumulation. The remaining volumes occur in Rogn Formation in the south extension and in the Garn Formation in the west. The main development target is the Rogn Formation in the main north-central accumulation.

The intrinsic reservoir energy of the Rogn reservoir is relatively low. The oil is highly undersaturated and hydrostatically pressured. The Rogn Formation is generally confined to the hydrocarbon-bearing area, except to the south where a significant water volume is present in the formation. The potential exists for possible natural water influx along faults in the center and west of the main Rogn north accumulation. However, extensive reservoir simulation studies have demonstrated that water injection will be required to maintain the pressure. This will therefore form the basis for the conceptual field development plan.

The high vertical permeability of the Rogn reservoir will enhance gravity segregation of oil and water, implying that production wells should be located crestally.

The Rogn reservoir sands have generally very high permeability and show an overall decline in permeability toward the base. This, plus the favorable water-oil mobility ratio (M = 0. 6), results in a high sweep efficiency by water flooding. As Unit I pinches out in the west and shales out in the east just below the oil-water contact, water cannot be injected into a Unit I waterleg in the east and west; here, injection would have to occur in the oil zone where injection capacity is low. Injectivity will be highest in the extreme north and south of the field, and instead of applying peripheral water injection, it is preferred to group the injection wells in the north and south.

There is no evidence of lateral permeability barriers in the Rogn Formation in the north-central accumulation. As the Rogn reservoir is thin (maximum closure is 45 m), early water coning is a potential problem wherever bottom water is present. However, only very small volumes of water are present in the crestal area such that the initial water cut is expected to be low.

The Rogn Formation in the southern lobe and the small accumulation in the Garn Formation are characterized by thin (less than 18 m) oil columns in contact with significant bottom-water legs. Production from these two accumulations will therefore be hampered by severe water coning and the oil recovery will be low.

The development plan consists of seven crestal production wells over a development region of 1.5 × 4.5 km, and six water injection wells, split in three-well groups at a distance of some 11 km from one another. The large areal extent of the field and its relatively shallow depth (1600 m) effectively places all of the injection well locations and one of the oil production well locations beyond the reach of a single central drilling location. The Draugen development will therefore be characterized by a significant subsea-related element. The field development plan utilizes (1) a centrally located, concrete, gravity-based, mono-column fixed platform for the production and processing of oil and gas and (2) subsea completed wells that will be tied back to the platform. The preferred oil transportation method is offshore loading into tankers via a floating loading platform. The oil production at plateau rate is planned at 90,000 bbl/day annual average.

ACKNOWLEDGMENTS

The authors wish to thank other members of Norske Shell's staff: Bruce M. Thomas who was working as area geologist for Mid-Norway at the time of the license application; Richard Birtles and Geir Engen, who carried out the seismic interpretation; and Cees J. van der Zwan, for his geological contributions and valuable discussions.

The authors also wish to express their appreciation to the Management of A/S Norske Shell and to other participants in License PL 093 of Block 6407/9—Den Norske Stats Oljeselskap (Statoil) A/S and BP Petroleum Development of Norway A/S—for granting permission to present this paper.

REFERENCES CITED

Boyles, J. M., and A. J. Scott, 1982, A model for migrating shelf-bar sandstones in Upper Mancos shales (Campanian), Northwestern Colorado: AAPG Bulletin, v. 66, n. 5, p. 491-508.

Brenner, R. L., 1978, Sussex sandstone of Wyoming—Example of Cretaceous offshore sedimentation: AAPG Bulletin, v. 62, n. 2, p. 181-200.

Dalland, A., D. Worsley, and K. Ofstad, 1988, A lithostratigraphic scheme for the Mesozoic and Cenozoic succession offshore mid- and northern Norway: Norwegian Petroleum Directorate Bulletin 4.

Van der Zwan, C. J., 1990, Palynostratigraphy and palynofacies reconstruction of the Upper Jurassic to Lowermost Cretaceous of the Draugen field, offshore mid Norway: Review of Palaeobotany and Palynology, v. 62, p. 157-186.

SUGGESTED READING

Bukovics, A., and P. A. Ziegler, 1985, Tectonic development of the Mid-Norway continental margin: Marine and Petroleum Geology, v. 2, n. 1, p. 2-22.

Bukovics, C., E. G. Cartier, N. D. Shaw, and P. A. Ziegler, 1984, Structure and development of the mid-Norway continental margin, *in* A. M. Spencer et al., eds., Petroleum geology of the north European margin: London, Graham and Trotman, p. 407-423.

Caselli, F., 1987, Oblique-slip tectonics Mid-Norway Shelf, *in* J. Brooks and K. Glennie, eds., Petroleum geology of north west Europe: London, Graham and Trotman, p. 1049-1063.

Van der Zwan, C. J., 1989, Palynostratigraphical principles as applied in the Jurassic of the Troll and Draugen fields, Offshore Norway, *in* Correlations in hydrocarbon exploration: Norwegian Petroleum Society, London, Graham and Trotman, p. 357-365.

Appendix 1. Field Description

Field name .... *Draugen field*
Ultimate recoverable reserves .... *428 × 10^6 barrels of oil*

Field location:

- **Country** .... *Norway*
- **Basin/Province** .... *Haltenbanken area*

Field discovery:

- **Year first pay discovered** .... *Upper Jurassic Rogn Formation 1984*
- **Year second pay discovered** .... *Middle Jurassic Garn Formation 1985*

Discovery well name and general location:

- **First pay** .... *6407/9-1, approx. 160 km northwest of Trondheim, Norway (64°20′N, 7°45′E)*
- **Second pay** .... *6407/9-4, approx. 3.2 km west-northwest of 6407/9-1*

Discovery well operator .... *A/S Norske Shell (all wells)*

IP: .... *8500 BOPD (1350 m^3)*

All other zones with shows of oil and gas in the field: .... *None*

Geologic concept leading to discovery and method or methods used to delineate prospect

Interpretation of seismic data tied to adjacent wells in the Haltenbanken area led to the discovery and the delineation of the prospect.

Structure:

Province/basin type ..... *Eastern edge of the main Norway-Greenland Sea rift; Bally 1141, Klemme III C*

Tectonic history

Rifting in Paleozoic to Mesozoic time, north-south normal faulting, several kilometers of sediments deposited. Passive continental margin from early Eocene. Progressive infilling of westward thickening wedge of post-rift sediments, little affected by faulting.

Regional structure

Draugen field is situated on the Trøndelag platform, an approximately 30 km wide, relatively undeformed monocline, separated by a regional southwest-northeast-trending fault zone from the Halten terrace.

Local structure

Draugen field is comprised of an extremely flat, north-south-trending, anticlinal structure with a maximal vertical closure of 45-50 m.

Trap:

Trap type(s)

Anticlinal trap and truncation trap associated with tilted fault block.

Basin stratigraphy (major stratigraphic intervals from surface to deepest penetration in field):

Chronostratigraphy	Formation	Depth to Top in m
Tertiary		
Cretaceous		*Top Cret. approx. 1500*
Upper Jurassic	*Spekk*	*Approx. 1570*
	Rogn	
Middle Jurassic	*Melke*	
	Garn	*Approx. 1650*
	Not	
	Ile	

DRAUGEN

Chronostratigraphy	Formation	Depth to Top in m
Lower Jurassic	*Ror*	*Approx. 1900*
	Tilje	
	Åre	*Approx. 2050*
Triassic	*Grey Beds*	

Reservoir characteristics:

Number of reservoirs *2*

Formations .... *Rogn and Garn Formations*

Ages .... *Rogn, Kimmeridgian; Garn, Bajocian-Bathonian*

Depths to tops of reservoirs .... *Rogn, 1596 m; Garn, 1625 m*

Gross thickness (top to bottom of producing interval) .... *Rogn, 0–45 m; Garn, 0–< 18 m*

Net thickness—total thickness of producing zones

Average .... *20 m*

Maximum .... *60 m*

Lithology

Main Rogn reservoir: unconsolidated, moderately sorted, fine sand containing 10-60% medium to very coarse grains; sands are massive, bioturbated, and occasionally cross-bedded

Porosity type .... *Intergranular porosity*

Average porosity *32%*

Average permeability .... *2-7 darcys*

Seals:

Upper

Formation, fault, or other feature .... *Kimmeridge Clay*

Lithology .... *Claystone*

Lateral

Formation, fault, or other feature .... *Kimmeridge Clay (bottom seal water or Heather Formation in some areas)*

Lithology .... *Claystone*

Source:

Formation and age .... *Kimmeridge Clay, Late Triassic*

Lithology .... *Claystone*

Average total organic carbon (TOC) *7%*

Maximum TOC *11%*

Kerogen type (I, II, or III) .... *II*

Vitrinite reflectance (maturation) .... $R_o = 1.1$ *(at kitchen center, approximately 20 km west of Draugen structural closure)*

Time of hydrocarbon expulsion .... 65×10^6 *years, Paleocene Danian (at kitchen center)*

Present depth to top of source .... *4000 m (at kitchen center)*

Thickness .... *100 m (at kitchen center)*

Potential yield .... *NA*

Appendix 2. Production Data

Field name .... *Draugen field*

Field size .... *20 × 6 km*

Proved acres .... *NA*

Number of wells all years *6*

Current number of wells 6
Well spacing .... Average 4 km
Ultimate recoverable .... 428 million bbl (67.5 × 10^6 m^3)
Cumulative production .... Nil
Annual production .... None
Present decline rate .... NA
  Initial decline rate .... NA
  Overall decline rate .... NA
Annual water production .... NA
In place, total reserves .... 1140 million STBO (180 × $10^6 m^3$)
In place, per acre foot .... NA
Primary recovery .... NA
Secondary recovery .... NA
Enhanced recovery .... NA
Cumulative water production .... Nil

DRAUGEN

Drilling and casing practices:

Amount of surface casing set .... 30-in. at 74 m below seabed
Casing program .... 20-in at 775 mss; 13⅜-in. at 1590 mss; 9⅝-in. at 1770 mss
Drilling mud .... Water-based mud
Bit program .... NA
High pressure zones .... NA

Completion practices:

Interval(s) perforated .... 12 m from top reservoir
Well treatment .... Internal gravel pack

Formation evaluation:

Logging suites ISF/LSS/SP/GR, LDL/CNL/CAL/GR, MSFL/DLL/GR/SP, CST, HDT, RFT, SWC; cores at reservoir level
Testing practices .... Gravel packed and acidized interval production test
Mud logging techniques .... NA

Oil characteristics:

Type .... Undersaturated oil
API gravity .... 40.1°
Base .... Paraffinic
Initial GOR .... 0.291 Mscf/bbl
Sulfur, wt% .... 0.14 ppm
Viscosity, SUS .... 2.52 cSt at 40°C, 1.84 cSt at 60°C
Pour point .... <-36°C
Gas-oil distillate .... NA

Field characteristics:

Average elevation .... 1610 m (depth of reservoir)
Initial pressure .... 16,500 kPa
Present pressure .... 16,500 kPa
Present gradient .... 10.25 kPa/m
Temperature .... 68°C
Geothermal gradient .... 0.05°C/m
Drive .... Injection water drive
Oil column thickness .... 40 m initially
Oil-water contact .... 1638 m subsea

Connate water *10%*
Water salinity, TDS .... *38 g/L*
Resistivity of water .... *0.10 ohm-m*
Bulk volume water (%) .... *NA*

Transportation method and market for oil and gas:

Oil by tanker; gas will be reinjected (probably into a deeper water-bearing reservoir).

Gullfaks Field—Norway
East Shetland Basin, Northern North Sea

SNORRE OLAUSSEN
Statoil
Stavanger, Norway

LENNART BECK
Statoil
Bergen, Norway

LARS M. FÄLT
Statoil
Stavanger, Norway

EIRIK GRAUE
KNUT G. JACOBSEN
Statoil
Bergen, Norway

OVE A. MALM
DEREK SOUTH
Statoil
Stavanger, Norway

FIELD CLASSIFICATION

BASIN: North Sea
BASIN TYPE: Rift
RESERVOIR ROCK TYPE: Sandstone
TRAP TYPE: Combination Unconformity Truncation and Tilted Fault Blocks
RESERVOIR ENVIRONMENT OF DEPOSITION: Alluvial Plain, Deltaic, Shallow Marine
RESERVOIR AGE: Jurassic
PETROLEUM TYPE: Oil

LOCATION

The Statoil-operated Gullfaks field is located within Block 34/10 at 61°N and 2°E in the Norwegian sector of the North Sea (Figure 1). The Gullfaks field covers an area of 51 km^2 with water depths ranging from 135 to 220 m, from southwest to northeast. The field lies in the central part of the East Shetland basin and represents the most positive element of the Tampen spur, which here limits the Viking graben along its western and northwestern edge (Figure 1). Apart from more complex tectonic structural development and shallower burial depth, the Gullfaks field is comparable with the neighboring giant Statfjord field (inset, Figure 1). In both cases the hydrocarbons, sourced from the Kimmeridgian "Hot Shale," the Draupne Formation, have accumulated in Lower and Middle Jurassic sandstones, with their main recoverable oil reserves in the Bajocian to Bathonian Brent Group. The Gullfaks field was discovered in 1978 and was brought into production toward the end of 1986. The recoverable oil reserves are estimated to be 230 $\times$ 10^6 Sm^3 (1450 MMBO). Peak production from three platforms is expected to be 24 $\times$ 10^6 Sm^3/year (150 MMBO/yr).

HISTORY

Pre-Discovery

The exploration of Block 34/10 started in 1974 with a conventional seismic survey. A 1 $\times$ 1 km seismic grid was collected and interpreted, and this was combined with information from nearby U.K. fields and the newly discovered Statfjord field (Figure 1). These discoveries were related to Middle and Lower Jurassic reservoir rocks on tilted fault blocks below the late Kimmerian (Cimmerian) unconformity. In contrast to the Statfjord field, it was clear from seismic profiles (Figure 2) that the Gullfaks field had suffered more deformation, expressed by intense faulting and steeper structural dip. In addition, geophysical data indicated that the prospective reservoirs on the Gullfaks structure were approximately 500 m shallower than the Statfjord field. After a license bidding round, Block 34/10 was awarded in June 1978 to the three Norwegian oil companies: Saga Petroleum, Norsk Hydro, and Statoil. Statoil was awarded operatorship. Esso and, later, Conoco were appointed as technical assistants to Statoil.

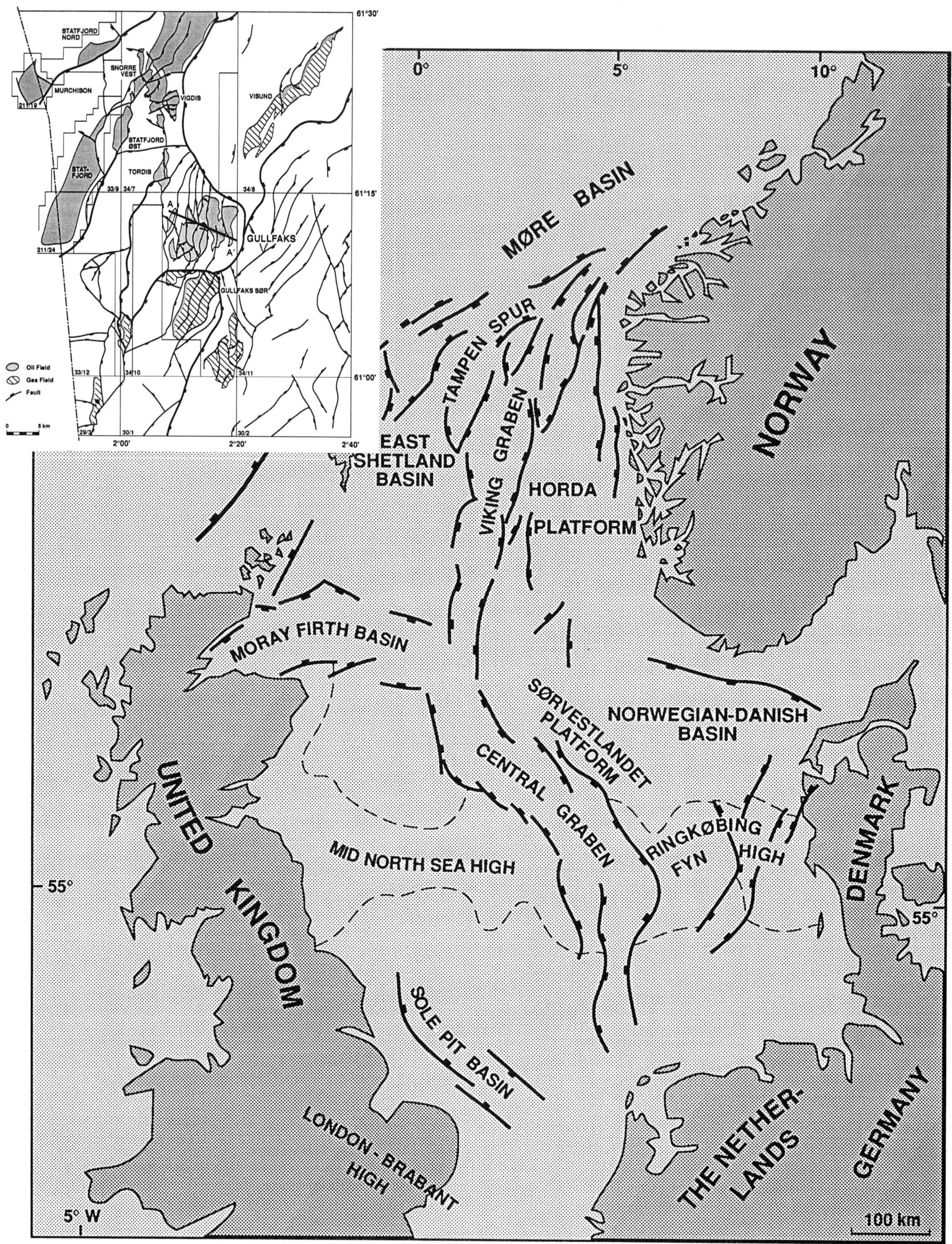

Figure 1. Location map of Gullfaks field in Block 34/10 (inset), with neighboring fields and generalized structural map of the northern North Sea. Line A–A′ in the inset shows location of seismic section in Figure 2. International boundary between Norway and U.K. shown by dash dot line in inset.

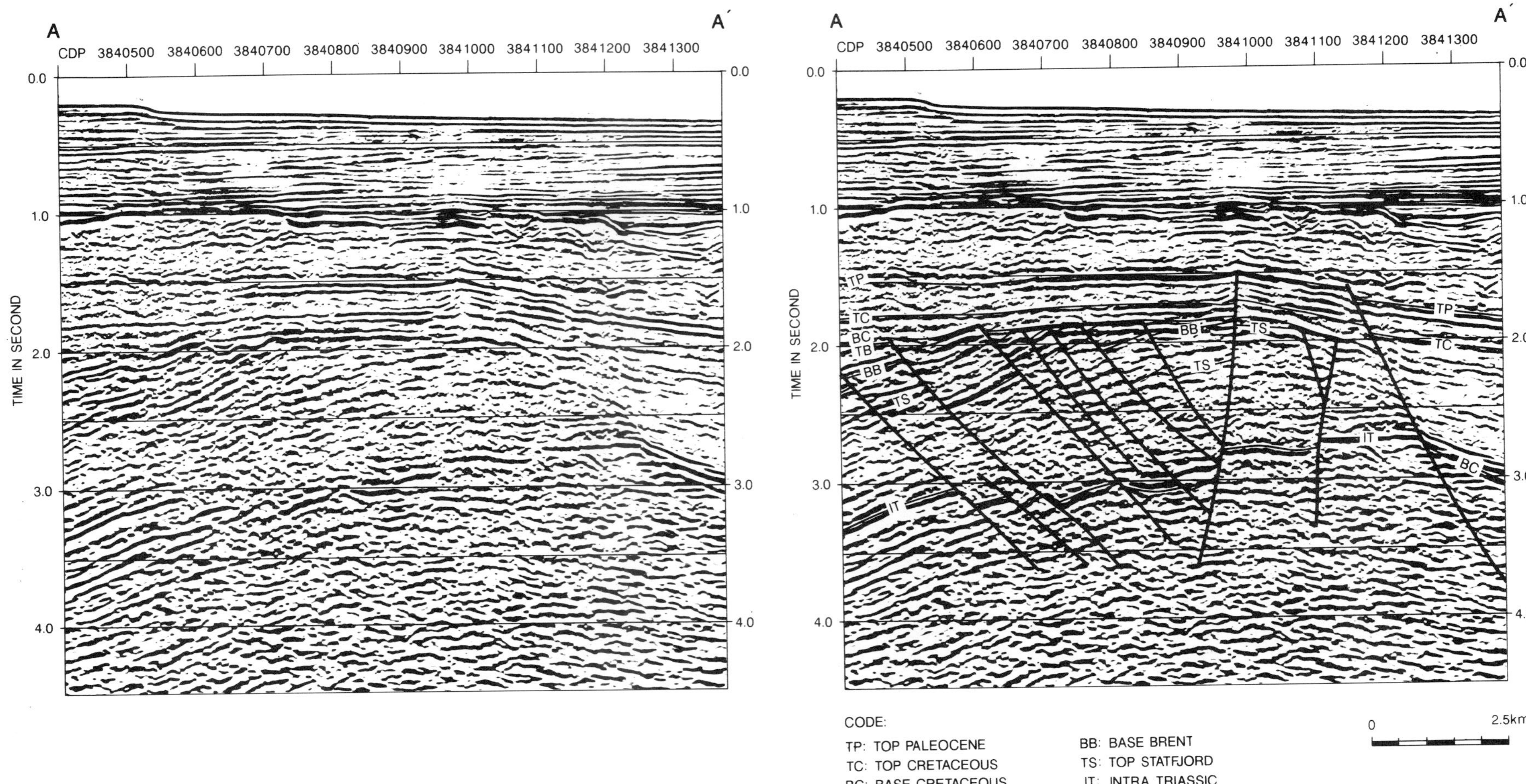

Figure 2. Interpreted and uninterpreted seismic lines. Line A–A′ in Figure 1 (inset) shows the location. (Modified from Erichsen et al., 1987.)

Before drilling, the following points had to be evaluated: (1) the possibility of absence of reservoir rocks as a result of nondeposition on a pre-existing high, (2) the possibility of erosion of sandstones during the late Kimmerian orogenic event, and (3) the possibility of poorly developed sandstones resulting from unfavorable deltaic settings. Erosion of sub-Upper Jurassic reflectors was recognized (Skarpnes et al., 1982).

Discovery

In July 1978 the first well, 34/10-1, was drilled downflank on a rotated fault block in the middle of the structure (Figure 3). It was successful and encountered 162 m of oil-filled sandstone in the Middle Jurassic Brent Group. The top of the reservoir was 50 m higher than predicted. The sandstone (Cook Formation) below the Brent Group (Figures 3 and 4) was water bearing. Porosities in the Brent Group sandstones ranged up to 34% while permeabilities were measured up to as much as 3 to 5 darcys. The reservoir temperature was 72°C and initial pressure was 310 bar. The hydrocarbons showed an oil gravity of 35° API and a gas-oil ratio of 200 Sm^3/Sm^3 (1.1 mcf/BO). Testing of the first well revealed that the friable condition of the highly porous sandstone in the Middle Jurassic Brent Group would be a problem for future test programs and that this would have to be taken into consideration in production planning. The second well was drilled into a separate and deeper structure in the southern part of the 34/10 block. This well encountered gas in the Brent Group and oil in the Lower Jurassic Statfjord Formation. This southern structure is now called the Gullfaks Sør field (Figure 1).

Post-Discovery

Subsequent activity in the block was directed toward further exploration and appraisal of the Gullfaks field. It was soon discovered that in favorable structural settings within the field, both the Rhaetian to Sinemurian Statfjord Formation and the Pliensbachian to Toarcian Cook Formation were oil bearing (Figure 4). Well 34/10-3 found an oil-water contact (OWC) within the Brent Group, well 34/10-7 discovered oil in the Cook Formation, while well 34/10-9 discovered oil both in the Brent Group and in the Cook Formation, with an OWC in the latter. Well 34/10-11 discovered an oil-bearing Statfjord Formation. Hydrocarbons have also been seen in the Upper Triassic Hegre Group in well 34/10-13, but this has, so far, been regarded as an unimportant reservoir. Oil shows have also been observed in Tertiary sandstones. A generalized sequence of the reservoirs in the Jurassic of the Gullfaks field is shown in Figure 5. Owing to tectonic complexity of the structure, the Gullfaks field has been difficult to interpret, and a three-dimensional seismic survey was carried out in 1979 to better define the field. Shallow biogenic or mixed thermogenic/biogenic gas in Neogene sandstones is common in the Gullfaks field area and has to be taken into account during drilling operations.

DISCOVERY METHOD

Early prospecting was based largely on seismic structural mapping. The potential of the Gullfaks area was highlighted by its geological similarity to neighboring discoveries. Uncertainty related to the subcrops beneath the unconformity in the tilted fault blocks was regarded as the main problem (Skarpnes et al., 1982). Prospect evaluation is briefly summarized below.

Source Rock

The low-velocity, organic-rich (5 to 10% TOC) shale of the Kimmeridgian to Ryazanian Draupne Formation (Figure 4) was postulated downdip west of the structure and in the downfaulted Viking graben to the east. Alternative source rocks were the black shales of the Dunlin Group and Cretaceous shales in the Viking graben.

Generation and Migration

Generally these processes are related to the strong subsidence in the central areas of the northern North Sea basins in the Paleocene and Eocene. Mapping of the mature Draupne Formation suggests three kitchen areas: south of the structure in the Viking graben, northeast of the structure in the Viking graben, and in the Møre basin (Figure 1).

Cap Rock

Upper Cretaceous shales and marls were considered an efficient seal in the area. However, the thinning of the Upper Cretaceous to 150 m over the structure was a risk element at Gullfaks.

Reservoir Rock

Paleogeographic reconstruction based on well information from the western part of the East Shetland basin indicated similarity to the Statfjord field. Shallower reservoir depths at Gullfaks and the potential for excellent reservoir conditions in the Statfjord Formation and the Brent Group as found in other fields in the area enhanced the prospect.

Suggested Recoverable Reserves

The "most probable" figure was estimated as 350 $\times 10^6$ Sm^3 (2200 MMBO), with an oil column of 300 m.

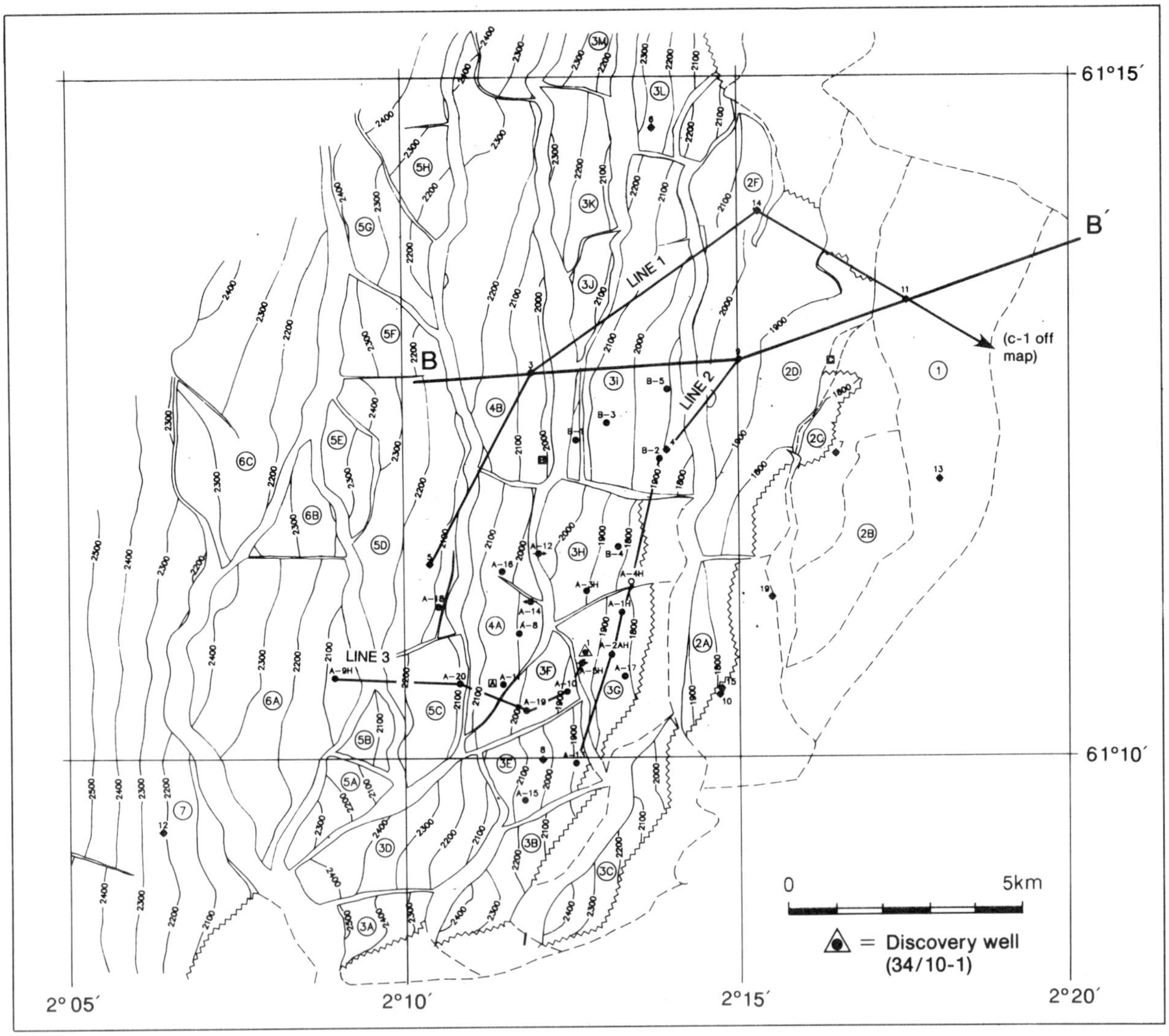

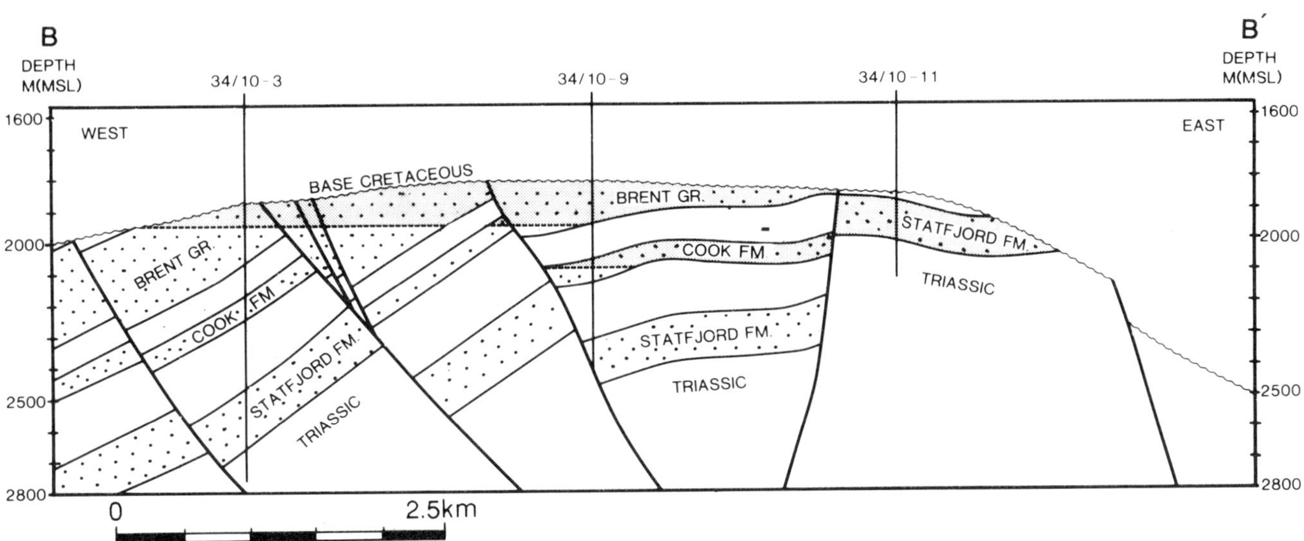

Figure 3. Structural map base Brent Group with well locations and cross section B-B′ (lower figure) through key wells. Shaded areas show productive zones. Contours in meters subsea. Lines 1, 2, and 3 show the locations of correlation cross sections by Figures 7, 11, and 14, respectively. (Modified from Petterson et al., 1990.)

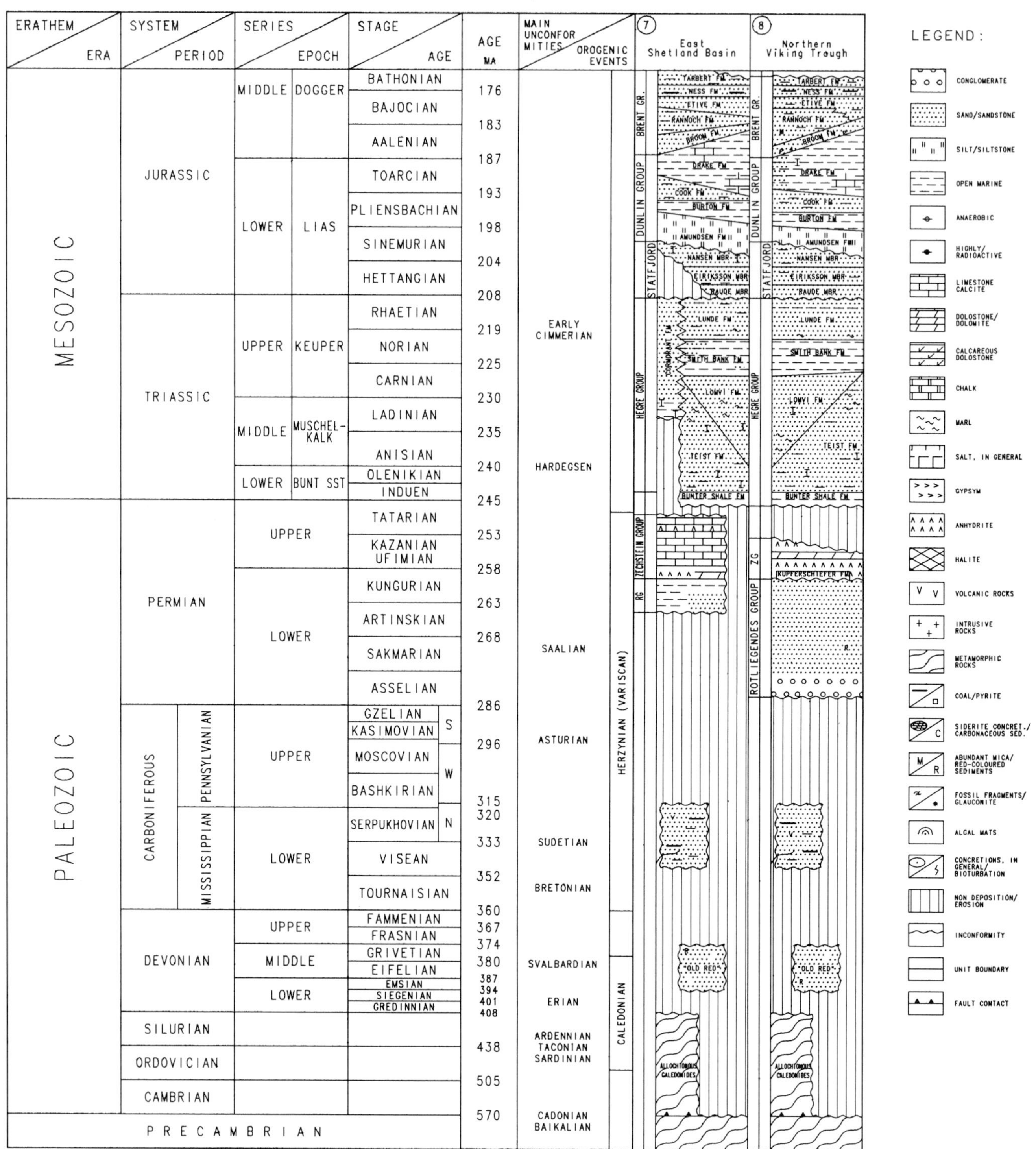

Figure 4. Generalized stratigraphy of the East Shetland basin (left column) and northern Viking graben (right column). (After internal Statoil report; Fjæran et al. written communication.)

ERATHEM / ERA	SYSTEM / PERIOD		SERIES / EPOCH		STAGE / AGE		AGE MA	MAIN UNCONFORMITIES / OROGENIC EVENTS
CENOZOIC	QUATERNARY		HOLOCENE				0.01	
			PLEISTOCENE				2	
	TERTIARY	NEOGENE	PLIOCENE		PIACENZIAN		3	
					ZANCLIAN		5	
			MIOCENE		MESSINIAN		6	ATTICAN
					TORTONIAN		11	
					SERRAVALLIAN		15	STYRIAN
					LANGHIAN		16	
					BURDIGALIAN		21	
					AQUITANIAN		23	SAVIAN
		PALEOGENE	OLIGOCENE		CHATTIAN		30	
					RUPELIAN		36	
			EOCENE		PRIABONIAN		40	PYRENEAN
					BARTONIAN		43	
					LUTETIAN		52	
					YPRESIAN		57	
			PALEOCENE		THANETIAN		63	LARAMIDE
					DANIAN		66	
MESOZOIC	CRETACEOUS		UPPER	SENONIAN	MAASTRICHTIAN		74	
					CAMPANIAN		84	
					SANTONIAN		87	
					CONIACIAN		88	SUB HERZYRIAN
					TURONIAN		91	
					CENOMANIAN		97	
			LOWER		ALBIAN		113	
					APTIAN		119	
					BARREMIAN		124	
				NEOCOMIAN	HAUTERIVIAN		131	
					VALANGINIAN		138	
					RYAZANIAN	BERRIASIAN	144	LATE CIMMERIAN
	JURASSIC		UPPER	MALM	VOLGIAN	TITHONIAN	150	
					KIMMERIDGIAN		156	
					OXFORDIAN		163	
			MIDDLE	DOGGER	CALLOVIAN		169	MID CIMMERIAN
					BATHONIAN			

ALPINE

(7) East Shetland Basin

NORDLAND GROUP: UTSIRA FM
HORDALAND GROUP: FRIGG FM
ROGALAND GROUP: BALDER FM, HEIMDAL FM, LISTA FM, VALE FM
SHETLAND GROUP: JORSALFARE FM, KYRRE FM, TRYGGVASON FM, BLODØKS FM, SVARTE FM
CROMER KNOLL GROUP: RØDBY FM, SOLA FM, RAN SST. ENH., MIME FM
VIKING GROUP: DRAUPNE FM, MAGNUS FM, HEATHER FM

(8) Northern Viking Trough

NORDLAND GROUP: UTSIRA FM
HORDALAND GROUP
ROGALAND GROUP: BALDER FM, HERMOD FM, SELE FM, HEIMDAL FM, LISTA FM, TY FM, VALE FM
SHETLAND GROUP: JORSALFARE FM, HARDRADE FM, KYRRE FM, TRYGGVASON FM, BLODØKS FM, SVARTE FM
CROMER KNOLL GROUP: RØDBY FM, SOLA FM, MIME FM, ÅSGARD FM
VIKING GROUP: DRAUPNE FM, SOGNEFJORDEN FM, HEATHER FM

LEGEND:

- CONGLOMERATE
- SAND/SANDSTONE
- SILT/SILTSTONE
- OPEN MARINE
- ANAEROBIC
- HIGHLY/RADIOACTIVE
- LIMESTONE CALCITE
- DOLOSTONE/DOLOMITE
- CALCAREOUS DOLOSTONE
- CHALK
- MARL
- SALT, IN GENERAL
- GYPSYM
- ANHYDRITE
- HALITE
- VOLCANIC ROCKS
- INTRUSIVE ROCKS
- METAMORPHIC ROCKS
- COAL/PYRITE
- SIDERITE CONCRET./CARBONACEOUS SED.
- ABUNDANT MICA/RED-COLOURED SEDIMENTS
- FOSSIL FRAGMENTS/GLAUCONITE
- ALGAL MATS
- CONCRETIONS, IN GENERAL/BIOTURBATION
- NON DEPOSITION/EROSION
- INCONFORMITY
- UNIT BOUNDARY
- FAULT CONTACT

Figure 4. Continued

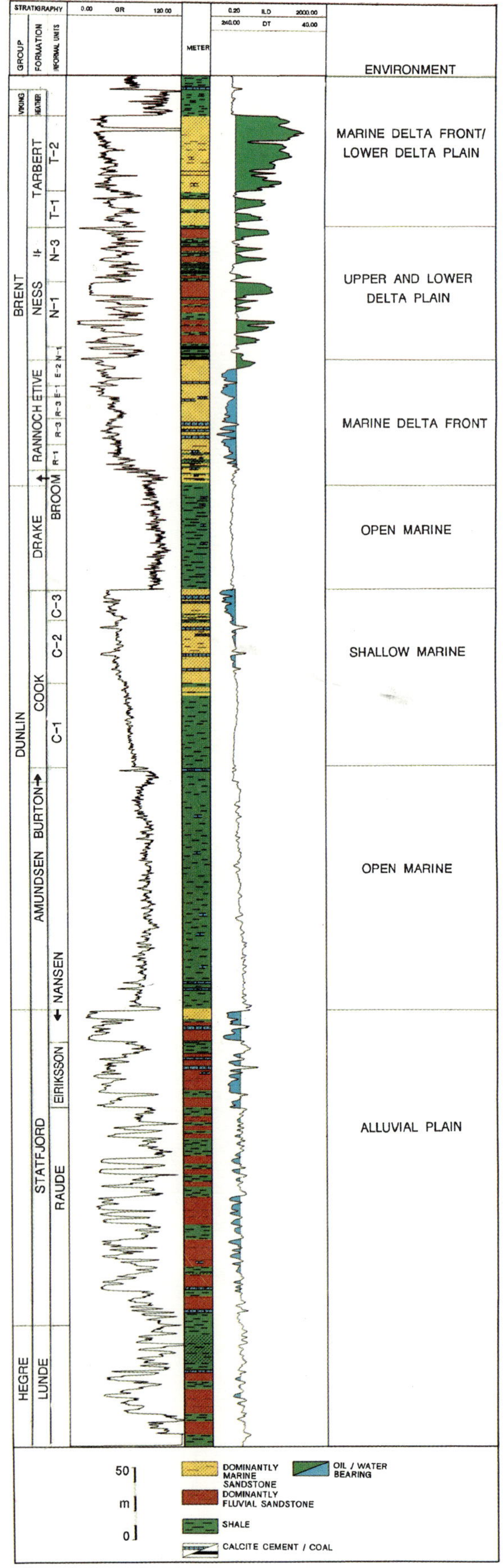

Figure 5. Type log showing general stratigraphy of the reservoir sequences in the Gullfaks field.

STRUCTURE

By middle Permian time, the major structural pattern was established in the North Sea, i.e., the Mid North Sea high (Figure 1) and rifting in the proto-Viking graben in the northern North Sea (Doré and Gage, 1987). In the East Shetland basin, syndepositional tectonism played an important role as a determinant of facies and thickness distribution of both Triassic and Jurassic rocks. During the Triassic, red bed deposition took place in a complex basin and range topography along the rift flanks. Major block faulting in the Early Triassic is expressed as horst and graben structures in the North Sea area. During the Permian and the Early Triassic, the eastern margin of the East Shetland basin and the primitive Viking graben were established as dominant structural elements (Nystuen et al., 1989). In the axial part of the Viking graben, a thickness of more than 3 km of Triassic continental deposits is estimated to be present (Lervik et al., 1989).

From Late Triassic to Early Cretaceous times a northwest to southeast extensional stress system, the Kimmerian phase, prevailed in the Viking graben (Ziegler, 1982). The three reservoir rock units in the Gullfaks field were deposited during the early and mid-Kimmerian phase and are regarded as pre-rift deposits. The northward-directed fluvial Rhaetian to early Sinemurian deposits (Statfjord Formation) probably represent the first indication of this tectonic influence on sedimentation in the northern North Sea basin. The Tethys and the Arctic seas were probably linked from south to north by the late Sinemurian transgression, seen in the northern North Sea (Ziegler, 1982), and marine shales and minor marine sands (the Cook Formation) covered most of the basins. The increased tectonic activity during the Early to Middle Jurassic transition is known as the mid-Kimmerian phase. Volcanic activity occurred in the North Sea area during this time (Malm et al., 1979).

From a structurally uplifted area in the central North Sea, regressive and transgressive clastics prograded northward into the subsiding basins of the northern North Sea. These clastics, known as the Brent Group, provide the major reservoir in Gullfaks and contain the bulk of the oil and gas reservoired in the area. Fault activity and rifting culminated during the Late Jurassic to Early Cretaceous, the late Kimmerian phase. Tectonics are expressed by normal faults overlying Triassic horsts and grabens in the East Shetland basin area (Karlson, 1986). Large-scale normal faulting of the Gullfaks structure is related to this Late Jurassic to Early Cretaceous tectonism, the Kimmerian and Austrian orogenic events. On both sides of the Viking graben, half-grabens were formed by rotated fault blocks. The Gullfaks structure was subsequently eroded at the time of the late Kimmerian and Austrian events. After these tectonic events, the East Shetland basin subsided.

Based on new seismic data, Fossen (1989) has subdivided the tectonism in the Gullfaks area into two phases, both related to Kimmerian extensional tectonics. A Middle to Late Jurassic extension was followed by a Late Jurassic to Early Cretaceous transpression in the Gullfaks field. The latter phase is interpreted as a response to dextral strike-slip movement along the northeast-southwest trend of the Tampen spur. Thrusts, reverse faults, and compressional folds were formed in this phase.

The marked north-south-oriented fault system, in which individual fault blocks vary from 0.5 to 1 km in width and with fault throws of the order of 50 to 200 m (Figure 2), is a typical feature of the Gullfaks field. Major boundary faults in the north, east, and west delineate the structure from the Tampen spur rim, the Viking graben, and the Gullfaks Sør structure. The Gullfaks field can be divided into three different structural regimes (Figure 3):

1. North-south-striking rotated fault blocks in the southwestern area (encircled 3, 4, 5, and 6 in Figure 3).
2. A central "graben" area (encircled 2 in Figure 3), interpreted as a result of "keystone" collapse (Sæland and Simpson, 1982).
3. An eastern horst structure (encircled 1 in Figure 3), which is bounded to the east by a major vertical fault with a throw on the order of 1500 m down to the Viking graben (Figures 2 and 3). The Brent Group is completely eroded over this eastern horst.

STRATIGRAPHY

The Mesozoic and Cenozoic of the northern Viking graben (Figure 4) is composed almost entirely of terrestrially derived siliciclastic deposits with less than 5% coals, limestones, and volcanics (tuff). The lithostratigraphical nomenclature given by Vollset and Doré (1984) is used in this paper.

The oldest rock sequence penetrated in the Gullfaks field is composed of Triassic continental deposits. Because of erosion (late Kimmerian unconformity), the Jurassic and Lower Cretaceous sequence can vary significantly in thickness. The Lower and Middle Jurassic fluvial/marine sequence of the Gullfaks field can reach a thickness of approximately 1 km. This sequence includes the reservoir units in the Gullfaks field. Upper Jurassic organic-rich shales (Draupne Formation) are normally absent over the structure but are more than 100 m thick downflank. These shales are the source rock for the accumulated hydrocarbons. Cretaceous shales and marls of the Shetland Group, which unconformably overlie the Jurassic sequence, are approximately 100 m thick in the area. These shales are the cap rock for the Gullfaks field. Tertiary sandstones and shales, including the seismic marker bed of the tuffaceous Balder Formation, are approximately 1.5 km thick in the Gullfaks area. Shallow gas and swelling clay in the Tertiary sequence are potential drilling hazards. Quaternary sediments may reach a thickness of 100 m.

TRAP

The Gullfaks field is a structural unconformity trap comprising many rotated and partly eroded fault blocks below the late Kimmerian unconformity. Lower and Upper Cretaceous shales and marls onlap and drape the structure and form the top and lateral seal. Minor amounts of biodegraded hydrocarbons having a Jurassic reservoir affinity, together with mixed biogenic and thermogenic gas, are found in the overlying Tertiary sandstones (Irwin, 1989). This suggests that minor leakage occurs along the major fault planes.

Reservoir

The Gullfaks field has three major sandstone reservoir units (Figures 3 and 5): the Rhaetian to Sinemurian Statfjord Formation, the late Pliensbachian to early Toarcian Cook Formation of the Dunlin Group, and the early Bajocian to earliest Bathonian Brent Group. All three reservoir units were deposited in the early- to mid-Kimmerian tectonic phase. This extensional phase, coupled with sea level changes, has been a major control on the thickness and distribution of the sands. The Brent Group is the most important reservoir unit. Each of the three reservoir units is capped by a marine offshore shale sequence. The stratigraphy, the depositional environments, and the reservoir characteristics of the three reservoir units are summarized below.

Statfjord Formation

The Statfjord Formation forms a reservoir primarily in the eastern parts of the Gullfaks field where this formation occurs structurally higher than in the rest of the field and where the Brent and Cook formations are extensively eroded (Figure 3). The Statfjord Formation contains recoverable oil reserves on the order of 25×10^6 Sm^3 (157 MMBO). It is composed of interbedded sandstones and mudstones, for the greater part representing alluvial environments (Figure 6).

Limited biostratigraphic evidence for the age of the Statfjord Formation is determined from samples from the formation; however, Rhaetian indicators are found in its lower parts and deposition extended into Sinemurian times. The formation, thus, spans the Triassic-Jurassic boundary.

The depositional basin of the primitive Viking graben and surrounding areas was bounded toward the Shetland platform in the west by major faults that were active during deposition and Statfjord Formation deposits have a limited extension west of this boundary. Apparently, syndepositional fault activity

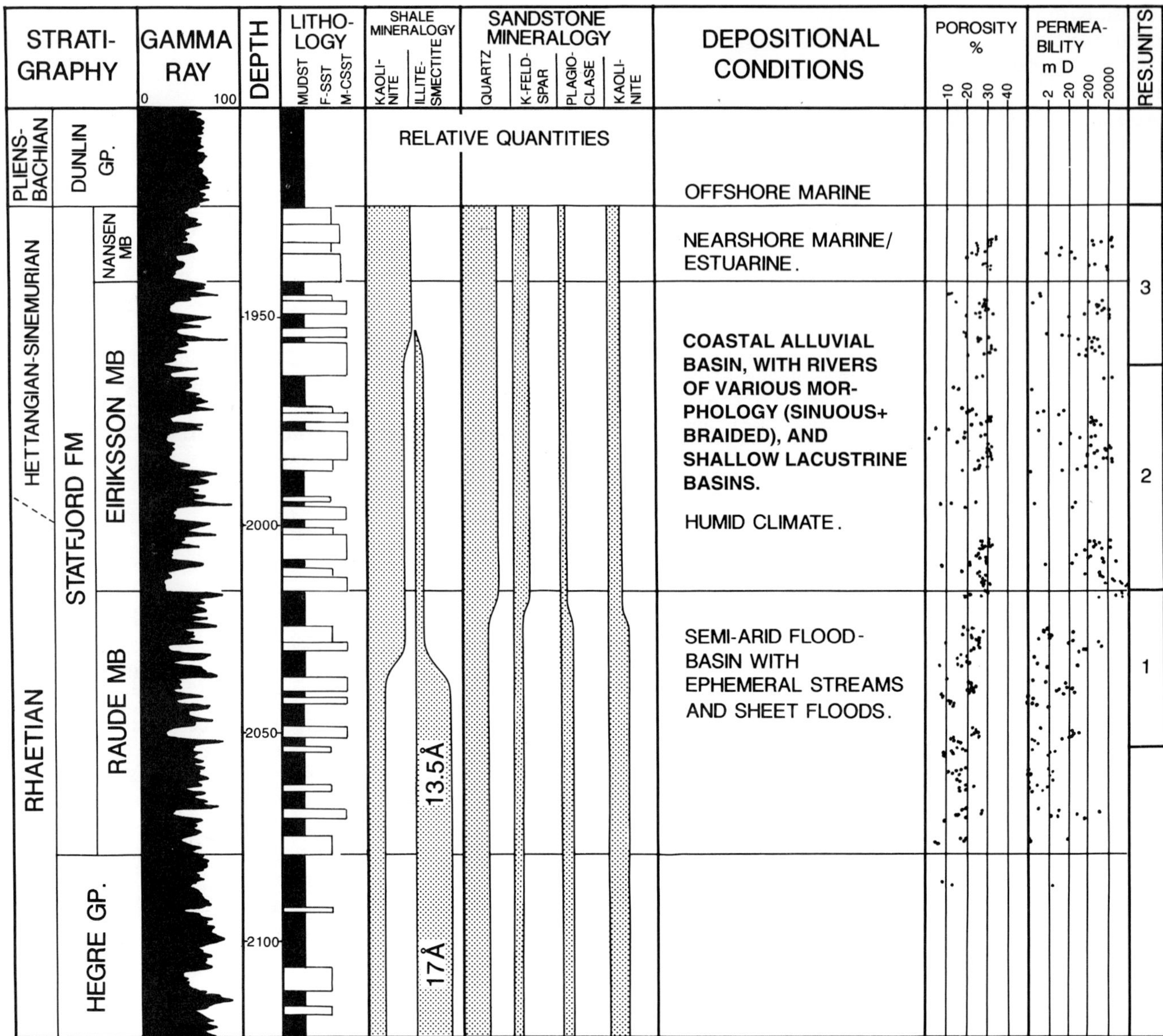

Figure 6. Summary of stratigraphy, mineralogic composition, and depositional conditions for the Statfjord Formation in the Gullfaks field.

influenced deposition and gave rise to the depositional cyclicity recognized in some areas (Røe and Steel, 1985; Falt, 1986). Thickness variations within the formation can also be partly ascribed to syndepositional fault-related tectonism and differential subsidence (Nystuen et al., 1989; Steel and Ryseth, 1991).

The alluvial basin in which the Statfjord Formation was deposited was connected to marine environments toward the north, but evidence of marine influence in the Gullfaks area is recognized only in the uppermost portions of the formation. The full thickness of the Statfjord Formation in the Gullfaks field is around 170 to 210 m, and three distinct members have been described (Vollset and Doré, 1984) (Figure 6).

Raude Member—The lowermost member, the Raude Member, rests upon the Triassic alluvial deposits of the Hegre Group. It generally reflects an upward-coarsening sequence and a thickening of sandstone beds (Figures 6 and 7). The interbedded mudstones are red and mottled red and green, with common root zones and frequent calcretes.

Sandstones of the Raude Member are interpreted to have formed by ephemeral streams, with sheet floods being an important process and channeling

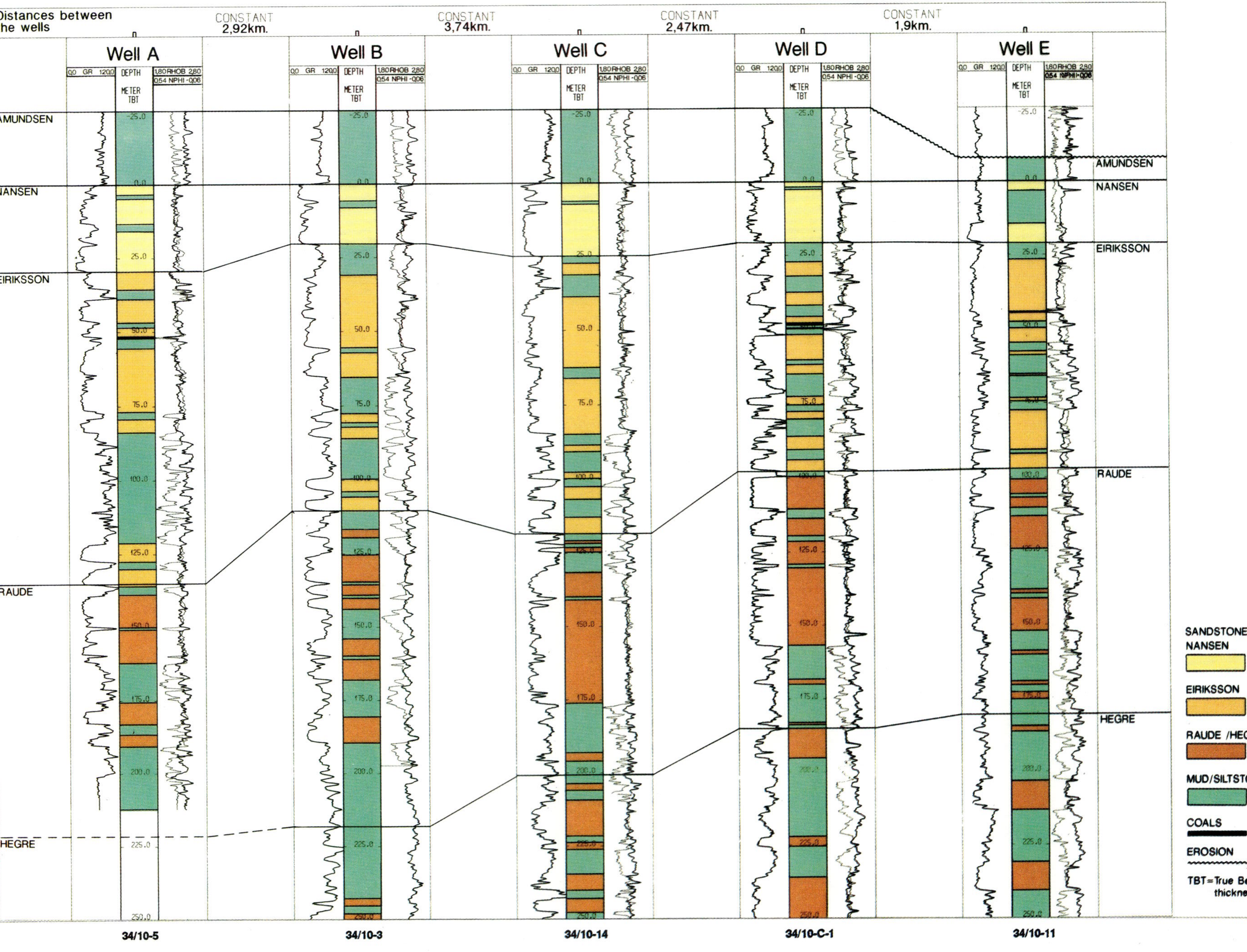

Figure 7. An example of reservoir correlation of the Statfjord Formation. See Figure 3 for well locations.

becoming more common upward through the sequence. In some wells in the Gullfaks field the number of fluvial channel sandstones in the Raude Member is higher than in other wells. This may reflect the location of fluvial channel belts, possibly with streams of a more permanent character.

Eiriksson Member—The Eiriksson middle member is volumetrically the most important reservoir interval. It is composed of generally thick (up to 10 to 15 m) sandstones that are mostly medium or coarse grained with an immature textural and mineralogical composition. The sandstones are interbedded with mudstones of variable thickness (Figure 7). The red coloration of the mudstones at the base passes upward into a gray and dark gray coloration. Root structures and other pedogenic phenomena in many of the mudstones indicate subaerial floodbasin conditions, but some dark laminated mudstones probably represent shallow lakes on the floodplain. Occasional thin coal beds occur.

The thicker sandstone units of the Eiriksson Member frequently exhibit single or multistory fining-upward patterns, and they were formed in fluvial channels and channel belts (Figure 8). Variations seen in channel sandstone characteristics and alluvial architecture probably reflect the fact that a range of fluvial styles is represented within the member, produced by the areal and temporal variability of stream types commonly seen in both recent and ancient alluvial basins.

As a hypothesis to support qualitative and quantitative reservoir modeling, it is assumed that most of the fluvial channel sandstones of the Eiriksson Member were formed by braided perennial or semiperennial streams. Toward the top of the member, channel sandstones formed by lower gradient, sinuous streams may also be identified. The multistory, multilateral alluvial architecture and the high sandstone-to-mudstone ratio in the Gullfaks field provides a reservoir unit with good sand-body intercommunication, although a few individual mudstones may act as efficient stratigraphic barriers to pressure and flow within the member.

A few of the thicker sandstone units of the Eiriksson Member exhibit coarsening-upward patterns (Figure 8), and these are suggested to have formed prograding distributary mouth bars in shallow lacustrine basins, possibly initiated by episodic tectonic downwarping (Fält, 1986).

Illite–smectite-dominated clay assemblages in the Raude Member are rapidly substituted by kaolinite-dominated assemblages in the Eiriksson Member (Figure 6). In addition, there is an upward increase of the mineralogical maturity of the sandstones through the Statfjord Formation. The quartz content increases, whereas the feldspar content diminishes upward. These changes, coupled with mudstone color changes and transition from calcrete soil profiles to more humid types, are best explained by climate changes (and evolving depositional processes) in combination with the possible existence of a significant hiatus at the Raude-Eiriksson boundary.

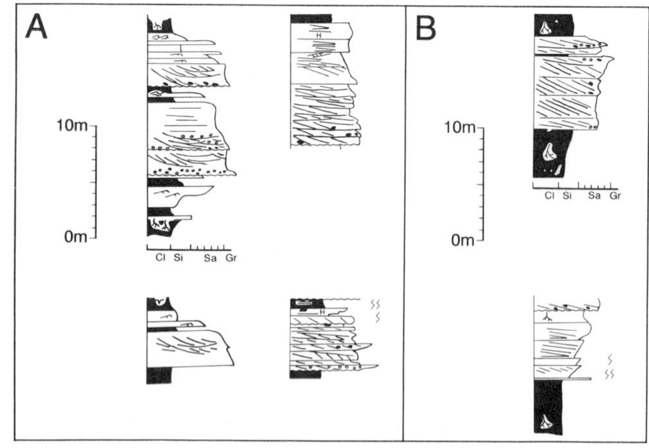

Figure 8. Examples from the Eiriksson Member of (A) single- and multistory fining-upward fluvial channel units and (B) coarsening-upward mouth bar deposits. For legend see Figure 13. Cl = claystone, Si = siltstone, Sa = sandstone, Gr = gravel.

Nansen Member—The Nansen Member, the uppermost member of the Statfjord Formation, is a relatively thin (20 to 30 m) unit (Figures 6 and 7).Trace fossils in intervals of fine-grained, micaceous sandstone indicate a marine-influenced depositional setting. In addition, the Nansen Member contains fining-upward channel deposits. It is interpreted to have formed in an estuarine setting with a strong fluvial influence during marine transgression in the late Sinemurian. Following the transgression, offshore fine-grained sediments of the Amundsen Formation, Dunlin Group, were deposited.

Reservoir Quality and Petrography of Sandstones in the Statfjord Formation—An example of reservoir correlation of the Statfjord Formation is shown in Figure 7. In the Gullfaks field, sandstones of the Statfjord Formation normally have good to excellent reservoir properties (Figure 9B), generally improving upward throughout the formation. Sandstone permeabilities are mostly in the range of 0.1 to 1.0 darcy and reach several darcys in the upper parts of the Eiriksson Member and in the Nansen Member. Sandstone porosities generally range from 20 to 30%. Intercommunication between individual Raude Member sandstone bodies is probably restricted, but interwell communication is expected to be effective in sandstone bodies of the Eiriksson and Nansen members.

The loose sandstones of the Statfjord Formation are mainly classified as subarkosic arenites. Reservoir quality is diminished by post-depositional alteration by compaction, kaolinitization of feldspar grains, and carbonate cementation. Because of the originally high content of feldspar and syndepositional or early diagenetic freshwater transformation of feldspar grains to kaolinite clusters, some sandstones have been easily deformed by compaction; e.g., loss of mechanical rock strength. Carbonate cementation (Figure 9A) is common, and although

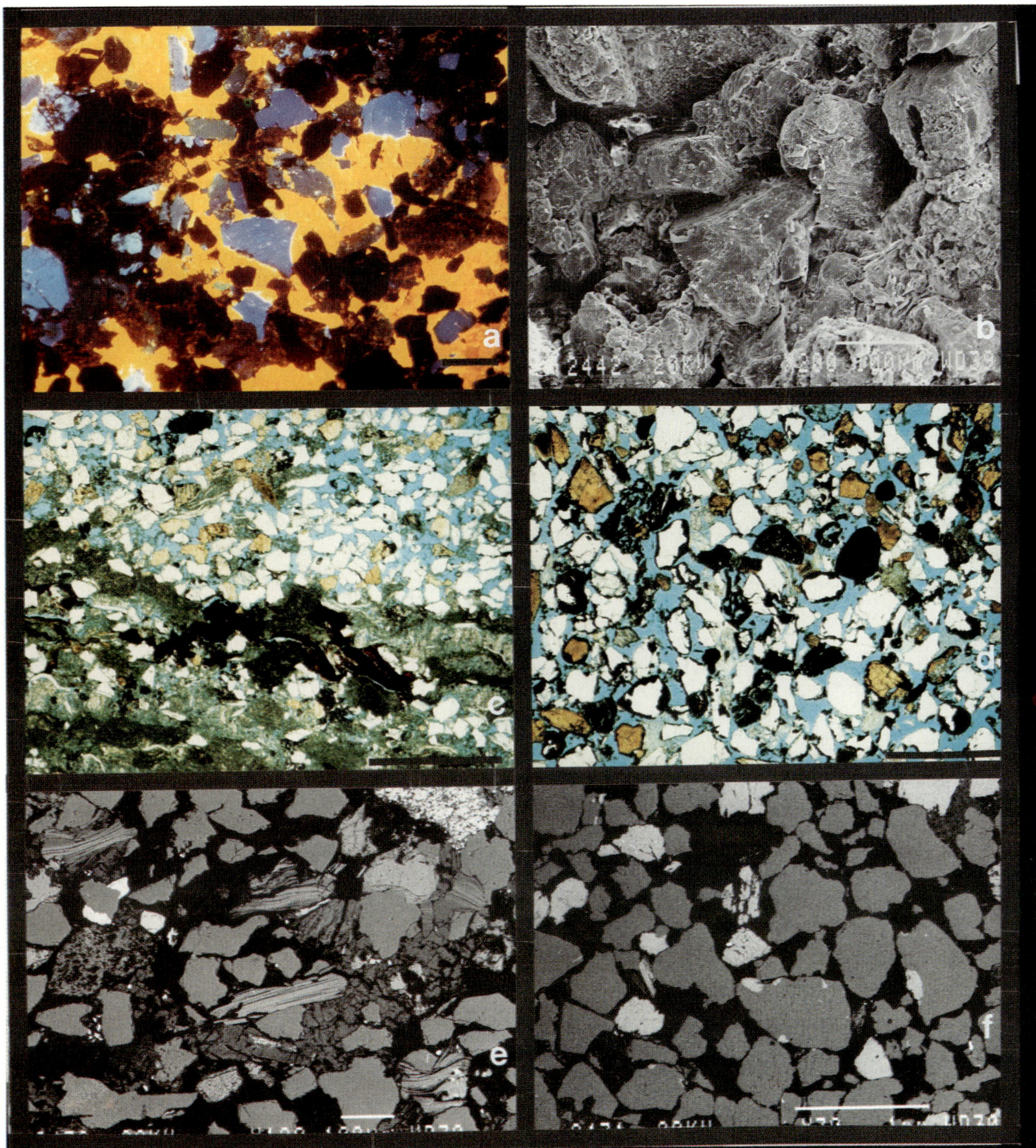

Figure 9. Petrographic characteristics of reservoir sandstones of the Gullfaks field. (A) Statfjord Formation carbonate-cemented coarse-grained sandstone photomicrograph. Luminescence of thin section shows carbonate cement as yellow, feldspar as blue, and quartz as dark brown. Other brown areas are porosity and clay minerals. Bar scale, 0.25 mm. (B) Scanning electron micrograph of Statfjord Formation fine-grained sandstone. Bar scale, 0.1 mm. (C) Thin section photomicrograph showing fine-grained sandstone and sandy intrapelsparite (lower part) from the Cook Formation. Blue color is porosity. Bar scale, 1 mm. (D) Cook Formation coarse-grained sandstone photomicrograph. Yellow color shows feldspar and blue color is porosity. Porosity varies from 35 to 38% and permeability in the range 3 to 5 darcys in this lithology. Black grains and coatings are pyrite. Bar scale, 1 mm. (E) Rannoch Formation back-scattered electron micrograph. Black, porosity; dark gray, kaolinite; medium gray, quartz; light gray, feldspar and mica; white, siderite. Bar scale, 0.1 mm. Fine-grained sandstone. Porosity and permeability are in the range of 30-35% and 1 to 2 darcys, respectively. (F) Etive Formation back-scattered electron micrograph. Mineral identification as in E. Note rock fragments and perthitic feldspars (upper right). Bar scale, 1 mm. Medium- to coarse-grained sandstone. Porosity and permeability are in the range of 35% and 5 to 10 darcys, respectively.

carbonate-cemented sandstone is not only associated with calcretes, the cement is probably redistributed carbonate from primary deposition within the formation. Lateral continuity of carbonate-cemented sandstone is regarded as only a minor production problem in these heterogeneous sandstone bodies.

Cook Formation

The Cook Formation is the only unit within the Dunlin Group that contains significant reservoir sandstones. Hydrocarbon reserves in the reservoirs have primarily accumulated in the eastern part of the Gullfaks field, with some minor accumulations in the central part. As in the Brent Group and Statfjord Formation, the formation shows increased erosional truncation eastward.

The reservoirs consist of bioturbated sandstones and interbedded sandstones and mudstones. The sedimentary source was located to the southwest, with a probable provenance area in the highlands surrounding the East Shetland basin (Mearns, 1989). The Cook Formation of the Lower Jurassic (Vollset and Doré, 1984) is dated from late Pliensbachian to early-middle Toarcian age in the Gullfaks area and contains 15×10^6 Sm^3 (94 MMBO) of recoverable reserves.

Bioturbated Sequence, Cook 2—This bioturbated, fine-grained sandstone sequence is fairly constant in thickness in the Gullfaks field, being normally around 50 m thick (Figure 10). The Cook 2 unit comprises coarsening-upward sequences, evident especially on resistivity and density logs. The upward-coarsening trends are an effect of shallowing cycles with improved reservoir quality upward. In the lower parts, these cycles exhibit moderate to high degrees of bioturbation, with sparse remnants of primary sedimentary structures. Upward, however, the sandstones are cleaner, less bioturbated, and contain poor to moderately preserved wavy bedding/hummocky cross stratification (Figure 10B, C). The lower cycle is much thicker than each of the two cycles above. Correspondingly, the lower part of the first cycle represents a deeper and lower energy depositional environment than the lower parts of the two cycles above.

The Cook 2 unit is interpreted to be deposited in a marine inner shelf environment. The relative sea level stand varied within each cycle from near the storm wave base in the lower part to somewhere below fair weather wave base at the top.

The Cook Formation, which on a large scale represents an upward-coarsening/shallowing sequence, may be subdivided sedimentologically into three major depositional units (Figure 10). Palynological and micropaleontological data indicate an upward transition from an open marine to a restricted marine environment. In the Gullfaks area the total thickness ranges from about 160 m in the southwest to 120 m in the northeast.

Laminated Sandstone and Mudstone Sequence, Cook 1—The lowermost, nonreservoir unit is a 50 to 65 m thick laminated sandstone, siltstone, and mudstone sequence (Figure 10). This unit is thickest in the central parts of the Gullfaks field (approximately 65 m thick) and thins toward the southwest and northeast. The base rests conformably on the Burton Formation (Figure 4) and consists of medium grayish-brown, laminated, very fine sandstones, siltstones, and mudstones (Figure 10A) . The sandstone content increases slightly toward the top. A few calcite-cemented horizons of less than a meter thick occur. The sedimentary structures and trace fossil associations indicate in general a quiet depositional setting. An open marine, middle shelf environment is suggested for this sequence.

Interbedded Sandstone and Mudstone Sequence, Cook 3—The upper sequence constitutes the best reservoir of the Cook Formation. The reservoir is very heterogeneous, with interbedded medium-grained sandstones and mudstones evident from MSF logs and cores (Figures 10D, E). The sandstones, which are locally very micaceous, have maintained a very porous and permeable texture, with porosities of generally 30 to 35% and permeabilities between 500 and 5000 md. This unit indicates, as does the bioturbated sequence, an upward-shallowing trend. The mudstone content decreases as the sandstone content correspondingly increases. Grain sizes are nearly constant throughout the interval. The unit is much more variable in thickness than the other Cook units, more than 45 m in the southwestern part of the field and decreasing to 20 m in the northeastern part. This reflects the more distal position of the interbedded sequence in the northern areas compared to the southern parts. While three upward-shallowing cycles are present in the south, only the upper cycle is developed in the north. The uppermost part of the upper cycle generally shows a massive or trough cross-bedded sequence (Figure 10F), with very good reservoir properties. Calcite-cemented horizons occur throughout this unit, but they are far less continuous than in the bioturbated unit. Similar relationships have been documented by Walderhaug et al. (1989).

The uppermost part of the unit seems to have been deposited in a more hospitable environment for the benthic fauna, as trace fossils appear locally at the top. A shallow marine, high-energy depositional environment is suggested for this unit.

Reservoir Quality and Petrography of the Sandstones of the Cook Formation—The Cook Formation shows improved reservoir quality upward. An example of reservoir correlation of the Cook Formation is given in Figure 11. The sandstones are classified as subarkosic to arkosic arenites and feldspathic graywackes. Mica may constitute 20% of the mineral composition in the finer-grained sandstones, especially in Cook 1. Glauconite, pyrite, and carbonate allochems are commonly observed in some beds. Zircon, tourmaline, and rutile occur as common accessory minerals. The feldspars show great

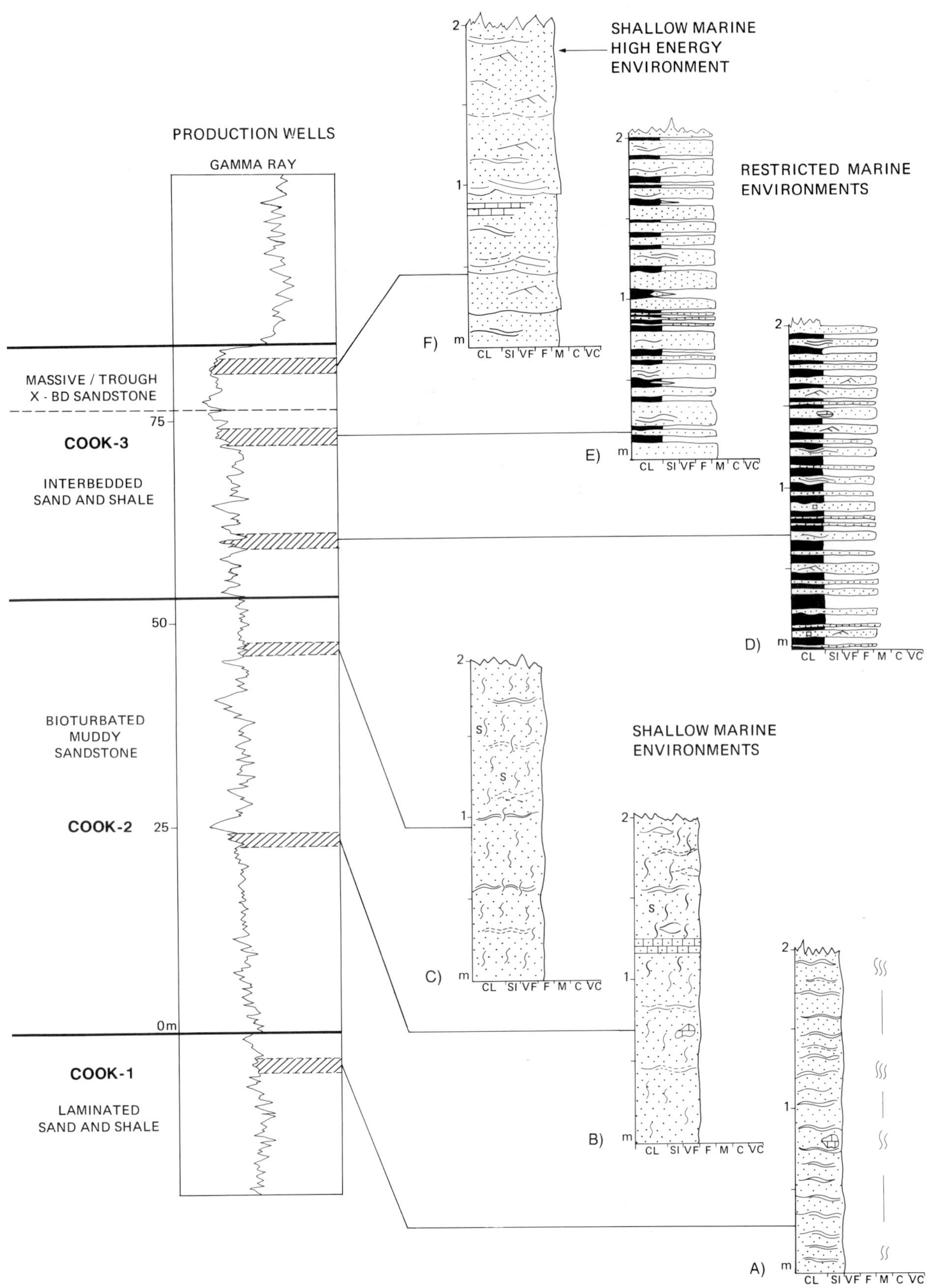

Figure 10. Facies sequences in the Cook Formation. (For legend see Figure 13.)

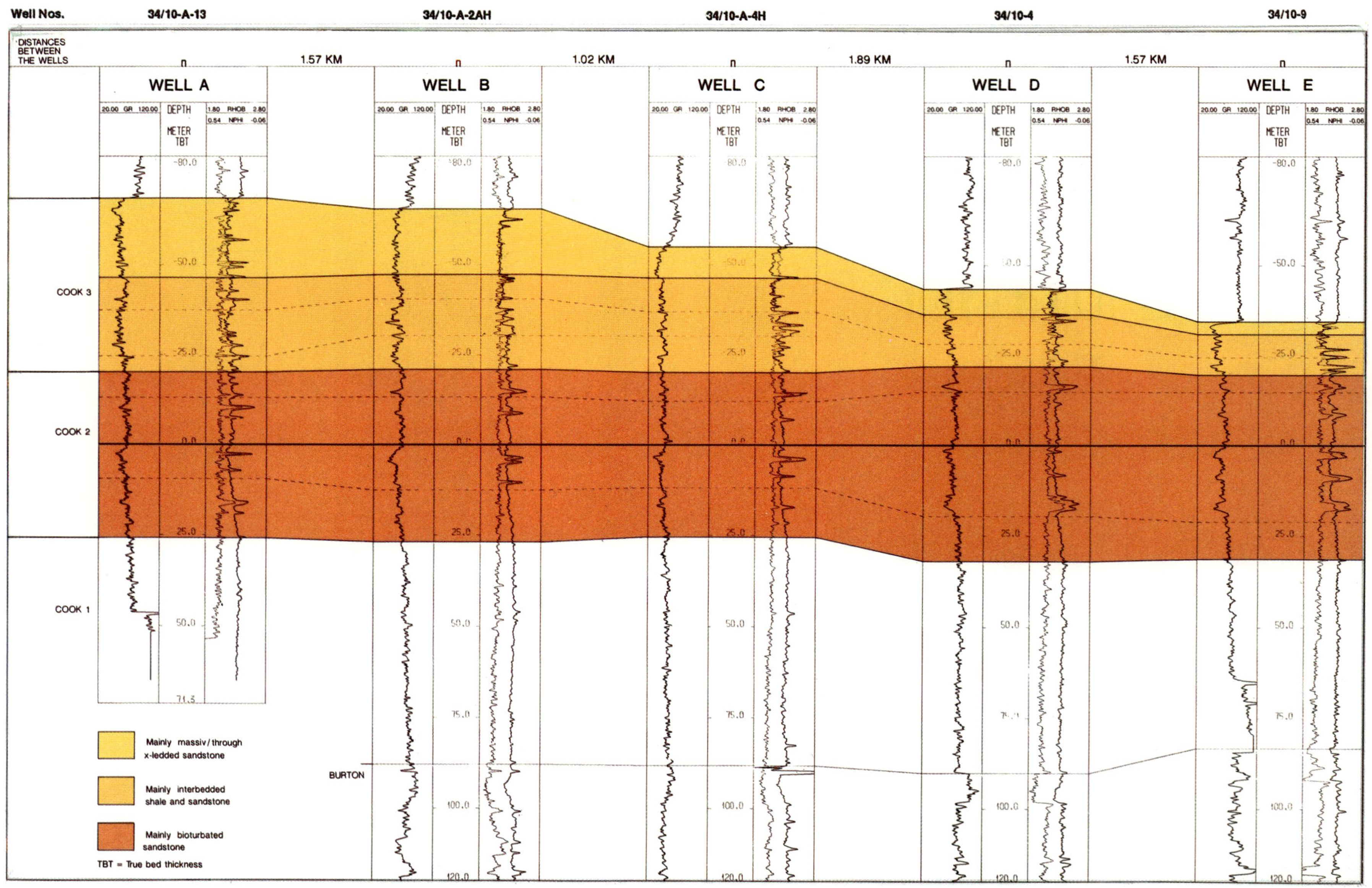

Figure 11. Example of correlation of the Cook Formation in the Gullfaks field. See Figure 3 for location of the correlation cross section.

variation, with the dominance of potash feldspars and fewer perthites and plagioclases. Like reservoir sandstones elsewhere in the Gullfaks field, the Cook Formation is loose, a factor that must be considered when designing production wells. Minor quartz and feldspar overgrowths are observed. Pyrite-coated grains (Figure 9D) constitute a striking feature of the Cook 3 unit sandstones.

Compaction has been the dominant post-depositional process, but it has had only minor to moderate effects on reservoir quality. The transformation of mica into kaolinite and the formation of carbonate-cemented layers have affected the reservoir quality. Up to 2 m thick carbonate-cemented beds occur in Cook 1 and Cook 2. Up to 25% of the rock column in Cook 1 has been observed as carbonate cemented (Figure 9C) in a few wells.

Some of the carbonate-cemented beds are suggested to be laterally continuous for hundreds of meters. A few of the carbonate-cemented beds in Cook 2 are classified as sandy intrabiosparites. These limestone beds are probably related to regional depositional episodes and are therefore potentially continuous over wide areas. They could thus act as vertical barriers to fluid migration. Probably most of the calcite cement has been sourced from shell debris. Both clastic kaolinite and diagenetic kaolinite are common in the finer-grained sandstones. Diagenetic kaolinite has reduced the permeability of the sandstones.

Brent Group

The Brent Group is the major reservoir unit in the Gullfaks field. The Brent Group has largely resulted from deposition related to a north-prograding deltaic coastline (Graue et al., 1987). In the Gullfaks field the Brent Group has been subdivided into two major deltaic depositional sequences (Figure 12). These are the lower Brent delta sequence, including the Broom, Rannoch, Etive, and Ness formations, and the upper Brent delta sequence, represented by the Tarbert Formation.

Prior to the northward advance of the Brent delta, the shoreline was east of the Horda platform (Graue et al., 1987). As stated above, uplift in the central North Sea resulted in a northward progradation into north-striking syndepositional, tectonic-subsiding basins. The provenance of the sands was thus the southern, eastern, and western margins of the basin. The lower Brent delta sequence, related to the main Brent delta progradation, was deposited during the Aalenian to mid-Bajocian rise in sea level (Haq et al., 1987) and is interpreted to be a wave-dominated to fluvial-dominated river delta system (Figure 12). The transition from the lower to upper Brent delta sequence is related to the time of Bajocian to early Bathonian transgressive events seen in the south, mid-, and north Norwegian shelves (Fält et al., 1989). The upper Brent delta sequence retreated southward during the latest Bajocian–earliest Bathonian deposition, and the Tarbert Formation represents a continued influx of sediments, deposited during an overall transgression that also partly resulted in a reworking of the underlying delta system deposits. The distribution and thickness of the Tarbert Formation was influenced by syndepositional tectonics resulting from shoreline retreat along active lineaments (Graue et al., 1987). Seaward of the shoreline, the marine shale of the Heather Formation was deposited, and further transgression of this sequence terminated Brent Group deposition. The Brent Group is approximately 300 m thick (Figure 13) in the Gullfaks field and has 190×10^6 Sm^3 (1195 MMBO) recoverable oil.

Lower Brent Delta—The lower Brent delta sequence comprises the four formations described and interpreted below.

Broom Formation—The lowermost lithostratigraphic unit in the Brent Group in the East Shetland platform area has been traditionally termed the Broom Formation (Figure 13) in the more southern area in the Viking graben. The formation in the Gullfaks field is developed as mudstones with thinly interbedded sandstone and gravel beds.

It is interpreted to represent storm as well as current and turbidite deposits in an offshore to prodeltaic environment. The Broom Formation is unimportant as a reservoir unit in the Gullfaks field.

Rannoch Formation—The Rannoch Formation (Figure 13) varies in thickness from 47 m in the southern part to 96 m in the northern part of the field and consists of well-sorted, micaceous, very fine to fine-grained sandstones. Plant remains are dispersed throughout the formation. The formation represents an overall upward-coarsening sequence which, in the Gullfaks field, can be subdivided into three reservoir units.

The lowermost unit, R-1, consists of alternating burrowed, lenticular to ripple laminated, flat-laminated, planar cross-laminated, and hummocky cross-stratified siltstones to very fine grained sandstones.

The two next units, R-2 and R-3, consist of micaceous very fine and predominantly fine-grained sandstones. R-2 is a monolithic unit of fine-grained sandstone with scattered carbonate-cemented intervals. The R-3 unit is more variable in grain size, mica content, and sedimentary structures. The R-2 and R-3 units are characterized by hummocky cross stratification, low-angle parallel lamination, and minor scour and fill structures. Ripple lamination and single burrows are locally developed in the R-3 unit.

The formation is interpreted as wave-dominated delta front deposits, grading from lower to upper shoreface.

Etive Formation—The Etive Formation (Figure 13) has a maximum thickness of 48 m in the central-southwestern part of the field, but decreases to about 15 m toward the south and the north. The formation is generally developed as an overall fining-upward

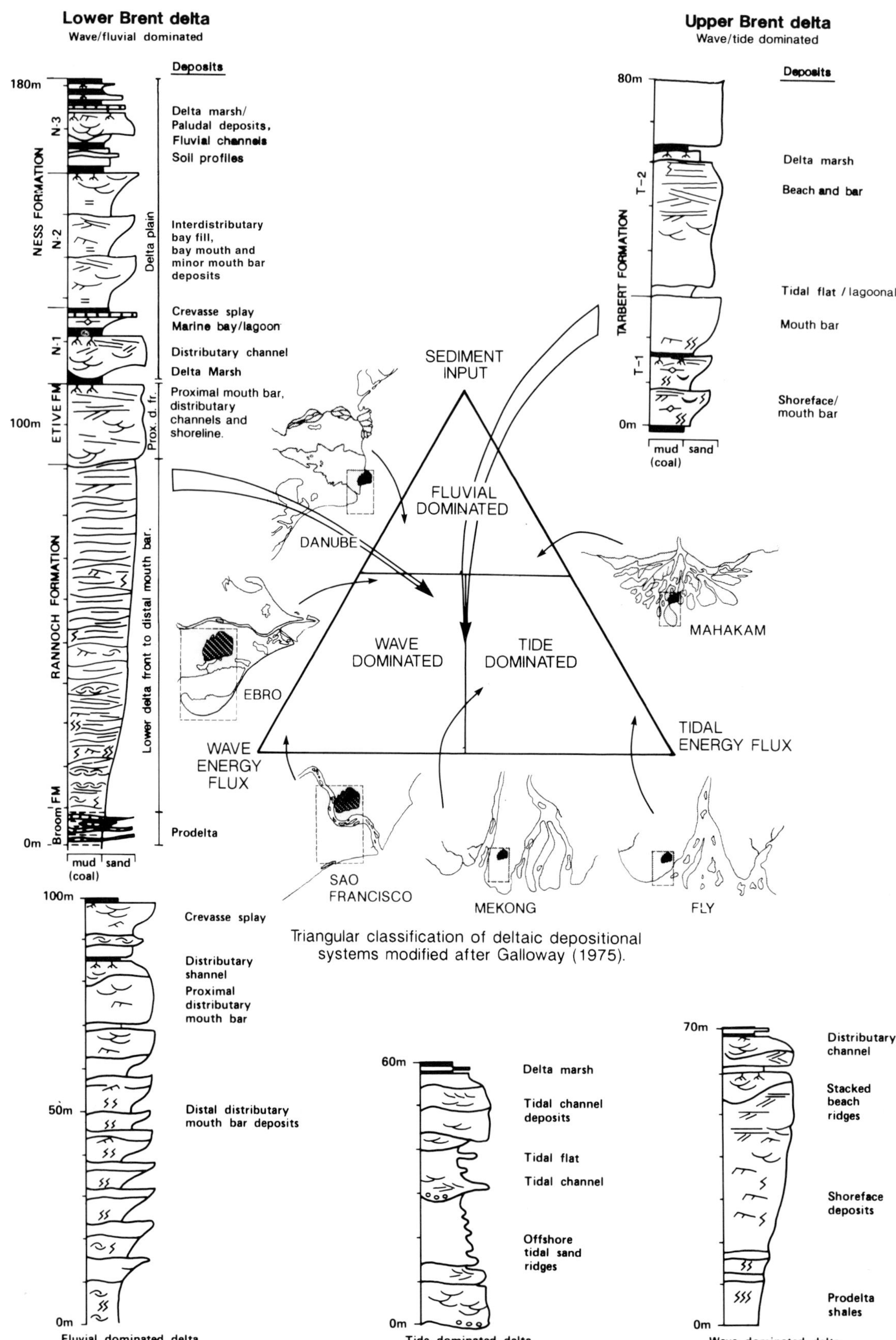

Figure 12. Lower and Upper Brent deltas compared with recent deltaic depositional systems. Stippled rectangles show areal comparison of recent deltas with area of the Gullfaks field in block 34/10. Note that even for a relatively small delta, such as the Ebro, a single bay or back barrier lagoon covers the entire areal extent of the Gullfaks field. (After Miall, 1979, and Galloway, 1975.)

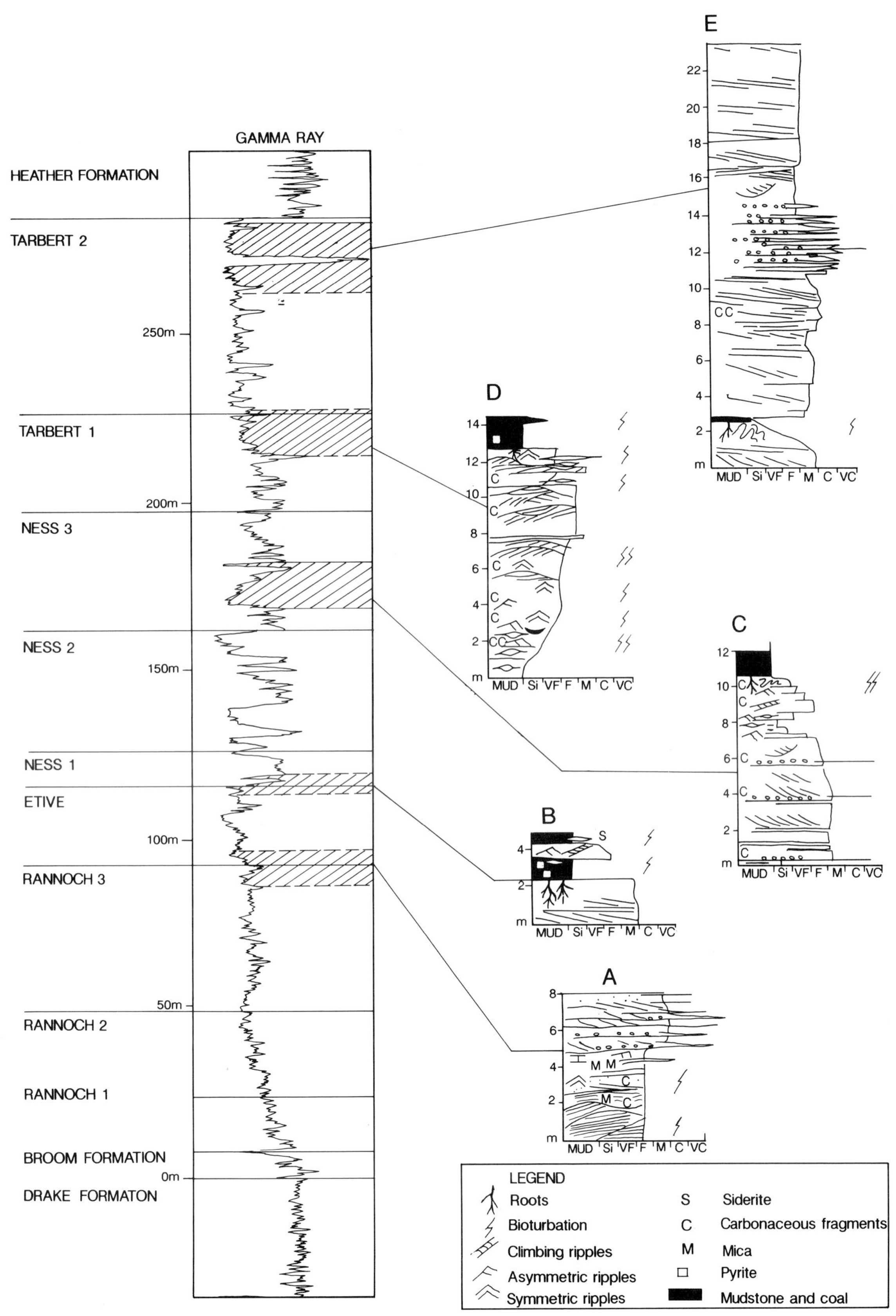

Figure 13. Facies sequences in the Brent Group.

sequence with medium- to coarse-, occasionally very coarse grained, pebbly sandstones in the lower part (Figure 13A) and medium- to fine-grained sandstones in the upper part. However, composite sequences of fining and coarsening upward do occur. Based on reservoir quality, the Etive Formation can be subdivided into two units, E-1 and E-2.

In the central part of the Gullfaks field, 3 to 5 m of the Rannoch Formation interdigitates with the lower part of the E-1 unit.

A barrier complex with migrating tidal channels has been suggested as a general sedimentological model for the Etive Formation (Budding and Inglin, 1981). A blending of distributary channels, mouth bar deposits, barrier islands, and shoreline deposition in a proximal delta front setting is suggested as the depositional environment for the formation in the Gullfaks field area.

Ness Formation—The Ness Formation (Figure 13) shows a general thinning from south to north (135 to 82 m) in the Gullfaks field and consists of interbedded coals, claystones, siltstones, and sandstones. The formation is subdivided into three main reservoir units, N-1, N-2, and N-3.

N-1 is the lowermost unit (5 to 35 m in thickness) and is dominated by coals, root zones (Figure 13B), underclays with siderite nodules, laminated clay- and siltstones, and minor ripple-laminated and cross-stratified medium- to fine- and sporadically coarse-grained sandstones.

N-1 represents lower delta plain environments containing swamp and lagoon deposits with crevasse splay and distributary channel sands. Because of the surrounding cohesive and erosion-resistant sediments the channels were probably fairly stationary.

The N-2 unit (34 to 56 m in thickness) is dominated by three to five coarsening-upward sequences grading from claystone to micaceous fine- to medium-grained sandstones. The sandstones are ripple-laminated, locally with mudstone drapes.

The N-2 unit represents bay fill, crevasse splays, and mouthbar deposits. Occurrence of marine palynomorphs shows that the unit was at times in communication with marine environments.

The N-3 unit comprises thick coal beds, mudstones, sandstones, and coarsening- as well as fining-upward sequences representing delta plain deposits with soil development, peat deposits, fluvial channels, and interchannel deposits (Figure 13C). Marine embayment infill is suggested for the upper part of the N-3 unit. Generally, grain size varies from medium sandstones to clay.

Upper Brent Delta: Tarbert Formation—The Tarbert Formation (Figure 13) in the Gullfaks field forms a succession, up to 115 m thick, of laterally continuous and correlatable units consisting of stacked sandstone bodies with interbedded mudstones, siltstones, and thin, often extensive coal beds. The overall high sandstone content and good porosity and permeability qualities make the formation an excellent reservoir.

The Tarbert Formation as a whole in the Gullfaks field area is interpreted to be a shallow marine sandstone sequence, deposited during a relative rise in sea level. The formation is subdivided into two units, T-1 (lower) and T-2 (upper).

The T-1 unit is interpreted to be mainly back-barrier lagoonal, delta marsh, and mouth bar deposits (Figure 13D), while unit T-2 is interpreted to consist mainly of foreshore and shoreface sequences, probably of beach and barrier island origin (Figure 13E).

Reservoir Quality and Petrography of the Sandstones of the Brent Group—The sandstones of the Brent Group show variable reservoir properties, with generally high porosity (30 to 40%) and an extreme variation in permeability from a few millidarcys to several darcys. As an example, the Rannoch Formation tends to show upward increase in porosity; in contrast, its permeability shows a very distinct increase upward. The porosity and permeability variations are related to the sedimentary facies of the Brent Group. The sandstone bodies are laterally continuous and correlatable in the Rannoch, Etive, and Tarbert formations. The Ness Formation and the lower part of the Tarbert Formation are vertically more heterogeneous. However, lateral communication of crevasse splays, bay mouth bars, and bay filling deposits and channels has proven lateral intercommunication between fault blocks of the Gullfaks field. Minor faults, below the resolution limit for seismic detection, are also important in their effect on the lateral and vertical intercommunication of reservoir bodies in the Brent Group. An example of reservoir correlation in the Brent Group in the Gullfaks field is given in Figure 14.

The Brent Group sandstones are loose and can be subdivided into two different types.

1. Medium- to coarse-grained sandstones composed of three major components: quartz, feldspars, and authigenic kaolinite. The sandstones are classified as quartz arenites to subarkosic arenites (Figure 9F). Rare, thin carbonate-cemented layers occur in these types of sandstones.
2. Very fine to fine-grained sandstones composed of five major components: quartz, feldspars, mica, authigenic and clastic kaolinite, and authigenic siderite (Figure 9E). These sandstones, generally called micaceous sandstones, are classified as subarkosic arenites to feldspathic graywackes. Carbonate-cemented layers are common in these sandstone types. Some of the carbonate-cemented beds in the Rannoch Formation are laterally continuous.

Accessory minerals in both sandstone types are zircon, tourmaline, epidote, hornblende, pyroxene, garnet, and Fe-Ti oxide. Shell fragments may occur in the Rannoch Formation. The Etive Formation and the upper part of the Tarbert Formation (T-2 unit) are examples of type 1 sandstones. The Rannoch

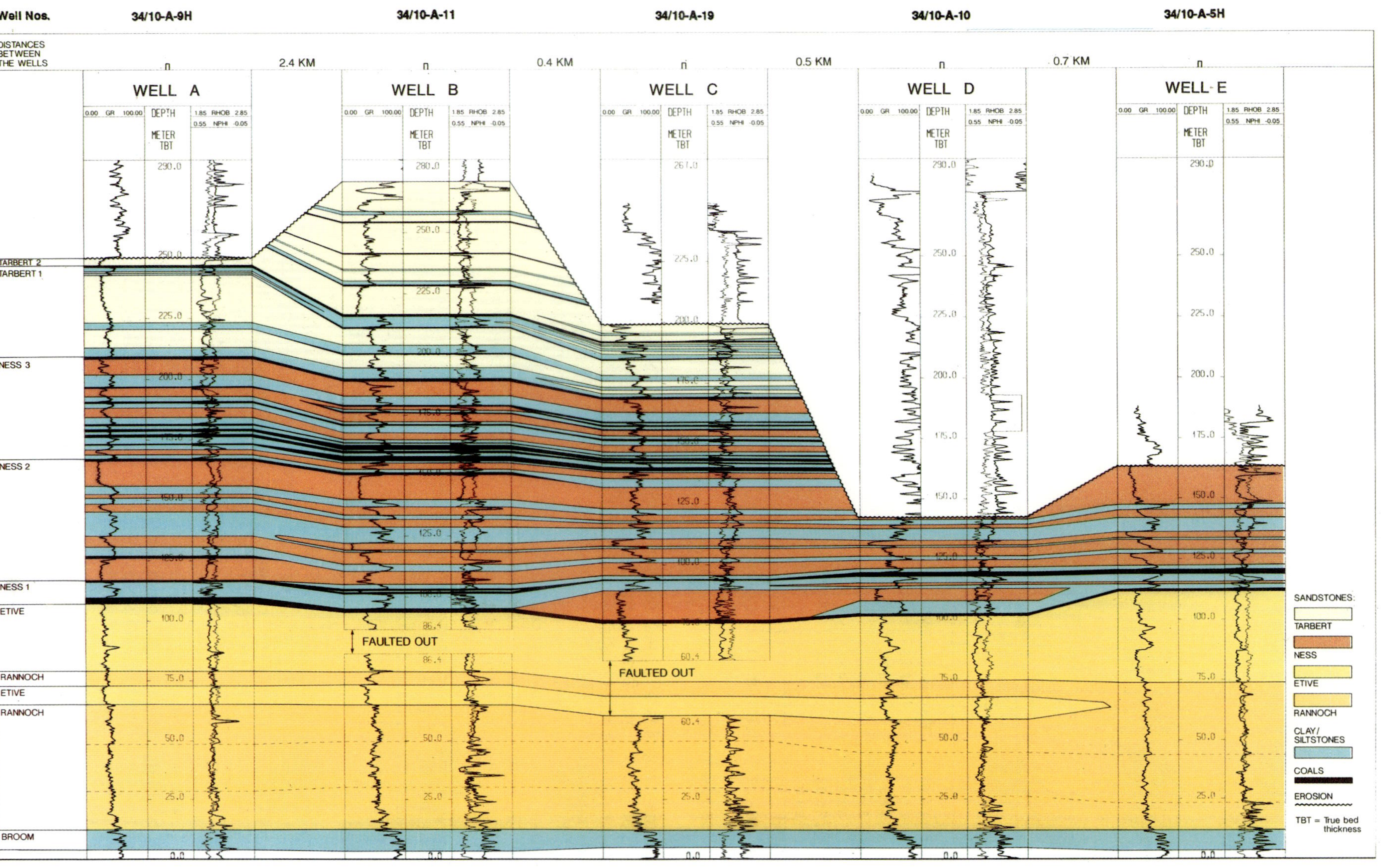

Figure 14. Example of correlation of the Brent Group in the Gullfaks field. See Figure 3 for the location of the wells.

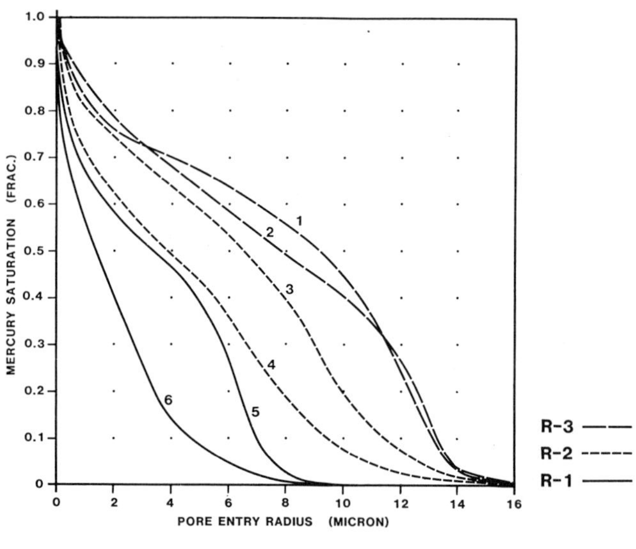

Sample	Porosity (%)	Permeability (mD)
1	37.9	1782
2	32.3	1105
3	36.6	993
4	33.1	422
5	35.0	295
6	29.8	54

Figure 15A. Pore-size distribution curves of six samples from the three units in the Rannoch Formation in well 34/10-4. See text for discussion.

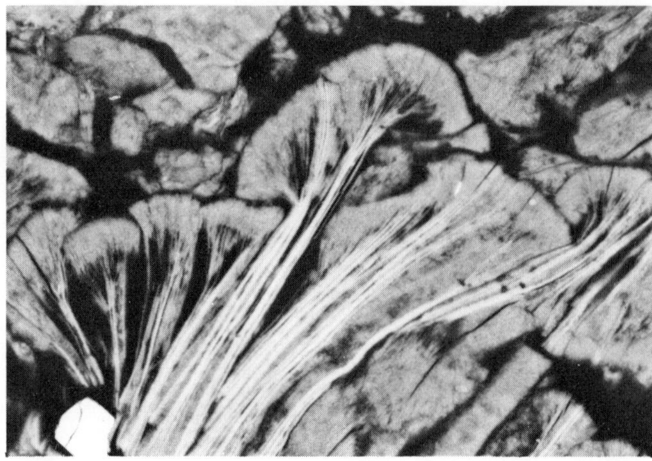

Figure 15B. Back-scattered microscopy picture showing mica (light gray) transformed to kaolinite (medium gray), creating microporosity. Note the fan character of the authigenic kaolinite. The Rannoch Formation bar scale, 10 mm.

Formation contains examples of type 2 sandstones, while a mixture of types 1 and 2 occur in the other formations and units. The major post-depositional alterations that have affected the primary porosity and permeability are mechanical compaction, dissolution of micas and feldspars accompanied by precipitation of kaolinite, and the formation of iron carbonates. Of these factors, kaolinite is probably the most important. Mechanical compactional effects in the Brent Group as well as in the Cook Formation are moderate to minor. The kaolinite and iron carbonate formation have an important influence on composition, micro-texture, porosity, pore size distribution, and permeability.

An example of the impact of kaolinite is seen in the overall coarsening-upward sequence of the Rannoch Formation. The Rannoch Formation generally has a high porosity (30 to 40%) throughout but shows a strong decrease in permeability downward. Figure 15A illustrates how the pore size distribution and permeability change down through the units R-3, R-2, and R-1 of the Rannoch Formation of well 34/10-4. Similar patterns are found in the Rannoch Formation in the other wells. By their S-shaped forms, the curves indicate a bimodal pore-size distribution. Most curves show a change toward steeper inclinations at a pore entry radius of approximately 2 microns. In unit R-3, approximately 20 to 30% of the total porosity is composed of pores with pore sizes less than 2 microns. In R-2 and R-1, 30 to 40% of the total porosity is composed of pores of less than 2 microns. In general 20 to 40% of the total porosity of the Rannoch Formation is composed of pores of less than 2 microns. This porosity is mainly found as micropores within the microcrystalline aggregates of kaolinite and siderite (Figures 9E and 15B).

Source

The Upper Jurassic "hot shales" of the Draupne Formation (equivalent to the Kimmeridge Clay Formation) are the principal source rocks for the hydrocarbon accumulation in the Gullfaks field. The Draupne Formation is 200 to 400 m thick in the deeper parts of the Viking graben. Organic geochemical studies of the Draupne Formation have revealed TOC values of 5 to 12% and high hydrocarbon index (HI) values (500 to 700 mg HC/gTOC). These suggest a dominantly sapropelic kerogen facies in the deeper part of the North Sea basin. (Thomas et al., 1985).

In addition to the Draupne Formation, the underlying Upper Jurassic Heather Formation probably contributes to the hydrocarbon accumulations in the northern North Sea basin. The Heather Formation contains 2 to 5% TOC and generally has values below 300 mg HC/gTOC. Detailed kerogen typing suggests that the Heather Formation generates mainly gas and light hydrocarbons. The potential may be higher in certain areas where it possesses liquid hydrocarbon potential (Thomas et al., 1985). However, the Heather Formation is subordinate to the Draupne Formation as a source for the Gullfaks area.

The Draupne Formation is currently mature for both oil and gas in the deeper parts of the basins adjacent to the Gullfaks structure. The secondary migration processes responsible for the filling of the Gullfaks field have been and remain poorly understood. However, application of petroleum reservoir

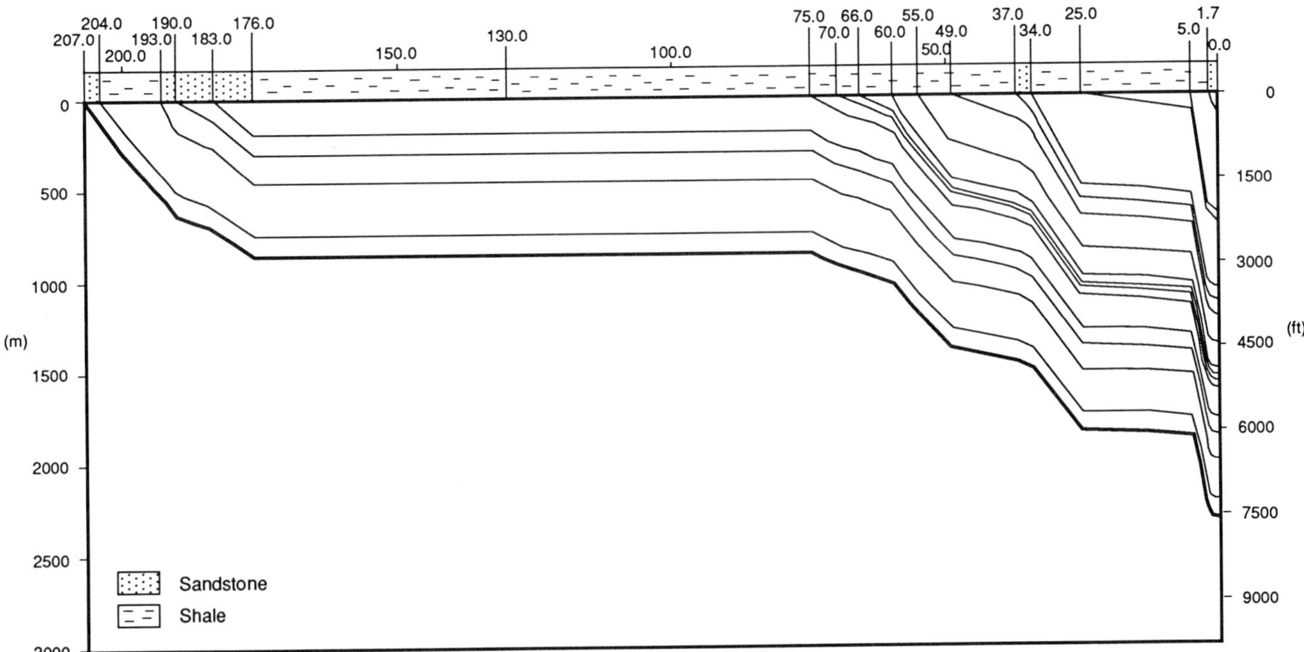

Figure 16. Burial history of the Gullfaks field based on 1-D modeling of well 34/10-1. The curves are not corrected for water depth.

geochemistry to the Gullfaks field has revealed geochemical trends within the field that indicate sourcing from two different directions. The Brent and Cook reservoirs in the western part of the field appear to have been sourced from the west. The suggested migration route is from the Oseberg kitchen through North Alwyn, Brent, Statfjord, and Snorre fields before entrapment in the Gullfaks reservoir sandstones. This migration and fill/spill route was described by Thomas et al. (1985) and Karlsson (1986), and their hypothesis seems likely on the basis of data from the adjacent fields. The eastern part of the field, comprising the Statfjord and Cook formations, is thought to have been filled from the east (Horstad et al., 1990).

Although not corrected for water depth, which is supposed to have been significant in the Late Jurassic and early Tertiary (hundreds of meters to 1 kilometer water depth), Figure 16 reflects the post-Triassic burial history of the Gullfaks field. Because of the mid-Cimmerian tectonic event, no sedimentary record is observed from mid-Jurassic to Late Cretaceous on the structural highs. In this time span, these areas were probably periodically exposed to meteoric water. The greater subsidence seen in the early Tertiary is also supposed to reflect the burial history of the kitchen area. Generation and migration of hydrocarbons is believed to have started in this or slightly earlier (latest Cretaceous) time.

Oil Characteristics

The characteristics resulting from biodegradation permits the hydrocarbons to be subdivided into two systems in the Gullfaks field. Oil in the Brent Group and in the Cook Formation in fault block 3 (western part of the field) (Figure 3) has been biodegraded and is characterized by a low concentration of normal alkanes (Figure 17). The hydrocarbons in fault blocks 1 and 2 (eastern part of the field) show no sign of biodegradation and have high concentrations of normal alkanes (Figure 17). Both oil gravity and PVT properties of the oil vary throughout the field. The initial pressure, temperature, and pressure gradient show values in the range 320 to 309 bar, 71.5 to 76.5°C, and 0.058 to 0.079 bar/m depth, respectively, at reservoir conditions. The oil gravity ranges from 27 to 39° API. Viscosity at bubble point ranges from 0.38 to 1.13 cp, and the gas-oil ratio from 200 to 75 Sm^3/Sm^3. It is suggested that the variation observed in the bulk properties of the oil is due to different levels of biological degradation in the reservoir.

EXPLORATION CONCEPTS

Regional seismic surveys that covered the 34/10 block indicated a geologic development similar to nearby discoveries. The exploration targets were Lower and Middle Jurassic sandstone reservoirs situated on rotated fault blocks below the Kimmerian unconformity. Well 34/10-1 was successfully drilled and hit oil in reservoirs of the Brent Group. However, the oil column was thinner than expected because the structure was not filled to spill point. New seismic indicated a complex structural development, and ineffective migration into some compartments was suggested. Appraisal and development wells have

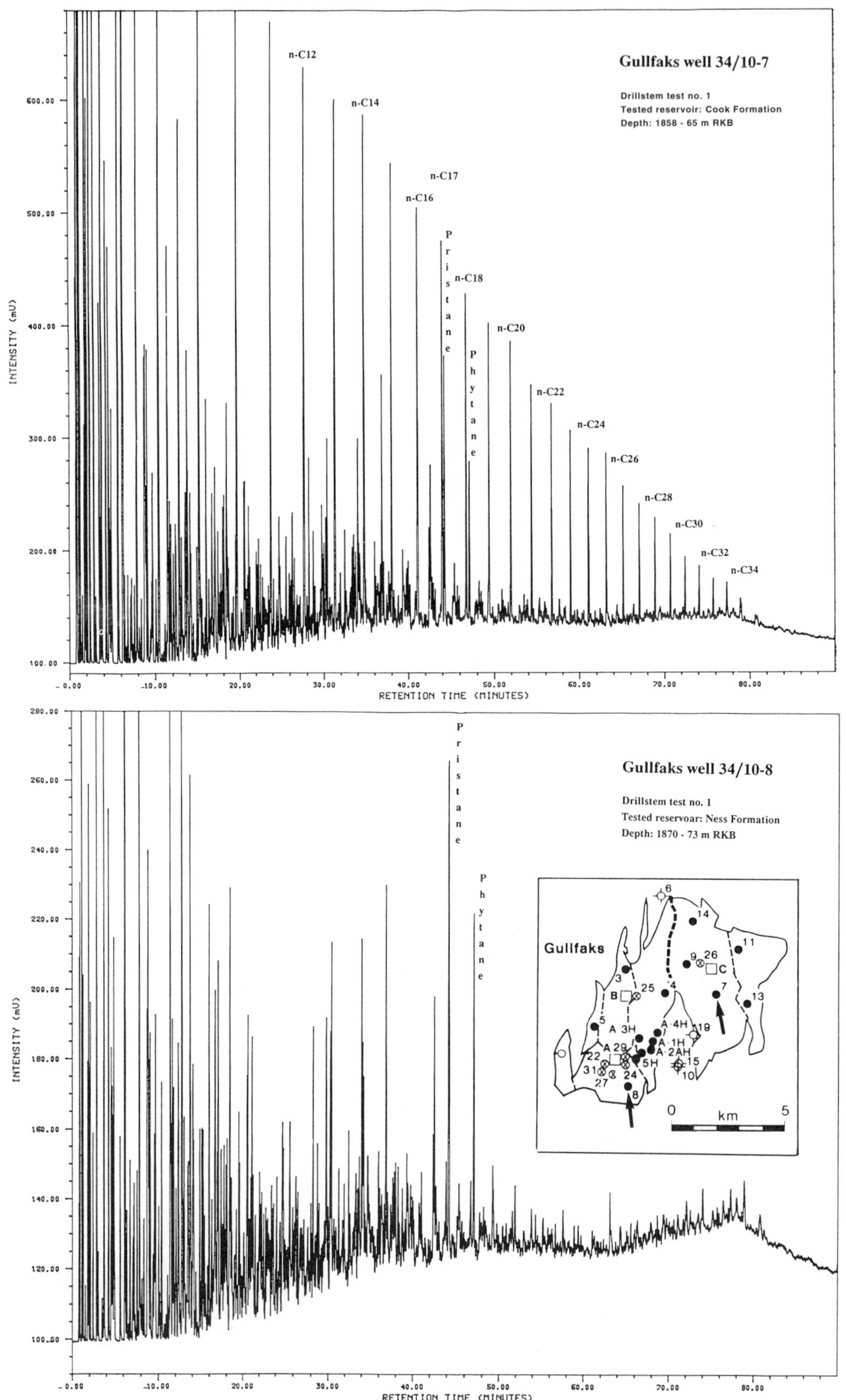

Figure 17. Chromatograms of biodegraded (well 34/10-8) and nonbiodegraded (well 34/10-7) oil of the Gullfaks field. Locations of the oil samples are shown in the map inset.

yielded unexpected results with respect to oil-water contacts and reservoir development.

In hindsight, it could be argued that more attention could have been paid to detailed tectonic modeling studies coupled with detailed seismic interpretation at an early stage during field appraisal.

ACKNOWLEDGMENTS

The authors wish to thank Statoil and the partners in the Gullfaks License, Norsk Hydro and Saga Petroleum, for permission to publish this field study.

REFERENCES CITED

Budding, M. C., and H. F. Inglin, 1981, A reservoir geological model of the Brent Sands in Southern Cormorant, *in* L. V. Ibling and G. D. Horbson, eds., Petroleum geology of the continental shelf of northwest Europe: London, Institute of Petroleum, p. 326-334.

Doré, A. G., and M. S. Gage, 1987, Crustal alignments and sedimentary domains in the evolution of the North Sea, northeast Atlantic margin and Barents Shelf, *in* Brooks and Glennie, eds., Petroleum geology of north west Europe: London, Graham & Trotman, p. 1131-1148.

Erichsen, T., M. Helle, J. Henden, and A. Rognebakke, 1987, Gullfaks, *in* A. M. Spencer, et al., eds., Geology of the Norwegian oil and gas fields: London, Graham & Trotman.

Fält, L.-M., 1986, Tectonic control on sedimentation in the Statfjord Formation, Statfjord Field, N. North Sea (abstr.): 12th International Sedimentological Congress, 24-30 August, Canberra, Australia.

Fält, L.-M., R. Helland, V. W. Jacobsen, and D. Renshaw, 1989, Correlation of transgressive-regressive depositional sequences in the Middle Jurassic Brent/Vestland Group megacycle, Viking Graben, Norwegian North Sea, *in* J. D. Collinson, ed., Correlation in hydrocarbon exploration: London, Graham & Trotman, p. 191-200.

Fossen, H., 1989, Indication of transpressional tectonics in the Gullfaks oil field, northern North Sea: Marine Petroleum Geology, v. 6, p. 22-30.

Galloway, W. E., 1975, Process framework for describing the morphologic and stratigraphic evolution of deltaic depositional systems, *in* M. L. Broussard, ed., Deltas, models for exploration: Houston Geological Society, p. 87-89.

Graue, E., W. Helland-Hansen, J. Johnsen, L. Lømø, A. Nøttvedt, K. Rønning, A. Ryseth, and R. Steel, 1987, Advance and retreat of Brent Delta systems, Norwegian North Sea, *in* Brooks and Glennie, eds., Petroleum geology of north west Europe: London, Graham & Trotman, p. 915-937.

Haq, B. U., J. Hardenbol, and P. R. Vail, 1987, Chronology of fluctuating sea levels since the Triassic: Science, v. 235, p. 1156-1167.

Horstad, I., S. Larter, H. Dypvik, P. Aagaard, A. M. Bjørnvik, P. E. Johansen, and S. E. Eriksen, 1990, Degradation and maturity controls on oil field petroleum heterogeneity in the Gullfaks field, Norwegian North Sea, *in* B. Durand and F. Behar, eds., Advances in organic geochemistry 1989, p. 497-510.

Irwin, H., 1989, Hydrocarbon leakage, biodegradation and the occurrence of shallow gas and carbonate cement (abstr.): Norwegian Petrological Society Conference on Shallow Gas and Leaky Reservoirs, Stavanger, Norway.

Karlsson, W. , 1986, The Snorre, Statfjord and Gullfaks oilfields and the habitat of hydrocarbons on the Tampen Spur, offshore Norway, *in* A. M. Spencer et al., eds., Habitat of hydrocarbons on the Norwegian continental shelf, NPF: London, Graham & Trotman, p. 181-197.

Lervik, K. S., A. M. Spencer, and G. Warrington, 1989, Outline of Triassic stratigraphy and structure in the central and Northern North Sea, *in* J. D. Collinson, ed., Correlation in hydrocarbon exploration: London, Graham & Trotman, p. 173-190.

Malm, O. A., H. Furnes, and K. Bjørlykke, 1979, Volcaniclastics of Middle Jurassic age in the Statfjord oil field of the North Sea: N. Jb. Geol. Paleont., v. 10, p. 607-618.

Mearns, E. W., 1989, Neodynium isotope stratigraphy of Gullfaks oil field, *in* J. D. Collinson, ed., Correlation in hydrocarbon exploration: London, Graham & Trotman, p. 201-215.

Miall, A. D., 1979, Facies models 5: deltas, *in* R. G. Walker, ed., Facies models: Geosci. Can. Reprint 1, p. 43-56.

Nystuen, J. P., R. Knarud, K. Jorde, and K. O. Stanley, 1989, Correlation of Triassic to Lower Jurassic, Snorre field and adjacent areas, northern North Sea, *in* J. D. Collinson, ed., Correlation in hydrocarbon exploration: London, Graham & Trotman, p. 273-290.

Petterson, O., A. Storli, E. Ljosland, and Ian Massie, 1990, The Gullfaks field geology and reservoir development, *in* Buller et al., eds., North Sea oil and gas reservoirs, Vol. II: London, Graham & Trotman.

Røe, S. L., and R. Steel, 1985, Sedimentation, sea level rise and tectonics at the Triassic-Jurassic boundary (Statfjord Formation), Tampen Spur, northern North Sea: Journal of Petroleum Geology, v. 8, n. 2.

Skarpnes, O., E. Briseid, and D. I. Milton, 1982, The 34/10 Delta prospect of the Norwegian North Sea: exploration study of an unconformity trap: AAPG Memoir 32, p. 207-216.

Sæland, G. T., and G. S. Simpson, 1982, Interpretation of 3-D data in delineating a subconformity trap in block 34/10, Norwegian North Sea: AAPG Memoir 32, p. 217-236

Steel, R. J., and A. Ryseth, 1991, The Triassic and Early Jurassic succession in the Northern North Sea: megasequence stratigraphy and intratriassic tectonics, *in* R. P. F. Hardman, ed., Tectonic movements responsible for Britain's oil and gas reserves: Geological Society Special Publication, London, Blackwell.

Thomas, B. M., P. Møller-Pedersen, M. F. Whitakker, and N. D. Shaw, 1985, Organic facies and hydrocarbon distributions in the Norwegian North Sea, *in* Thomas et al., eds., Petroleum geochemistry in exploration of the Norwegian Shelf: London, Graham & Trotman, p. 3-26.

Vollset, J., and A. G. Doré, 1984, A revised Triassic and Jurassic lithostratigraphic nomenclature for the Norwegian North Sea: Norwegian Petroleum Directorate Bulletin, v. 3, p. 1-53.

Walderhaug, O., P. A. Bjørkum, and H. M. Nordgård Bolås, 1989, Correlation of calcite-cemented layers in shallow marine sandstones of the Fensfjord Formation in the Brage field, *in* J. D. Collinson, ed., Correlation in hydrocarbon exploration: London, Graham & Trotman, p. 367-375.

Ziegler, P. A., 1982, Geological atlas of western and central Europe: SIPM, Elsevier, Amsterdam.

Appendix 1. Field Description

Field name .... *Gullfaks field*
Ultimate recoverable reserves .... *230 × 10^6 m^3 (1.45 MM bbl) oil*

Field location:

- **Country** .... *Norway*
- **Basin/Province** .... *North Sea*

Field discovery:

- **Year first pay discovered** .... *Middle Jurassic Brent Group 1978*
- **Year second pay discovered** .... *Lower Jurassic Cook Formation 1980*
- **Year third pay discovered** .... *Lower Jurassic Statfjord Formation 1980*

Discovery well name and general location:

- **First pay** .... *Block 34/10, well 34/10-1*
- **Second pay** .... *Block 34/10, well 34/10-7*
- **Third pay** .... *Block 34/10, well 34/10-11*

Discovery well operator .... *Statoil*

IP:

- **First pay** .... *Brent, 1050 m^3 (6600 bbl) per day of oil*
- **Second pay** .... *Cook, 265 m^3 (1670 bbl) per day of oil*
- **Third pay** .... *Statfjord, 400 m^3 (2520 bbl) per day of oil*

All other zones with shows of oil and gas in the field:

Age	Formation	Type of Show
Pliocene	*Nordland Group*	*Gas*
Oligocene/Eocene/Paleocene	*Rogaland and Hordaland Group*	*Gas*
Sinemurian/Pliensbachian	*Amundsen Formation*	*Oil*
Norian/Rhaetian	*Lunde Formation*	*Oil*

Geologic concept leading to discovery and method or methods used to delineate prospect

Evaluation based on regional geology and analogies with similar structures (e.g., Statfjord field) in the region gave high hopes for the Gullfaks structure. Initial exploration was based on the interpretation of a 1 × 1 km two-dimensional seismic grid.

Structure:

Province/basin type .... *North Sea/Viking graben; Bally 1211, Klemme III A*

Tectonic history

Extensional tectonics; rift basin. Large-scale normal faulting in relation to Late Jurassic and Early Cretaceous tectonism, the main Kimmerian orogenic event.

Regional structure

The field is located in the middle of the East Shetland basin on the western flank of the Viking graben.

Local structure

Rotated fault block with 10–15° westerly dip.

Trap:

Trap type(s)

The Gullfaks field is a structural unconformity trap comprising many rotated and partly eroded fault blocks.

Basin stratigraphy (major stratigraphic intervals from surface to deepest penetration in field):

Chronostratigraphy	Formation	Depth to Top in m
Pleistocene/Pliocene/Paleocene	*Nordland Group*	*135 MSL*
Cretaceous	*Shetland Group*	*1640 MSL*
Jurassic	*Brent/Dunlin groups*	
	Statfjord Formation	*1720 MSL*
Triassic	*Hegre Group*	*2000 MSL*

Reservoir characteristics:

Number of reservoirs *3*
Formations .... *Brent Group (BG) with five formations; Cook Formation (CF); Statfjord Formation (SF)*
Ages .... *BG, Aalenian to Bathonian; CF, Pliensbachian to Toarcian; SF, Rhaetian to Sinemurian*
Depths to tops of reservoirs .... *BG, 1720 m; CF, 1780 m; SF, 1865 m*
Gross thickness (top to bottom of producing interval) .... *BG, 230 m; CF, 310 m; SF, 185 m*
Net thickness—total thickness of producing zones
Average .... *BG, 160 m; CF, 220 m; SF, 120 m*
Maximum .... *BG, 180 m; CF, 250 m; SF, 135 m*
Lithology
Brent Group: fine to coarse, micaceous to subarkosic arenites with mudstones and coal beds and minor calcite-cemented beds
Cook Formation: fine to medium, micaceous, subarkosic arenite with carbonate-cemented beds
Statfjord Formation: fine to coarse subarkosic arenites with mudstones, coal beds, and minor carbonate beds
Porosity type .... *BG, CF, SF, intergranular porosity*
Average porosity .... *BG, 30%; CF, 25%; SF, 26%*
Average permeability .... *BG, 1000 md; CF, 500 md; SF, 1000 md*

Seals:

Upper
Formation, fault, or other feature .... *Unconformity with overlying shales and marls*
Lithology .... *Mudstones/marls*
Lateral
Formation, fault, or other feature .... *Faults*
Lithology .... *Mudstone*

Source:

Formation and age .... *Draupne Formation, Kimmeridgian/Volgian*
Lithology .... *Mudstone*
Average total organic carbon (TOC) *7%*
Maximum TOC *12%*
Kerogen type (I, II, or III) .... *II*
Vitrinite reflectance (maturation) .... *NA*
Time of hydrocarbon expulsion .... *Early Tertiary, possibly latest Cretaceous*
Present depth to top of source .... *3800 m*
Thickness .... *200–400 m in the kitchen area*
Potential yield .... *NA*

Appendix 2. Production Data

Field name ... *Gullfaks field*

Field size:

- **Proved acres** ... *51 km^2*
- **Number of wells all years** ... *140*
- **Current number of wells** ... *13 exploration wells; 45 production/injection wells (as of 1/90)*
- **Well spacing** ... *400–800 m*
- **Ultimate recoverable** ... *230 million Sm^3 (1450 MM bbl) oil*
- **Cumulative production** ... *62.9 million Sm^3 (396 MM bbl) oil (as of 1/92)*
- **Annual production** ... *19.9 million Sm^3 (125 MM bbl) oil (1991)*
- **Present decline rate** ... *NA*
  - **Initial decline rate** ... *NA*
  - **Overall decline rate** ... *NA*
- **Annual water production** ... *6.3 $\times$ 10^6 m^3 (40 MM bbl)*
- **In place, total reserves** ... *522.1 million Sm^3 (3283 MM bbl) oil*
- **In place, per acre foot** ... *~325 bbl/ac-ft*
- **Primary recovery** ... *230 million m^3 (1450 MM bbl) oil*
- **Secondary recovery** ... *NA*
- **Cumulative water production** ... *13.6 $\times$ 10^6 m^3(86 MM bbl) (as of 1/92)*

Drilling and casing practices:

- **Amount of surface casing set** ... *1800–1850 m*
- **Casing program** ... *32-in. conductor, 20-in., 13⅜-in., and 9⅝-in. casing, 7-in. liner*
- **Drilling mud** ... *Water-based above, oil-based in the reservoir*
- **Bit program** ... *Conventional/stratapax with mud motor*
- **High pressure zones** ... *Below 1400 m*

Completion practices:

- **Interval(s) perforated** ... *Jurassic sandstones (see Appendix 1)*
- **Well treatment** ... *Gravel packing where sandstones are loose*

Formation evaluation:

- **Logging suites** ... *DITE/BHC/LDT/CNL/EPT/NGT/AMS, OBDT/GR, RFT/GR*
- **Testing practices** ... *NA*
- **Mud logging techniques** ... *Full mud logging services in key wells, well control purposes in other wells*

Oil characteristics:

- **Type** ... *Naphthenic*
- **API gravity** ... *29°*
- **Initial GOR** ... *BG, 100 to 199 Sm^3 m^3 (500 to 1117 ft^3/bbl); CF, 75 to 155 Sm^3 m^3 (421 to 870 ft^3/bbl); SF, 163 Sm^3 m^3 (915 ft^3/bbl)*
- **Sulfur, wt%** ... *0.45 (average high)*
- **Viscosity, SUS** ... *19cSt (20°C) average (BG, 1.11; CF, 1.13; SF, 0.43 cp at bubble point)*
- **Pour point** ... *–30°C/–45°C*
- **Gas-oil distillate** ... *NA*

Field characteristics:

- **Average elevation** ... *1850 m (approx.) subsea*
- **Initial pressure** ... *BG, 310 bar; CF, 311 bar; SF, 320 bar (variable in fault blocks)*

Present pressure .... *300 bar*
Pressure gradient .... *BG, 0.075 bar/m; CF, 0.079 bar/m; SF, 0.079 bar/m*
Temperature .... *BG, 72°C; CF, 76.5°C; SF, 76°C*
Geothermal gradient .... *0.032°C/m*
Drive .... *Water drive (with water and gas injection)*
Oil column thickness .... *250 (average)(BG, 250 m; CF, 290 m; SF, 180 m)*
Oil-water contact .... *BG, 1947 m; CF, 2090 m; SF, 2043 m*
Connate water .... *Variable*
Water salinity, TDS .... *44,000 mg/L*
Resistivity of water .... *0.168 ohm-m (at 21.6°C)*
Bulk volume water (%) .... *NA*

Transportation method and market for oil and gas:

Gas via subsea pipelines to European market (Emden); oil is loaded into tankers at the field.

Piper Field—U.K.
Outer Moray Firth Basin, North Sea

CONRAD E. MAHER
IGGE-SIRS Ltd.
Aberdeen, Scotland

H. RICHARD H. SCHMITT
Occidental International Exploration and Production Company
Bakersfield, California

SIMON C. H. GREEN
Occidental Petroleum (Caledonia) Ltd.
Aberdeen, Scotland

FIELD CLASSIFICATION

BASIN: North Sea
BASIN TYPE: Rift
RESERVOIR ROCK TYPE: Sandstone
RESERVOIR AGE: Jurassic
PETROLEUM TYPE: Oil
TRAP TYPE: Tilted Fault Block
RESERVOIR ENVIRONMENT OF DEPOSITION: Nearshore to Shallow Marine

LOCATION

The Piper field is in U.K. North Sea Block 15/17, 110 mi (177 km) northeast of Aberdeen, Scotland (Figures 1A and 1B). It is situated on a shelf on the northern margin of the Witch Ground graben (WGG) in the Outer Moray Firth basin (Figure 1B). The WGG is a northwesterly trending graben that developed in the Late Jurassic, branching off from the intersection of the north-south-trending Viking and Central grabens, 50 mi (80 km) southeast of Piper (Figure 1A).

Nearby oil fields include Claymore, Tartan, Scapa, Highlander, Scott, Galley, Petronella, Chanter, Rob Roy, and Ivanhoe. These fields have Upper Jurassic sandstone reservoirs except for Scapa, which produces from Lower Cretaceous sandstones. The Claymore field also produces from Lower Cretaceous sandstones.

The Piper field is ranked number 271 among the world's giant fields (Carmalt and St. John, 1986), and its ultimate recovery will be almost 1 billion bbl of oil and gas liquids.

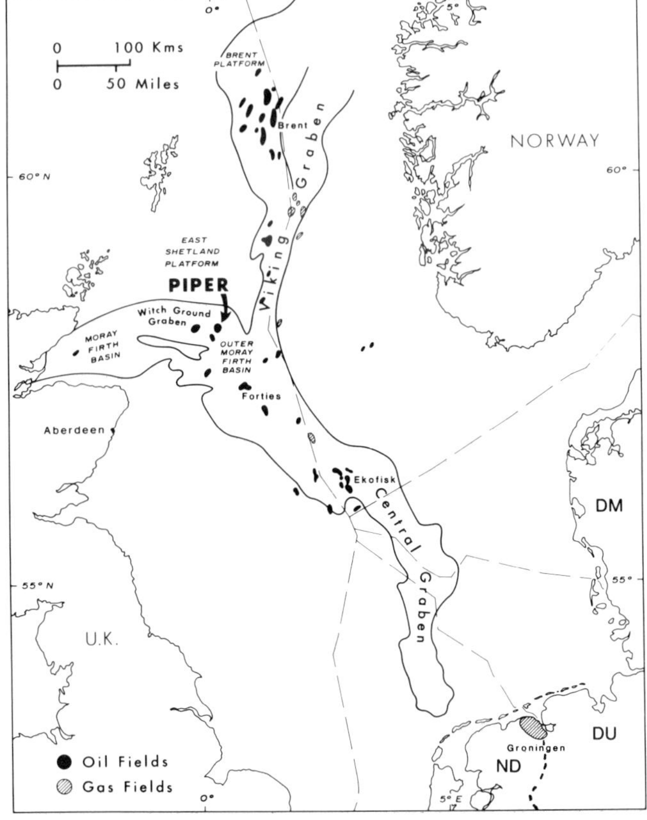

Figure 1A. The location of Piper and other fields in the Central, Viking, and Witch Ground grabens of the North Sea, showing national sectors. DM, Denmark; DU, Germany; ND, Netherlands.

HISTORY

Pre-Discovery

Occidental began regional seismic and geological studies of the petroleum potential of the onshore and

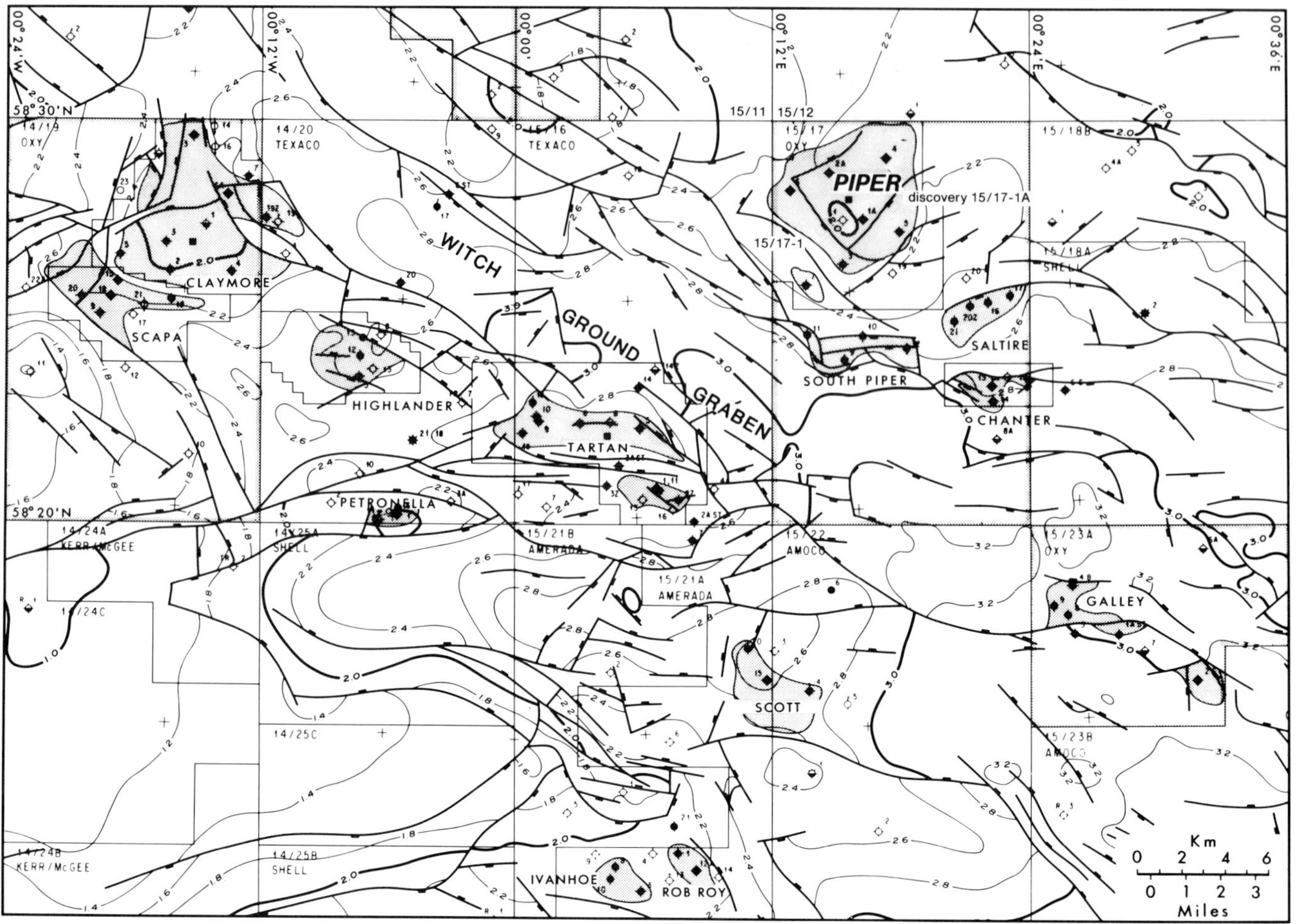

Figure 1B. Two-way time structure map of the Witch Ground graben at base of the Cretaceous unconformity level, showing the locations of license blocks and important exploration and appraisal wells. Contour interval, 0.2 m sec.

offshore United Kingdom in 1969 in anticipation of a forthcoming licensing round by either the United Kingdom or Norway. At that time only one major oilfield, Ekofisk, had been discovered (Figure 1A). This offshore field produces from the Danian-Maastrichtian Chalk (Figure 2).

From an initial interest in the onshore basins and with the knowledge of good source rock from the Kimmeridge Clay Formation, 500 mi (804 km) of spec seismic coverage of the Central graben was obtained. This was subsequently traded for five other spec shoots in the North Sea. One of these covered the Outer Moray Firth basin and the area of the subsequent Piper discovery. With 1500 mi (2400 km) of seismic data, the geophysicists and geologists were able to make a good structure map of the Central graben, Outer Moray Firth basin, and the Viking graben.

Interest intensified following the discovery of the Forties field in 1970 (Figure 1A). This field produces from Tertiary sandstones. By early 1971, it had become apparent that the United Kingdom would be the next to offer production licenses. As a result, work was concentrated on the U.K. sector. Occidental as operator formed a consortium with Getty Oil International (England) Ltd., Allied Chemical (Great Britain) Ltd., and Thomson Scottish Associates. In the period January-August 1971, more than 19,000 line miles (30,400 km) of seismic data in the U.K. North Sea between 56° and 62°N were evaluated.

In July 1971, prior to license application, the Occidental Group contracted for the building of a semi-submersible rig that would be capable of winter drilling in the northern North Sea. In addition, a drill ship was contracted to take advantage of favorable summer weather should a license be awarded to the Consortium in the spring of 1972.

Applications for petroleum production licenses in the U.K. 4th licensing round were submitted in August 1971, and the Occidental Group was granted six blocks in March 1972. Three of these blocks—14/19, 15/11, and 15/17 (Figure 1B)—were in the Outer Moray Firth basin and each contained at least one large feature with dip closure at the base of Tertiary or the base of Cretaceous.

The Sonda 1 drillship spudded the 15/11-1 well (Figure 1B) in May 1972, a little more than one month after the block had been awarded. The primary target

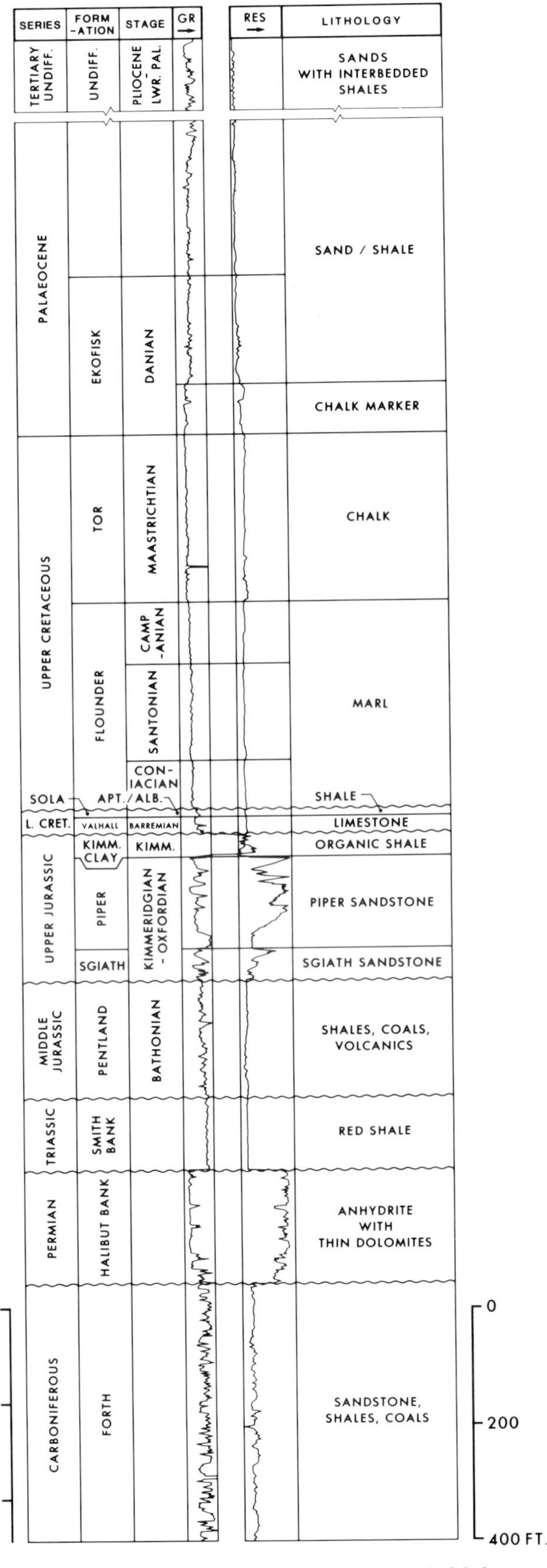

Figure 2. Characteristic log response and thickness of formations in Piper field.

was the Tertiary, where sandstones were known to be productive in the Forties field. The well was the first to be drilled in the Outer Moray Firth basin and tested a large anticlinal structure mapped at the base of the Tertiary. It found a relatively thick sequence of Tertiary sandstones, Cretaceous chalks, and Upper Jurassic sandstones but encountered no oil.

After the 15/11-1, a well was drilled on a large anticlinal feature in Block 14/19 (Figure 1B). Again the Tertiary sandstones were the primary target. The 14/19-1 well encountered oil in thin sandstones of either the Jurassic or Lower Cretaceous and minor amounts of oil in tight Triassic sandstones. Oil was recovered by wireline test from two horizons in the Permian Halibut Bank Formation dolomites.

The Sonda 1 then moved to Block 15/17 to test another large structure. This well was planned to test the Upper Jurassic sandstones with the Tertiary sandstones as a secondary objective. The 15/17-1 had to be abandoned in September 1972 at 1500 ft (457 m), due to anchoring and shallow casing problems. With the onset of winter weather, the drillship was released.

Discovery

The semi-submersible *Ocean Victory* arrived at the 15/17-1A location (Figures 1B and 3) in November 1972. This location is 0.5 mi (0.8 km) east and slightly downdip from the original 15/17-1 location. The location was moved downdip because of concern that the Jurassic sandstones could be eroded from the crest of the structure as had already been seen in the 14/19-1 well.

On 22 December 1972, the 15/17-1A well encountered 192 ft (58 m) of porous, permeable, oil-bearing sandstone of Late Jurassic age between 7523 and 7729 ft (2293 and 2356 m) subsea. In January 1973, the well produced 36° API low sulfur oil at 5266 BOPD through a 2-in. (5.08 cm) choke from 143 ft (44 m) of perforated zone between 7526 and 7693 ft (2294 and 2345 m) subsea.

Post-Discovery

Following the initial discovery well, an appraisal drilling program commenced in order to delineate the field. Three more wells, 15/17-2, -3, and -4, were drilled on the 15/17 block, and a bottom hole contribution was made to the Burmah 15/12-1 well drilled just north of the 15/17 block boundary (Figure 3). The 15/17-2 well, 1.6 mi (2.6 km) northwest of 15/17-1A, encountered the same reservoir section as 15/17-1A between 7855 and 8114 ft (2394 and 2473 m) subsea. Production testing was conducted through a 2-in. (5.08 cm) choke and flowed oil at rates of 15,257 and 16,873 BOPD, respectively, from the perforated intervals 8074 to 8114 ft (2461 to 2473 m) subsea and 7942 to 8038 ft (2421 to 2450 m) subsea.

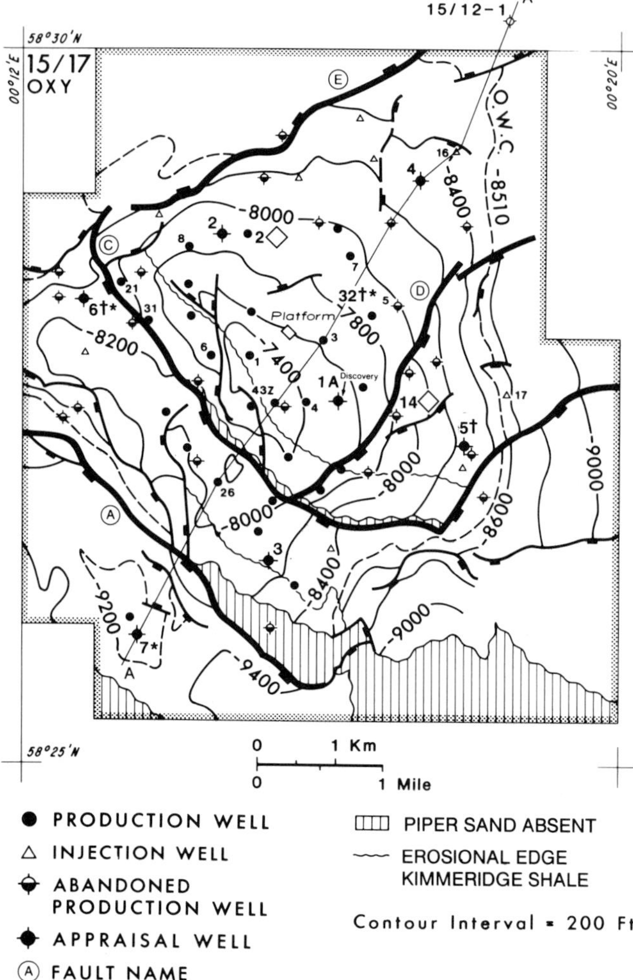

Figure 3. Top Piper sandstone depth structure map of Piper field. ◇, type logs shown by Figures 5A and 5B, wells 15/17-P2 and 15/17-P14. *, core photos and thin sections, Figures 17, 18, and 19 for wells 15/17-6, 15/17-7, and 15/17-P32. †, log porosity vs. core porosity and permeability, Figures 20 and 21, wells 15/17-5 and 15/17-8; also gamma ray-dual lateralog from well 15/17-P32, Figure 22.

The 15/17-3 well was drilled on the downthrown side of a normal fault 1.4 mi (2.2 km) south of 15/17-1A. This well found the same Jurassic sandstones between 8131 and 8246 ft (2478 and 2537 m) subsea. It was production tested through a 2-in. (5.08 cm) choke from the intervals 8131 to 8211 and 8228 to 8243 ft (2478 to 2503 and 2508 to 2512 m) subsea at a rate of 15,509 BOPD. The upper part of the Jurassic sandstone in this well was missing, a circumstance interpreted to be the result of Late Jurassic or Early Cretaceous erosion.

The 15/12-1 well was then drilled by the Burmah Group. It encountered thick, wet reservoir sandstones. The 15/17-4 well was then drilled updip of 15/12-1 and established an oil-water contact at 8510 ft (2594 m) subsea.

Detailed seismic coverage was shot after the initial discovery. This resulted in a spacing between seismic lines of approximately 2460 ft (750 m). The data, together with the new velocity and geological data from the appraisal wells, helped to define the field limits, and a single production platform was planned.

Faults of varying displacement were mapped within the field limits, and some of these faults offset the reservoir sands (Figure 4). It was therefore considered necessary to drill further appraisal wells to determine if the oil-water contact was the same for the entire field and if the faults were likely to affect production. At the same time, field limits would be more precisely defined and would allow for better placement of the production platform.

Wells 15/17-5 and 15/17-6 confirmed that the Piper field could be fully developed from one centrally located platform. Both encountered the same oil-water contact as 15/17-4 (8510 ft subsea). The 15/17-7 well was drilled to the southwest and across a major fault (now designated the "A" fault) in late 1973. The Jurassic sandstones were present but significantly deeper and only partially filled, with an oil-water contact at 9200 ft (2804 m) subsea.

The discovery and appraisal program had established an oil column of approximately 1200 ft (366 m) within a layered reservoir (Figure 5) covering an area of 7350 ac (29.75 km^2) and containing 1.4 billion STBOIIP. A single steel platform containing 36 well slots was centrally located over the field in 474 ft (144 m) of water in June 1975 and made ready for production drilling by October 1976.

The P1 production well was spudded on 10 October 1976 and established commercial production on 7 December 1976 at more than 30,000 BOPD, restricted by 5½-in. (14 cm) tubing. Development progressed steadily, and the P7 well was completed in April 1977 producing more than 50,000 BOPD restricted by 7-in. (17.8 cm) tubing (Figure 3).

From field characteristics, nearby well data, and the availability of accurate pressure data, it was apparent that natural water influx was occurring in Piper field. This aquifer drive was calculated by material balance to be approximately 250,000 bbl/d. By mid-1977, it was obvious that reservoir pressure could not be sufficiently maintained with natural water influx at projected reservoir production rates of 250,000 to 300,000 BOPD. In late 1977, the first injection wells, P16 and P17, were drilled to supplement the natural water influx, and injection commenced in early 1978 (Figure 3). These and subsequent injectors halted the decline in reservoir pressure as production rates greater than 300,000 BOPD were sustained (Figure 6).

In 1980, following a formal request to the U.K. government, the field rate was reduced to allow for selective completion and injection and more efficient reservoir management (Figure 6). Further improvements in recovery were also made following the installation of a gas lift system in 1977 and high volume, submersible pumps in 1982. Selective completion of both injectors and producing wells has

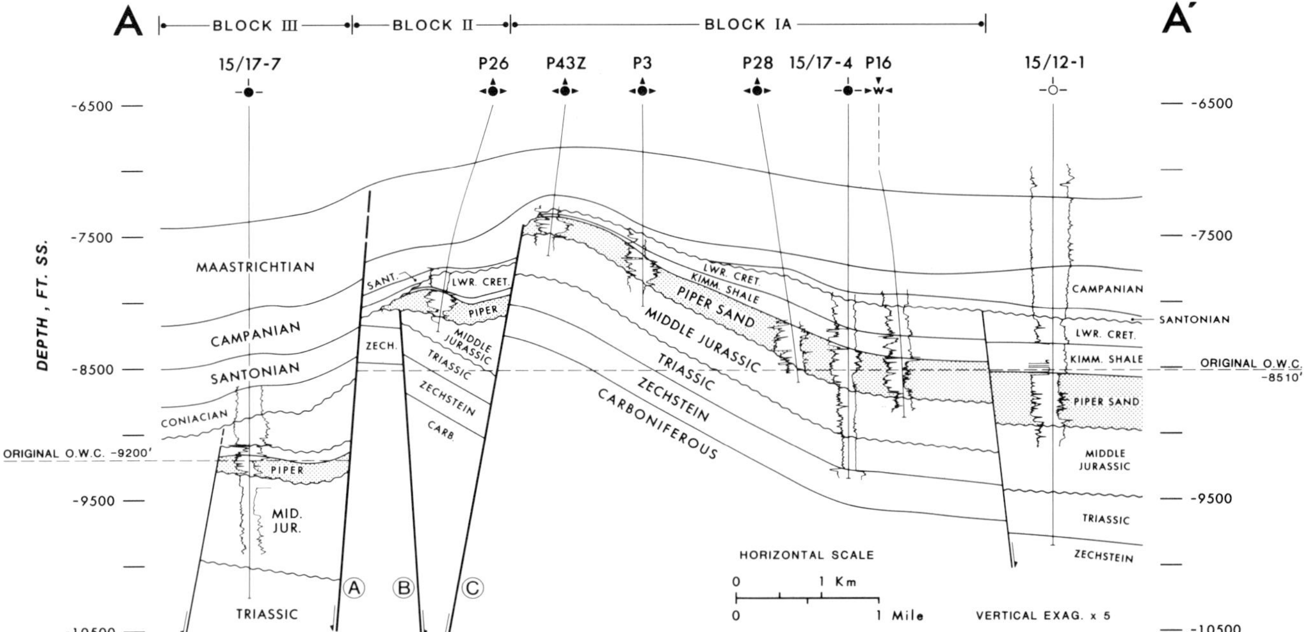

Figure 4. Structural cross section running southwest-northeast through Piper field, demonstrating the northeasterly fault block rotation and the structural configuration. See Figure 3 for location.

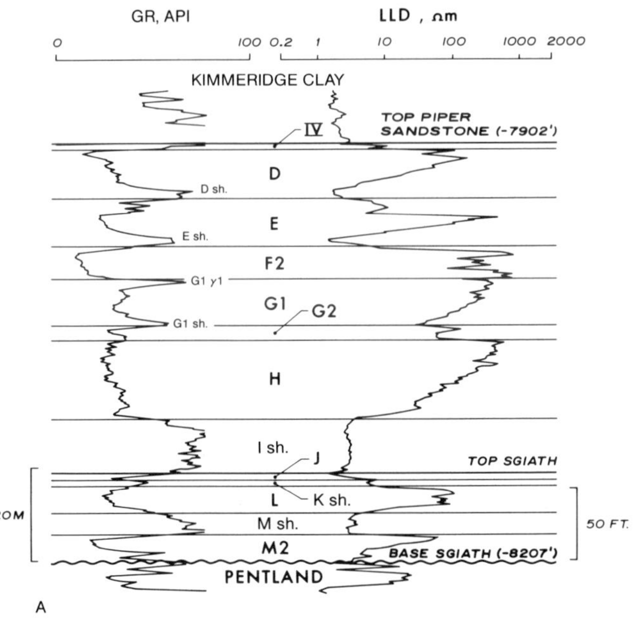

Figure 5. (A) and (B) show type logs for wells 15/17-P2 and 15/17-P14 for the western and eastern areas of the Piper field, respectively. (See Figure 3 for well locations.)

been practiced in order to balance the natural influx and injected water with offtake from each layer within the reservoir. The combination of these management techniques is expected to lead to a recovery factor of approximately 70%.

Initial ultimate recovery for the Piper field was put at 618 MMBO, but subsequent reservoir performance confirmed this to be conservative. There has been a steady increase in the estimated ultimate recovery to an accepted figure of 952 MMBO. Production exceeded 800 MMBO on 25 October 1987, and the field was still producing at a rate of 120,000 BOPD at the time of the disaster on 6 July 1988. Production is scheduled to resume from a new platform in mid-1992.

The discovery of Piper encouraged further drilling activity in the Outer Moray Firth area leading to the discovery of the Claymore, Tartan, Scott,

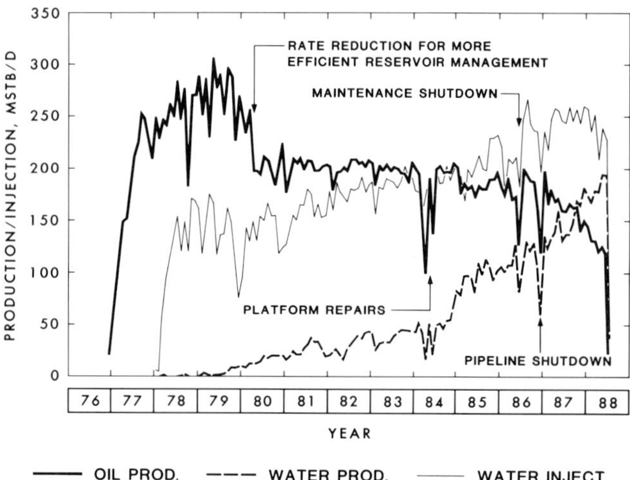

Figure 6. Graph showing Piper field's performance history, 1976–1988. Production averaged over 250,000 BOPD for the first three years but had declined to approximately 120,000 BOPD by mid-1988, 62% watercut, at the time of the disaster. Water production exceeded oil production in 1987.

Petronella, Highlander, Scapa, Rob Roy, Ivanhoe, and Chanter fields. Several additional discoveries have yet to be declared commercial (Figure 1B).

DISCOVERY METHOD

Then

Prior to the application for acreage that contains the Piper field, the Occidental exploration team had acquired and interpreted over 19,000 line miles (30,400 km) of seismic data covering large areas of the U.K. northern North Sea. The prospectivity of this region was considered to be good, based on the knowledge of existing source rocks and reservoir sandstones cropping out onshore and on the fact that two large fields had already been discovered (Ekofisk and Forties). A Mesozoic discovery that turned out to be the Brent field was also rumored (Figure 1A).

A large number of potentially prospective structures were mapped within the U.K. 4th Round acreage. Seismic data quality was good, and it was a reasonable assumption that nearly every operator in the United Kingdom had comparable structural interpretations. However, some of the major oil companies had the advantage of having already drilled wells through the Mesozoic sandstones and had good data on reservoir quality, source rock, and caprock. A key point in the Occidental Group's acquisition strategy was therefore to avoid head-on competition with other larger and more knowledgeable operators for the structures in the Northern Viking graben. Instead, the virgin Outer Moray Firth basin blocks were nominated as their first choice.

As explained above, following the successful application for the acreage, the initial two wells drilled in this virgin basin, 15/11-1 and 14/19-1, had Tertiary sandstones as their primary objective. With the integration of the results of these wells, the primary objective of the Piper discovery well, 15/17-1A, were Upper Jurassic sandstones.

Now

With today's exploration methods, Piper field would be easily mapped and discovered. Since the discovery of the Piper field, the industry has become very much more knowledgeable about extension-type basins and the stratigraphy, origin, migration, and entrapment of oil and gas within them.

It would be easy to map Piper-type structures and, perhaps, with the quality of the seismic data available, to recognize that the reservoir sandstone and caprocks are present. It still is not possible to tell which undrilled structures contain oil or gas. Today, Occidental's first choice structure in block 15/11 still might be drilled first, as it was in 1972, and be found dry. Although not definitely proven by the three wells on this block, it is now thought that the 15/11 structure is dry because it remained too high during the Cretaceous. Consequently, the Piper sandstones were onlapped by the Maastrichtian chalk, which is not a caprock in this area. Indeed, the chalk came to within a few feet of onlapping the sandstone at Piper, in which case the Piper structure would also have been dry.

STRUCTURE

Tectonic History

The North Sea and northwest Europe were part of the landmass until the late Permian (Ramsbottom, 1978). At this time, a shallow epicontinental gulf covered most of the North Sea basin area and resulted in widespread evaporite deposition. In the southern North Sea, thick salt deposits of the Halibut Bank Formation (Zechstein) were laid down. In the Piper area, 100 ft (30 m) of interbedded dolomite and anhydrite containing an assemblage of Upper Permian palynomorphs were overlain by 300 ft (91 m) of massive anhydrite containing a few dolomite beds up to 20 ft (6 m) thick (Figure 2).

The Halibut Bank Formation overlies a thick sequence of Lower to Middle Carboniferous sandstones and shales. It in turn is overlain by extensive sections of nonmarine shale and claystone of Triassic age. Drilling in the Claymore field has confirmed structural movement on northwest-southeast-trending faults during the deposition of the Triassic. In the Piper area, the Triassic red shales are overlain by a nonmarine sequence of siltstones, shales, lignites, tuffs, and basalt flows with occasional thin-bedded freshwater limestones of Middle Jurassic age.

In the Claymore area, and possibly the Piper area, the end of the Triassic was marked by a period of erosion. Middle Jurassic sediments in the Piper area were also partially eroded prior to deepening of the basin and deposition of the widespread, lower deltaic plain sediments of the Sgiath Formation (Maher and Harker, 1987).

There was a major transgression in the late Oxfordian over the whole of the Outer Moray Firth basin area. This is represented by the "I shale" in the Piper field (Figure 5). Following this transgression, the shallow marine Piper sandstones were deposited.

After the deposition of the Piper sandstones, there was major extension and extensive deepening of the Witch Ground graben (WGG). The graben had been initiated as early as Triassic times and subsidence and tilting occurred during the deposition of the Piper and Sgiath sandstones. Consequently, wells within the graben have thicker Piper and Sgiath sandstone sections than do wells on the northern margin of the WGG. The deepening of the graben and rotation of shelf areas away from the graben margin resulted in erosion of Piper and Sgiath sandstones along fault scarps and the graben margin. These sediments were probably redeposited as gravity flows within the graben.

Near the end of the Jurassic, further basin extension occurred, accompanied by a widespread transgression that deposited the organic-rich shales of the Kimmeridge Clay Formation over the Outer Moray Firth basin area. The Piper structure remained high throughout the Early Cretaceous as a result of fault block rotation and a fall in sea level. This resulted in local erosion of Jurassic sediments and nondeposition of Cretaceous sediments. Only a thin, "condensed" sequence of Lower Cretaceous deposits is present in the Piper field.

During the Late Cretaceous, regional subsidence and a rise in sea level resulted in the progressive onlap of the Piper structure by marls and chalks of Santonian, Campanian, and Maastrichtian age (Figure 4). By the end of the Cretaceous, tectonic movements had largely ceased and a thick sequence of Tertiary sands and clays was deposited throughout the area. The burial history of the Piper reservoir is summarized in Figure 7.

Regional Structure

The regional structure is an extensional graben trending northwest, away from the intersection of the Viking and Central grabens (Figures 1A and 1B). Extensional subsidence of the WGG basin may have taken place during the Permian and Triassic, but the major phases of graben formation occurred during the Late Jurassic and Early Cretaceous. The graben may have developed as a reactivation of Hercynian features.

Local Structure

The Piper field lies on the northern margin of the WGG. Four significant faults are present within the Piper field (Figures 3 and 8). Most of the fault movement occurred after the deposition of the Piper sandstones, but some, notably along the "D" fault, was syndepositional. The field comprises three major tilted fault blocks, dipping gently to the northeast and slightly folded along the northeast-southwest axis. The main Piper field (blocks IA and IB) dips at about 8° to the northeast (Figures 3, 4, and 9).

STRATIGRAPHY

The oldest sediments penetrated in Piper field are Lower Carboniferous Coal Measures of the Forth Formation (Figures 2 and 10). They comprise a thinly interbedded deltaic sequence of sands, shales, and coals. Upper Permian sediments, which unconformably overlie the Carboniferous, represent evaporite deposits of the Zechstein sea. Fluvio-lacustrine sedimentation followed with argillaceous Triassic red beds of the Smith Bank Formation. No Lower Jurassic rocks are present due to nondeposition or erosion. The Middle Jurassic consists of Rattray Formation volcanics and Pentland Formation alluvial to marginal-marine argillaceous clastics and coals. During the Late Jurassic transgression, the interbedded sands, shales, and coals of the paralic Sgiath Formation were overlain by shallow marine sands and shales of the Piper Formation. These in turn were overlain by anoxic marine shales of the Kimmeridge Clay Formation. Synrift deposition was brought to a close during the Early Cretaceous when sandy marls and limestones of the Valhall Formation onlapped the Piper structure. Diminishing tectonic activity into the Late Cretaceous was marked by postrift hemipelagic deposition of marls, limestones, and chalks. Clastic sedimentation returned in the Tertiary, represented by a thick sequence of sands and shales.

The Piper reservoir is composed of two formations within the Upper Jurassic Humber Group. These are the Oxfordian Sgiath Formation (Harker et al., 1987) and the Upper Oxfordian to Kimmeridgian Piper Formation (Deegan and Skull, 1977) (Figure 10). Originally the section now assigned to the Sgiath Formation was dated as Callovian (Maher, 1981), but this has now been revised (Turner et al., 1984).

The Sgiath sandstones are interbedded with shales and coals up to 10 ft (3 m) thick. Immediately overlying the Sgiath is a silty, sandy, bioturbated shale that grades upwards into a series of stacked, shallow marine sands. This sequence comprises the Piper Formation.

These sands extend over hundreds of square miles on the shelf in which the Piper field is located and across the WGG. They shale out on the west side of the Claymore field (Boote and Gustav, 1987).

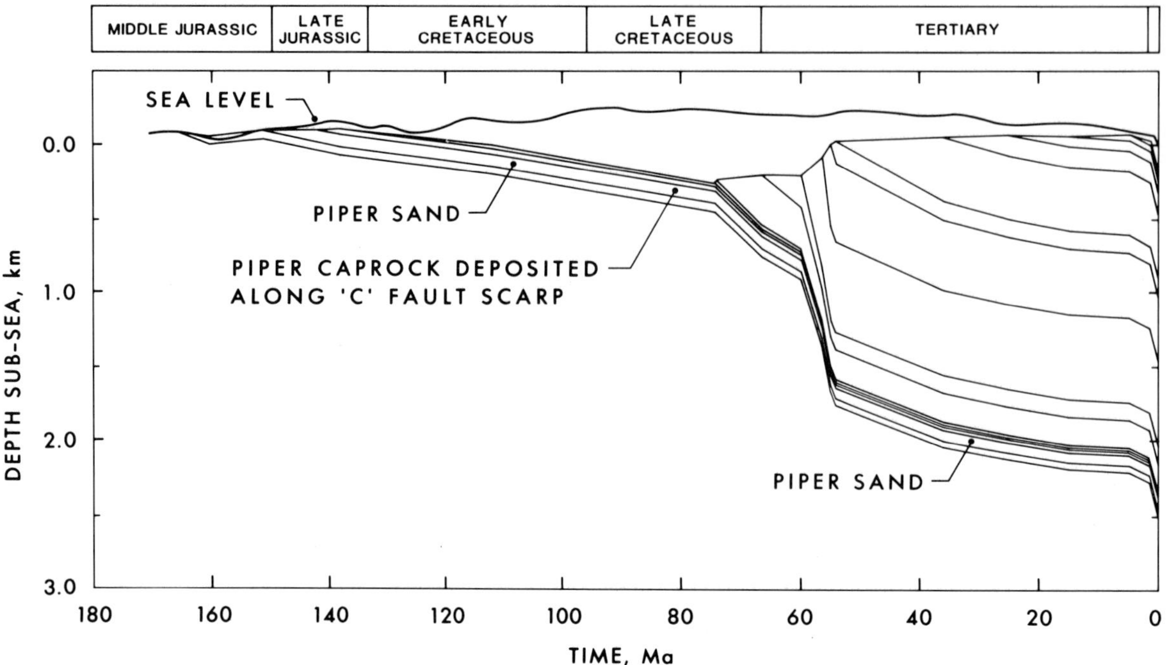

Figure 7. Typical burial history curve for Piper field. The period of most rapid burial was during the inferred Late Cretaceous to early Tertiary oil maturation and migration phase.

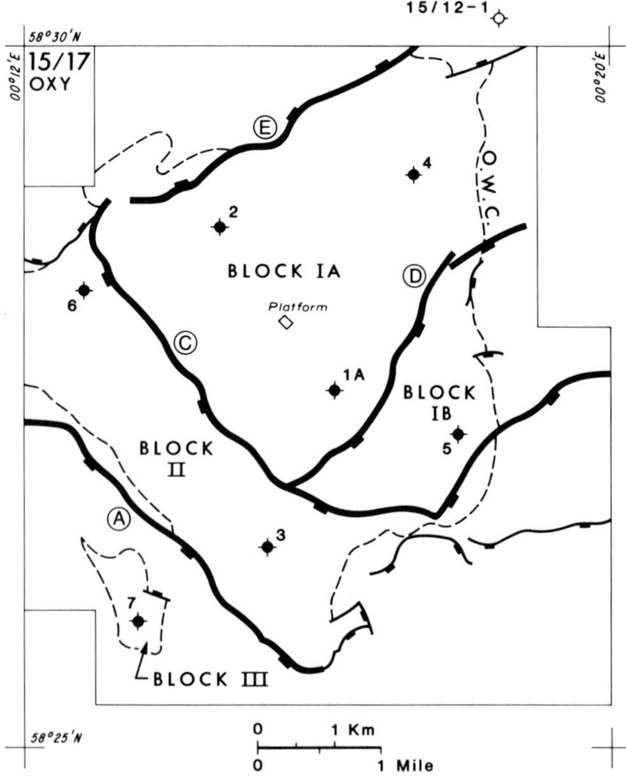

Figure 8. Outline map of Piper field showing principal faults and appraisal well locations.

The source rock for the Piper field is the organic-rich shale of the Kimmeridge Clay Formation that directly overlies the Piper sandstone over most of the Piper field. It is widespread throughout the Outer Moray Firth basin and the Central and Viking grabens.

TRAP

Trap Type

The Piper field is a series of three major tilted, folded fault blocks. It is the gentle folding about a northeast-southwest axis together with the drape associated with the northeast-southwest Caledonian fault trends that provides the critical closure to the northwest and southeast (Figures 1B and 3). To the northeast, dip closure results from the gentle tilting of the shelf away from the WGG (Figures 1B, 3, and 4). To the southwest, closure is provided by a major northwest-southeast fault, the A fault (Figure 8). The four-way closure is mapped from seismic data and has been verified by appraisal and development drilling.

The vertical and lateral seals are shales of the Kimmeridge Clay Formation. Along the "C" fault where these shales are eroded, onlapping Campanian marls form the seal (Figure 4).

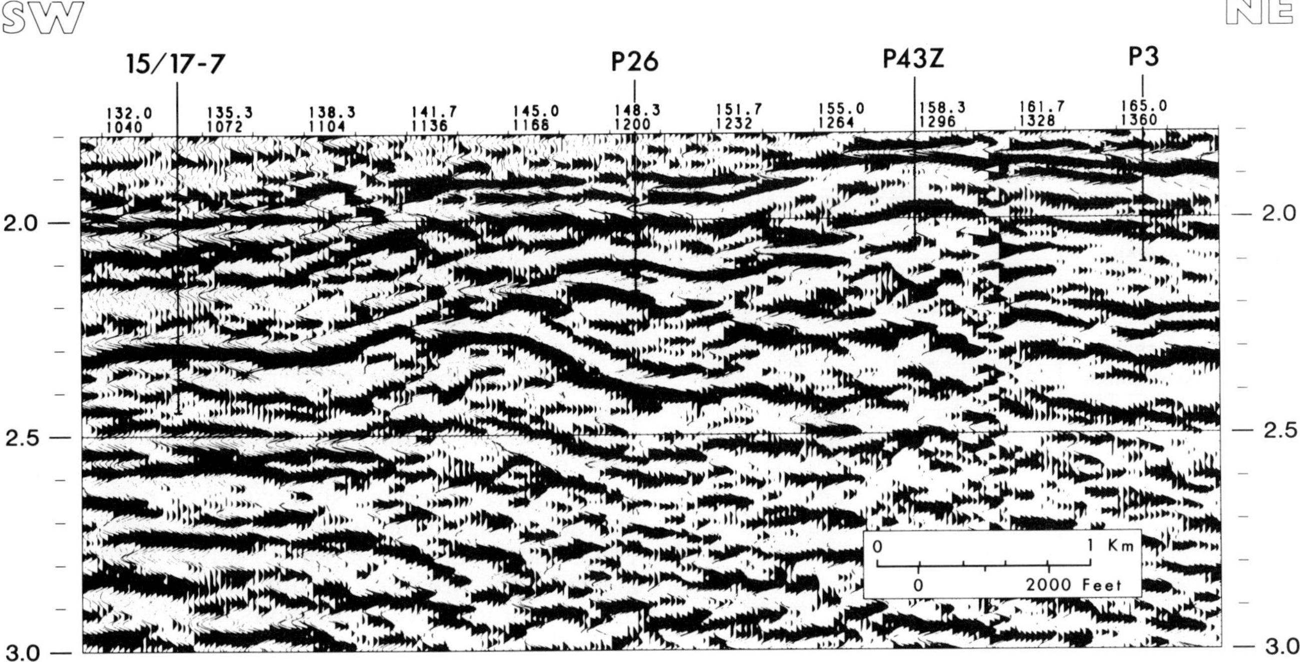

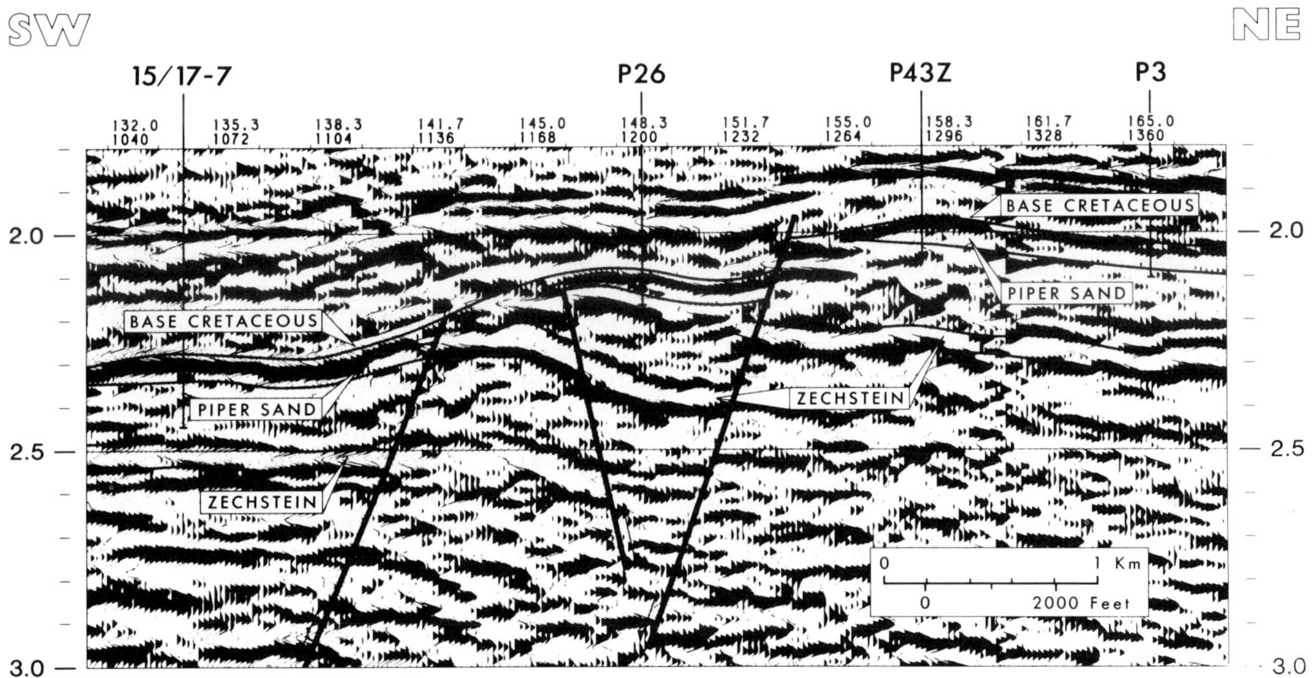

Figure 9. Uninterpreted and interpreted southwest-northeast seismic line through Piper field. Two-way time in seconds. The section illustrates key seismic horizons along the southern half of line A–A′ in Figures 3 and 4.

The oil-water contact for the main part of the field (blocks I and II) is at 8510 ft (2594 m) subsea and has equalized across all of the faults. The small four-way dip closure to the southwest (block III) has a deeper oil-water contact at 9200 ft (2804 m) subsea but contains only a minor amount of oil.

Although the structural closure of the blocks occurred at the end of the Jurassic, the trap for the main part of the Piper field was not formed until the Campanian marls onlapped the Piper sandstones along the "C" fault some 88 million years after the Piper sandstones had been deposited.

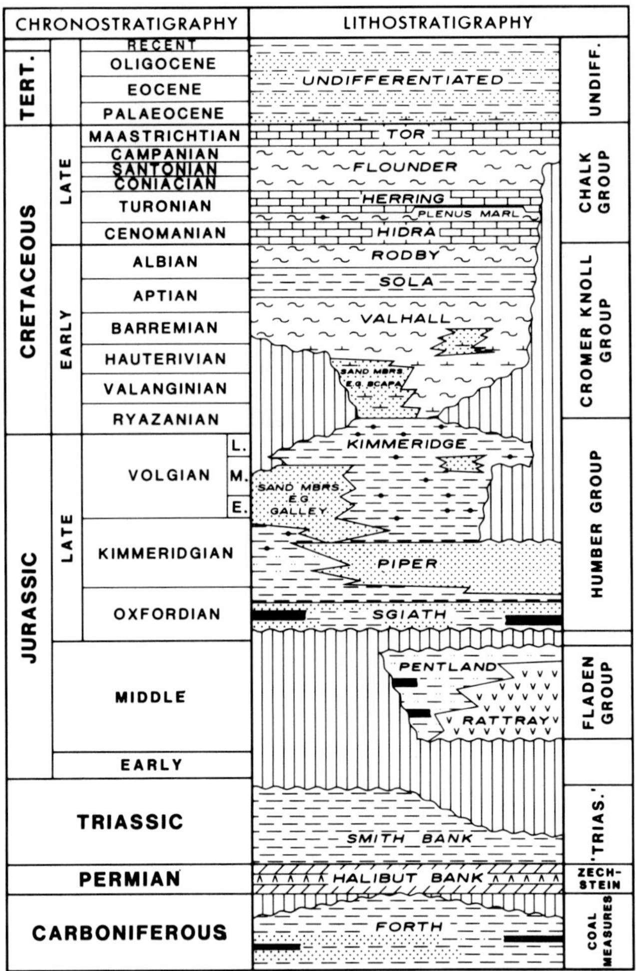

Figure 10. Stratigraphic column relating chronostratigraphy to lithostratigraphy in the Witch Ground graben area. Thicknesses are not to scale.

Reservoir

The Piper field is a layered reservoir. In many cases, the layers are sandstone on sandstone and show little variation in porosity. However, the big differences in permeability between these layers necessitate careful reservoir management to achieve high recovery. Those layers with the highest resistivities are where water break-through occurs first (Figure 5).

Two production wells, P2 and P14 (Figure 5), define the Piper reservoir subdivisions used informally within the Occidental Group. The P2 well is used for the western and the P14 well for the eastern half of the field. The subdivisions are changed from those defined by Maher (1979) following a major re-correlation exercise carried out in 1986 (Figure 11). The distribution of the various significant reservoir subdivisions within the Piper Formation and the net sandstone and net oil sandstone isochores for both the Piper and the Sgiath formations are shown in Figures 12A through 16E.

The Sgiath is comprised of five divisions as follows: from base to top, the M sandstone, M shale, L sandstone, K shale, and J sandstone (Figure 5). Throughout the field the Sgiath rests unconformably on the Middle Jurassic Pentland Formation, and the upper boundary is a conformable contact with the Piper Formation.

The Piper Formation comprises 12 major subdivisions and subsidiary shale units, commencing with the field-wide "I shale" and ending with a reworked zone (Unit IV) at the top. The P2 and P14 wells (Figure 5) do not exhibit the complete sequence because the uppermost A and B sandstones are only developed downflank along the eastern margin of the field (Figures 4 and 13A). (The A and B sandstones may be equivalent to the Galley sandstones [Figure 10] deposited to the east and southeast of Piper field.)

The above reservoir subdivision is based on log character correlation using primarily the GR, DLL, FDC, and CNL. Correlation on a field-wide basis is aided by the presence within the Piper Formation of three radioactive horizons in the G and H sandstones, termed Glγ1, Glγ2, and Hγ, respectively (Figures 12A, 16A, and 16B), and thin shales at the base of the D and E sandstones (Figures 13D and 14B). Within the Sgiath, field-wide correlation of the subdivisions is relatively straightforward, but the Piper Formation exhibits a greater variability from east to west across the field as shown by the stratigraphic correlation sections (Figure 11). Wells in the western half of the field are relatively easy to correlate while those in the eastern half are more difficult, as is correlating between east and west.

Sgiath lithology west of the "D" fault (Figure 8) consists of coal, silty bioturbated shales (M shale, K shale), and planar and trough cross-bedded, rippled, medium- to coarse-grained, poorly to well-sorted sandstones with occasional bioturbated surfaces (J and L sandstones). East of the "D" fault the sandstones are medium to very coarse grained, cross-bedded, moderately to poorly sorted, with numerous bioturbated surfaces overlain by coarser-grained sandstones (Maher, 1979). Body fossils are absent from the sandstones, but the shales contain a sparse fauna including heterodonts and oyster fragments. The coal beds, which occur in the basal M unit, are discontinuous and variable in thickness. Throughout the formation there are abundant plant and coal fragments and minor pyrite. The sandstones are moderately to poorly cemented with calcite.

The Piper reservoir units consist of a series of stacked sandstones organized in overall coarsening-upward cycles. Individual units may be fine to medium grained, medium to coarse grained, and coarse grained to pebbly. Sorting is variable, from poor to excellent and occasionally bimodal. Bioturbation is pervasive except in the coarsest-grained units and has destroyed many of the primary sedimentary structures. Planar and trough cross-lamination, ripple-drift cross-lamination, and pebble-lined scour surfaces are the most common of the remaining sedimentary structures. Calcite doggers

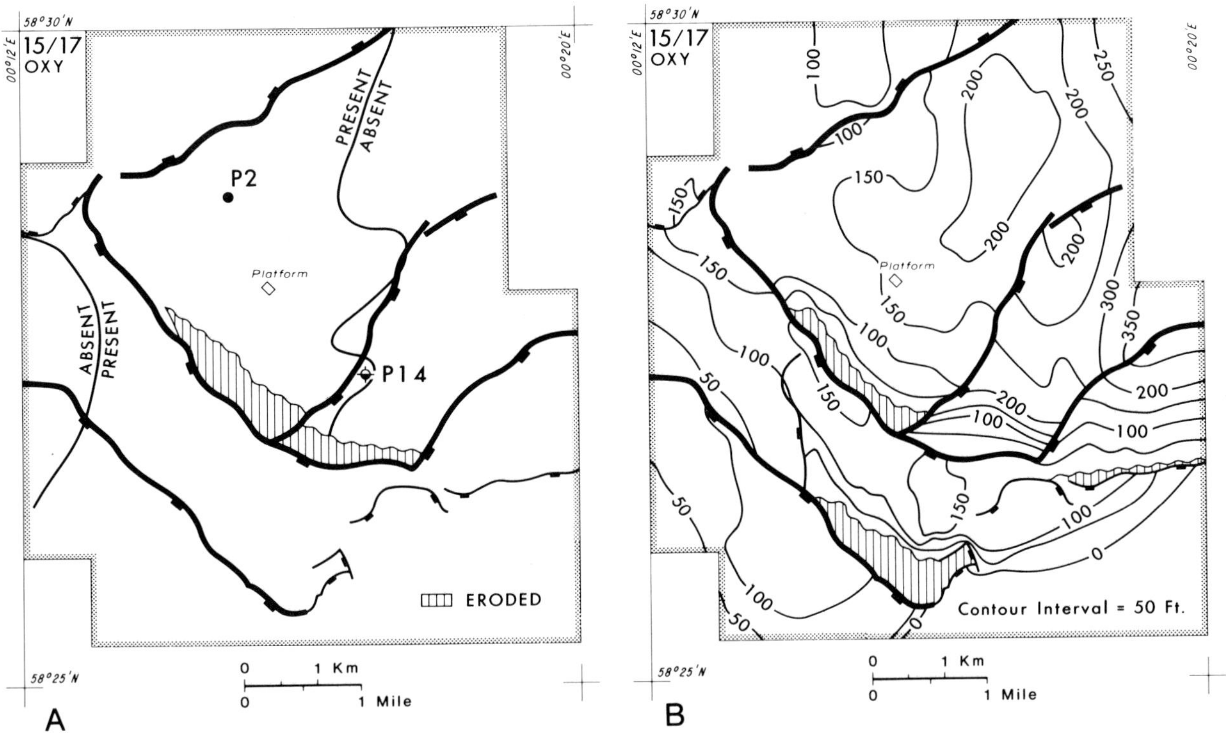

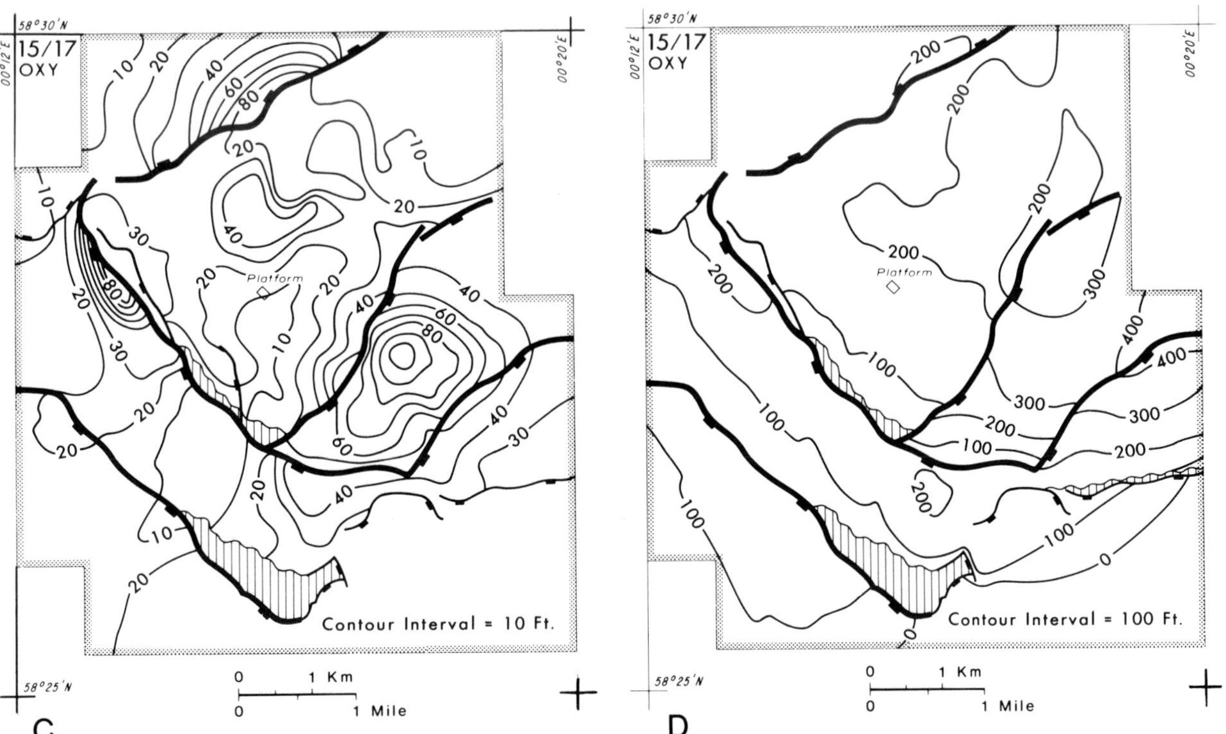

Figure 12. (A) Areal distribution of the G1γ1 radioactive marker in Piper field. (B) Net Piper sandstone isochore illustrating thickening of the sandstone to the east and northeast. (C) Net Sgiath sandstone isochore. (D) Combined net sandstone isochore for the Piper and Sgiath.

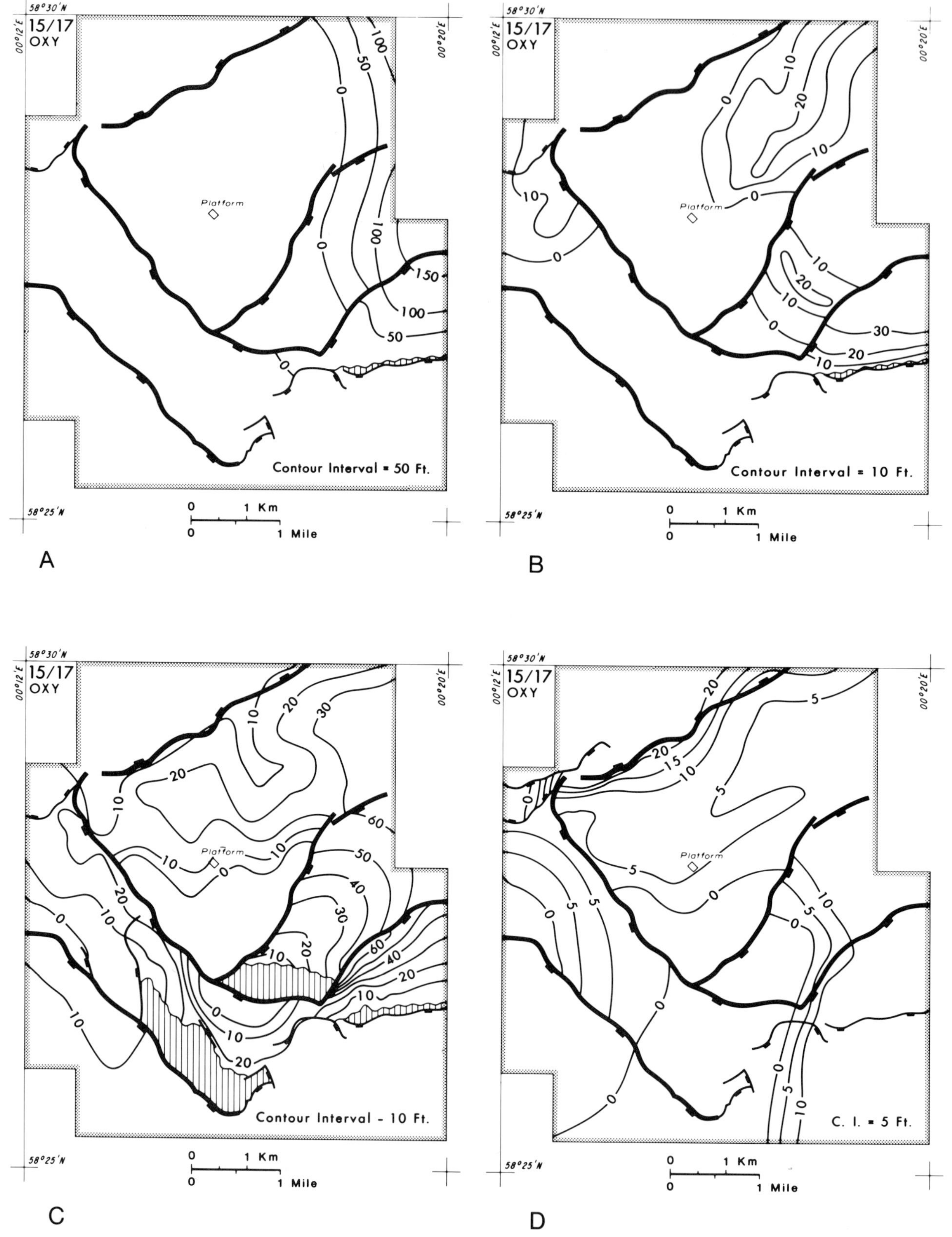

Figure 13. (A) Gross isochore for units A and B. These units are restricted to the eastern flank of the field and are now considered likely equivalents to the Galley sandstones. (B) Gross isochore for unit C, restricted to down-flank areas of the field. (C) Gross isochore for unit D, the uppermost reservoir unit with field-wide extent. As with all of the units, erosion has occurred along the southwestern leading edge of each of the main fault blocks. (D) Gross isochore for the D shale, the upper of three shales with significant extent within the Piper Formation in Piper field.

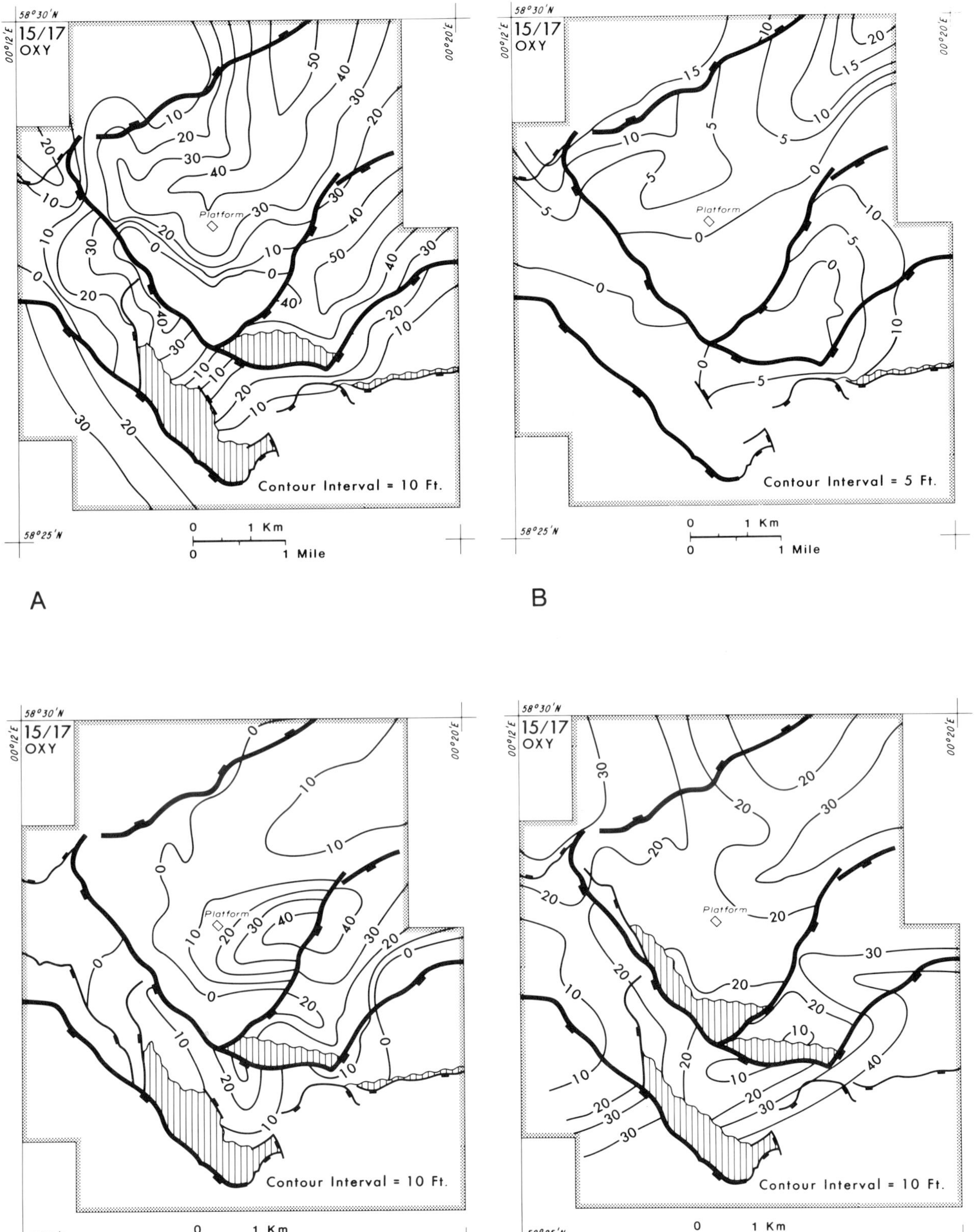

Figure 14. (A) Unit E gross isochore. (B) Gross isochore for the E shale, the second of the three important shales within the Piper Formation. (C) Unit F1 gross isochore. (D) Unit F2 gross isochore.

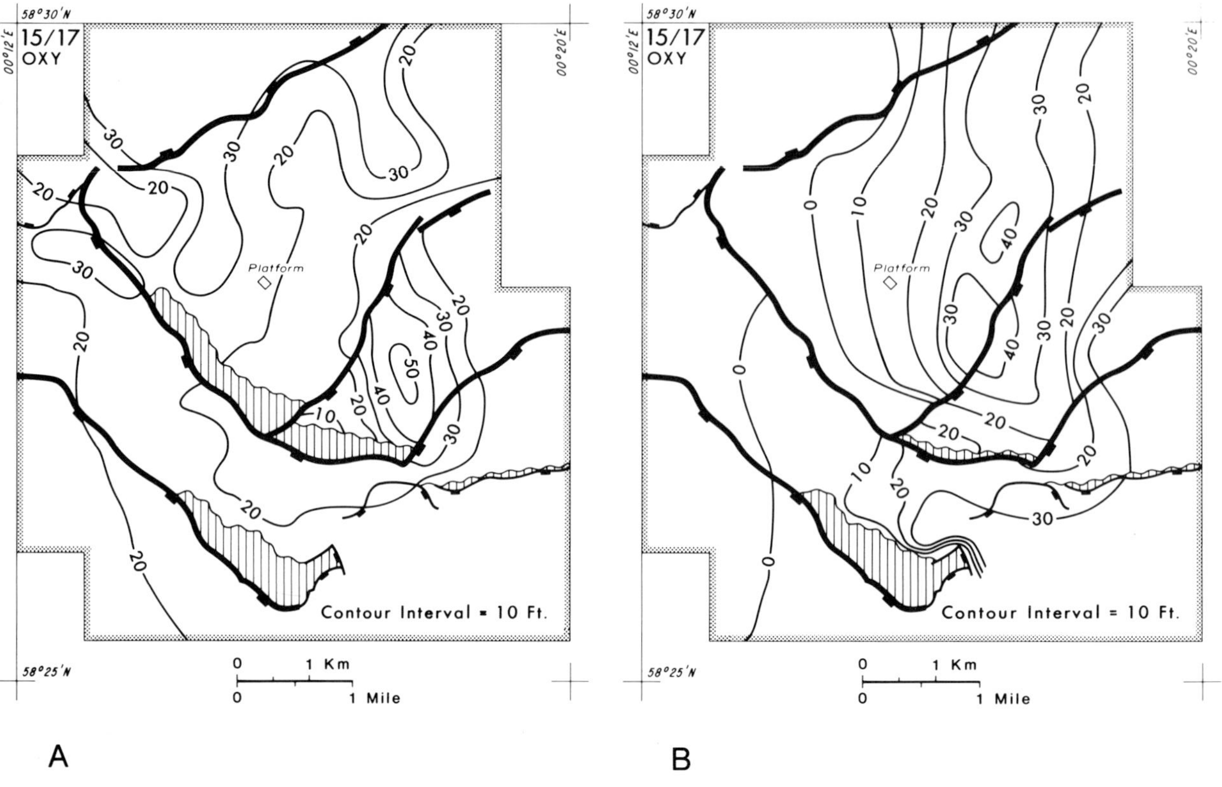

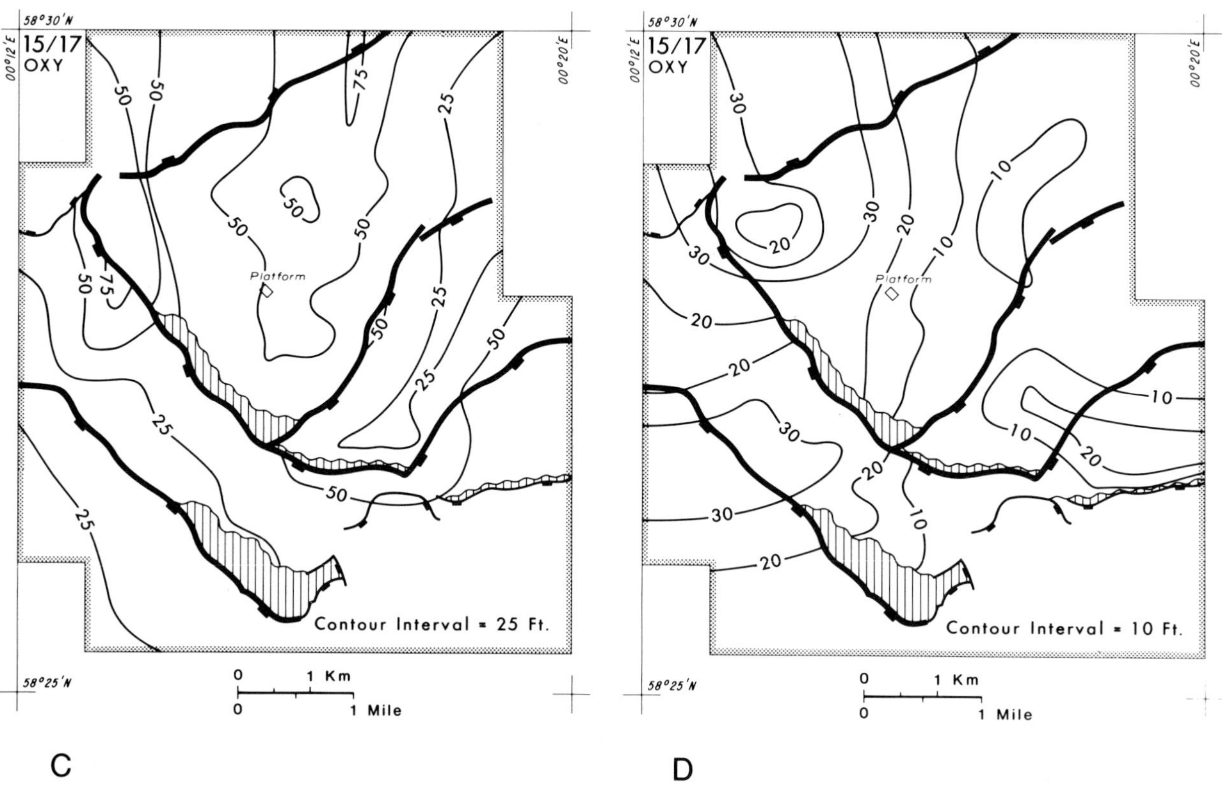

Figure 15. (A) Unit G1 gross isochore. (B) Unit G2 gross isochore. (C) Unit H gross isochore. (D) Gross isochore for the "I shale," which is the basal member of the Piper Formation. It is the lowest of the three significant shales within the Piper Formation and is regionally extensive throughout the Witch Ground graben.

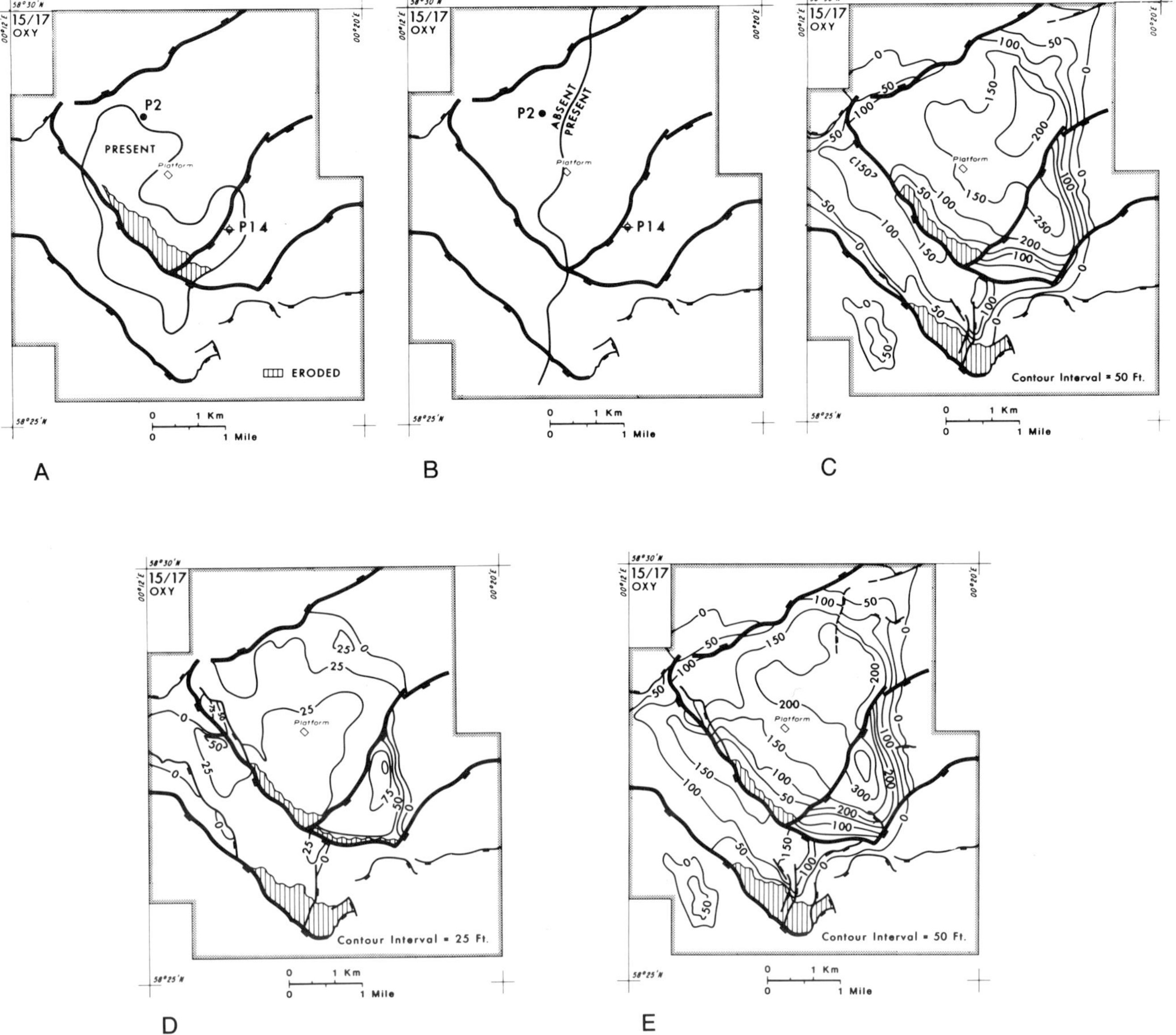

Figure 16. (A) Distribution of the G1γ2 radioactive marker, which is confined to the central part of the field. (B) Distribution of the Hγ radioactive marker, which is present only across the eastern side of the field. (C) Piper net oil sandstone isochore. (D) Sgiath net oil sandstone isochore. (E) Combined Piper and Sgiath net oil sandstone isochore.

occur sporadically but are more common in the E sandstones. Shell debris is common in the H and E sandstones, and pyrite nodules have been noted in the H sandstones. The sandstones are friable overall and are poorly cemented with minor calcite cement and by quartz overgrowths. Core recovery in some units is poor because of their unconsolidated nature.

Most macrofauna observed in cored sections are preserved in the shaley-silty units of the "I shale," E shale, and D shale. Bioturbation in these units is ubiquitous together with shells and shell debris (oysters, bivalves), plant debris, and occasional coal clasts. On a field-wide basis, the lithology of the "I shale" remains constant, but both the D and E shales become increasingly silty and sandy up structure until they eventually disappear as identifiable nonreservoir units.

Figure 17 shows core photographs of typical Piper reservoir sandstones from wells 15/17-6 and P32. Figure 18 illustrates thin sections from the E and F1 units in P32, and Figure 19 shows SEMs from the F2 unit in well 15/17-7 and the J unit in well 15/17-6.

The Sgiath is interpreted as a lower delta plain to interdistributary bay sequence deposited in three regressive cycles that correlate from the Claymore field in the west, through the 15/11 block and into Piper field. Deposition occurred over a surface of very low relief where small fault displacements and local topography were gradually infilled by a succession

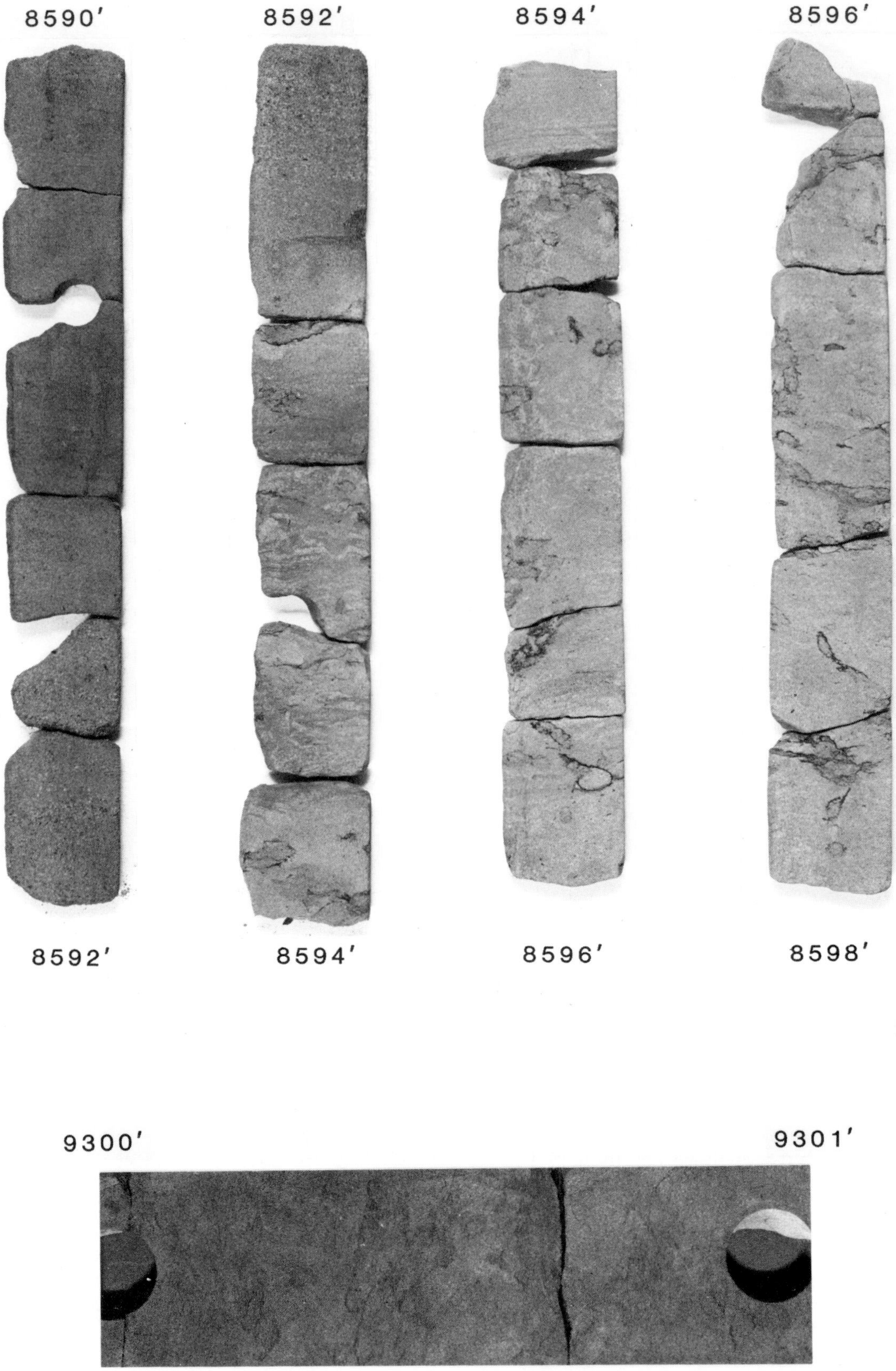

8590′ 8592′ 8594′ 8596′

8592′ 8594′ 8596′ 8598′

Figure 17. Core from well 15/17-6 (Figure 3). The boundary between the coarse-grained F2 sandstone with 6 darcy permeability and the underlying fine-grained bioturbated G1 sandstone with 1 darcy permeability is at 8592 ft 9 in. The lower photograph of core from well 15/17-P32 (Figure 3) shows bioturbated H sandstone immediately below foreshore and lag deposits containing radioactive monazite. Measured porosity is 26% and permeability 1130 md.

Figure 18. Two photomicrographs from well 15/17-P32 (Figure 3). The upper one shows fine- to medium-grained F1 sandstone at 7913 ft (2412 m) subsea with 26% porosity and permeability of 1910 md. The lower one shows fine- to medium-grained E sandstone at 7909 ft (2411 m) subsea with 26% porosity and 1380 md permeability.

Figure 19. The upper SEM photograph shows coarse-grained well-sorted F2 sandstone at 9212 ft (2808 m) subsea in well 15/17-7 (Figure 3). Porosity is 26% and permeability 5 darcys. The lower SEM photograph shows medium to coarse-grained J sandstone at 8633 ft (2631 m) subsea in well 15/17-6 (Figure 3). Porosity is 26% and permeability 2 darcys.

of minor alluvial and deltaic cycles, periodically abandoned during brackish transgressions (Boote and Gustav, 1987). Figure 12C illustrates the net sandstone distribution for the Sgiath, composed largely of the J and L sandstones. The M sandstone is discontinuous and much thinner. Limited core data indicate a more marine influence in the east. The permeability within the Sgiath sandstones varies significantly both laterally and vertically. The better quality sandstones are concentrated in block IB in the east of the field. Permeabilities range from 200 md to 4 darcys, with an average of 1200 md. Permeability tends to be higher near the top of the sandstones, although this is not universally true.

The Piper Formation was deposited in a shallow marine wave dominated delta environment and comprises a series of stacked sand bodies. They are interpreted as lower to upper shoreface and foreshore deposits. As previously described, the units of the Piper Formation can be correlated with various degrees of difficulty on a field-wide basis using log character. The changes seen in the log character of the sandstone units across the field are related to transitions between a lower shoreface environment in the west and an upper shoreface and foreshore environment in the east. Within most sedimentary cycles, the permeability and grain size show an increasing-upward trend that reflects the transition from lower to higher energy. This transition is reflected in both the resistivity and gamma ray profiles in many wells; e.g., the H sandstone cycle (Figures 5, 20, 21, and 22).

Other units within the Piper show similar trends, notably the D, E, and G1. Where this trend is poorly developed or absent, such as in the F1 unit, only the upper shoreface and foreshore deposits of a cycle have been preserved. Permeabilities within a unit can vary from 500 md at the base to 10 darcys at the top. Lateral permeability variations within individual units are insignificant compared to the large vertical changes observed.

The porosity of both the Piper and Sgiath Formation sandstones is almost entirely intergranular and has been considerably modified by diagenesis during burial. The most significant effect of burial diagenesis was porosity reduction by quartz overgrowths on many of the originally moderately to well-rounded sand grains. Detailed studies on the Piper

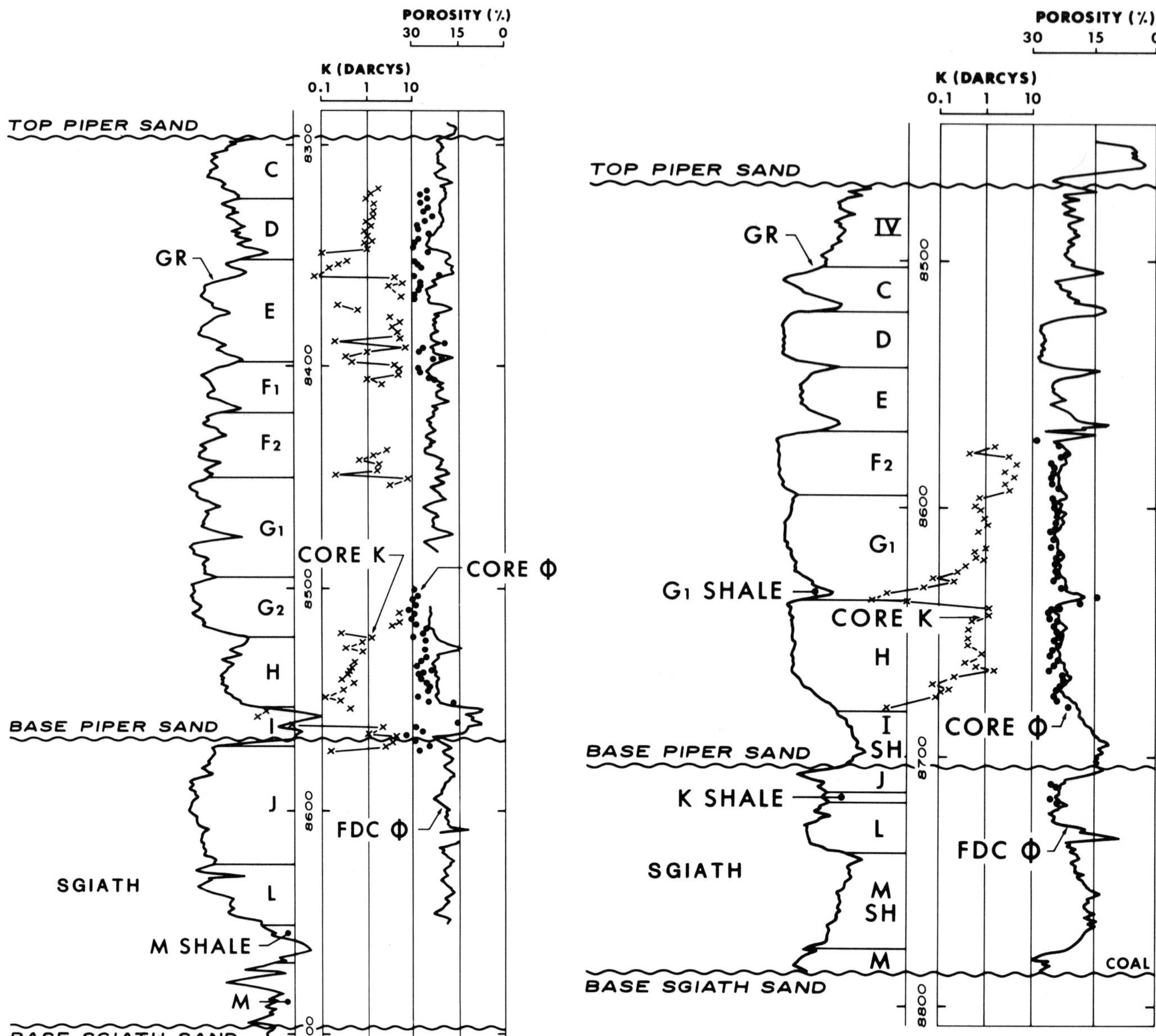

Figure 20. Comparison of log porosity with core porosity and permeability in well 15/17-5 (Figure 3). The lack of core measurements from the F and G units is indicative of poor core recovery from these poorly consolidated sands. (Depths are measured depths.)

Figure 21. Comparison of log porosity with core porosity and permeability in well 15/17-6 (Figure 3). (Depths are measured depths.)

sandstone (Burley, 1986) have shown that the pore network displays many textural characteristics typical of secondary porosity (Schmidt and McDonald, 1979). However, it is possible that the pore geometry at Piper is principally modified intergranular porosity.

The combination of these diagenetic effects has resulted in a field-wide average porosity for both the Sgiath and Piper sandstones of 24%. A relationship between current depth of burial and porosity is observed such that porosity is higher at the crest than downflank. Some studies have noted slightly lower intergranular porosities in the water zone due to late stage cement phases that precipitated subsequent to oil entrapment (Burley, 1986).

The Piper sandstones are neither oil-wet nor even partially oil-wet. In the coarse-grained, highly permeable sandstones, the smallest pores that contain the irreducible water constitute only about 1% of the rock. The wetting layer on the surface of the quartz is only three molecular layers (10^{-3} microns) thick. The specific inner surface area of these very permeable sands has been measured at ± 0.1 m^2/cm^3. Thus, the wetting layer occupies a

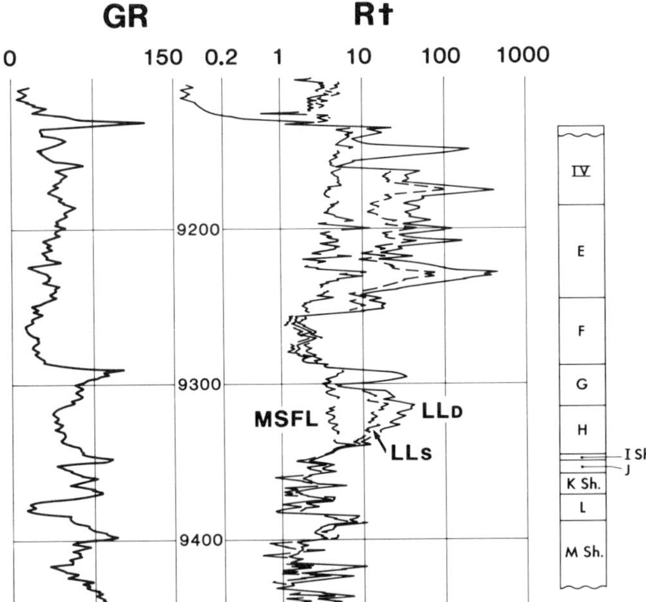

Figure 22. Gamma ray and dual laterolog from well 15/17-P32 (Figure 3). The upward-increasing resistivity profile in the H sandstone is a characteristic log response in coarsening-upward units. (Depths are measured depths.)

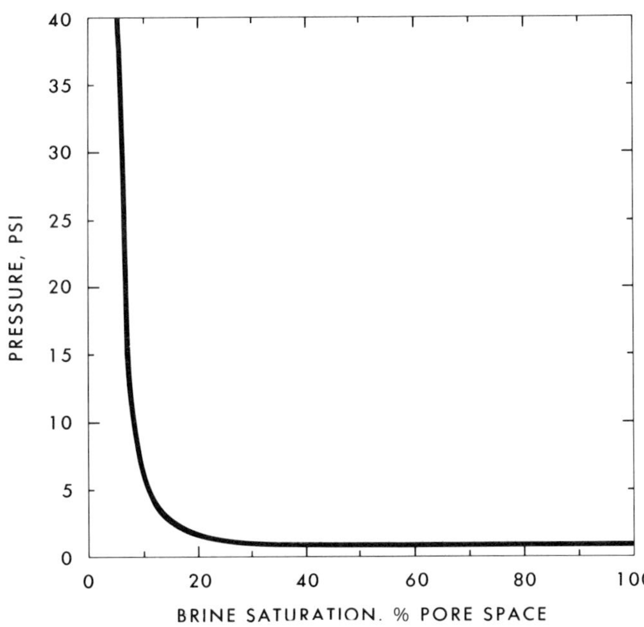

Figure 23. A typical capillary pressure curve for Piper sandstone in Piper field. In the more permeable units, the water saturation is often less than 10% within 20 ft (6 m) of the oil-water contact.

volume equal to 10^{-9} m (0.1 m^2)/cm^3. That is 10^{-10} m^3/cm^3 or

$$\frac{10^{-4}\ \text{cm}^3\ \text{water}}{\text{cm}^3\ \text{rock}} \quad \text{or } 0.01\ \%\ \text{of the bulk volume.}$$

This is true whether the rock is near the oil-water interface or 1000 ft (305 m) higher in the reservoir. However, as the capillary pressure increases upwards within the oil column, oil displaces water out of increasingly smaller pores and the lowest irreducible water saturations are found at the highest elevation in the rocks with the highest permeability. The permeability is a function of pore throat size, which is affected more by grain size than by sorting.

Extremely high resistivities (2000 ohm-m or greater) are observed within the high permeability zones of the Piper reservoir (Figure 5). These resistivities are associated with very low irreducible water saturations that result from the high permeability, thick oil column, and minor percentage of small pores present within the pore network. Wells drilled close to the original oil-water contact have demonstrated that the irreducible water saturation increases in a perfectly normal manner for a water-wet reservoir. In the very permeable sands, the irreducible water saturation is less than 10% within 20 ft (6 m) of the oil-water contact. On the crest of the structure, these same sands have connate water saturations as low as 1%. Figures 23 and 24 show typical capillary pressure and relative permeability curves for oil and water within the Piper reservoir.

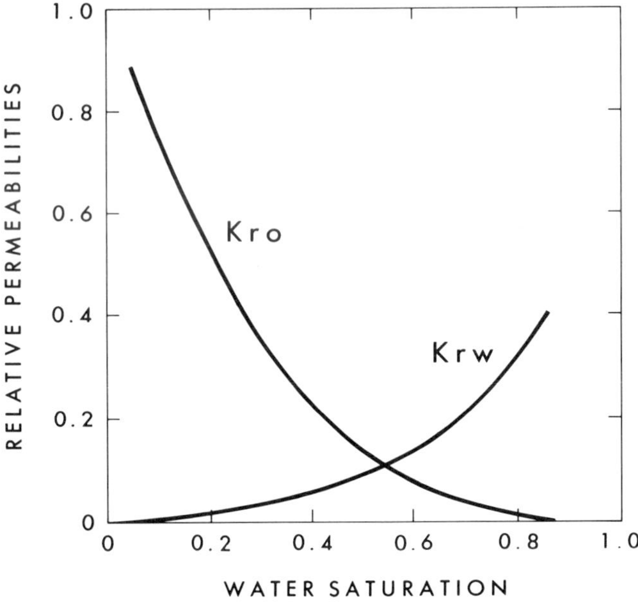

Figure 24. Typical highly favorable oil-water relative permeability curves for Piper sandstone in Piper field.

Producing horizons within the main Piper field fall within the depth window 7300 to 8510 ft (2225 to 2594 m) subsea, while the subsidiary block III accumulation falls between 9100 and 9200 ft (2774 and 2804 m) subsea. The maximum net pay within the Piper field is 375 ft (114 m), with an average of 140 ft (43 m). Piper oil is 36° API, low sulfur crude with a GOR of 430 SCF/bbl. Figure 25 shows a representative whole oil gas chromatogram for Piper crude.

Both the Piper and Sgiath sandstones have exceptionally good flow characteristics. Faults within the Piper field are nonsealing unless complete sand offset occurs (e.g., the "C" fault, Figure 4). The principal effect of the nonsealing faults is to juxtapose zones of differing permeability, which is an aid to more effective sweep. Barriers to vertical fluid flow are provided by three widespread shales (D, E, and "I shale," Figures 5, 13D, 14B, and 15D). The "I shale" is field-wide and serves to separate the Piper sandstones from the Sgiath sandstones. The D and E shales, where present, subdivide the Piper but pinch out and are absent over the crest of the structure.

The flow characteristics of the individual reservoir layers are governed by the permeability distribution. As previously mentioned, the permeability distribution is largely a function of the energy of the depositional environment and associated grain size and sorting (McCubbin, 1982). Many of the sandstones are extensively bioturbated, but this has a limited effect on reservoir performance owing to the absence of fine-grained, low-energy laminae and good sorting of the sandstones.

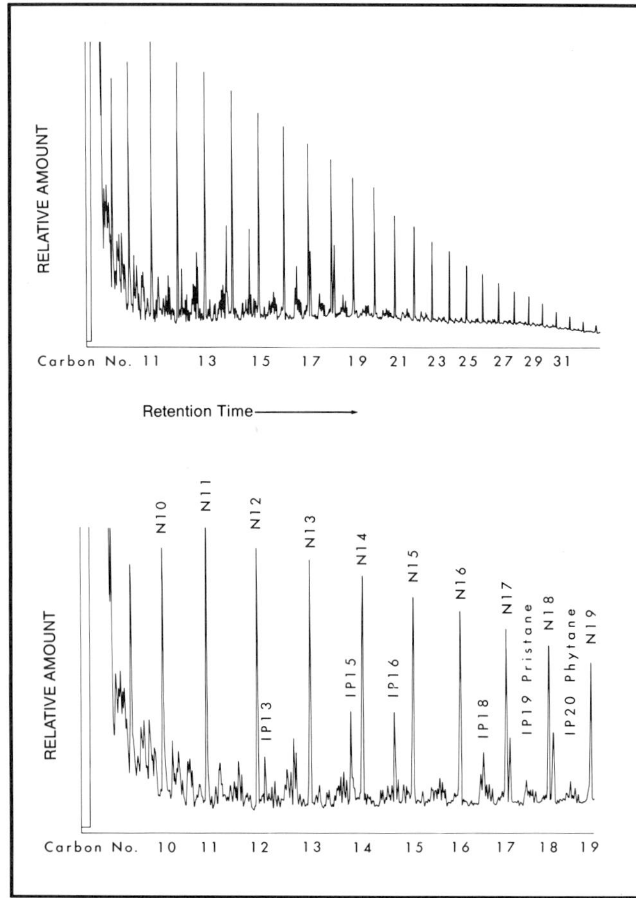

Figure 25. Piper field crude oil/gas chromatogram. Some of the better known components are labeled on the enlarged section in the lower part of the diagram.

The P1 well came on stream at 30,000 BOPD restricted by 5½-in. (14 cm) tubing. The P7 well came on stream at 50,000 BOPD restricted by 7-in. (17.8 cm) tubing. After the field rate was reduced in 1980 to improve reservoir management, wells were generally produced at rates of less than 25,000 BOPD. The primary consideration in the development of the field was to ascertain the reservoir pressure above the bubble point of 1600 psi required to flow high watercut wells after a shutdown.

From the very high production rates in the appraisal wells and the production geology studies, it was clearly understood that a relatively large spacing of more than 300 ac (1.2 km^2) would be adequate for excellent sweep. This has been confirmed by reservoir performance.

Both producers and injectors were selectively completed to provide a balanced offtake for individual reservoir units. This ensured a controlled water advance in each of the layers. There was some water override in the more permeable sandstones such as the F2, but the rate reduction in 1980 effectively minimized this problem.

In 1977, a gas lift system was installed, and this was supplemented in 1982 by the use of high-volume submersible pumps that allowed economic production of wells with watercuts as high as 98%. A 3D reservoir simulation of the field was initiated in 1982 and regularly updated and refined as development drilling and production performance provided additional information (King et al., 1983).

The reservoir has been effectively managed, and approximately 70% of the oil-in-place will ultimately be recovered by supplementing natural water influx with selective water injection.

Faults

Faults in the Piper field were recognized with the use of 2D seismic data. The data were high quality, but, while faults of quite small magnitude could be mapped, only base of the Cretaceous and the Permian Halibut Bank (Zechstein) Formation were clearly recognizable. Those faults having expression at both base of the Cretaceous and Zechstein levels were certain to affect the Piper reservoirs. However, some easily recognizable Zechstein faults may also extend up into the reservoir without offsetting the base of the Cretaceous. In the absence of reliable top and base sandstone reflectors, sandstone thickness variations could only be confirmed by drilling.

In 1982, 3D seismic was acquired over the Piper field area and much of the 15/17 block. This resulted in better definition of the fault pattern and helped in the location of late stage and difficult development wells. However, many of the small faults can still only be inferred by extrapolation upwards from the much deeper Halibut Bank Formation.

The 3D seismic data also enabled the interpretation and recognition of the northwest corner of the field as a very complex intersection of the Cimmerian or Witch Ground graben trend with the Caledonian fault trend. This intersection has formed a complex fault zone along the northern section of the "C" fault from P31 to where it intersects the northeast–southwest-trending "E" fault (Figures 3 and 8). In this corner of the field, two wells, P21 and P31, drilled prior to the 3D survey, penetrated the Piper sandstones on the upthrown side of the "C" fault, but both had been intended as water injection wells on the downthrown side.

The best example of a syndepositional fault in the Piper field is the "D" fault (Figure 8), which has significantly thicker Piper and Sgiath sandstones on the downthrown side (Figures 12B, 12C, and 12D).

Source

The organic-rich shales of the Kimmeridge Clay Formation are the source for the Piper oil. The most likely migration path is out of the Witch Ground graben, south of the field, and updip into the Piper reservoir.

The migration could not have taken place before the Campanian marls onlapped the Piper sandstones and formed the seal along the "C" fault during Late Cretaceous times (Figures 4 and 8). Most evidence points to an early Tertiary maturation and migration phase, coinciding with the period of most rapid burial of the Piper reservoir (Figure 7).

The richness of the Kimmeridge shales varies from 4 to 16% TOC, comprising mainly type II kerogen (Cayley, 1986). The potential yield is 0.3 barrels per cubic meter of source rock, assuming an average TOC of 8%. At 12,000 ft (3658 m) subsea, the source rock is just within the oil window, as indicated by the observed vitrinite reflectance of 0.6 (Tissot and Welte, 1978). Deeper in the graben, the maturation level would be even higher.

EXPLORATION CONCEPTS

Regional Play

The regional play is one of tilted fault blocks on the edge or within an extensional graben (Figures 1A and 1B). Similar plays exist wherever extensional grabens are present, such as the Central and Viking grabens of the North Sea, the Gulf of Suez, West Africa, and Brazil.

General Application of Geologic Parameters

The Piper field is by far the most productive oil field in the Outer Moray Firth basin. Other fields produce from the Piper and Sgiath sandstones, but they have been buried to greater depths and have therefore suffered a greater reduction in reservoir quality due to burial diagenesis and are much smaller accumulations.

As mentioned previously, faulting does not pose a significant problem in the Piper field. However, those fields actually in the graben margin are much more difficult to develop and produce owing to structural complexity caused by fault offsets.

The Piper field has good four-way closure as mapped with reflection seismic data near the top of the Piper sandstone. It is at a favorable depth to allow full development from a single, centrally located platform (Figure 26). It has a large aquifer of similarly high quality sandstones in good pressure communication.

In common with other highly permeable sandstone reservoirs that have strong natural water drive and large aquifer support, the Piper field will have a high recovery factor (Arps et al., 1967). This is helped by the layering of the reservoir, which allows selective completion and efficient reservoir management.

Lessons

Exploring today for another field like Piper would probably be done in very much the same way. Very likely, more detailed seismic would be shot over the mapped closures and tilted fault blocks to improve stratigraphic interpretation of lithology and cap rock prior to application for the acreage.

Three-dimensional seismic would be shot soon after one or two appraisal wells had been drilled. In Piper, the persons responsible did an excellent job of placing the original appraisal wells. In most fields, however, the appraisal program could be improved with the use of 3D seismic data. It would also have been useful to have had the 3D seismic to detail some of the more complex fault intersections within the field and for the planning of the initial development wells.

The Piper field has reiterated that caution should be exercized in the use of the gamma ray log when there are radioactive elements present in the reservoir. The radioactive sandstones in Piper appear as shales on the gamma ray and on the TDT logs (Figures 5 and 11). This initially led to an underestimation of oil-in-place by some parties.

In any field where significant vertical variations in permeability are present, selective completion of both injectors and producers should commence from the beginning of field development. This, combined with balanced injection and offtake by layer, will result in improved reservoir management and higher ultimate recovery.

Piper is a good example of how field development and reservoir management can benefit from close cooperation between geological, engineering, and drilling staff. Occidental, as operator, has always followed a policy of organizing reservoir management

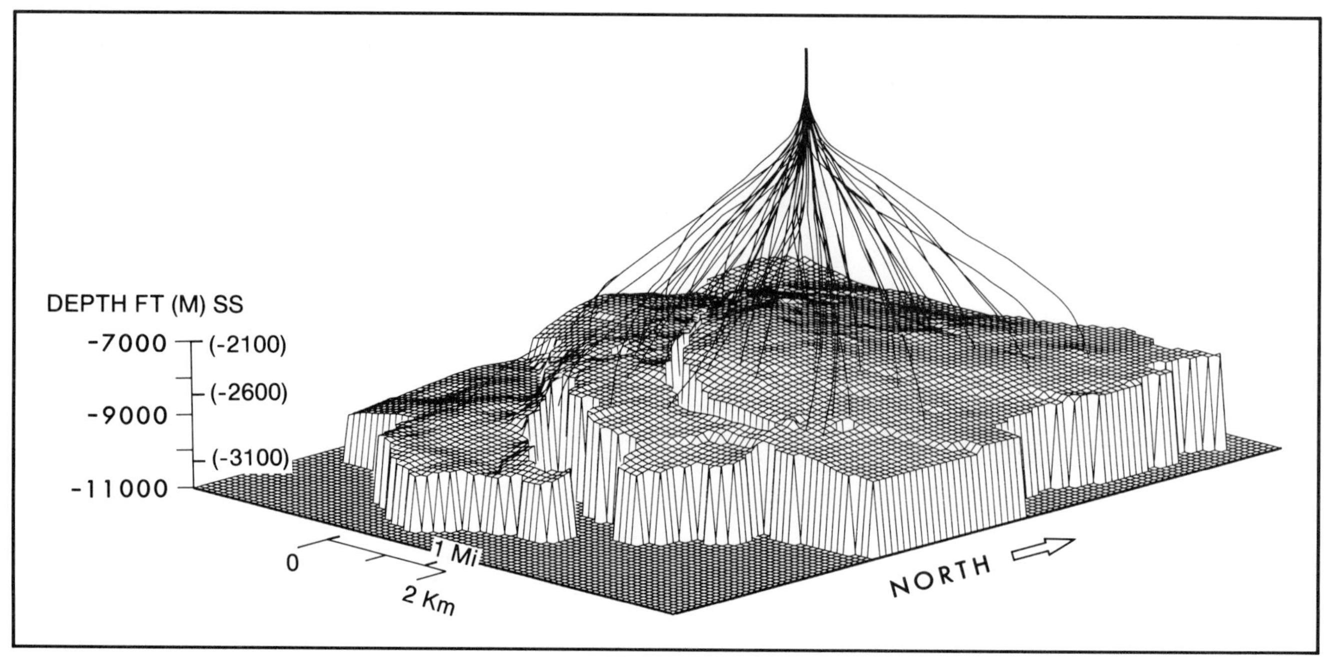

Figure 26. A 3D projection of the Piper field from the southeast showing the top Piper sandstone and well tracks. The entire field was within drilling reach of the single centrally located platform.

and development personnel on a team basis, resulting in effective integration of data.

ACKNOWLEDGMENTS

The authors wish to thank Occidental Management and their Piper Operating Group Partners for permission to publish this paper. The interpretations expressed in this paper are entirely the responsibility of the authors.

REFERENCES CITED

Arps, J. J., F. Brons, A. F. von Everdingen, R. W. Buchwald, and A. E. Smith, 1967, A statistical study of recovery efficiency: API Bulletin 14D, 33 p.

Boote, D. R. D., and S. H. Gustav, 1987, Evolving depositional systems within an active rift, Witch Ground graben, North Sea, *in* J. Brooks and K. Glennie, eds., Petroleum geology of north-west Europe: London, Graham and Trotman Publishing Company, p. 819–833.

Burley, S. D., 1986, The development and destruction of porosity within Upper Jurassic reservoir sandstones of the Piper and Tartan fields, Outer Moray Firth, North Sea: Clay Minerals, v. 21, p. 649–694.

Carmalt, S. W., and B. St. John, 1986, Giant oil and gas fields, *in* Future petroleum provinces of the world: AAPG Memoir 40, p. 11–54.

Cayley, G. T., 1987, Hydrocarbon migration in the central North Sea, *in* J. Brooks and K. Glennie, eds., Petroleum geology of north-west Europe: London, Graham and Trotman Publishing Company, p. 549–555.

Deegan, C. E., and B. J. Scull, 1977, A standard lithostratigraphic nomenclature for the central and northern North Sea: London, H.M.S.O., I.G.S. Report 77/25, 35 p.

Harker, S. D., S. H. Gustav, and L. A. Riley, 1987, Triassic to Cenomanian stratigraphy of the Witch Ground graben, *in* J. Brooks and K. Glennie, eds., Petroleum geology of north-west Europe: London, Graham and Trotman Publishing Company, p. 809–818.

King, P. A., T. Robinson, A. A. Abdullah, D. Waldren, and D. Brindle, 1983, A six year simulated-history match of the Piper field, *in* Proceedings of the Offshore Europe Conference: Society of Petroleum Engineers (UK) Paper 11898, 20 p.

Maher, C. E., 1979, Piper oil field, *in* Giant oil and gas fields of the decade: 1968–1978: AAPG Memoir 30, p. 131–142.

Maher, C. E., 1981, The Piper oil field, *in* Petroleum geology of the continental shelf of north-west Europe: 2nd Conference Proc. Inst. Petroleum, p. 358–370.

Maher, C. E., and S. D. Harker, 1987, The Claymore oil field, *in* J. Brooks and K. Glennie, eds., Petroleum geology of north-west Europe: London, Graham and Trotman Publishing Company, p. 835–845.

McCubbin, D. G., 1982, Barrier-island and strand plain facies, *in* Sandstone depositional environments: AAPG Memoir 31, p. 247–279.

Ramsbottom, W. H. C., 1978, Permian, *in* W. S. McKerrow, ed., The ecology of fossils, an illustrated guide, p. 164–193.

Schmidt, V., and D. A. McDonald, 1979, Texture and recognition of secondary porosity in sandstones, *in* P. A. Scholle and P. R. Schluger, eds., Aspects of Diagenesis: SEPM Special Publication 26, p. 209–225.

Tissot, B. P., and D. H. Welte, 1978, Petroleum formation and occurrence: New York, Springer-Verlag, p. 538.

Turner, C. C., P. C. Richards, J. L. Swallow, and S. P. Grimshaw, 1984, Upper Jurassic stratigraphy and sedimentary facies in the central Outer Moray Firth basin, North Sea: Marine and Petroleum Geology, v. 1, p. 105–117.

SUGGESTED READING

Conner, D. C., and D. G. Kelland, 1976, Piper field, UK North Sea interpretive log analysis and geologic factors, Jurassic reservoir sands: Log Analysis, v. 17 p. 12–21.

Conner, D. C., K. E. Peterson, and J. J. Williams, 1975, Piper oil field, North Sea: fault-block structure with Upper Jurassic beach/bar reservoir sands: AAPG Bulletin, v. 59, p. 1585–1601.

Dunlop, K. N. B., D. Waldren, and E. T. de Boer, 1982, Modelling the Piper field using layer cross flow at faults, *in* Proceedings of the 6th Symposium on Reservoir Simulation: Society of Petroleum Engineers of the AIME, p. 539–551.

King, P. A., 1980, Piper field: operational aspects of drilling/completions workovers/data acquisition and well performance, *in* Proceedings of the European Offshore Petroleum Conference, v. 1, p. 19–32.

Naylor, D., R. M. Pegrum, and G. Rees, 1975, Geology of the north-west European continental shelf: London, Graham Trotman Dudley Publishing Company, v. 2, p. 225.

Wiebe, M., 1980, Piper field: Geology, *in* Proceedings of the European Offshore Petroleum Conference, v. 1, p. 7–18.

PIPER

Appendix 1. Field Description

Field name .... *Piper field*

Ultimate recoverable reserves .... *952 million bbl*

Field location:

Country .... *United Kingdom*

State .... *U.K. Block 15/17, offshore North Sea*

Basin/Province .... *Outer Moray Firth basin*

Field discovery:

Year first pay discovered .... *Layered sandstones of the Oxfordian-Kimmeridgian Sgiath and Piper formations 1972*

Discovery well name and general location:

First pay .... *15/17-1A 110 mi (177 km) northeast of Aberdeen, Scotland*

Discovery well operator .... *Occidental Petroleum (Caledonia) Ltd.*

IP:

First pay .... *30,000 bbl/day restricted by 5½-in. tubing*

All other zones with shows of oil and gas in the field:

Age	Formation	Type of Show
Paleocene	*Undifferentiated*	*8° API oil*

Geologic concept leading to discovery and method or methods used to delineate prospect

Basin analysis of all offshore North Sea basins. Available well data were confined to southern North Sea gas province; Argyll, Ekofisk, and Forties were the only significant oil fields and little information was available. Approximately 8000 mi (12,892 km) of reflection seismic data were acquired and numerous closed structures were mapped. Jurassic sands and Kimmeridge oil shale were known from onshore basins. Scout data indicated that a major Mesozoic discovery had been made in the northern North Sea.

Structure:

Province/basin type .... *Bally 1211, Klemme IIB/IIIA*

Tectonic history

Gentle downwarping, folding, and faulting occurred in a cratonic basin in the Permian. Uplift and gentle folding followed in the Triassic and the Witch Ground graben (WGG) was initiated. Intermittent fault movement occurred through Late Jurassic. Major fault movement and development of WGG at the end of the Jurassic was followed by regional subsidence in Cretaceous and southeasterly tilting and subsidence in the Tertiary.

Regional structure

The Piper field is situated on a shelf on the northern margin of the WGG.

Local structure

Tilted, folded, fault blocks plunge northwest-southeast from 4 to 10°. The main block dips to the northeast at 8°.

Trap:

Trap type(s)

Anticlinal. Vertical and lateral seals are shales. Onlapping marls seal eroded fault scarps and the crestal area of the structure.

Basin stratigraphy (major stratigraphic intervals from surface to deepest penetration in field):

Chronostratigraphy	Formation	Subsea Depth* in ft (m)
Recent-Eocene	*Undifferentiated*	*1800 (550)*
Paleocene	*Undifferentiated*	*3350 (1020)*
	Ekofisk	*6740 (2055)*

Upper Cretaceous	*Tor*	*6860 (2050)*
	Flounder	*7400 (2255)*
Lower Cretaceous	*Sola*	*7450 (2270)*
	Valhall	*7455 (2277)*
Late Jurassic	*Kimmeridge Clay*	*7490 (2285)*
	Piper	*7525 (2295)*
	Sgiath	*7660 (2335)*
Middle Jurassic	*Pentland*	*7730 (2360)*
Triassic	*Smith Bank*	*8310 (2535)*
Permian	*Halibut Bank*	*8525 (2600)*
Carboniferous	*Forth*	*8920 (2720)*

**Approximate depth at a location on structure near the production platform.*

Reservoir characteristics:

Number of reservoirs *1*
Formations .... *Piper and Sgiath formations sandstones*
Ages .... *Late Jurassic*
Depths to tops of reservoirs .... *7250 ft (2211 m) subsea*
Gross thickness (top to bottom of producing interval) .... *115–453 ft (35–138 m)*
Net thickness—total thickness of producing zones
Average .... *155 ft (47 m)*
Maximum .... *335 ft (102 m)*
Lithology
Fine- to coarse-grained, predominantly medium- to coarse-grained, moderate to well-sorted, subrounded quartz sand and sandstone
Porosity type .... *Intergranular modified by quartz overgrowths*
Average porosity *24%*
Average permeability .... *4000 md*

Seals:

Upper
Formation, fault, or other feature .... *Kimmeridge Clay and Flounder formations*
Lithology .... *Shale, marl*
Lateral
Formation, fault, or other feature .... *Dip closure with Kimmeridge Clay above*
Lithology .... *Shale*

Source:

Formation and age .... *Kimmeridge Clay Formation*
Lithology .... *Organic-rich black shale*
Average total organic carbon (TOC) *8%*
Maximum TOC *15%*
Kerogen type (I, II, or III) .... *I*
Vitrinite reflectance (maturation) .... $R_o = 0.6$
Time of hydrocarbon expulsion .... *Early to middle Tertiary*
Present depth to top of source .... *11,900 ft (3630 m)*
Thickness .... *955+ ft (290 m)*
Potential yield .... *0.32 bbl/m³**

**Total potential yield approximately equal to 32 × 10⁹ bbl for WGG to Central graben based on 320 km² area and 300 m thickness; about 15% effective expulsion and entrapment required to equal STOIP in WGG area*

Appendix 2. Production Data

Data current to 1987. The disaster in July 1988 caused the field to be shut in. Production is scheduled to resume in mid-1992 from a new platform.

Field name ... *Piper field*

Field size:

- **Proved acres** ... *7350 ac (2974 ha)*
- **Number of wells all years** ... *60*
- **Current number of wells** ... *37*
- **Well spacing** ... *300 ac*
- **Ultimate recoverable** ... *1000 million bbl oil*
- **Cumulative production** ... *800 million bbl oil as of 22 October 1987*
- **Annual production** ... *60 million bbl*
- **Present decline rate** ... *30% (field has just commenced decline)*
- **Initial decline rate** ... *Nil*
- **Overall decline rate** ... *Nil*
- **Annual water production** ... *60,000 bbl/d*
- **In place, total reserves** ... *1390 million bbl oil*
- **In place, per acre foot** ... *1500 bbl/ac-ft*
- **Primary recovery** ... *800 million bbl oil*
- **Secondary recovery** ... *
- **Enhanced recovery** ... *

**Water injection has been used since the first year of production to supplement aquifer influx and maintain field pressure more than 1000 psi above the bubble point. Gas-lift, high-volume submersible pumps and infill drilling have been used to improve recovery.*

- **Cumulative water production** ... *170 million bbl*

Drilling and casing practices:

- **Amount of surface casing set** ... *900± ft (274 m)*
- **Casing program**

20-in. 900 ± ft (274 m) (400 ft [122 m] below seabed); 13⅜-in. to 3200± ft (976 m) subsea (top upper Paleocene sand); 9⅝-in. into Valhall above top Piper sand (7100–8500 ft [2166–2592 m] subsea); 7-in. liner to total depth

- **Drilling mud** ... *Gel polymer in 12¼-in. hole, KCl polymer through reservoir*
- **Bit program** ... *17½-in. hole (Tricone); 12¼-in. 70/30 Stratapax/Tricone; 8½-in. 50/50 Stratapax/Tricone*
- **High pressure zones** ... *None*

Completion practices:

- **Interval(s) perforated** ... *Selective completion of producing and injection wells to balance offtake for each sand member*
- **Well treatment** ... *Some injection wells acidized*

Formation evaluation:

- **Logging suites** ... *ISF-sonic-GR, dual LL-MSFL-GR, CNL-FDC-GR, HDT-GR, CBL, TDT, PCT*
- **Testing practices** ... *RFT in separate sand units; wells flowed through test separator weekly*
- **Mud logging techniques** ... *Full mud logging capabilities from Upper Cretaceous to TD; 10 ft sampling*

Oil characteristics:

- **Type** ... *Naphtheno-paraffinic*
- **API gravity** ... *37°*
- **Initial GOR** ... *446 ft^3 gas/bbl oil*
- **Sulfur, wt%** ... *1*

Viscosity, SUS .... *38.7 cs at 50°C*
Pour point .... *−6°C*
Gas-oil distillate .... *NA*

Field characteristics:

Average elevation .... *8000 ft (2440 m) subsea*
Initial pressure .... *3700 psi at −8100 ft (255.3 bar at 2470 m)*
Present pressure .... *3000 psi (207.0 bar)*
Pressure gradient .... *0.46 psi/ft (10.3 kPa/m)*
Temperature .... *175°F (95°C)*
Geothermal gradient .... *0.017°F/ft (0.031°C/m)*
Drive .... *Natural water plus water injection*
Oil column thickness .... *1262 ft (385 m)*
Oil-water contact .... *8512 ft (2596 m) subsea*
Connate water .... *1± to 10±%*
Water salinity, TDS .... *75,000 ppm*
Resistivity of water .... *0.045 at 175°F (95°C)*
Bulk volume water (%) .... *0.25 to 2.5%*

Transportation method and market for oil and gas:

Pipeline to Flotta Terminal in Orkney Islands and offshore loading to tankers; gas to MCP07 by pipeline and pipeline to St. Fergus, Scotland

Ramadan Field—Egypt
Gulf of Suez Basin

A. SHAWKY ABDINE
WAFIK MESHREF
A. NABIL SHAHIN
PAUL GAROSSINO
SALAH SHAZLY
Gulf of Suez Petroleum Company
Cairo, Egypt

FIELD CLASSIFICATION

BASIN: Gulf of Suez
BASIN TYPE: Rift
RESERVOIR ROCK TYPE: Sandstone
TRAP TYPE: Tilted Fault Block
RESERVOIR AGE: Carboniferous and Cretaceous
PETROLEUM TYPE: Oil
RESERVOIR ENVIRONMENT OF DEPOSITION: Braided Stream, Deltaic, and Shallow Marine
TRAP DESCRIPTION: Complex of tilted fault blocks with truncated reservoirs sealed above by shale, limestone, and evaporite

LOCATION

Ramadan field covers some 2850 ac (11.5 km^2; 4.5 mi^2) in the offshore, central province of the Gulf of Suez basin, some 174 km (108 mi) to the southeast from Suez City (Figure 1). The field is estimated to contain several hundred million barrels of oil as ultimate recoverable reserves, making it the fourth largest oil field in Egypt. The field is located on a gulf parallel (clysmic trending) tilted fault block system, generated during two pulses of rifting. Within the same area, and situated on similar structures, are the July and El-Morgan oil fields which, together with Ramadan field, account for over 39% of the oil reserves in Egypt. The field is operated by the Gulf of Suez Petroleum Company (GUPCO), which is a joint venture between the Egyptian General Petroleum Corporation (EGPC) and AMOCO Egypt Oil Company (a subsidiary of AMOCO Production Company).

HISTORY

Pre-Discovery

In 1964, aeromagnetic and seismic surveys were conducted over several concessions assigned to the Pan American UAR Oil Company (now AMOCO Egypt Oil Company). Two mappable seismic horizons "M" and "C" were identified (Brown, 1978). These coincide, respectively, with the upper and lower limits of the Miocene evaporite section. Several structural anomalies were mapped using these horizons (Figure 2). In 1965, the well El Tor-1 (later renamed Morgan-1) was spudded on the "C" structure and became the discovery well for El-Morgan field. Later that year, the well Alef-1 was spudded on the crest of the "A" structure (Figures 3 and 4), reaching a total depth of 3784 m (12,412 ft). It was abandoned in the Nubia "A" (Cretaceous) sandstones with no hydrocarbon indications in that zone. In 1966, the "J" structure (Figure 2) was tested by the well J-1, which encountered uncommercial quantities of oil within the lower Miocene Nukhul Formation and the pre-Miocene Cretaceous sands of the Senonian, Turonian, and Cenomanian (Figure 5). Two additional wells were drilled on the structure before the well GS 311-1, later renamed July-4 (J-4), was spudded in May 1973 and encountered the lower Rudeis (lower Miocene) sand reservoir of July field.

Discovery

In May 1974 the well GS 303-1 (Figure 2), renamed Ramadan-1 (R-1), was spudded by GUPCO. The well R-1 ultimately encountered 415 m (1361 ft) of net oil pay extending from 3018 to 3674 m (9900–12,050 ft) across several pre-Miocene formations. Within these, 338 m (1110 ft) of pay was recorded within the Nubia "C" massive sandstones, 55 m (180 ft) within the Senonian, Turonian, and Cenomanian sandstones (Matulla, Wata, and Raha formations,

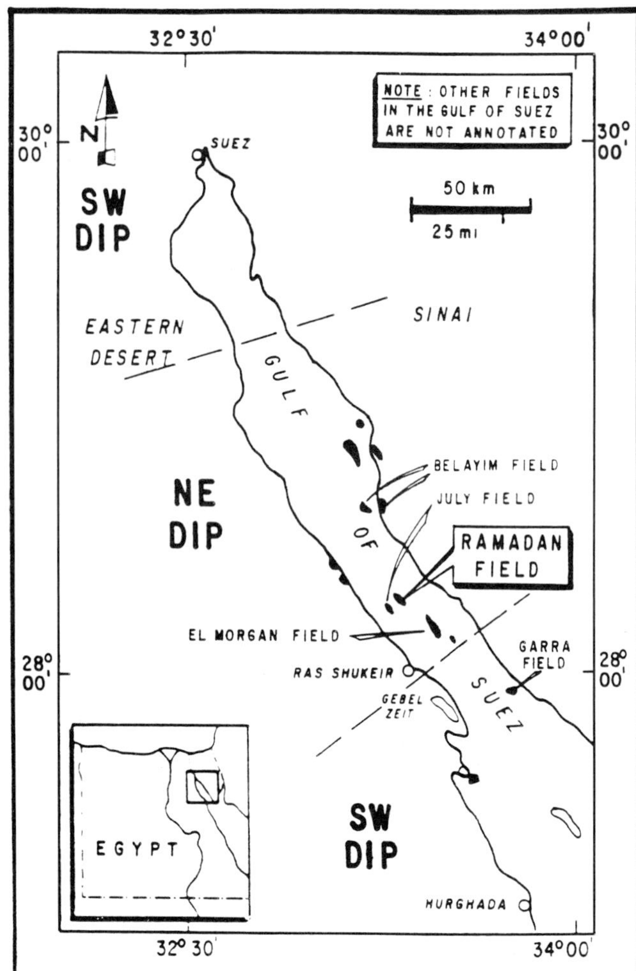

Figure 1. Ramadan field is located in the central Gulf of Suez in a northeasterly dipping province that separates two southwesterly dipping provinces.

respectively), and 22 m (71 ft) from within the early Cenomanian Nubia "A" sandstone. Ramadan's main pay zone, the Nubia "C" sandstone, was put on production in December 1974 at an initial rate of 25,000 barrels of oil per day (BOPD).

Post-Discovery

In April 1983, production from within the Nezzazat Group (Senonian, Turonian, and Cenomanian age) was tested at a rate of 2422 BOPD. Production from the Nubia "A" sandstone was first tested in November 1986 at a rate of 3200 BOPD.

To date, 51 wells have been drilled over the field area, resulting in a cumulative production of 360 million barrels of oil (MMBO). This drilling has delineated the northern, eastern, and western limits of the field. The southern limit has yet to be reached by the development program, though recent mapping has identified its potential location.

DISCOVERY METHOD

Just to the east of July field, interest was being regenerated in the "A" structure (Figure 2). Although Alef-1 (Figure 2) was a dry hole, dipmeter data indicated a northeast dip in the pre-Miocene section which, if present far enough to the west, could have resulted in a commercial accumulation of hydrocarbons structurally high to the Alef-1 well location. In 1971 and 1972, additional seismic data were obtained which, together with much improved seismic processing techniques, allowed the interpretation of deeper pre-salt structure. Seismic processing also verified that Alef-1 had been drilled downdip on a northeasterly dipping pre-Miocene fault block. The bounding fault of the structure was subsequently mapped at about 3.2 km (2 mi) farther west from the Alef-1 well location (Figures 4A and 4B).

STRUCTURE

Tectonic History

Following the layer-cake deposition of the pre-Miocene sediments, the entire area was subjected to the first phase of rifting during the Oligocene, accompanied by uplifting and tilting of faulted blocks and followed by Oligocene–Miocene crestal erosion. This was followed by a transgressive phase that resulted in the deposition of the Miocene clastics and evaporites. A second phase of rifting started during the late Miocene, resulting in the rejuvenation of the older faults.

Geologic Evolution

Initially the Ramadan pre-Miocene structure was shown bounded by only one fault system (Figure 6) extending from south to north of the field. Recently it has been discovered that two fault systems, of very different ages and histories, combine to form the field-bounding fault complex resulting in three fault zones of unique structural style (Figure 7). In the "Northern Zone," the two systems "A" and "B" are separate, with the eastern fault "B" bounding the field. Over the central "Transfer Zone," a portion of the throw from the "B" system was transferred through a series of cross-elements to the "A" system, which trends toward the field from the northwest. In the "Southern zone," the two fault systems have combined to form the western-bounding fault "AB" of the field. Since the geologic evolution of the northern and southern zones of the field have been so different, a separate analysis has been generated for each.

Evolution of the Northern Zone

Prior to the late Eocene, the entire Ramadan area was covered by a flat-lying blanket of sediments from the Carboniferous Nubia sandstones through the

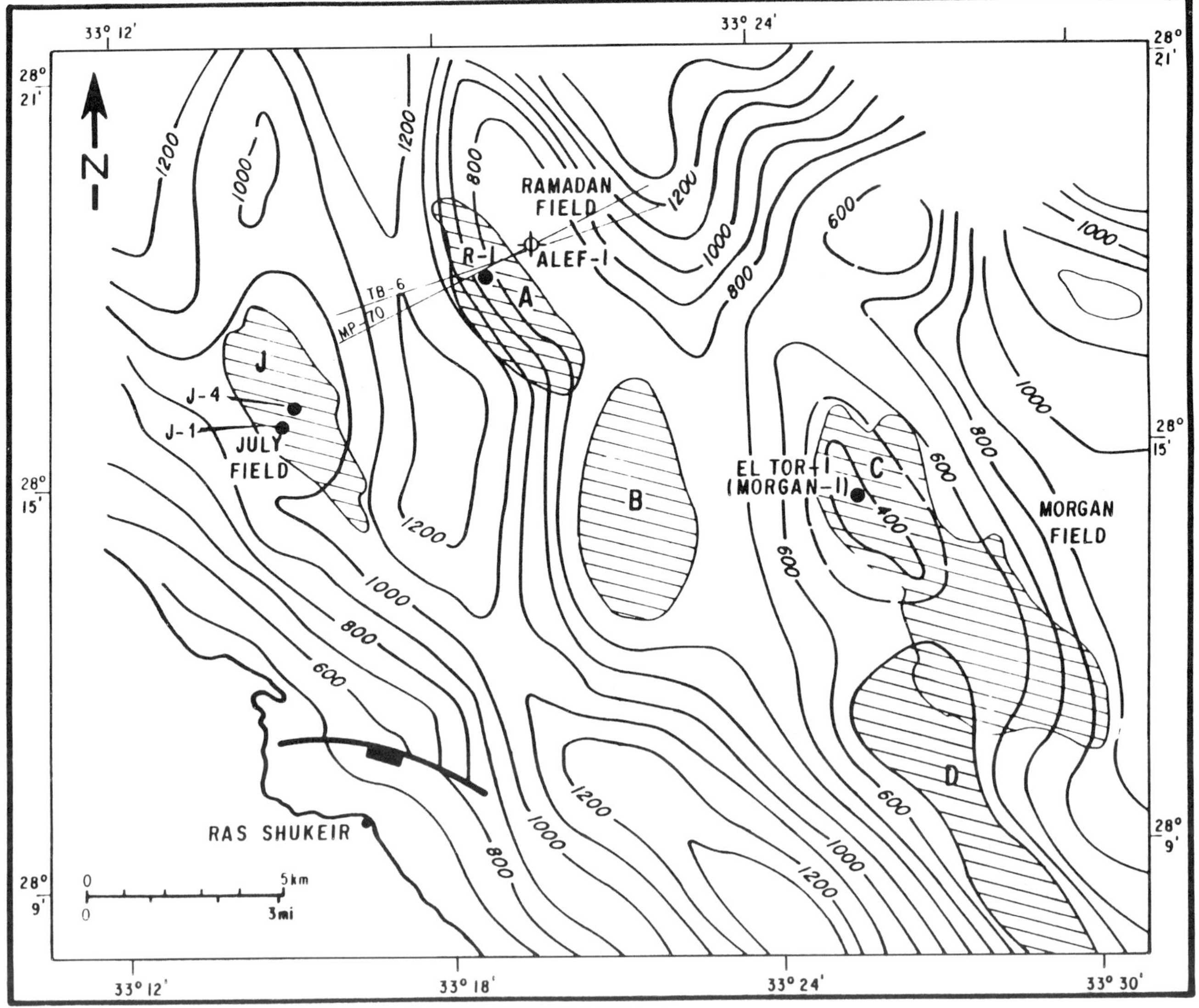

Figure 2. The "M" horizon (see Figure 3B) seismic structure map represents the top of the Miocene evaporites. Several structural anomalies are shown at A, B, C, D, and J structures (hachured). Contours are in meters below sea level. Contour interval is 100 m (328 ft).

Eocene Thebes Formation (Figure 8-1). The sedimentary record from the upper Eocene to the lower Miocene is missing in Ramadan, making it possible that rifting could have begun over this area as early as late Eocene. Oligocene-age volcanics are present to the north of this area, indicating that the initial phase of rifting occurred when the major Clysmic (Gulf of Suez trend) "A" fault system and many associated cross-elements (faults that are normal to Clysmic) were formed (Figure 8-2). The "A" fault system incurred in excess of 610 m (2000 ft) of displacement, imparting a 10° dip to the upthrown pre-Miocene section. During the 15 to 20 million years available from late Eocene to early Miocene, crestal erosion occurred along this fault system (Figure 8-3). The cross-elements that formed during the initial rift event tend to have maximum throw near the crests of upthrown rocks, with diminishing throw downdip and terminating in regionally dipping pre-Miocene strata.

The basal Miocene Nukhul Formation onlapped the pre-existing pre-Miocene structure some 3.5 km (2.2 mi) northeast from fault "A" (Figure 8-4). As about 30 m (100 ft) of Nukhul sediments are present in the well R-19 (Figure 7) at the bounding fault on the first block south of this zone, a cross-element (normal to the Gulf trend) must have existed near the crest of fault "A" with a magnitude in excess of 457 m (1500 ft), down to the south. This fault separated the 10° dip of the northern block from the 4° dip over the rest of the field and diminished to the east where the pre-Miocene dip was relatively constant. An incipient "B" fault system may have been present at this time separating the 10° dip regime, upthrown to the bounding fault, from the approximately 8° regional dip to the east. As Miocene

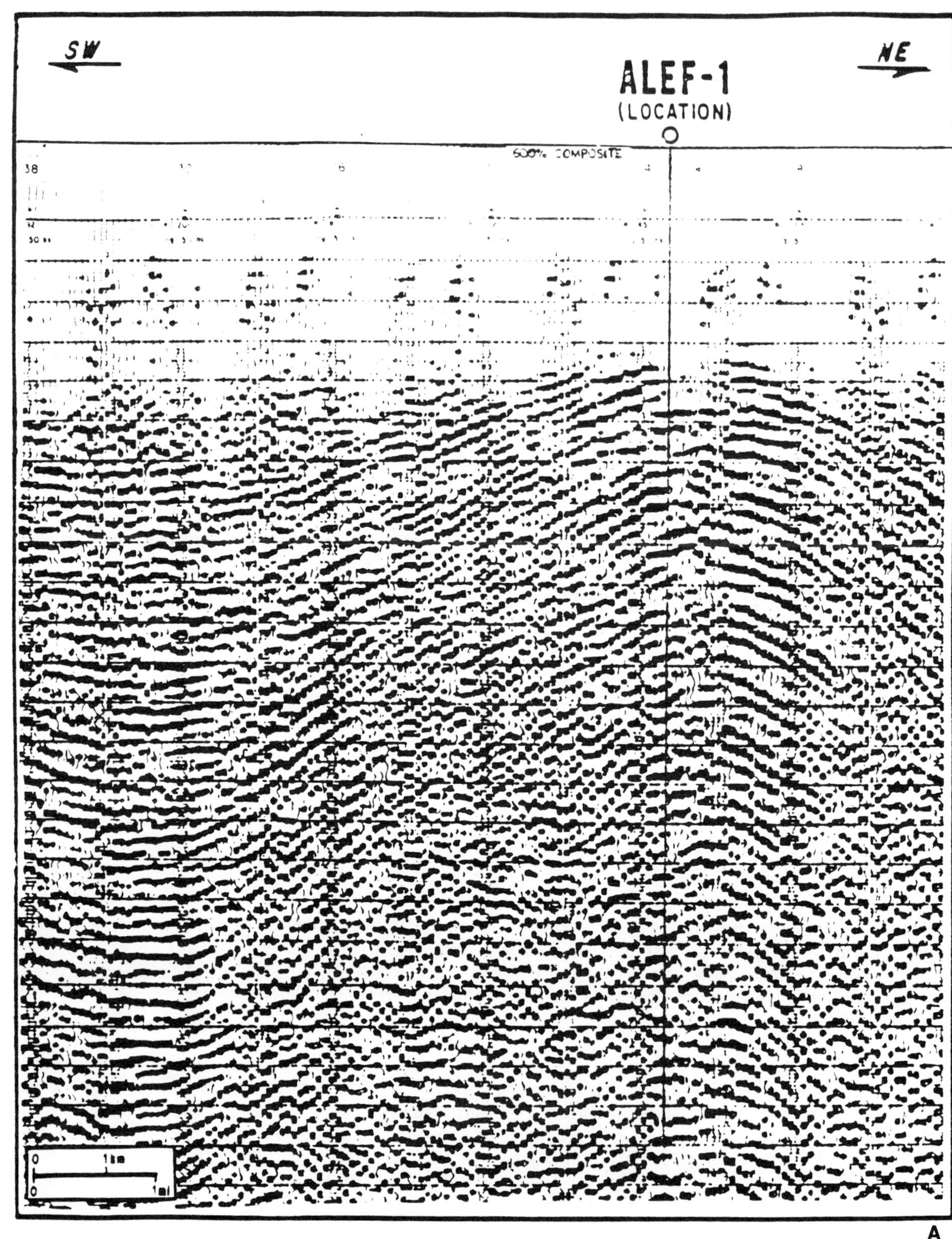

Figure 3. Seismic line TB-6 (see Figure 2 for location), drill line for well Alef-1 to test a crestal position of an upthrown side of a Clysmic (Gulf of Suez trend) normal fault on the "A" structural anomaly (Figure 2). (A) Uninterpreted. (B) Interpreted.

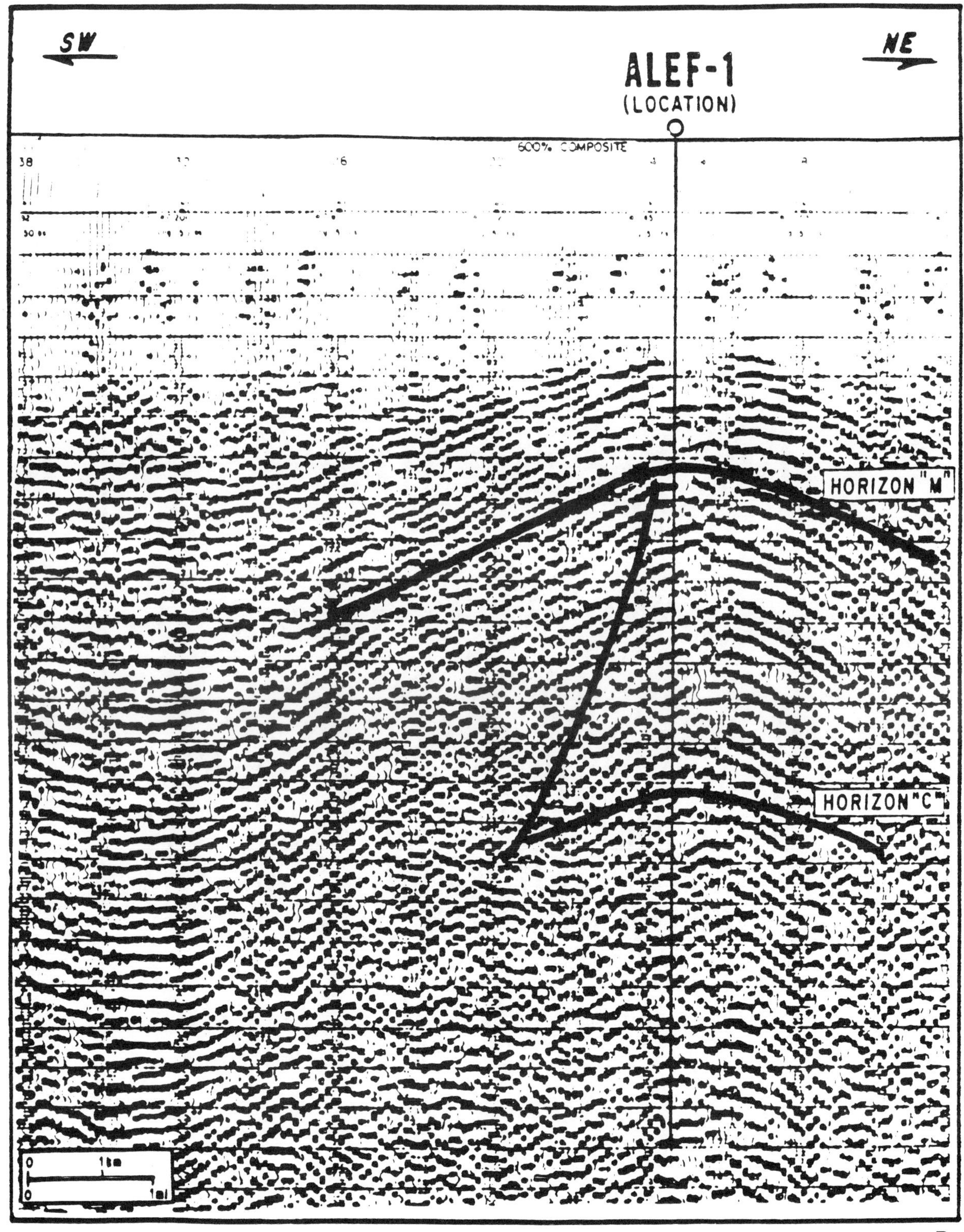

B

Figure 3. (Continued)

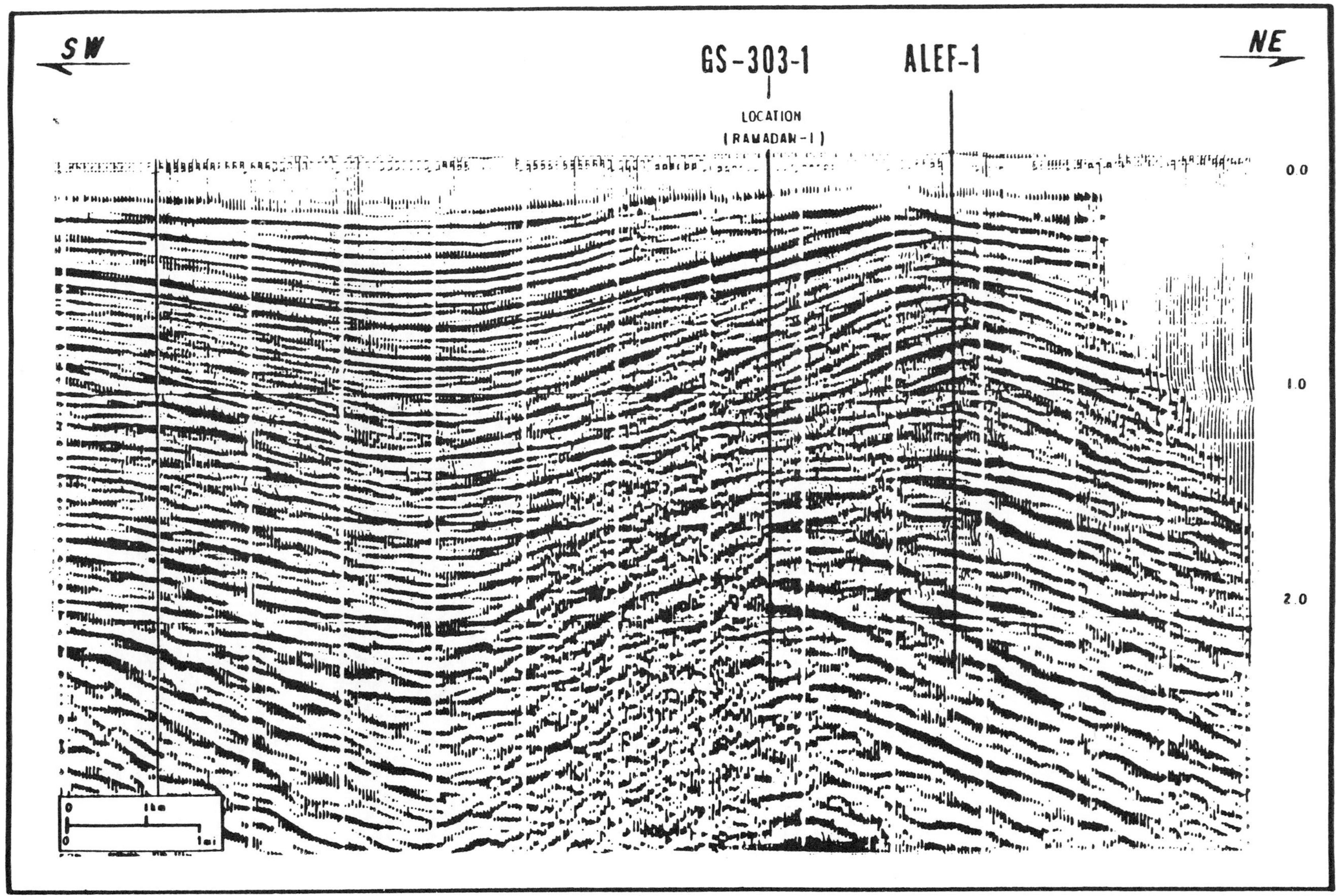

Figure 4. Seismic line MP-70 (see Figures 2 and 7 for location), drill line for well GS 303-1 (the discovery well, Ramadan-1), showing that well Alef-1 was drilled downdip on a northeasterly dipping pre-Miocene fault block. (A) Uninterpreted. (B) Interpreted (1974 processing and interpretation).

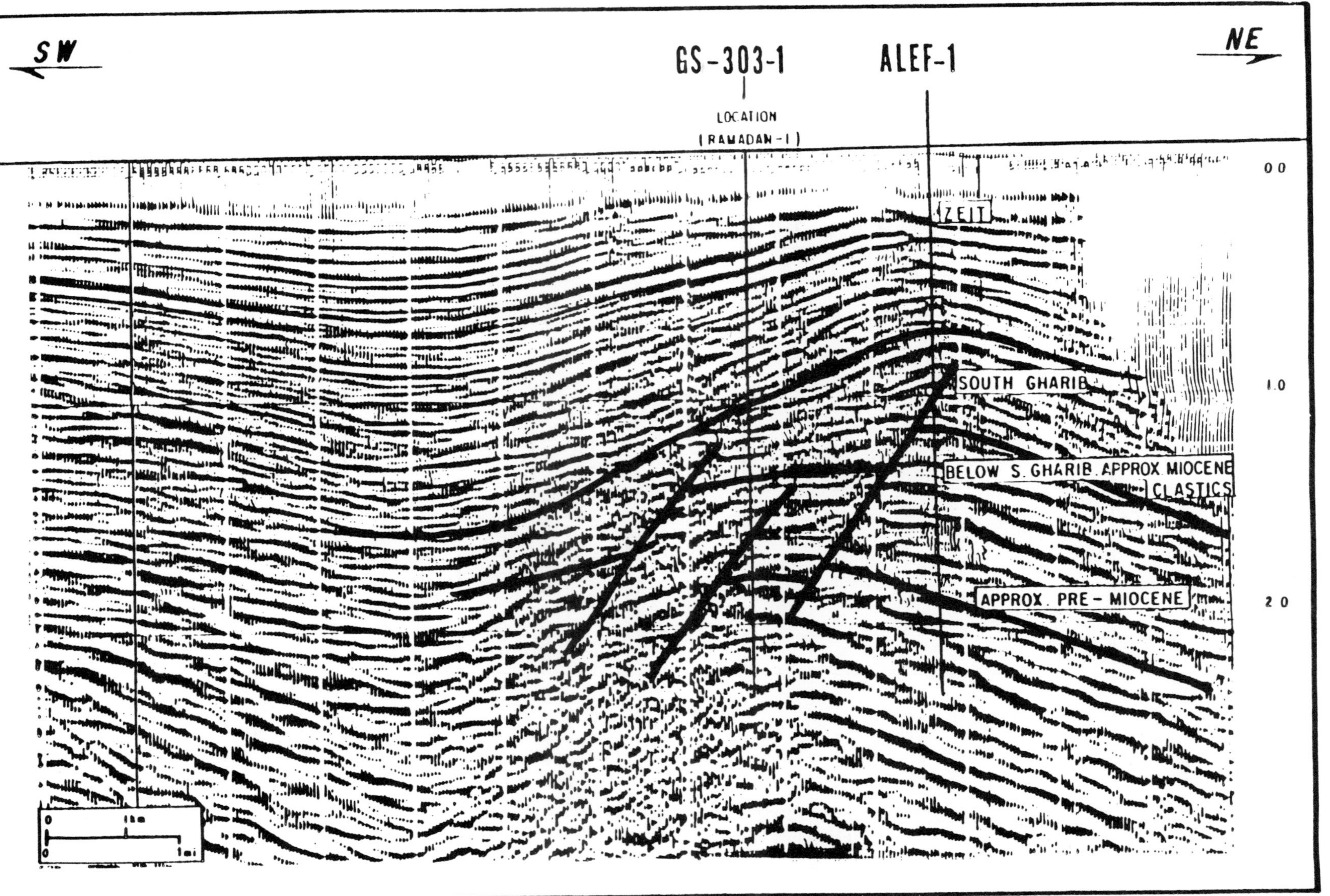

Figure 4. (Continued)

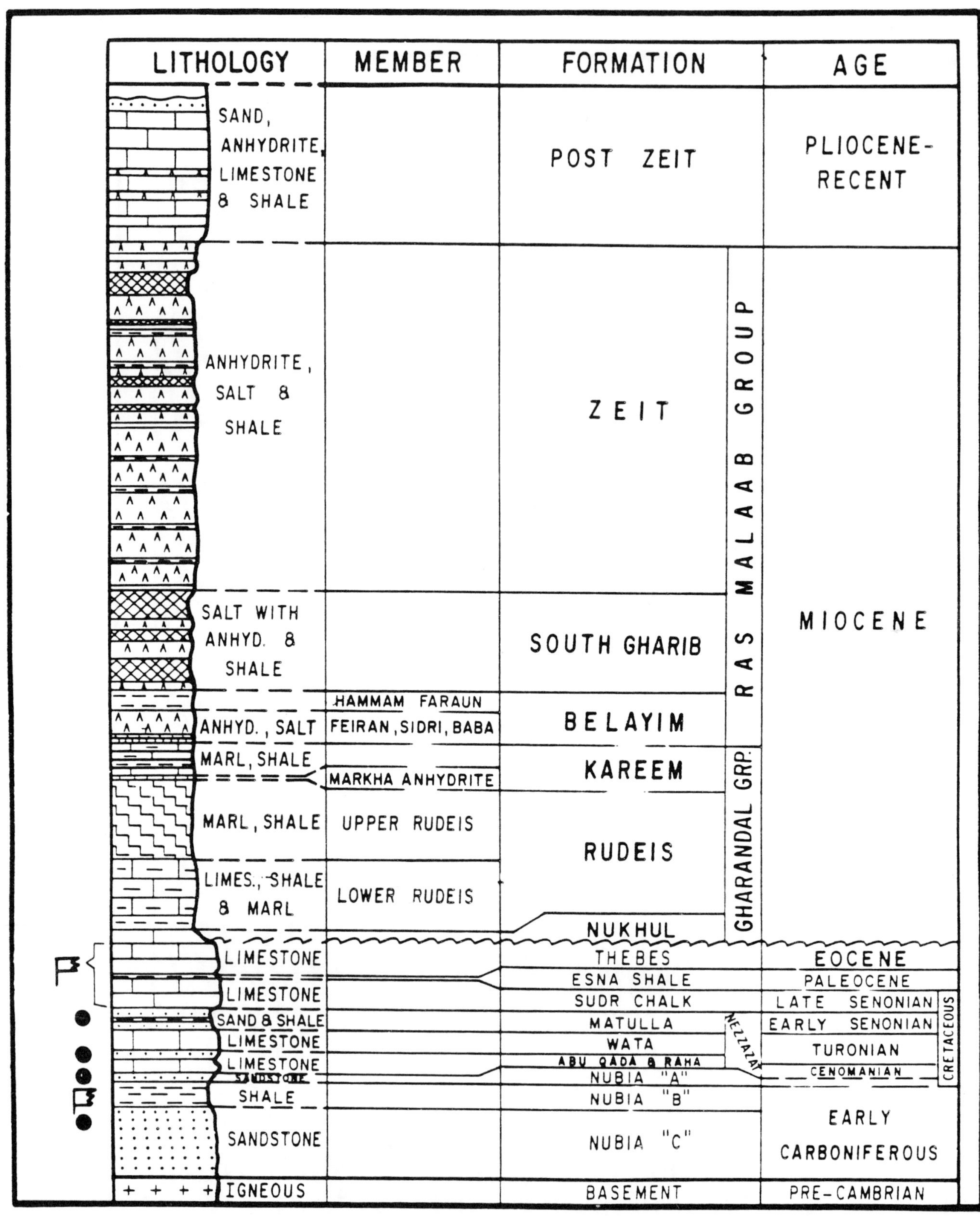

Figure 5. Stratigraphic column of Ramadan field area, typical of the central Gulf of Suez. Reservoir rocks (solid circles) are the limestones and sandstones of the Nezzazat Group and the Nubia "A" and "B." Possible source rocks (flags) are the organically rich limestones and shales of the Eocene Thebes Formation, the Paleocene Esna shales, the Campanian Sudr Formation, and the Carboniferous Nubia "B."

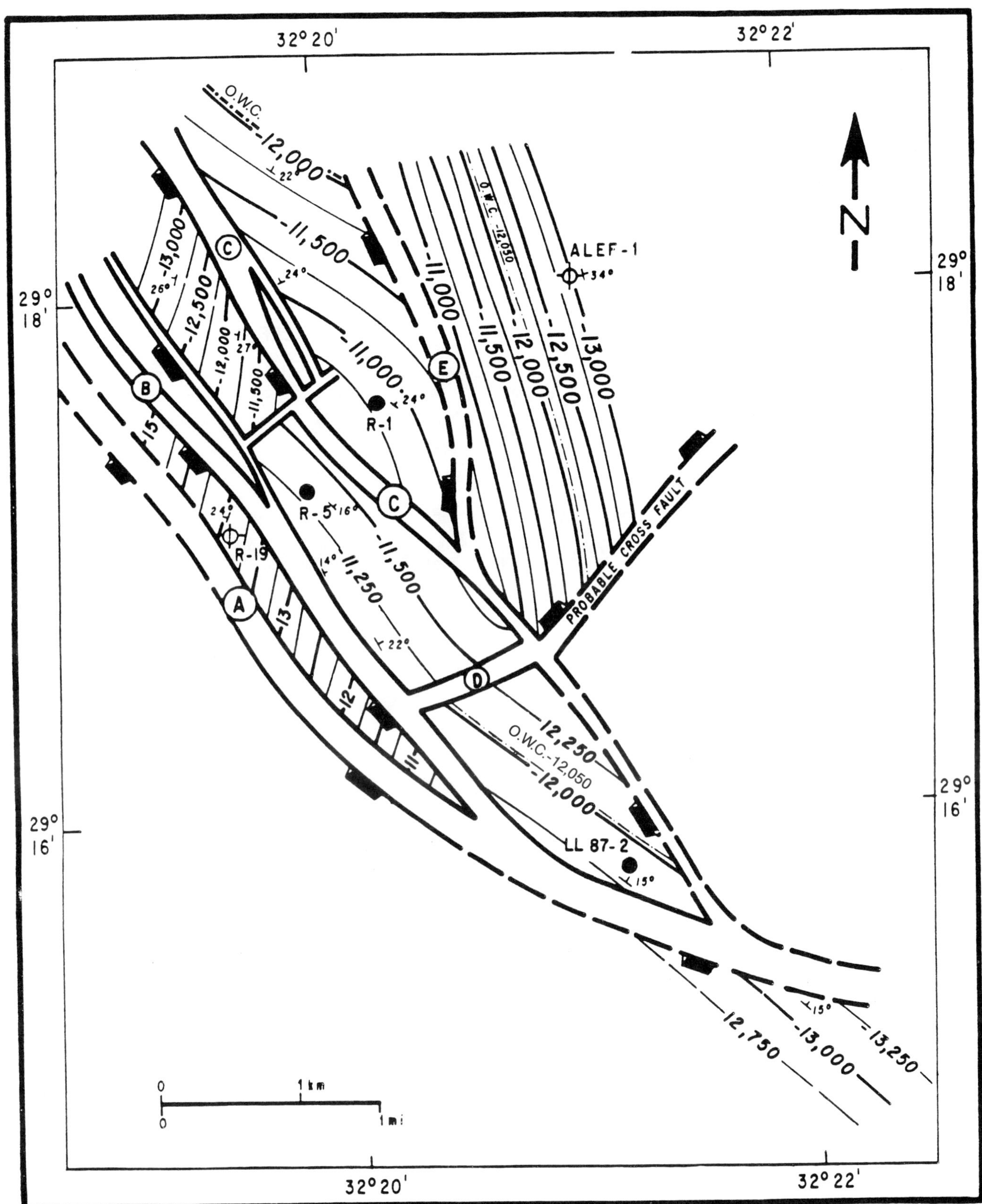

Figure 6. Top of the Nubia "C" structure contour map of Ramadan field (Brown, 1978) showing that the pre-Miocene faulted tilted block is bounded from south to north by one fault system "A." This system was initiated during the main rifting as shown by the fact that it is parallel to the Gulf of Suez trend (clysmic). Oil-water contact (OWC) is -12,050 ft (-3073 m). Contour interval is 250 ft (76.2 m).

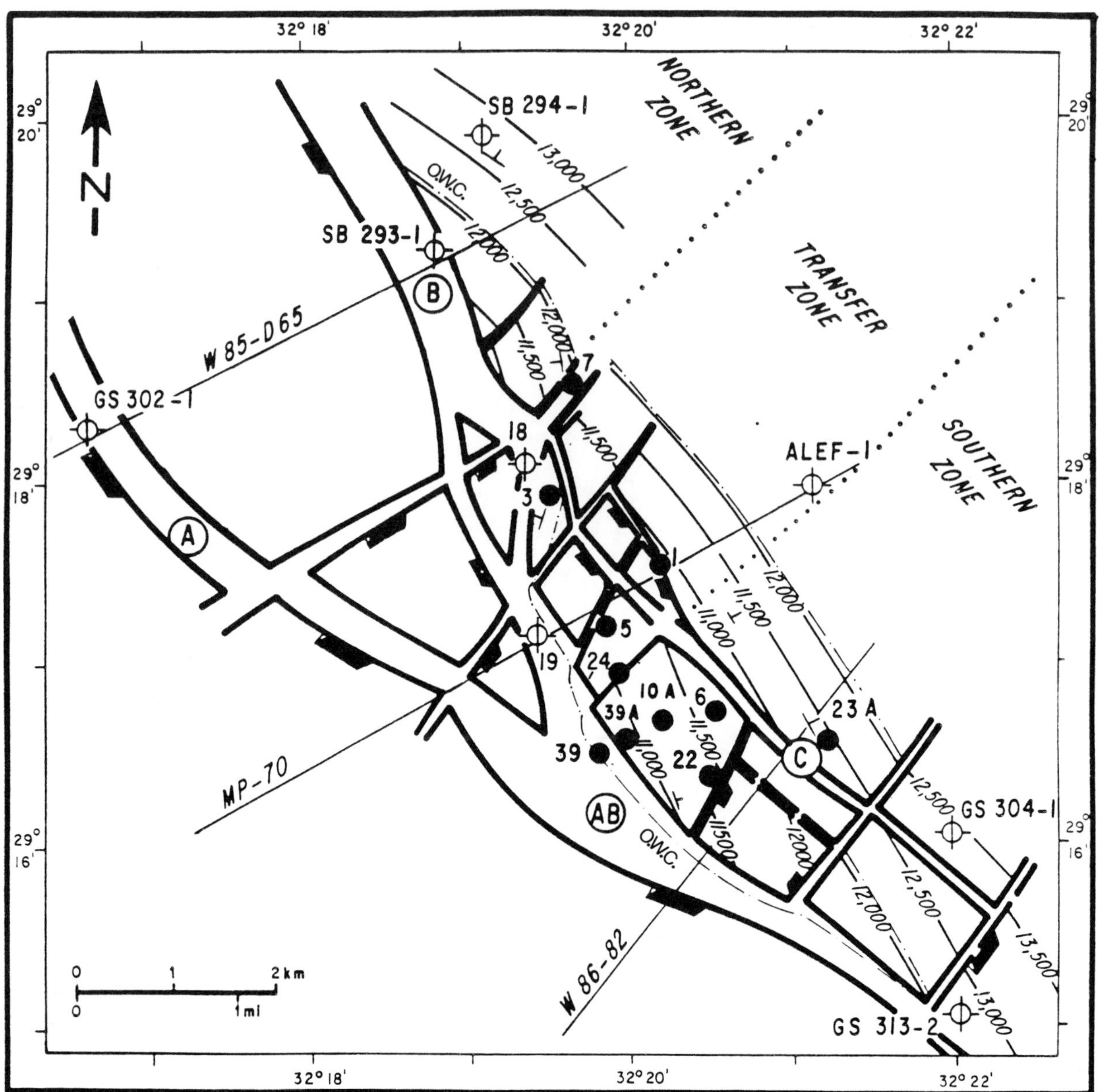

Figure 7. Top of the Nubia "C" structure contour map showing that the pre-Miocene faulted and tilted block of the Ramadan field is bounded by two major fault systems (A and B) of different ages (rifting and pre-rifting, respectively). The two systems combine in one system (AB) toward the south via a Transfer Zone that separates the Northern Zone from the Southern Zone with markedly different histories. Contour interval is 500 ft (152.4 m). OWC is -12,050 ft (-3073 m).

sedimentation continued, the first sediments to completely cover the upthrown block were laid down sometime between deposition of the Miocene clastics (Rudeis Formation) and the Markha anhydrites of the Kareem Formation (Figure 5). Sometime after deposition of the Belayim Formation, a second pulse of rifting occurred, resulting in an additional 10° rotation of the block formed between the "B" and "A" systems and allowing for over 610 m (2000 ft) of displacement on the "B" fault. It was during this rotation that the Miocene fault system formed over fault "B," which extended up to but not completely through the Belayim Formation (Figure 8-5). The Belayim surface was seen to drape over the underlying structure.

Evolution of the Southern Zone

As in the northern zone, prior to late Eocene the stratigraphic section was basically flat-lying (Figure 9-1). Sometime during late Eocene the first pulse of rifting occurred, resulting in the initiation of about 550 m (1800 ft) displacement on the "A" fault system and probably the initiation of the incipient "C" fault system. The pre-Miocene dip of the upthrown block

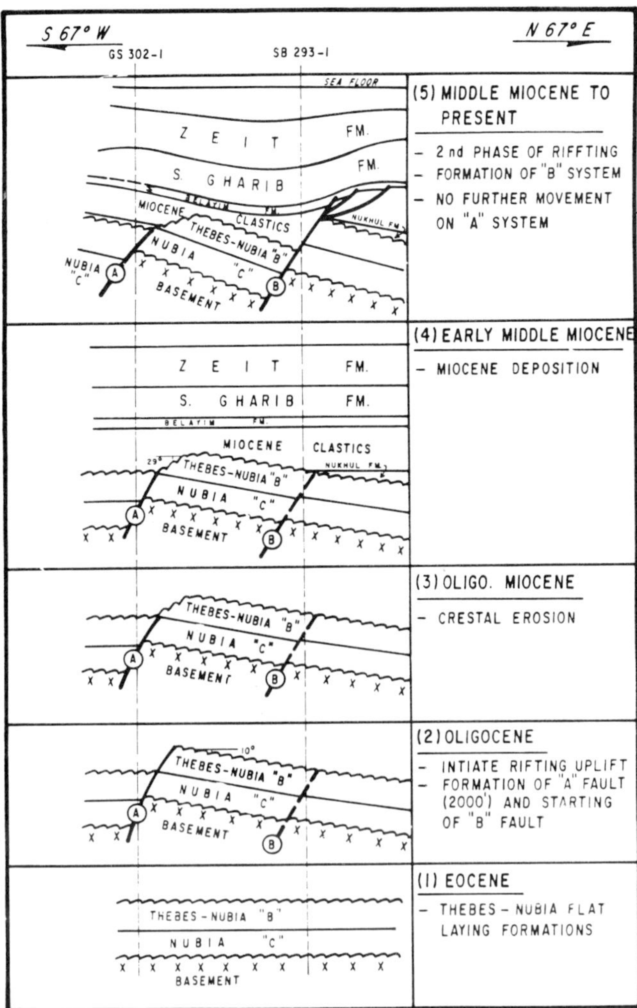

Figure 8. The geologic evolution of the Northern Zone of Ramadan field. No post-Oligocene movement is associated with fault A, so there has been no deformation of the overlying Belayim Formation in this zone.

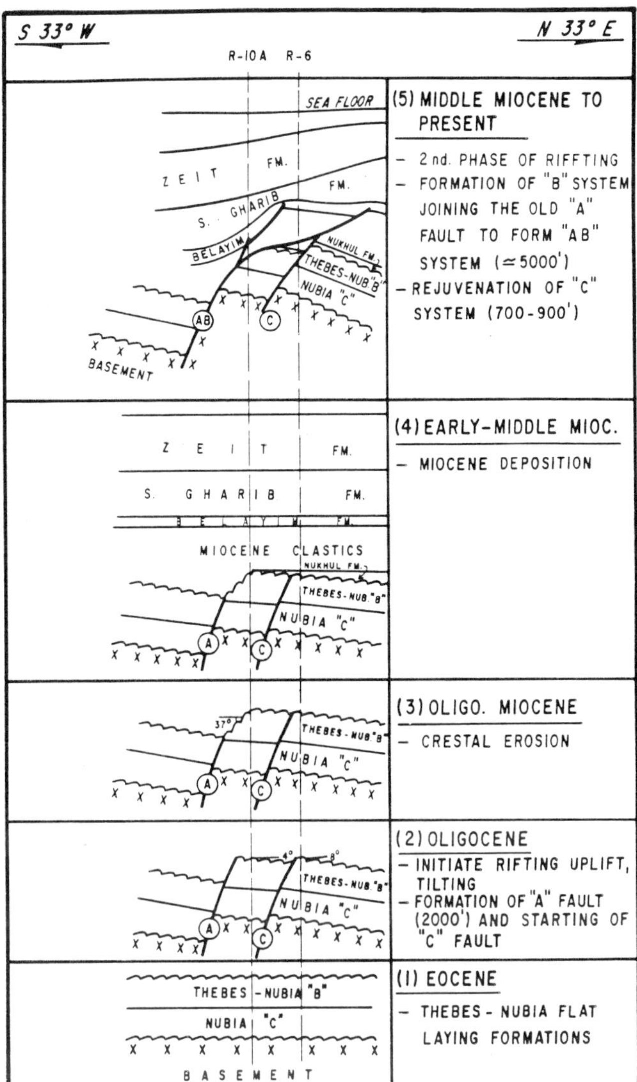

Figure 9. The geologic evolution of the Southern Zone of Ramadan field. Here the rejuvenation of fault A has resulted in a significant structural displacement at the Belayim surface.

was in the order of 4° toward the northeast to the west of the "C" fault and as much as 8° to the east (Figure 9-2). In the time elapsed between the formation of the "A" system and the deposition of the lower Miocene sediments, crestal erosion along the "A" system resulted in the development of a pre-Miocene erosional scarp at that location. The angle of that scarp surface with respect to the horizontal would have been 37° (Figure 9-3).

During the early Miocene, the Nukhul Formation was deposited over the block, thickening over any existing cross-elements. The rest of the Miocene section, through the Belayim and possibly including some of the South Gharib Formation, was laid down (Figure 9-4). During the second phase of rifting, the "A" and "C" fault systems were rejuvenated. These systems each received a portion of the throw attributed to the "B" system of northern Ramadan. During this phase, an additional 10° of block rotation occurred over the western block of the field, resulting in a final scarp angle of 27° and a bounding fault angle of about 54° with respect to the horizontal (Figure 9-5).

During this second rotational phase, a Miocene fault system formed in much the same fashion as in the Northern Zone. The existence of the underlying pre-Miocene scarp allowed for a detachment surface to develop, over which the overlying Miocene sediments could have moved, thus resulting in the present-day fault relationship across that surface.

Present-Day Structural Setting

The following describes the present-day structural setting in each of the three recognized zones: Northern Zone, Transfer Zone, and Southern Zone.

Northern Zone Structure

The structure of the Northern Zone is dominated by the presence of two separate fault systems, "A" and "B" (Figure 7). The "B" system forms the main bounding fault of Ramadan field, while the "A" system bounds a more westerly block. The historical relationship of these two systems can best be discussed through a detailed examination of the schematic geo-seismic cross section (Figure 10). The line of section is constructed along seismic line W85-D65 (Figure 11), tying the wells GS 302-1 and SB 293-1 and intersecting both faults "A" and "B."

Fault "A" is a two fault system associated with at least 649 m (2130 ft) of missing section. The uppermost fault displays 67 m (220 ft) of missing section between the Matulla and Wata formations. The lower fault separates the Nubia "A" and Nubia "C" sandstones with a minimum of 584 m (1916 ft) of missing section.

The "B" fault system, encountered in the well SB 293-1, is associated with a minimum of 869 m (2850 ft) of missing section between the Miocene Rudeis Formation and the sands of the upthrown Nubia "C" sandstone.

The block between the "A" and "B" fault systems, as mapped on the pre-Miocene unmigrated time structure map (Figure 11), is considered to be continuous. The dip on this block, shown as 18° toward the northeast, is controlled geologically by the conformable basement contact and the "B" fault throw, estimated from the SB 293-1 wellbore data. Modeling of this dip fits well with the seismic data.

An angular unconformity is evident in the seismic data along the top of this pre-Miocene block (Figure 12). This is consistent with geologic data from the SB 293-1 wellbore where an intra-Miocene isopach is some 107 m (350 ft) thinner than the comparable isopach upthrown to fault "B," indicating onlap of the Miocene sediments to the pre-existing pre-Miocene structure. Further support for onlap is present on the Nukhul Formation isopach map (Figure 13), which shows the Nukhul Formation to thicken to the east.

Evidence to support the relative ages of the two fault systems is present in the relationship of the position of the present-day Belayim surface to the underlying pre-Miocene structural configuration. On the seismic data, a very reliable Belayim event is present between the two well control points, providing excellent control on the Belayim structure. Although both the underlying "A" and "B" fault systems display an excess of 610 m (2000 ft) of offset,

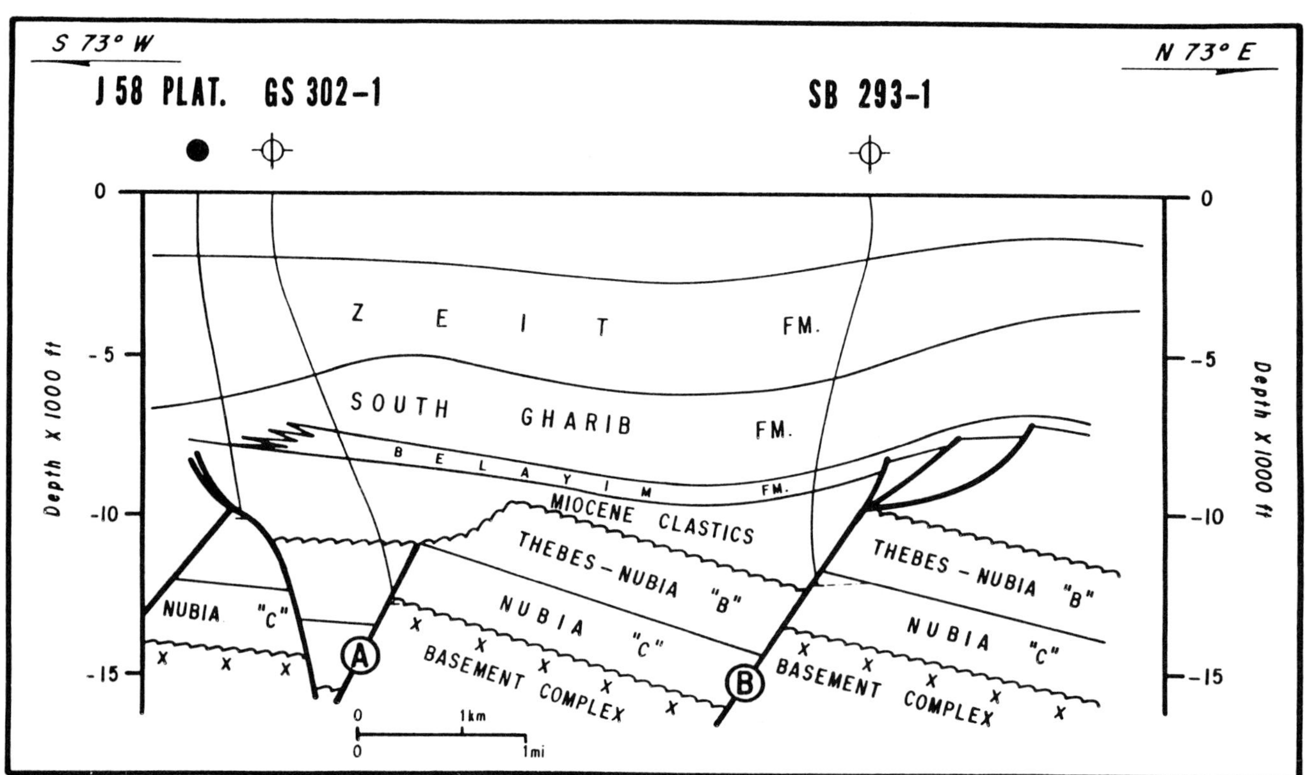

Figure 10. The geo-seismic cross section along seismic line W88-D65. Movement on the B fault system bounding northern Ramadan field has caused a significant structural flexure of the overlying Belayim Formation. No such event has occurred over fault A. This, together with the onlap of the Miocene clastics, indicates that the block to the east of fault A was a pre-existing "high" later buried during Miocene sedimentation. Movement on fault A probably pre-dated Miocene sedimentation and certainly pre-dated the major displacement along the fault B system.

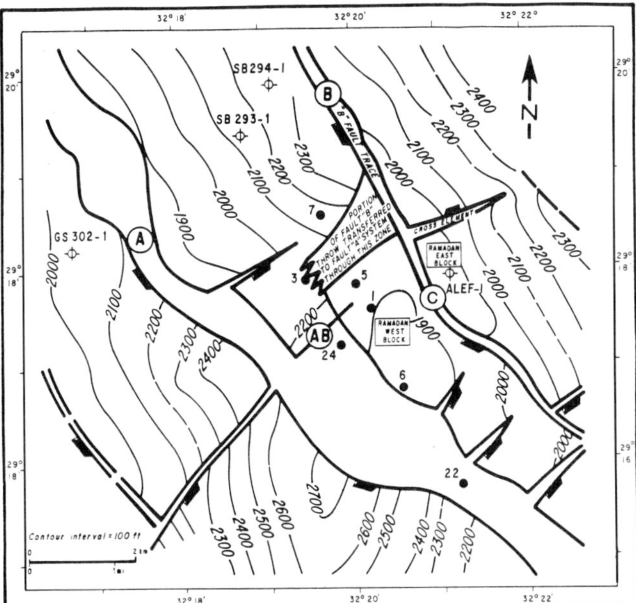

Figure 11. Pre-Miocene unmigrated time structure map showing that in the Northern Zone the block between the "A" and "B" fault systems is undisturbed, while in the Transfer Zone it is characterized by the presence of several cross-elements through which a portion of the throw of the "B" fault system is transferred to the "A" system; in the Southern Zone the "A" and "B" fault systems have come together to form the "AB" system. The remaining flow of the "B" system is accounted for by movement along the "C" fault.

the Belayim above the two faults behaves quite differently. Over fault "B," the Belayim horizon drapes some 610 m (2000 ft) from the crest of the Ramadan feature to the synclinal axis between July and Ramadan fields, whereas there is no expression whatsoever at the Belayim horizon of fault "A." Therefore, movement along fault "A" must have predated deposition of the Belayim Formation, whereas major motion along fault "B" had to occur after Belayim deposition. The Miocene onlap, discussed above, points to the existence of fault "A" prior to any Miocene sedimentation. Furthermore, any attempt to reconstruct the stratigraphic section above the basement requires the Belayim horizon to be some 610 m (2000 ft) higher in the section than observed over fault "A."

Transfer Zone Structure

This portion of the field is characterized by the presence of numerous cross-elements, through which a portion of the throw associated with the "B" fault system is transferred to the "A" system, forming the "AB" bounding fault system of southern Ramadan field. The remaining throw on the "B" system is accounted for by movement along the "C" fault system that trends into the "B" system of northern Ramadan and separates the eastern and western productive blocks of the field. A corner is formed in this zone where the "B" fault system intersects a series of cross-elements in which three downthrown blocks are trapped and where northwest dip is developed as a consequence.

This zone is typified by cross section MP-70 (Figure 14), named for the coincident seismic line (Figure 4) oriented along an azimuth of 58°. It traverses the central portion of the field, joining wells R-19A and Alef-1. In this section, the "A," "B," and "C" fault systems are cut by wells R-19A, R24-27A, and R1-29. Fault "A" is cut by the R-19A wellbore within the Miocene clastics, where 442 m (1450 ft) of section is missing. In well R24-27A, 976 m (3200 ft) of section is missing at the pre-Miocene contact, which represents a combination of components from both the "A" and "B" fault systems. The "C" fault system was encountered in the R1-29 wellbore where a total of 210 m (700 ft) of section are missing in the pre-Miocene section at the intersections with two faults. The "C" fault is terminated by a low-angle Miocene normal fault, which indicates that the "C" system was there prior to that Miocene structural failure. The small block defined between these two faults in the R1-29 wellbore displays 18° of westerly dip, indicating block rotation within the fault zone. Between the "A" and "B" fault systems, well R-19A encountered a block displaying 20° westerly dip. This block has been caught up between the "A" and "B" fault systems and rotated toward the west.

The Belayim Formation is shown to drape, not fault, over the underlying structure. This conclusion was based on seismic mapping of the Belayim horizon where no evidence of a fault was found. Geologic evidence for drape can be recognized in many crestal wells where a measured thickness of 845 m (2772 ft) of unrepeated Belayim section was penetrated, which display formation dips from 35° to 45° toward the west. When this information is combined with the well R-19A data, which show 18° to 20° of dip toward the southwest off the western flank of the structure, it becomes clear that no fault is required structurally at the Belayim level. There is a slight thickening of the Hammam Faraun and Feiran members on the downthrown side of the system that could be associated with growth during the later stages of Belayim deposition. However, it appears that the major movement affecting the Miocene section occurred after deposition of Belayim Formation.

The 61 m (200 ft) displacement fault associated with the "B" fault in the R24-27A and R24-28 wells is consistent with the displacement of the fault zone resulting from the proximity of the "A" and "B" fault systems. As these systems converge, the number of faults associated with the bounding fault is seen to increase. In well R6-39 to the south, five faults were associated with this zone.

Southern Zone Structure

It is within the Southern Zone that the "A" and "B" fault systems are merged to form the "AB" bounding fault (Figure 15). The resulting throw

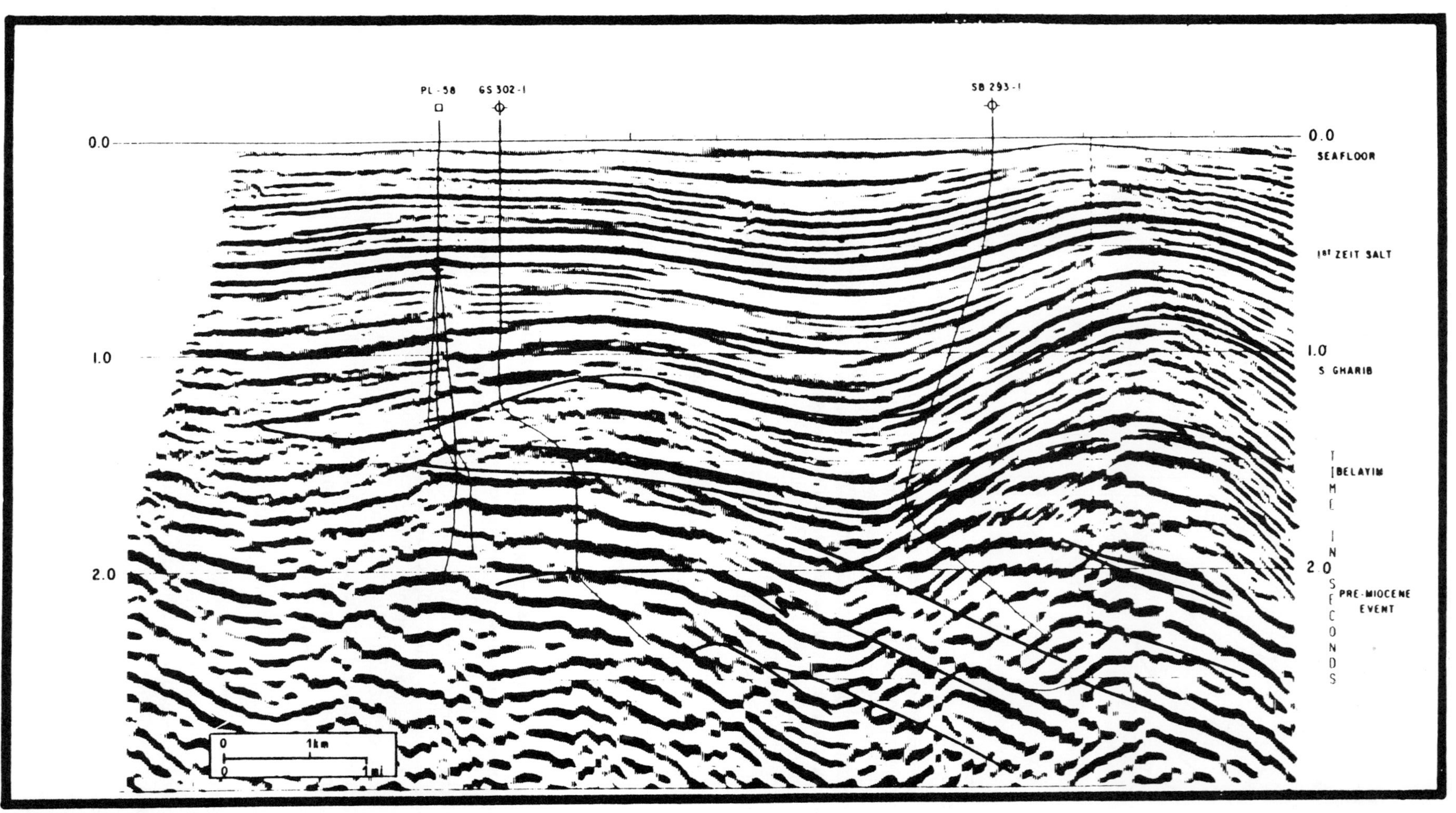

Figure 12. Interpreted unmigrated seismic line W88-D65 showing over 400 milliseconds of structural drape at the SB293-1 location in the Northern Zone of the Ramadan field and ramp dip up to the GS302-1 location over the fault A system.

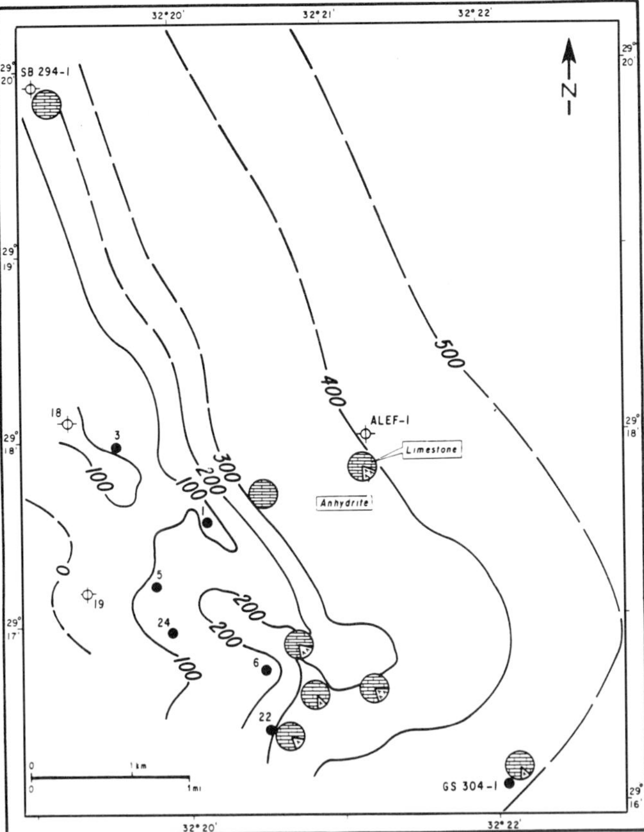

Figure 13. The basal Miocene Nukhul Formation isopach map. The Nukhul Formation is shown to thicken to the east, supporting an onlap on a northeasterly dipping pre-Miocene block. Contour interval is 100 ft (30.5 m).

within the system is in excess of 1524 m (5000 ft), as seen in well R6-39.

The pre-Miocene subcrop map (Figure 16) over Ramadan field shows a distinct difference between that of the Southern Zone, bounded by the "AB" system, and that of the remainder of the field. Over the Southern Zone, the pre-Miocene subcrop has a much wider expression in plan, indicating a lower angle with respect to the horizontal on the pre-Miocene surface. This is consistent with the assumption that the "A" system pre-dated Miocene sedimentation and that a significant erosional scarp was formed at its upthrown edge (Figure 9-5). From well control over this zone, the scarp is known to dip between 24° and 28° and to strike near 125°. The bounding fault has been measured at 54° along an azimuth of 127°. The real challenge over the Southern Zone of the field is to predict the location of the intersection of the bounding fault with the pre-Miocene erosional scarp, since this intersection defines the apex of the trap containing the Nubia "C" reserves.

GENERAL STRATIGRAPHY

The stratigraphic succession of Ramadan field is summarized in Figure 5. As noted above, the sandstones of the Nezzazat Group (Cenomanian to lower Senonian), Nubia "A" (Lower Cretaceous), and Nubia "C" (Lower Carboniferous) act as reservoirs while, as it will be shown in the geochemistry section below, the limestones and shales of the Thebes (Eocene), Esna (Paleocene), and Sudr (Campanian) formations and the Nubia "B" (Lower Carboniferous) shales display good source potential. In this field, the Miocene section from the Nukhul through Zeit formations is considered the ultimate seal for the underlying pre-Miocene reservoirs.

TRAP

Ramadan field is a complex of tilted fault blocks with reservoirs truncated and sealed by overlying impermeable beds. The field is contained in a northeasterly dipping regime (Figures 11 and 12) bounded to the west by a Clysmic trending fault system with between 854 and 1677 m (2800–5500 ft) of throw. The southern and northern limits of the field are also structurally controlled by northeasterly trending oblique-slip faults. The updip seals for all reservoirs are impermeable shales, limestones, and evaporites of Miocene age. The Nubia "C" reservoir is sealed vertically by the Nubia "B" massive shales, while the vertical seal for the Nubia "A" and Nezzazat reservoirs is probably massive limestone of the Sudr Formation.

Reservoir Characteristics and Facies

In Ramadan field, additional production has been established from the sandstones of the Nubia "A" and Nezzazat Group (Matulla, Wata, and Raha, respectively), but primary production is still mainly from the Nubia "C" sandstone.

Work is just beginning toward a better understanding of the Nezzazat and Nubia "A" reservoirs. Neither reservoir has yet been mapped nor adequately examined from a petrographical or petrophysical standpoint to provide enough detail to support a drilling process. The Nubia "C" reservoir, on the other hand, has received more attention, thereby allowing a relatively comprehensive reservoir description.

The factors affecting permeability within the Nubia "C" reservoir have been studied by GUPCO. These factors include the presence of authigenic clay minerals (Figure 17), asphaltic matter (Figure 18), and silica overgrowths (Figure 19) within the pore space, as well as the effects of compaction. In the central field, the presence of authigenic clay minerals has greater effect on porosity and permeability than does the presence of asphaltic matter. This is due

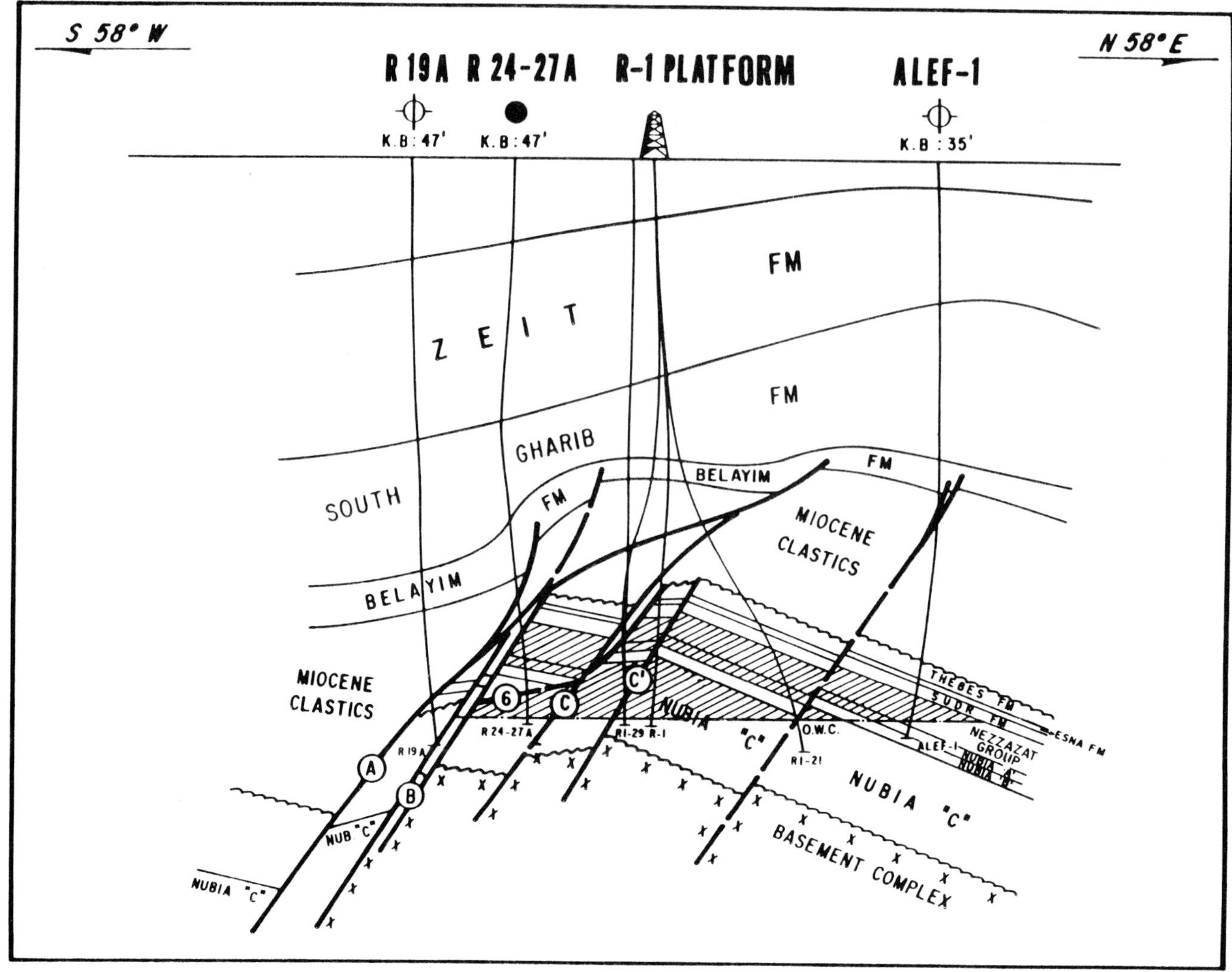

Figure 14. Schematic cross section on seismic line MP-70 (Figure 4), through the Transfer Zone in Ramadan field, shows the complexity of the structure and the drape of Belayim Formation over the underlying structure. Oil-bearing intervals are hachured.

to the overall decrease in the amount of asphaltic matter from the north to the south of the field. In the north, asphaltic matter can represent as much as 30% of the total rock volume.

The Nubia "C" reservoir section has been subdivided into three zones based on permeability, porosity, and lithology (Figure 20). The uppermost zone (Zone 1) is characterized by the presence of uncorrelatable shale units within a massive sandstone section. It ranges in thickness from 30 to 76 m (100–250 ft). The permeability of this zone varies as a function of the amount of asphaltic matter present. This asphaltic matter affects mainly the northern portion of the field. The intermediate zone (Zone 2), composed of massive sandstone with very minor discontinuous shale interbeds, displays good permeability and porosity (Figure 20) over a thickness of 76 to 210 m (250–690 ft) of section. The lowermost interval (Zone 3), although made up of massive sandstone, displays an overall increase in clay and asphalt content resulting in an overall decrease in permeability and porosity.

Through petrophysical analysis it has been shown that the net cut-off within the reservoir occurs when (1) porosity drops below 11%, (2) permeability decreases to less than 10 md, or (3) the percentage of clay in the rock matrix becomes greater than 9%.

Through correlation of lithology and core permeability, 10 md of permeability correlates to a combination of clay and asphaltic matter in the rock matrix of greater than 14% of the total (Figures 21 and 22). This is probably closer to reality than the petrophysical clay estimate since the gamma-ray tool is not sensitive to kaolinite. Petrographic work (Figure 23) also has shown that 10 md of core permeability is highly correlatable with a bulk density of 2.45 g/cm^3 throughout the reservoir.

The Nubia sandstones are characterized by (1) fining-upward, large-scale, high-angle cross-bedding; (2) the presence of carbonaceous fragments; and (3)

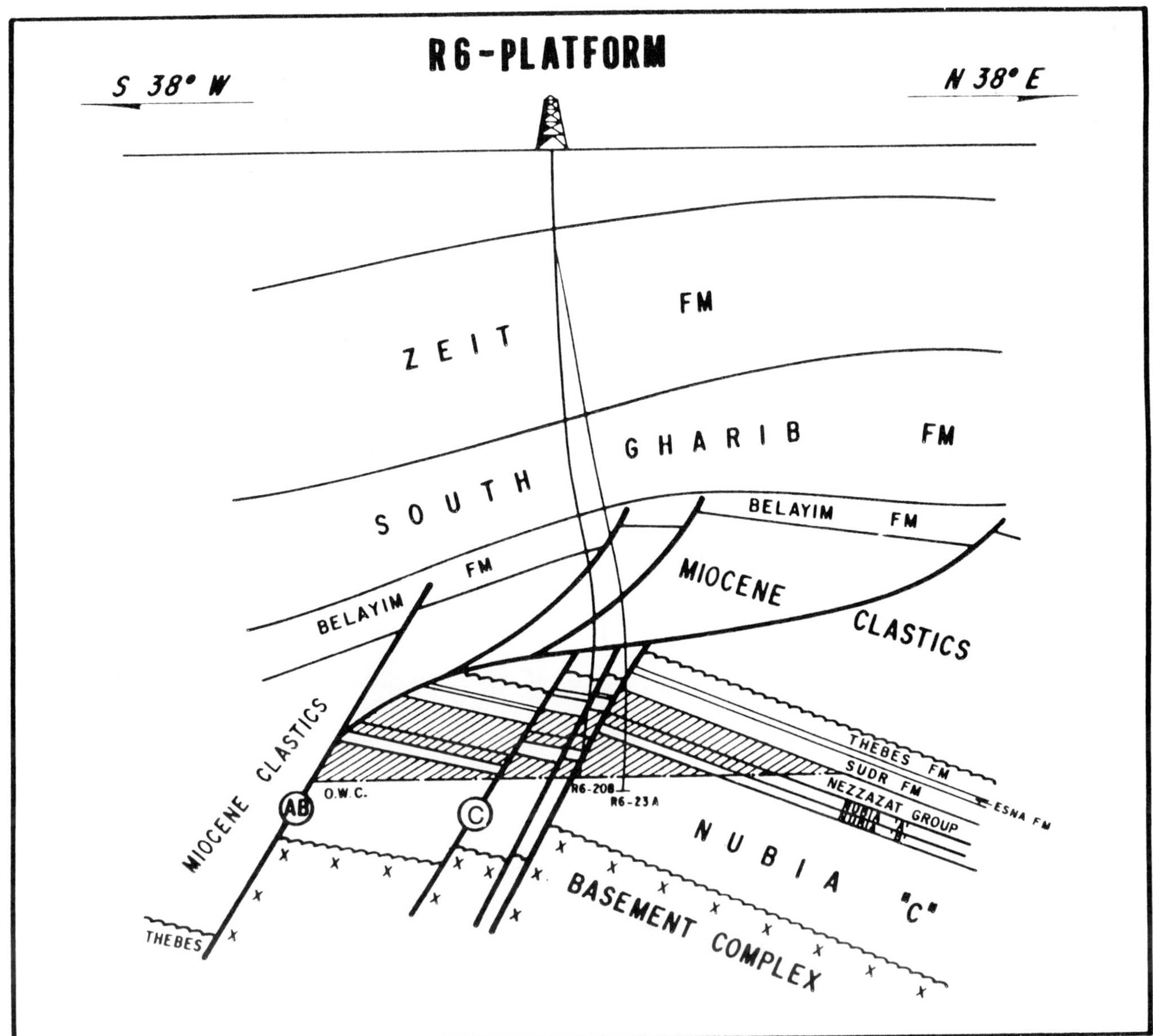

Figure 15. Schematic cross section on seismic line W86-82 through the Southern Zone (Figure 7) of the Ramadan field. Here the "A" and "B" fault systems have come together to form "AB" bounding system with noticeably less drape of Belayim Formation over the underlying structure. Oil-bearing intervals are hachured.

the absence of marine fauna. The Nubia sands are believed to have been deposited in braided streams that were widespread and that intersected to produce sheet sand with lenses of clayey and silty impermeable sandstones characteristic of overbank and levee environments. The frequently observed black and reddish shale interbeds may be attributed to flood plain origin (Hassouba, personal communication).

Source, Maturation, and Migration

Ramadan field geochemistry is based on GUPCO's evaluation of potential source rocks in the field and the maturation modeling of the hydrocarbon-generating basin in the vicinity of the field.

Potential Source Rocks

Total organic carbon determinations in weight percent (TOC wt.%) of ditch samples (Figure 24) indicate intervals of variable good to excellent organic content within the Thebes, Esna, and Sudr formations and in the Nezzazat Group. Results of "Rock Eval" pyrolysis indicate that these intervals show variable capabilities for generating oil and minor gas (mainly from type II kerogen). Intervals of fair organic content within the lower Miocene clastics were tested

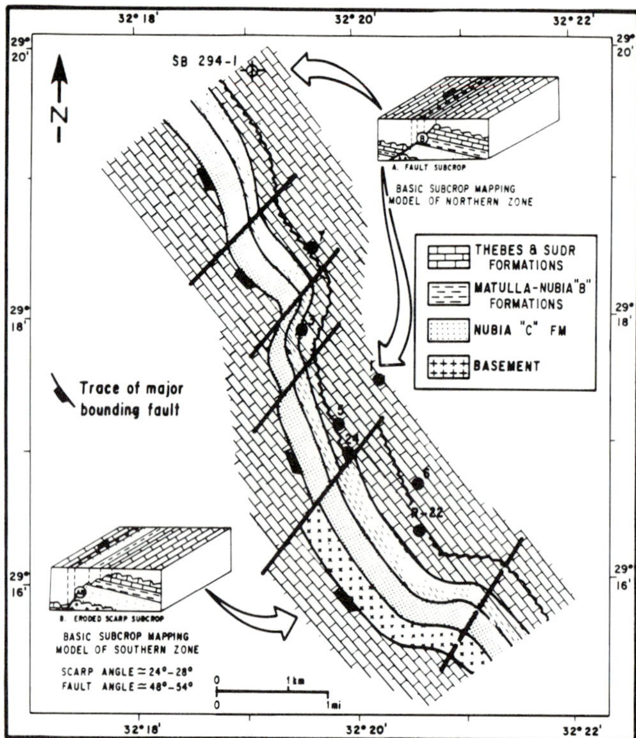

Figure 16. Pre-Miocene subcrop map over Ramadan field shows that the Southern Zone bounded by "AB" system has a much wider expression in plan, indicating a lower-angle pre-Miocene surface than in the remainder of the field.

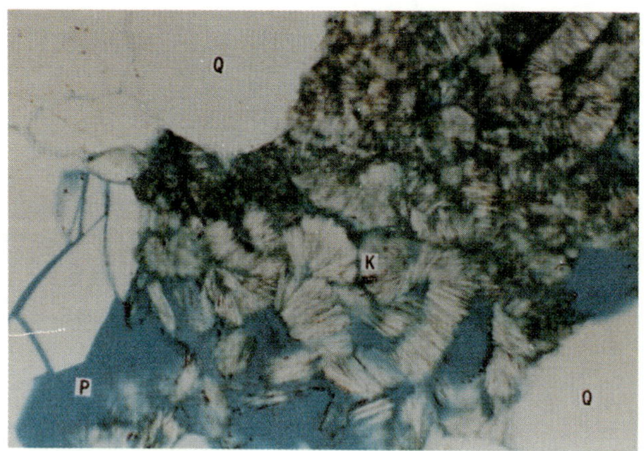

Figure 17. Thin-section showing fair quality reservoir. The rock is composed mainly of quartzose sandstone, "Q," with kaolinite, "K," partially filling the pore spaces, "P." ×110.

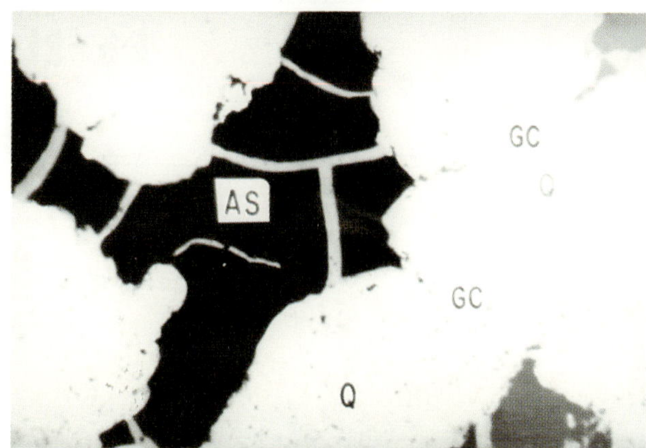

Figure 18. Thin-section showing cracked asphaltic matter "AS" produced by sudden cooling of asphaltine when core reached the surface. "GC" indicates grain-to-grain contact. ×65.

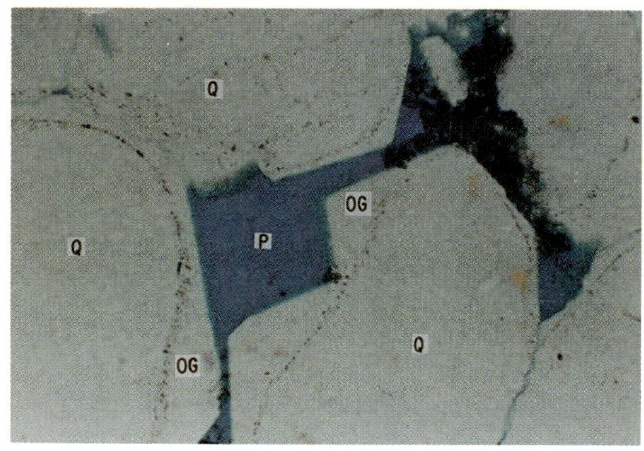

Figure 19. Thin-section showing rectangular porosity, "P," produced by development of silica overgrowth, "OG," in optical continuity with quartz grains, "Q." Note at the upper right corner the development of authigenic kaolinite clays that absorb asphaltic matter. ×80.

and found to possess poor to fair generating potential from gas-prone type III kerogen.

The Eocene Thebes Formation displays an average TOC of 1.8 wt.% (maximum 1.95 wt.% TOC), type II kerogen, a vitrinite reflectance (R_o %) of 0.68, and an average thickness of 122 m (400 ft). The Paleocene Esna Formation, a shale of 15 m (50 ft) average thickness, has a TOC of 0.85 wt.%, type III kerogen, and R_o of 0.69%. The Campanian Sudr Formation has an average TOC of 1.57 wt.% (maximum 3.03 wt.%), contains both type I and II kerogen, has an R_o of 0.71%, and averages 91 m (300 ft) in source potential thickness. The Matulla Formation has an average TOC of 0.97 wt.% (maximum 1.08 wt.%), contains types II and III kerogen with an R_o of 0.77%, and an average thickness of 30 m (100 ft) of source potential.

Maturation and Migration

An explorationist generally is interested more in *when* and *where* the oil was expelled and migrated than in the beginning of generation. Therefore, in this work, the oil window is defined as the depth interval between the depth of peak generation (not

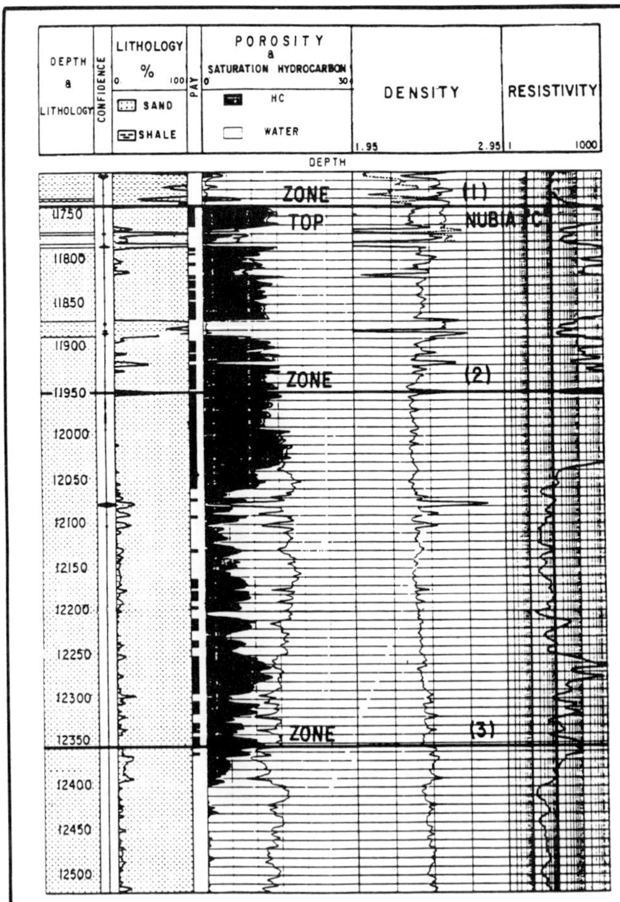

Figure 20. Electric log response of the three zones comprising the Nubia "C" reservoir in well R5-30. Subdivision into three zones is based on permeability, porosity, and lithology.

at the onset of generation) and the depth of the oil floor (the depth below which liquid oil may not generate). A thermal burial history model was generated for the South Belayim trough based on the Lopatin (1971)–Waples (1980, 1985) method. The South Belayim trough (Figure 25) is located downdip to the north and east from Ramadan field and considered to be the source of Ramadan oil. The R_0 % data were used to adjust the magnitudes of local tectonic events and erosions in the model. The geothermal gradient employed is based on bottom-hole temperatures corrected for loss of heat during mud circulation (Figure 25).

In the field, the geothermal gradient is about 1.6°F/100 ft (28.9°C/km). In the basinal model (Figure 26), a lower geothermal gradient of 1.3°F/100 ft (23.7°C/km), as derived from the geothermal gradient map (Figure 25), was employed. The uplift and erosion are replaced by a hiatus (or a continuous period of low sedimentation rate). The model suggests that the pre-Miocene formations of Nezzazat Group reached the stage of peak hydrocarbon generation and began

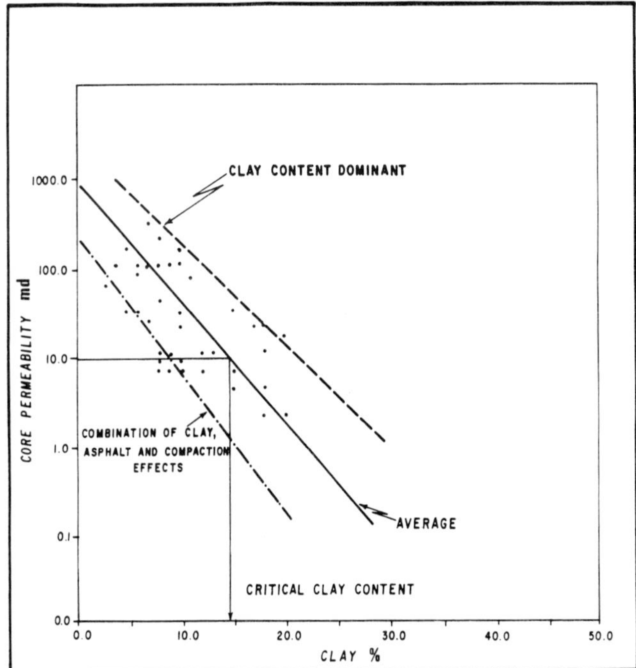

Figure 21. Crossplot of permeability vs. clay content of examined samples from the cored well R1-34, Ramadan field.

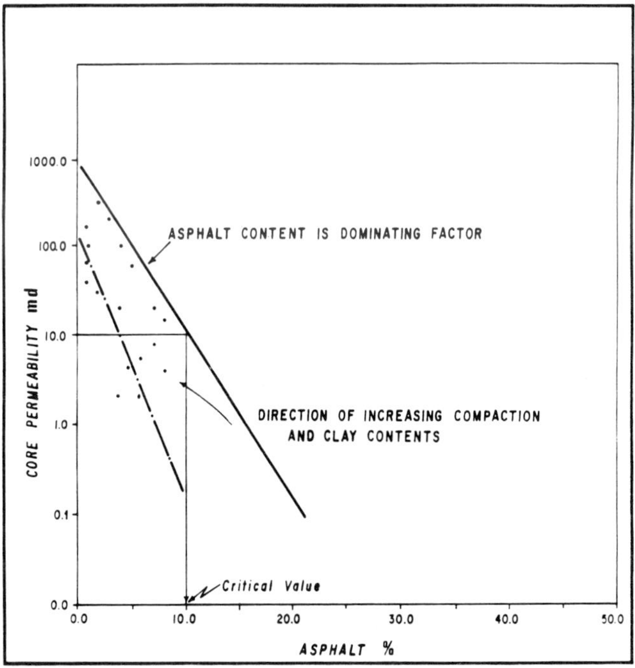

Figure 22. Crossplot of permeability vs. asphaltic content of examined core samples from well R1-34, Ramadan field.

to expel oil some 8 Ma, very close to the average timing of migration in the Gulf of Suez (Shahin, 1988).

Such timing post-dates the deposition of thousands of feet of Miocene evaporites that must have

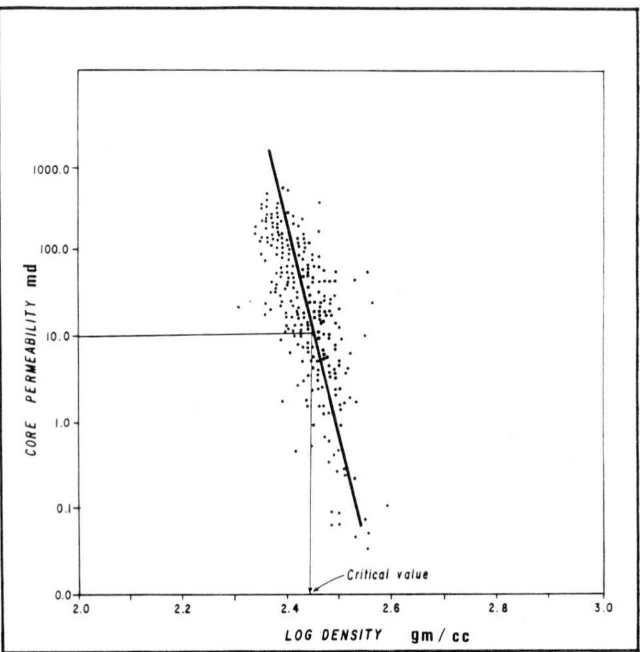

Figure 23. Relation between permeability and density of cored interval from well R1-34. Ramadan field core permeability of 10 md is highly correlatable with a bulk density of 2.45 g/cm³ throughout the reservoir.

accumulated as the basin subsided. These evaporites, as well as providing the burial depth necessary for kerogen maturation and hydrocarbon generation, provided the ultimate seal in the Gulf of Suez. Presently, the pre-Miocene source intervals are shown to have passed through the peak generation stage (at $R_o = 0.8\%$). The Nubia "B" shales are presently overmature as they are shown in Figure 26 below the oil floor.

The present-day depth of peak oil generation is 4938 m (16,200 ft), while the end of oil generation (oil floor) is shown at 5822 m (19,100 ft).

The term *spent source rock* may be applied here in describing the Nubia "B" shales at present, whereas the Matulla, Wata, Sudr, Esna, and Thebes formations remain "active" within the hydrocarbon generation limits. Similar conclusions pertaining to the timing of oil expulsion and migration and the pre-Miocene sediments as the main source in the northern and central Gulf of Suez have been reached by Shahin and Shehab (1984), Chenet et al. (1984), Ungerer et al. (1986), RRI (1986), and Shahin (1988).

The northeasterly dip and the northern bounding cross-element, which brings into juxtaposition the Nubia "C" sandstone of Ramadan field and the pre-Miocene sediments to the northeast, provides the most likely oil migration pathway into the structure. The southwestern corner of the mapped structure shows a potential spill point at 3659 m (12,000 ft) where the Nubia "C" sand comes in contact with the Nezzazat Group on the downthrown side. The original OWC observed in the field was 3674 m (12,050 ft).

Field Characteristics

Oil produced in Ramadan field is paraffinic-naphthenic (Tissot and Welte, 1984) and non-biodegraded (Figure 27). It is 30.8° API, 0.88 wt.% sulfur, and has an estimated oil viscosity of 0.5 cp. Oil analyses have shown that trace amounts of what may be pyrobitumen are consistently present.

High temperature melting-point wax is also present in all analyzed samples in sufficient quantity to form a sludge in the bottom of the sample containers. These waxes, although insoluble at room temperature, may be soluble at reservoir temperatures, 295°F (146°C). Based on carbon isotopes vs. optical rotation plots, all Ramadan oils are classified as being of one type. A similar conclusion was reached by Rohrback (1982).

Production began December 1974 at a rate of 25,000 BOPD, no water, and a solution GOR of 500 ft³/bbl. The initial pressure within the field was 5869 psi. As the field was further developed, the production rate increased (Figure 28) to a peak rate of 130,000 BOPD in January 1979 from 13 wells. The average oil column thickness within the Nubia "C" sandstones was 305 m (1000 ft) with a connate water content of 7.8%. Oil production remained fairly constant at rates near 100,000 BOPD until 1981 when production began a steady decline. Drilling of additional wells along with some successful water shutoff workovers performed in 1982 and 1983 arrested this decline for a short period of time. However, in 1984 the field again began a decline in production and reached a low of 30,000 BOPD in November 1986.

Three new wells have been recently completed in the west block of the field and the oil production rate of the field has been increased close to 50,000 BOPD. It has also been recognized that the majority of the oil produced from the Nubia "C" reservoir has come from Zone 2. Presently its pressure is 4250 psi with a gradient of 0.3 psi/ft at a temperature of 146°C (295°F). Production tests have verified that when Zones 2 and 3 are co-mingled, Zone 2 is responsible for over 90% of the observed production. This appears to be due to the overall lower permeability of Zone 3. It has been recognized that within Zone 3, however, there are intervals with reservoir parameters equivalent to those of Zone 2 and that, through a selective perforation program, production from Zone 3 may be realized at a commercial rate.

EXPLORATION AND DEVELOPMENT CONCEPTS

Ramadan structure is a scarp resulting from crestal erosion of the upthrown edge of tilted pre-Miocene fault blocks during Oligocene. Tilting of the uplifted pre-Miocene fault blocks was revealed by a three-arm dipmeter survey obtained from Alef-1 and by the seismic interpretation of a deep pre-salt structure.

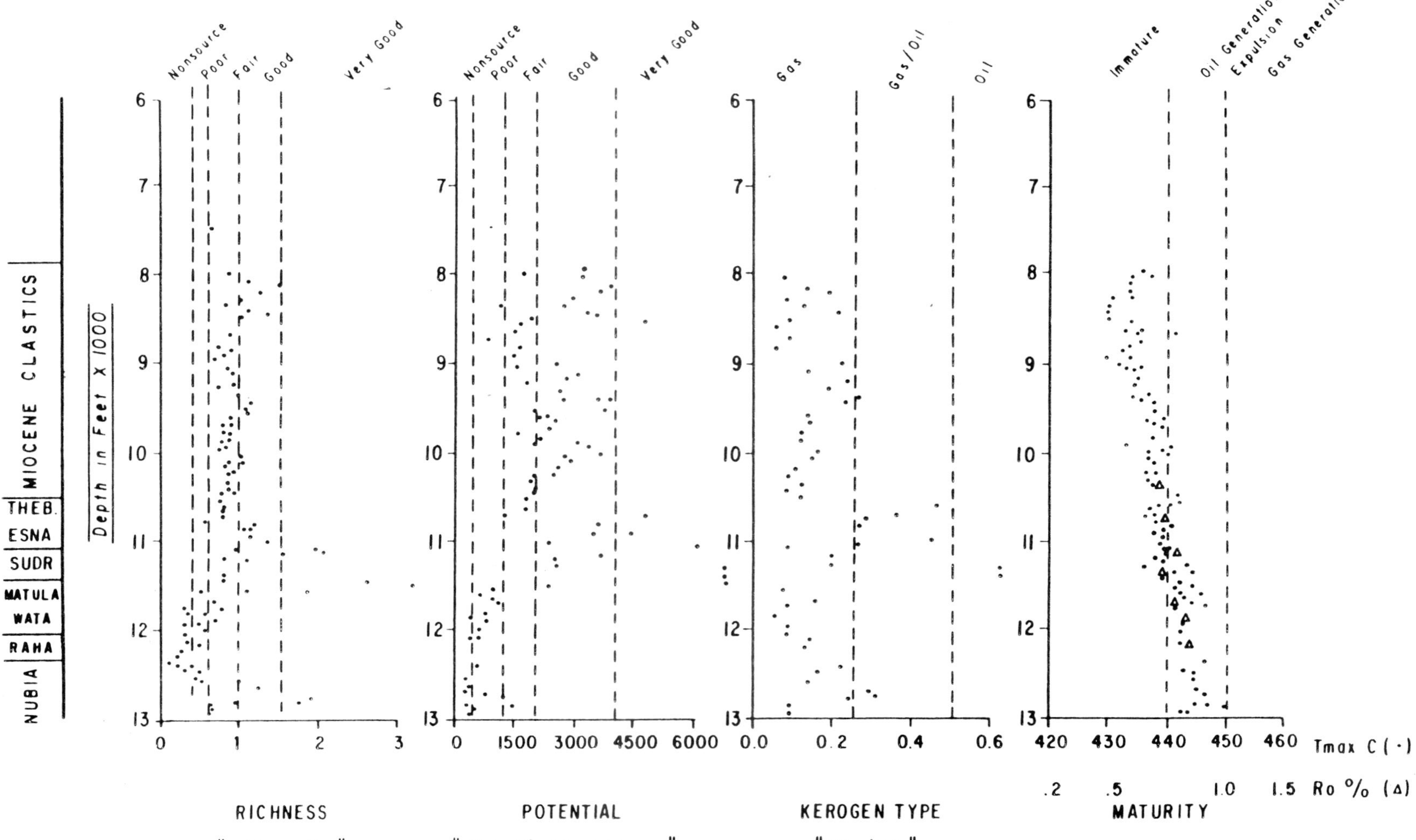

Figure 24. The Campanian lower Sudr Formation and intervals within the Thebes, Esna, Matulla, Wata, and Nubia formations are potential source rocks in the vicinity of Ramadan field as revealed by TOC wt.% and the pyrolysis-derived genetic potential, Hydrogen Index/Oxygen Index, T_{max}, as well as R_o measurements.

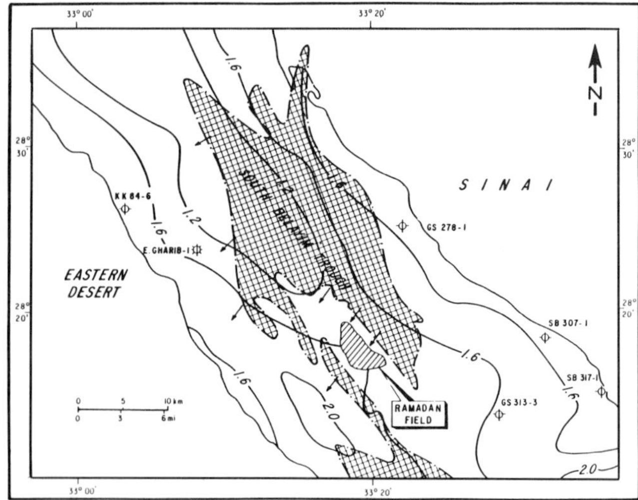

Figure 25. Geothermal gradient map suggesting the South Belayim trough, located downdip to the north and east from Ramadan field, as the likely generating basin for the Ramadan structure. Arrows represent potential migration pathways, generally updip from the generating basin. Contour interval is 0.4°F/100 ft.

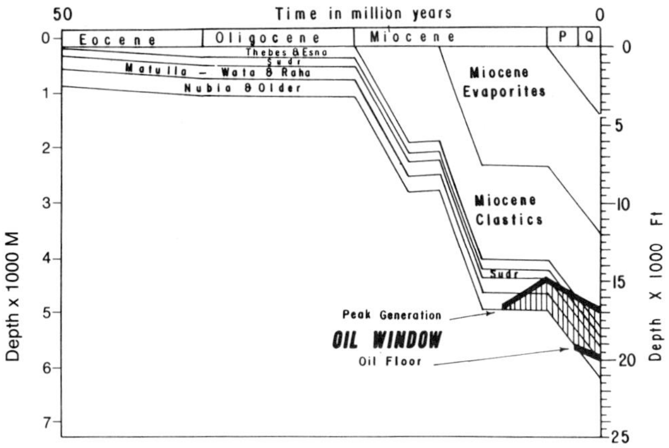

Figure 26. Thermal burial history model of the South Belayim trough (Figure 25), which is considered to be the source of Ramadan oil. The pre-Miocene source rocks reached the stage of peak generation after deposition of the Miocene evaporites, the ultimate seal in the Gulf of Suez.

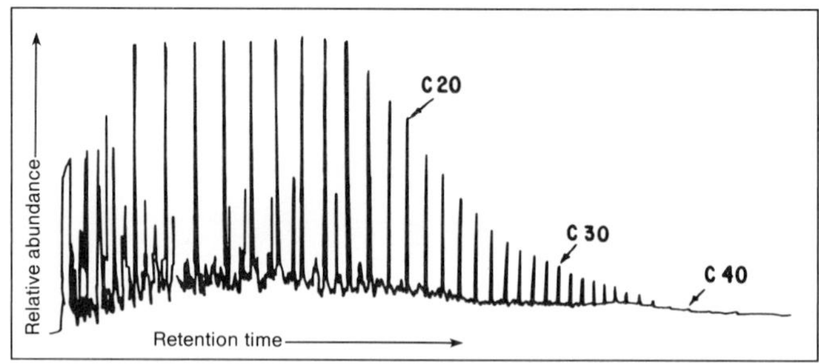

Figure 27. A representative gas chromatogram of Ramadan field oil, revealing its paraffinic-naphthenic, non-biodegraded, and mature nature.

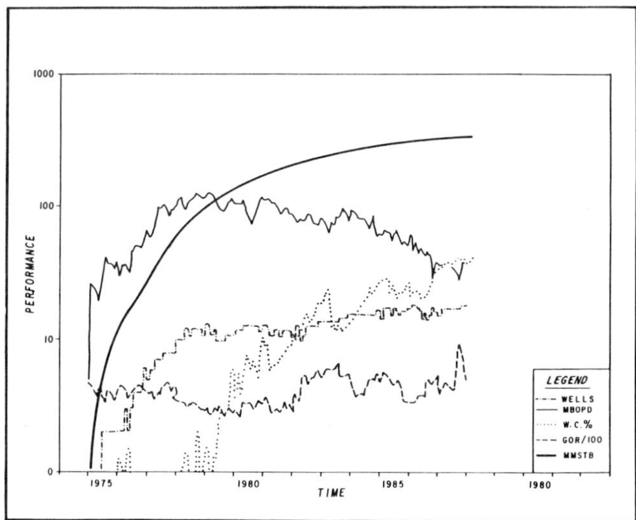

Figure 28. Nubia "C" reservoir production performance, Ramadan field. MBOPD, thousand barrels of oil per day; W.C.%, water cut %; GOR, gas-oil ratio in SCFG/100 per STB (stock tank bbl of oil); MMSTB, million stock tank barrels of oil, cumulative.

Reserves are defined by the intersection of the main bounding fault with the pre-Miocene erosional scarp. Adopting the concept of eroded scarp has allowed a westward repositioning of the interpreted updip limit of the Nubia "C" reservoir, resulting in an increase in the estimate of recoverable reserves in excess of 50 million barrels of oil.

Ramadan field has been produced under a strong water-drive, primary recovery mechanism. Significant water production first began in mid-1979 and has continued to increase as various intervals are flooded out. However, a continuing workover program to plug off watered-out intervals while perforating additional oil pay intervals has been successful in restricting the increase in water production. Future performance will be characterized by increasing water-cuts as intervals are watered out. Although the potential still exists for water shutoff workovers, this practice will decrease water production and increase oil production only temporarily.

Drilling will continue to the south and southwest until all remaining reserves within the Nubia "C" reservoir are identified. Infill drilling will continue in order to optimize areal sweep. As water production from the Nubia "C" increases, many wells will be recompleted in the secondary and tertiary reservoirs of the Nubia "A" and Nezzazat sections. The field may be a candidate for an enhanced oil recovery (EOR) program.

ACKNOWLEDGMENTS

The Egyptian General Petroleum Corporation (EGPC) and AMOCO permitted the publication of this paper.

The study was prepared under supervision of A. Shawki Abdine (Chairman of the Board), Wafik Meshref (Exploration General Manager), and Safi Wasfi (Chief Geologist), Gulf of Suez Petroleum Company (GUPCO).

The study was edited by A. Nabil Shahin and compiled by Paul Garossino and Salah Shazly.

In addition to cited literature, several unpublished GUPCO reports were the basis for the Ramadan field study. These reports are authored by A. Abdine, A. Badawi, M. Bahr, R. Brown, M. Fathy, P. Garossino, M. Hamawy, H. Hammouda, A. Hassouba, H. Hataba, H. Mansour, A. Mokhtar, W. Meshref, M. Sarawy, A. Shahin, B. Torkelson, and S. Shazly.

The manuscript was reviewed by H. Dalton, A. Gad, A. Hassouba (who also provided reservoir petrology and environment), H. McDowell, R. Nelson, J. Jenner, T. Russell, and T. Thompson. Thanks are extended to AAPG reviewers E. Beaumont, J. Lowell, and W. Brownfield for their critical review and valuable comments. We are, however, responsible for any negative qualities that remain.

Technical assistance was provided by A. Salem, M. Morsi, M. Aman, A. Badawi, M. Laboudy, H. Mostafa, and G. Moustafa.

REFERENCES CITED

Abdine, A. S., 1981, Egypt's petroleum geology: good grounds for optimism: World Oil, v. 92, p. 99-112.

Brown, R. N., 1978, Ramadan oil field: Egyptian General Petroleum Corporation, 5th International Exploration Conference.

Chenet, P., J. Letouzey, and E. Zaghloul, 1984, Some observations on the rift tectonics in the eastern part of the Suez Rift: Egyptian General Petroleum Corporation, 7th International Exploration Conference, Cairo.

Lopatin, N. V., 1971, Temperature and geologic time as factors in coalification: USGS translation from Akad. Nauk. SSR. IZV. Geol. Series no. 3, p. 95-106.

Rohrback, B. G., 1982, Crude oil geochemistry of the Gulf of Suez: Egyptian General Petroleum Corporation, 6th International Exploration Conference, Cairo.

RRI (Robertson Research International), 1984, The Gulf of Suez area, stratigraphy, petroleum geochemistry, and petroleum geology, Consultant Report: Egyptian General Petroleum Corporation.

Shahin, A. N., 1988, Oil window in the Gulf of Suez, Egypt (abs.): American Association of Petroleum Geologists Bulletin, v. 72, n. 8, p. 1024-1025.

Shahin, A. N., and M. M. Shehab, 1984, Petroleum generation, migration and occurrence in the Gulf of Suez, offshore South Sinai: Egyptian General Petroleum Corporation, 7th International Exploration Conference, Cairo.

Tissot, B. P., and D. H. Welte, 1984, Petroleum formation and occurrence: New York, Springer-Verlag, 537 p.

Ungerer, P., P. I. Chenet, I. Moretti, A. Chiarelli, and J. L. Ondin, 1986, Modeling of oil genesis and migration in the southern part of the Suez Rift, Egypt: Egyptian General Petroleum Corporation, 8th International Exploration Conference, Cairo.

Waples, D. W., 1980, Time and temperature in petroleum formation, application of Lopatin's method to petroleum exploration: American Association of Petroleum Geologists Bulletin, v. 64, n. 6, p. 916-926.

Waples, D. W., 1985, Geochemistry in petroleum exploration: Boston, International Human Resources and Development Corporation, 232 p.

Appendix 1. Field Description

Field name .... *Ramadan field*

Ultimate recoverable reserves .... *NA*

Field location:

Country .... *Egypt*

Basin/Province .... *Gulf of Suez*

Field discovery:

Year first pay discovered .... *Nubia "C" sandstone, discovered and on production 1974*

Year second pay discovered .... *Nezzazat Group, discovered 1974, first production 1983*

Third pay .... *Nubia "A" sandstone, discovered 1974, first production 1986*

Discovery well name and general location:

First pay ..... *GS 303-1, lat. 28°17′31″N, long. 33°18′52″E; 174 km (108 mi) to the southeast of Suez City; 31 km (19 mi) east of Ras Shukheir base*

Second pay .... *Same as discovery well*

Third pay .... *Same as discovery well*

Discovery well operator .... *GUPCO*

Second pay .... *GUPCO*

Third pay .... *GUPCO*

IP in barrels per day and/or cubic feet or cubic meters per day:

First pay .... *25,000 BOPD (well GS 303-1, December 1974)*

Second pay .... *2420 BOPD (well R4-12, April 1983)*

Third pay .... *3200 BOPD (well R3-25, November 1986)*

Geologic concept leading to discovery and method or methods used to delineate prospect

Prior to the discovery well GS 303-1 (R-1), Alef-1 was drilled in October 1965 on the crest of the Miocene evaporite sequence (Brown, 1980; Abdine, 1981). This well was abandoned in Nubia "A" Formation without any hydrocarbon indications. Additional seismic data obtained in 1971–1972 allowed the interpretation of the deeper pre-salt structure. In addition, the three-arm dipmeter obtained from Alef-1 well, although poor in quality, indicated a northeasterly dip in the pre-Miocene strata. All above led to the subsequent decision to drill well GS 303-1, which is updip from Alef-1. Well GS 303-1 was later called Ramadan-1.

Structure:

Province/basin type .... *Bally III, Klemme IIIA; Gulf of Suez*

Tectonic history

After the deposition of the pre-Miocene sediments, the entire area was subjected to the first phase of rifting during the Oligocene, accompanied by uplifting and tilting and followed by an Oligocene–Miocene crestal erosion. This was followed by a Miocene transgressive phase that resulted in the deposition of the Miocene clastics and evaporites. A second phase of rifting started during the late Miocene, resulting in the rejuvenation of the older faults.

Regional structure

Ramadan oil field is located in the central province of the Gulf of Suez. This province is characterized by northeast dip regime bounded to the north and south by southwesterly dipping provinces.

Local structure

Northeast-dipping pre-Miocene faulted tilted blocks bounded from the west by northwest-trending fault of 152 to 610 m (500–2000 ft) average throw.

Trap:

Trap type(s) .... *Structural, faulted tilted blocks*

Basin stratigraphy (major stratigraphic intervals from surface to deepest penetration in field):

Chronostratigraphy	Formation	Depth to Top in ft (m)
Middle Miocene	*Zeit*	*1400 (427)*
Middle Miocene	*South Gharib*	*4800 (1463)*
Middle Miocene	*Belayim*	*6650 (2027)*
Middle Miocene	*Kareem*	*7100 (2165)*
Lower Miocene	*Rudeis*	*7600 (2317)*
Lower Miocene	*Nukhul*	*9100 (2774)*
Eocene	*Thebes*	*9170 (2796)*
Paleocene	*Esna Shale*	*9540 (2909)*
Upper Cretaceous	*Sudr*	*9570 (2918)*
Upper Cretaceous	*Nezzazat Group*	*9930 (3027)*
Lower Cretaceous	*Nubia "A"*	*10,600 (3232)*
Paleozoic	*Nubia "B"*	*10,700 (3262)*
Paleozoic	*Nubia "C"*	*10,841 (3305)*

Reservoir characteristics:

Number of reservoirs *3*

Formations .... *Nubia "C," Nubia "A," and Nezzazat Group (Matulla, Wata, and Raha formations sandstones)*

Ages .... *Paleozoic, Early and Late Cretaceous, respectively*

Depths to tops of reservoirs .... *Nubia "C," 3475 m (11,400 ft); Nubia "A," 3400 m (11,160 ft); Nezzazat Group, 3195 m (10,480 ft)*

Gross thickness (top to bottom of producing interval) .... *1000 ft (304.8 m)*

Net thickness—total thickness of producing zones

Average .... *750 ft (228.6 m)*

Maximum .... *900 ft (274.3 m)*

Lithology

Nubia "C": fine- to medium-grained, occasionally coarse-grained sandstone with kaolinite matrix

Nubia "A": same as Nubia "C"

Nezzazat Group: very fine to fine sandstone, moderately sorted, with glauconite and some pyrite crystals

Porosity type .... *Primary intergranular*

Average porosity *15%*

Average permeability .... *100 md*

Seals:

Upper

Formation, fault, or other feature .... *Sudr limestone, Nubia "B" shale*

Lithology .... *Massive limestones and shales*

Lateral

Formation, fault, or other feature .... *Fault*

Lithology .... *NA*

Source:

Formation and age .... *Thebes (Eocene), Sudr (Campanian), Matulla (early Senonian–Turonian)*

Lithology .... *Thebes (limestone), Sudr (brown limestone), Matulla (shales and marls)*

Average total organic carbon (TOC) .... *Thebes, 1.8%; Esna, 0.9%; Sudr, 1.6%; Matulla, 1.0%*

Maximum TOC .... *Thebes, 1.95%; Esna, 0.85%; Sudr, 3.03%; Matulla, 1.08%*

Kerogen type (I, II, or III) .... *Thebes, II; Esna, III; Sudr, II, I; Matulla, II, III*

Vitrinite reflectance (maturation) .... $R_o = 1.02$ *at 12,800 ft (3900 m)*

Time of hydrocarbon expulsion ... *10 Ma (middle Miocene)*
Present depth to top of source ... *9000 ft (2740 m) avg.*
Thickness ... *Thebes, 400 ft (122 m); Esna, 50 ft (15 m); Sudr, 300 ft (91 m); Matulla, 100 ft (31 m)*
Potential yield ... *NA*

Appendix 2. Production Data

Field name ... *Ramadan field*

Field size (data given for Nubia "C"; not available for Nubia "A" or Nezzazat Group):

- Proved acres ... *2851*
- Number of wells all years ... *NA*
- Current number of wells ... *40*
- Well spacing ... *40*
- Ultimate recoverable ... *NA*
- Cumulative production ... *360 million bbl*
- Annual production ... *13.6 million bbl*
- Present decline rate ... *25%*
  - Initial decline rate ... *Nil*
  - Overall decline rate ... *25%*
- Annual water production (1987) ... *8.8 million bbl*
- In place, total reserves ... *NA*
- In place, per acre-foot ... *NA*
- Primary recovery ... *NA*
- Cumulative water production ... *50 million bbl (as of February 1988)*

Drilling and casing practices:

- Amount of surface casing set ... *395 ft*
- Casing program ... *24 in. to 350 ft; 13⅜ in. to 3500 ft; 9⅝ in. to 10,150; 7 in. to TD*
- Drilling mud ... *Sea water salt saturated, oil-base mud*
- Bit program ... *Varies*
- High pressure zones ... *Miocene clastics*

Formation evaluation:

- Logging suites ... *BHC, GR, FDC, CNL, and DLL*
- Testing practices ... *Typically production tested*
- Mud logging techniques ... *Basic online logging for development wells and pressure detection inline logging for exploratory wells*

Oil characteristics:

- Type ... *Paraffinic-naphthenic*
- API gravity ... *30.8°*
- Base ... *NA*
- Initial GOR ... *500*
- Sulfur, wt% ... *0.88*
- Viscosity, SUS ... *0.5 cp*
- Pour point ... *NA*
- Gas-oil distillate ... *NA*

Field characteristics:

- Average elevation ... *11,500 ft*
- Initial pressure ... *5870 psi*
- Present pressure ... *4250 psi*

Pressure gradient *0.3*
Temperature .... *295°F (146°C) (surface 80°F; 27°C)*
Geothermal gradient .... *1.6°F/100 ft (29.2°C/km)*
Drive .... *Natural water influx*
Oil column thickness .... *1000 ft*
Oil-water contact .... *12,050 ft subsea*
Connate water *7.8%*
Water salinity, TDS .... *NA*
Resistivity of water .... *NA*
Bulk volume water (%) .... *NA*

Transportation method and market for oil and gas:
Submarine pipelines from offshore platforms to Ras Shukheir facility (land tanks) then pumping and shipping through the port by shipping facilities.

RAMADAN

Rio Itaúnas Field—Brazil
Espirito Santo Basin, Southeastern Brazil

ALCIDES PALHARES, JR.
Petrobrás/Dexes
São Mateus, Brazil

HAMILTON D. RANGEL
ANTONIO M. F. de FIGUEIREDO
Petrobrás/Depex
Rio de Janeiro, Brazil

FIELD CLASSIFICATION

BASIN: Espirito Santo
BASIN TYPE: Passive Margin
RESERVOIR ROCK TYPE: Sandstone
RESERVOIR AGE: Cretaceous
PETROLEUM TYPE: Oil
TRAP TYPE: Horst Block
RESERVOIR ENVIRONMENT OF DEPOSITION: Transitional Fluvial to Shallow Marine
TRAP DESCRIPTION: Faulted horst block with multiple pays

LOCATION

Rio Itaúnas field is located in the Espirito Santo basin, Espirito Santo State, southeastern Brazil. The basin comprises an area of about 3000 km^2 onshore and about 10,000 km^2 offshore out to the bathymetric contour of 200 m, but the basin extends beyond that into deeper waters of the South Atlantic Ocean (Figure 1).

Exploration of the basin was initiated in the mid-1950s. Since then 825 wells have been drilled. Up to now 33 hydrocarbon accumulations have been discovered in the onshore area and one offshore. The onshore fields have an ultimate recoverable reserve of 72.4 million barrels of oil (MMbbl) and a cumulative production of 57.8 MMbbl of oil, while the offshore field (Cacão) has ultimate recoverable reserves of 12.6 MMbbl with 10.4 MMbbl cumulative production. The basin holds a known ultimate recoverable gas reserve of 127.6 bcf and a cumulative production of 48.3 bcf, of which 21.6 bcf ultimate recoverable and 11.3 bcf cumulative production are from the offshore Cacão field.

The oil fields of the basin are in four geologic provinces (Figure 1): the São Mateus platform in the northern part of the basin, the Fazenda Cedro paleocanyon in the central part of the basin, the Regência platform to the south, and the Regência paleocanyon still farther southward. The Rio Itaúnas field is located in the northern part of the 1500 km^2 São Mateus platform. (Other papers in the *Atlas of Oil and Gas Fields* describe the offshore Cacão field associated with the Fazenda Cedro canyon and the onshore Lagoa Parda field in the Regência canyon.)

Eighteen hydrocarbon accumulations, including the Rio Itaúnas field, have been discovered on the São Mateus platform. The Rio Itaúnas field produces from reservoirs in the Mariricu Formation Mucuri Member sandstones of Aptian age that were deposited in a transitional fluvial to shallow marine environment (Figures 2 and 3). Ultimate recoverable oil reserves are 5.4 MMbbl.

HISTORY

Pre-Discovery

Geophysical prospecting initiated exploration in the mid-1950s, and in 1959 Petrobrás started effective exploration drilling with a stratigraphic test, 2-CB-2-ES, which was dry but which obtained oil shows in the well cuttings. This location was based on Bouguer gravity mapping from a semi-detail gravimetric survey using station spacing of approximately 700 m in a 15 to 20 km polygon. Gravity interpretation was used to site the well, because in the absence of outcrops in the Espirito Santo basin, surface geology was not helpful and because adequate reflection seismic profiles were not available at that

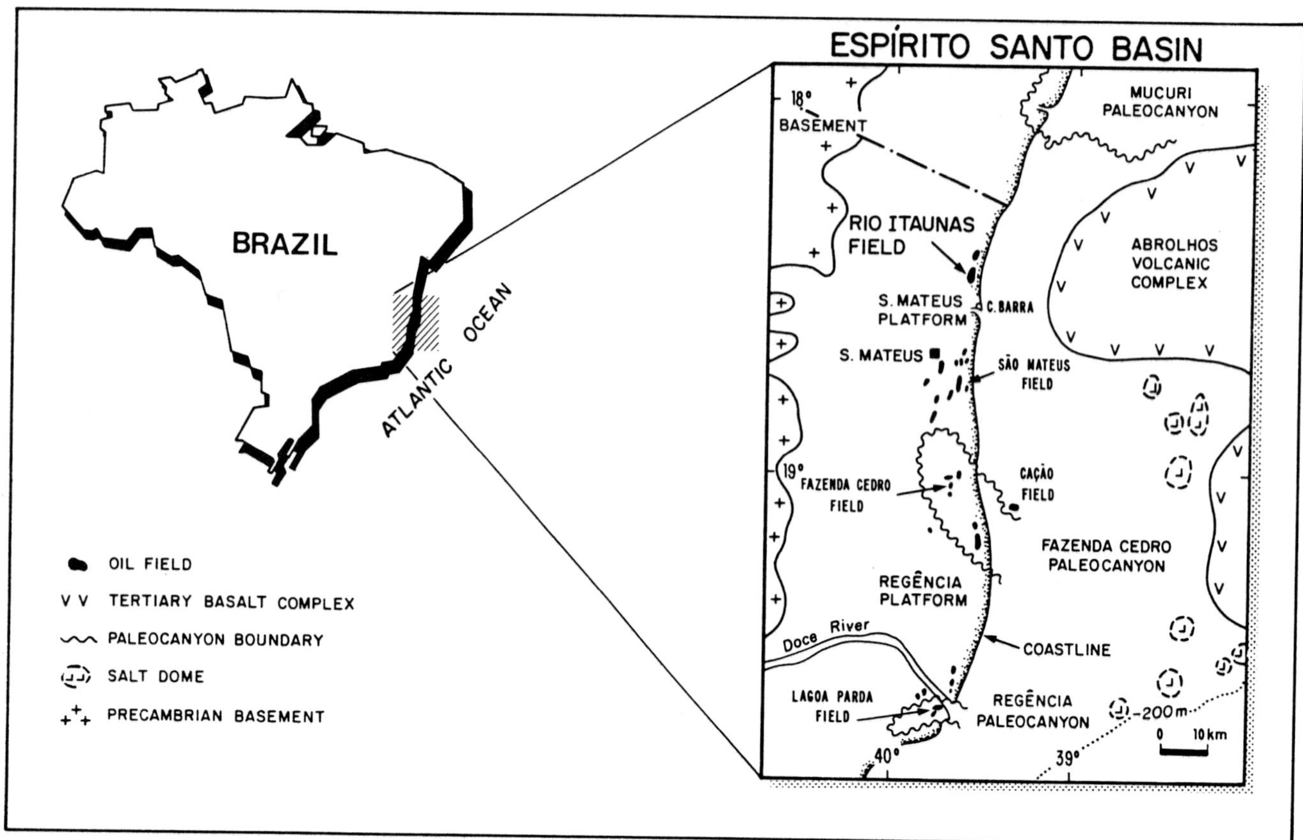

Figure 1. Location map of Espirito Santo basin and Rio Itaúnas field.

time. The original purpose of the well was to investigate the sedimentary sequence of the Espirito Santo basin. The first discovery in the São Mateus platform took place in 1969 by the 1-SM-1-ES wildcat. This was also the first oil strike in the Espirito Santo basin. An important trend of prospective structures aligned northeast was later confirmed.

Discovery

Rio Itaúnas field was discovered in 1977 by the wildcat 1-RI-1-ES, drilled by Petrobrás. It is the northernmost field of the São Mateus platform. The wildcat tested a structural high detected by seismic data at the top of the Aptian anhydrite bed of the Mariricu Formation, the sealing rock of the area (Figure 4). This anhydrite bed is found throughout Espirito Santo basin and other basins along the Brazilian coast; its deposition is associated with the opening of the South Atlantic Ocean between Brazil and Africa.

Reservoir rocks are sandstones of the Aptian Mucuri Member, which were detected at a depth of 1175 m, with a gas and oil net pay of 14 m and 5 m, respectively. The well was completed in 1977. Other hydrocarbon accumulations similar to Rio Itaúnas have been found in the area.

Post-Discovery

Following the discovery of Rio Itaúnas oil and gas accumulations, seven other exploratory wells were drilled by Petrobrás, two of them oil and gas producers, one of them subcommercial, and the rest dry. All these wells aimed to delineate the accumulation discovered by 1-RI-1-ES.

One of the successful appraisal wells, 4-RI-5-ES, located 4 km north of 1-RI-1-ES, discovered an adjacent oil pool (in Albian Barra Nova Formation/ Regência Member carbonate reservoir rocks, which occur just above the Aptian anhydrite bed) also included in the Rio Itaúnas field. Development of the field started in 1978, a year after the discovery well, and a total of 70 development wells have been drilled (Figure 5). Eighteen other accumulations were encountered in the São Mateus platform after the discovery of São Mateus, Rio Preto, and Rio Itaúnas fields.

DISCOVERY METHOD

The discovery well, 1-RI-1-ES, was based upon reflection seismic data, whose quality was only sufficient to detect the anticline associated with the

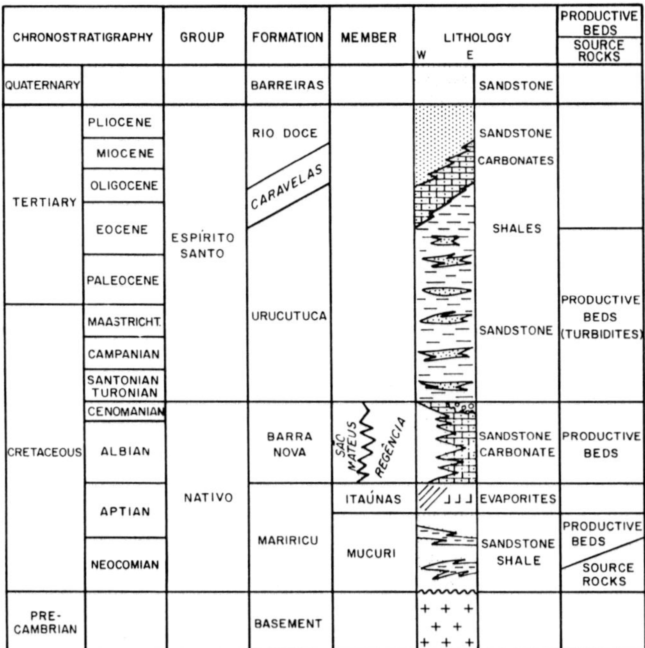

Figure 2. Espirito Santo basin stratigraphic column.

field. The stratigraphy of the São Mateus platform province was already known by previous wells drilled in the region, such as 1-SM-1-ES and 2-CB-2-ES.

Acquisition and reprocessing of reflection seismic data of the area, using new parameters, allowed much better resolution of the structural style and of reservoir thickness prediction in this province. This was important. The reservoir rock is positioned between the anhydrite bed and the Precambrian basement and can be detected only in seismic data with good resolution. In the past, a few exploratory wells were drilled on structural highs without encountering reservoir rocks

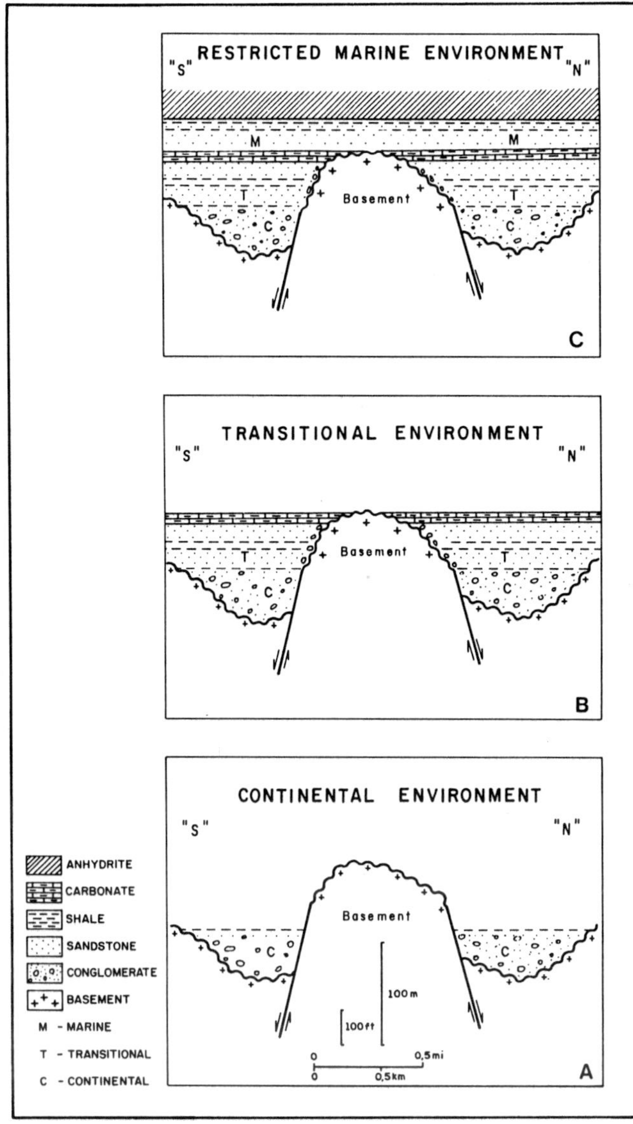

Figure 3. Schematic depositional environment evolution of Mucuri Member at Rio Itaúnas area.

STRUCTURE

Espirito Santo basin structural framework is similar to other East Brazilian marginal basins whose principal features reflect continental tectonic events inherent in plate separation and continental drift processes.

Tectonic deformations of the basin are associated with tensional effects caused by the African and South American plates' separation, such as normal faults, the formation of a gentle horst and graben structural style, tilt of the basin, and halokinetic faults. The basement is a dominantly extensive regional monocline dipping to the southeast, whereas the main lineament pattern is controlled by tension faults with strikes dominantly north-northeast to south-southwest.

São Mateus platform province is the most stable area of the basin. It dips gently to the southeast and is devoid of faults with large downthrow. Although it was also covered by evaporites during Aptian time, the deposition occurring with the entrance of the sea during the African and South American separation, its stratigraphy, predominantly characterized by coarse, terrigenous-clastic sediments, is that of a stable platform area.

The structural style of the field, as are most fields of the area, is associated with a large number of faults with small downthrow, generally less than 100 m, that strike predominantly north-northeast to south-southwest (Figures 4 and 5). These faults affected the Precambrian basement and were later reactivated after the Aptian evaporite deposition, creating trap conditions for the fields encountered in the platform.

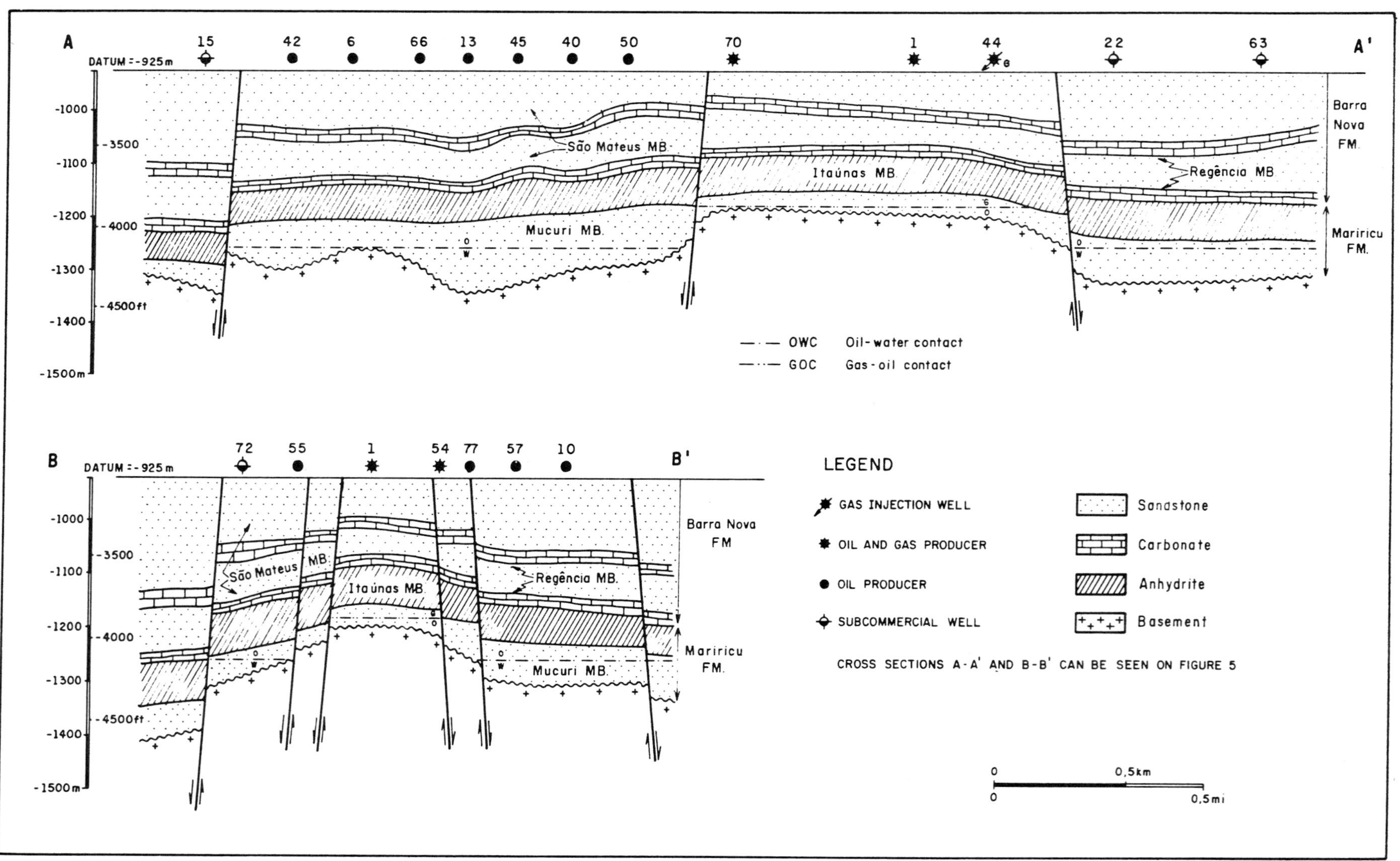

Figure 4. Dip and strike structural cross section of Rio Itaúnas field. See Figure 5 for the locations of the cross sections.

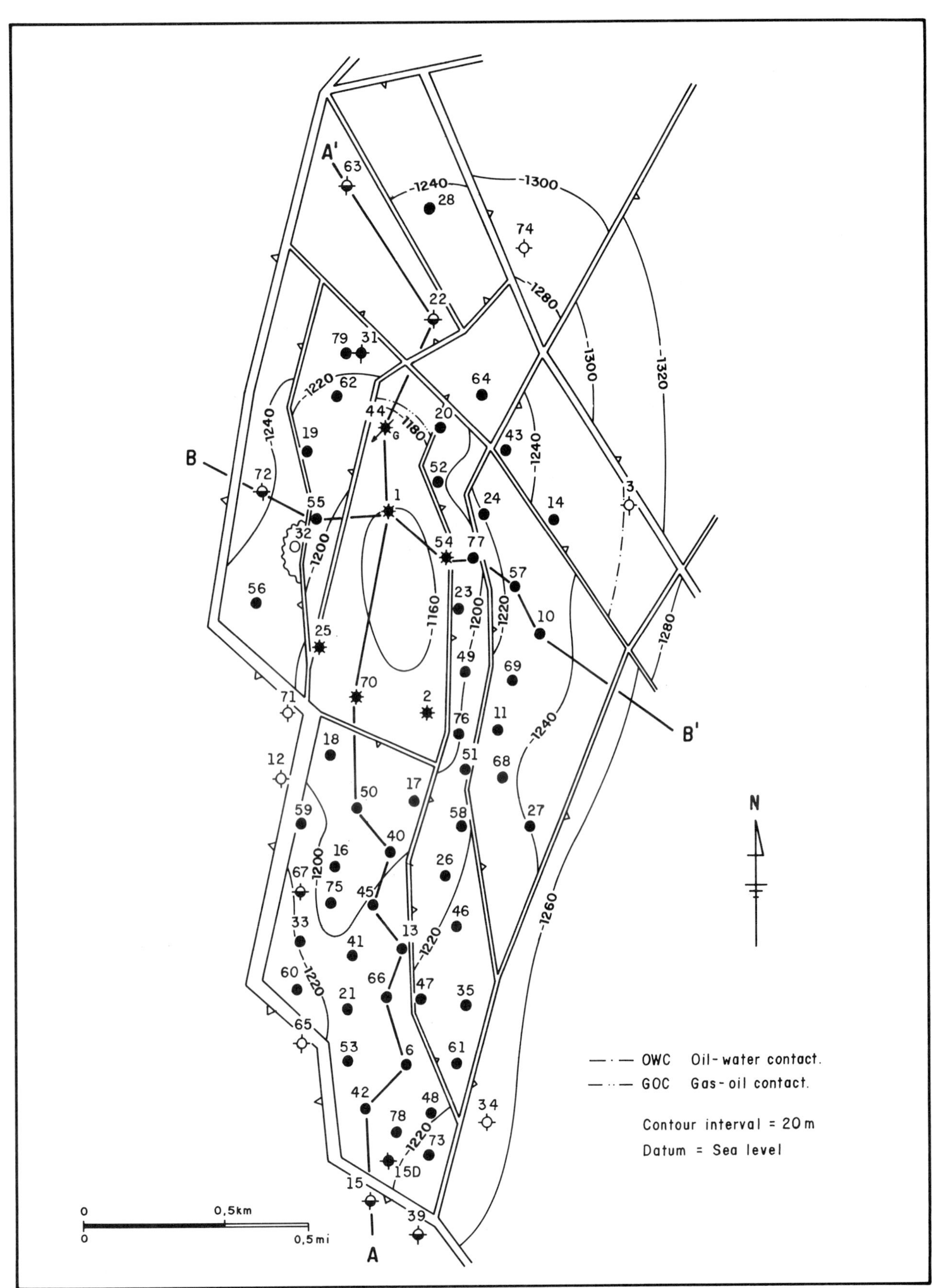

Figure 5. Structural map on the top of zone 1 sandstone reservoir, Mucuri Member.

STRATIGRAPHY

Espirito Santo basin sedimentary column was based initially on biostratigraphic standards and later on lithostratigraphic concepts made by Asmus (1971). The stratigraphic column presently used divides the sedimentary rocks of Espirito Santo basin into two groups: the Nativo Group and the Espirito Santo Group (Figure 2).

Nativo Group

The Nativo Group discordantly overlies the Precambrian basement and contains sediments from Lower Cretaceous upward through the Cenomanian, comprising the Mariricu and Barra Nova formations.

The Mariricu Formation Mucuri Member is characterized by continental sediments deposited in alluvial (Figure 3A), fluvial, and lacustrine (Figure 3B) environments. These sediments formed conglomerate and sandstone reservoirs interbedded with organic-rich shales. The upper portion of the Mucuri Member sandstones was deposited under shallow marine conditions. These sands have better permeabilities and porosities and constitute the main reservoir rocks of the Rio Itaúnas field (Figure 3C).

The Mariricu Formation Itaúnas Member evaporitic layers, which generally act as seals for the reservoirs, occur at the top of the Mariricu (Figures 2 and 3).

The Barra Nova Formation is associated with a coastal sandstone fan delta system (São Mateus Member) that is interfingered with the carbonate platform sediments (Regência Member), evincing the South Atlantic initial opening (proto-oceanic stage). The Regência Member carbonates just above the anhydrite bed are the reservoir rocks in the northern area of Rio Itaúnas field.

A sequence of basic rocks comprises a lithostratigraphic unit called Abrolhos Formation. These basalt layers are interbedded with Lower Cretaceous to Oligocene sediments.

Espirito Santo Group

The Espirito Santo Group includes sediments deposited from Turonian to Quaternary times. Three basic lithostratigraphic units comprise this group: Urucutuca Formation (not deposited in the Rio Itaúnas field area), associated with shales and thin turbidite sandstone beds deposited in deep sea environment; Caravelas Formation, open-marine, shallow platform carbonates deposited in a regressive environment; and Rio Doce Formation, comprised of sandstones and rare mudstone layers, representing the proximal facies of the Urucutuca Formation (Figure 2).

TRAP

In the Rio Itaúnas field, as well as over other parts of the São Mateus platform province, tectonic activity resulted in normal faults and in basin tilt. Faults are associated with two different stages: pre-Itaúnas Member evaporites diastrophism, related to normal faulting (rift phase) and post evaporites isostatic marginal adjusting (drift) phase. All petroleum accumulations discovered in the São Mateus platform are associated with reactivated antithetic faults of the rift phase, striking north-northwest to south-southeast preferentially. These faults create westward trap conditions for hydrocarbon accumulation (Figure 4) and also result in grabens and semi-grabens with stratigraphic traps against basement highs in these structurally low areas.

The Rio Itaúnas field is a structural trap associated with antithetic faults with 100 m throw, striking westward and southwestward, bounding the field on the west and south. Normal synthetic faults and regional dip bound the structure eastward, and the northern side is limited by normal faults that strike northwest to southeast, resulting in an entrapment area of about 4.5 km^2 (Figure 5). The Itaúnas Member anhydrite is the sealing bed of the field.

Main faults of the area are associated with the rift phase. These were later reactivated during the beginning of the drift phase. There is evidence that the faults dipping to the basin, related to the rift stage, were more active than the antithetic faults, which were intensely reactivated after the evaporitic phase, producing the conditions for entrapment.

RESERVOIR

The Rio Doce Formation (at surface), the Caravelas Formation (-650 m), the Barra Nova Formation (-680 m), the Mariricu Formation (-1100 m), and the Precambrian basement (-1330 m) constitute the major stratigraphic column in the Rio Itaúnas field area (Figure 2).

The main producing reservoirs in the Rio Itaúnas field are the Mariricu Formation Mucuri Member sandstones and conglomerates that occur at a depth of about 1175 m. The Mucuri Member is divided into three producing zones, of which zone 1 is the main producing interval (Figures 5 and 6). Zone 1 presents a gas cap on its apex (Figure 7).

The Mucuri Member in the Rio Itaúnas field area is a transgressive sequence constituted from bottom to top by continental sediments (subzones 3D and 3C), transitional sediments (subzones 3B, 3A, and zone 2), and restricted marine sediments (subzones 1C, 1B, and 1A) (Figure 3). The reservoir rocks contain very little fine allogenic matrix sediments but there is abundant calcite cement and authigenic clay minerals, especially illite, smectite, and kaolinite. The Rio Itaúnas field zonation is based on

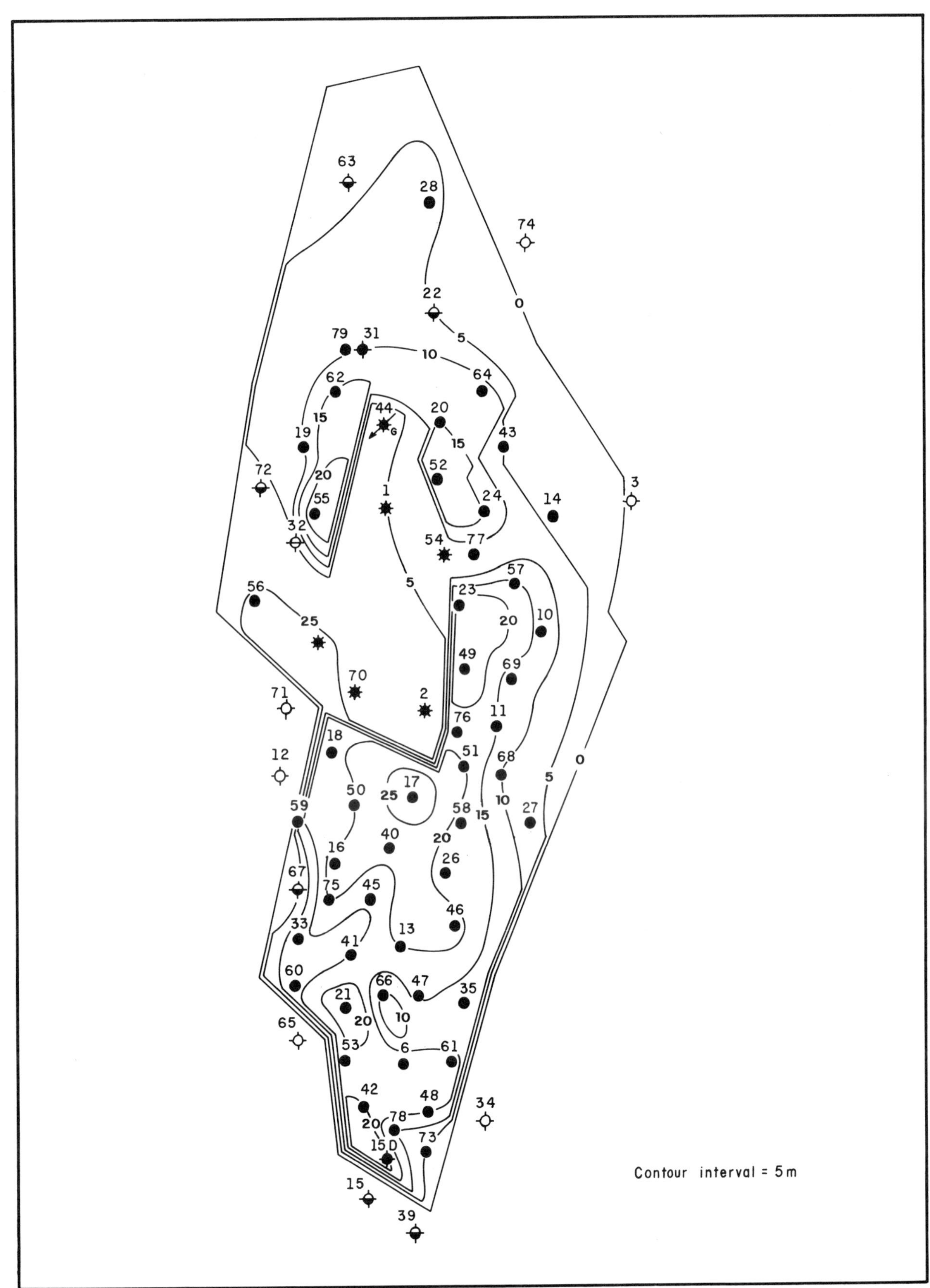

Figure 6. Net oil sandstone map of zone 1 sandstone reservoir, Mucuri Member.

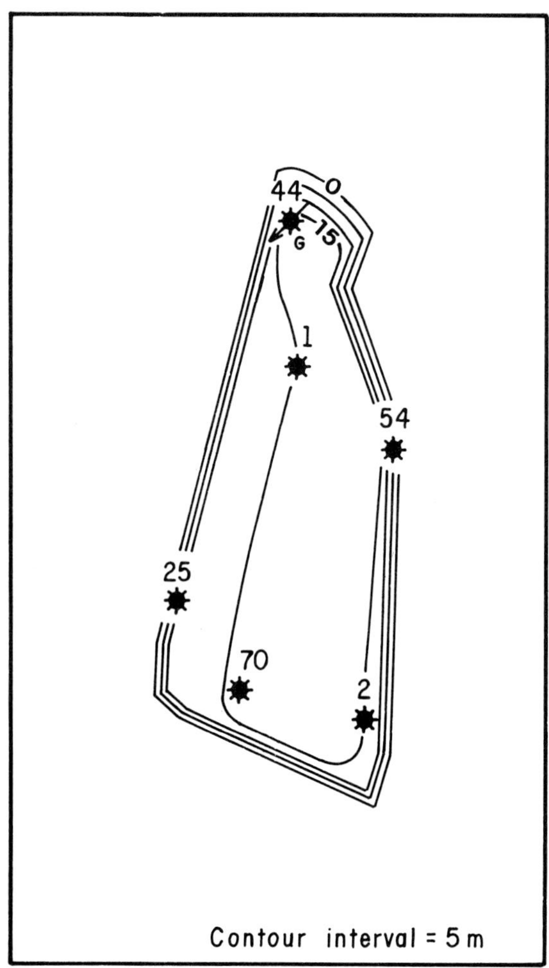

Figure 7. Net gas sandstone map of zone 1 sandstone reservoir, Mucuri Member.

The field reservoir does not show natural fractures, but induced fracturing in zone 1 has effectively stimulated intervals with unfavorable original reservoir conditions.

Zones 1, 2, and 3 have an average thickness of about 20 m, 5 m, and 75 m, respectively. The oil column in the Mucuri Member sandstones is 75 m, and zone 1 (main reservoir) has an average net pay of 14 m. Original reservoir pressure was 126 kgf/cm^2 (at -1250 m) and presently is 70 kgf/cm^2.

The Mucuri Member reservoir oil is mature and degraded, with gravity between 15° and 18° API. Viscosity, from PVT analysis reconstituted data, reaches 45 cp at reservoir conditions, and at surface conditions, it varies from 570 to 1610 cp at 100°C for different samples. Original gas/oil ratio is 38 m^3/m^3 (PVT). Gas cap gas gravity is 0.607 and is composed of methane (93.1%), ethane (2.79%), and nitrogen (2.65%).

Gas cap expansion is the main producing mechanism and water drive is secondary. Reservoir pressure maintenance, begun in February 1984, is by gas injection into the gas cap. In the same year development activity was rapid, increasing the oil production of the field (Figure 8). Part of the gas reinjected in the gas cap flows back toward the southern part of the field. Effective anhydrite layer and fault pattern seals prevent the gas from flowing toward the northern, eastern, and western areas of the field.

Initially, the reservoir was drained by wells spaced at 300 to 400 m. From June 1983, after drilling of development well 7-RI-40-ES, well spacing was changed to 200 m, resulting in more favorable exploitation. Field development is now nearly complete, with a total of 79 wells.

lithostratigraphic correlation from well and pressure data obtained by RFT (Repeat Formation Test).

The sandstone and conglomerate present secondary porosity that is associated with the diagenetic sequence related to compaction, secondary growth of quartz, and substitution of calcite cement (which generated good permeability and porosity). Sampling is difficult in the best reservoir zones due to low compaction or poorly cemented intervals, so petrophysical data are available only in the more cemented intervals, not truly reflecting reservoir conditions.

Reservoir porosity based on petrophysical data reaches values up to 30%, with an average of 15 to 20%, and permeability between 10 and 400 md, reaching values up to 2250 md. The interpretation of quantitive data and the geological model explain the performance of the productive wells; $\eta\phi S_o$ (net pay × porosity × oil saturation) mapping of zone 1 (main reservoir) is the best instrument of correlation with the productive capacity of each well.

SOURCE ROCK AND BURIAL HISTORY

Geochemical analysis of Espirito Santo basin indicates that upper Neocomian and Aptian shales of the Mariricu Formation constitute excellent source rocks for hydrocarbons, with a dominance of type II organic matter (Figure 2). These source rocks reach the maturation window for oil along an extensive area in the present submerged region of the basin. Hydrocarbon index is high and the hydrocarbon source rock potential shows values higher than 5 kg HC/ton. The burial history of the area is shown in Figure 9.

On the São Mateus platform, Aptian shales are immature and upper Neocomian shales are absent. Maturity is reached in the adjacent continental platform where source beds are more than 2000 m deep. Therefore, hydrocarbons must have migrated at least as far as 10 km.

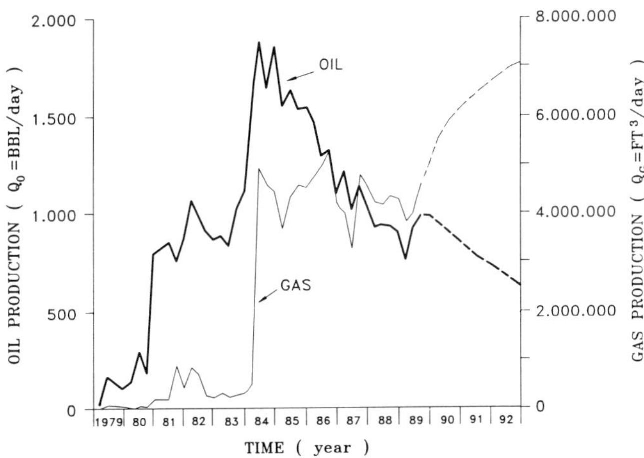

Figure 8. Oil and gas production, history and forecast, Rio Itaúnas field.

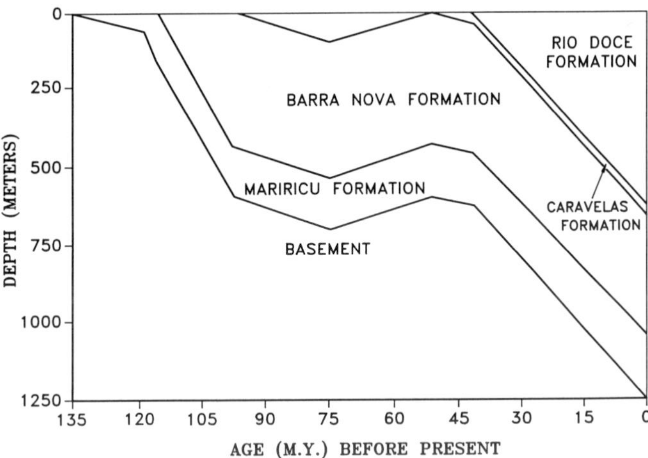

Figure 9. Burial history diagram, Rio Itaúnas field.

EXPLORATORY CONCEPTS AND LESSONS

The São Mateus platform province has the largest number (18) of hydrocarbon fields of the provinces of the basin and is second to Regência paleocanyon province in the volume of recoverable oil, having an ultimate recoverable reserve of 20.5 MMbbl, of which Rio Itaúnas field accounts for 5.40 MMbbl.

In spite of being a very prolific area relative to volume of oil in place, the negative aspects of the São Mateus platform are the small dimensions of the accumulations, the poor reservoir quality of some reservoir lithologies, and the low API gravity values of the oil.

A well-defined seismic reflection is associated with Aptian anhydrite, which seals the producing reservoirs of the São Mateus platform. Gentle structures at this level control commercial hydrocarbon accumulations, while steep structures, generally associated with prominent Precambrian basement highs, are linked to poorly developed, unfavorable reservoir conditions, resulting in dry holes.

Stratigraphic traps in the structurally low areas adjacent to these highs, with thicker and better reservoir conditions, are being researched. Although wildcat 1-RI-1-ES found only a 35 m thick Mucuri Member sandstone at the top of the structure, thicker sand and better reservoir conditions were found in the 7-RI-13-ES, an adjacent and lower area with an average reservoir thickness of about 110 m.

ACKNOWLEDGMENTS

We thank Petróleo Brasileiro S.A. (Petrobrás) for permission to publish this paper. Unpublished Petrobrás studies by R. J. Alves, B. Ansaloni, L. Arantes, C. H. R. Cunha, E. A. C. Gueiros, U. M. Maino, D. Milanez, F. Nepomuceno Filho, J. L. Pantoja, L. A. F. Trindade, and M. M. Silva and conversations with N. Aurich, A. Z. N. de Barros, V. Q. Lima, G. A. V. da Rosa and many other Petrobrás colleagues were very helpful in preparing this paper.

REFERENCES

Asmus, H. E., J. B. Gomes, and A. C. B. Pereira, 1971, Integracão geológica regional de Bacia do Espirito Santo (Regional Geologic Integration of Espirito Santo Basin), *in* XXV Congresso Brasileiro de Geologia, São Paulo, SBG, v. 2, p. 235–252.

Estrella, G. O., M. R. Mello, P. C. Gaglianone, R. L. M. Azevedo, K. Tsubone, E. Rossetti, J. Concha, and I. M. R. A. Brüning, 1984, The Espirito Santo Basin (Brazil) source rock characterization and petroleum habitat, *in* G. Demaison, and R. J. Murris, eds., Petroleum geochemistry and basin evolution: American Association of Petroleum Geologists Memoir 35, p. 253–271.

Morgan, W. S., 1975, Heat flow and vertical movements of the crust, *in* A. G. Fisher and S. Judson, eds., Petroleum and global tectonics: New Jersey, Princeton University Press.

Appendix 1. Field Description

Field name .... *Rio Itaúnas field*

Ultimate recoverable reserves .... *5.4 MMbbl*

Field location:

- **Country** .... *Brazil*
- **State** .... *Espirito Santo*
- **Basin/Province** .... *Espirito Santo basin/São Mateus platform*

Field discovery:

- **Year first pay discovered** .... *Mucuri Formation 1977*
- **Year second pay discovered** .... *Caravelas Formation 1978*
- **Third pay** .... *Barra Nova Formation (Regência Member) 1979*

Discovery well name and general location:

- **First pay** .... *Rio Itaúnas No. 1, 5 km northwest of Itaúnas River mouth*
- **Second pay** .... *Rio Itaúnas No. 2, 4.5 km northwest of Itaúnas River mouth*
- **Third pay** .... *Rio Itaúnas No. 5, 9 km northwest of Itaúnas River mouth*

Discovery well operator .... *Petróleo Brasileiro S.A. (Petrobrás)*

- **Second pay** .... *Petróleo Brasileiro S.A. (Petrobrás)*
- **Third pay** .... *Petróleo Brasileiro S.A. (Petrobrás)*

IP in barrels per day and/or cubic feet or cubic meters per day:

- **First pay** .... *150 BOPD (Mariricu Formation)*
- **Second pay** .... *Zero (viscous oil)*
- **Third pay** .... *180 BOPD (Barra Nova Formation)*

All other zones with shows of oil and gas in the field:

Age	Formation	Type of Show
Albian	*Barra Nova (São Mateus Member)*	*Whole cuttings*

Geologic concept leading to discovery and method or methods used to delineate prospect, e.g., surface geology, subsurface geology, seeps, magnetic data, gravity data, seismic data, seismic refraction, nontechnical:

Wildcat 1-RI-1-ES tested a structural high mapped by seismic profiles at the top of the Aptian anhydrite bed, which acted as an excellent seal for the underlaying Mucuri Member sandstone reservoir.

Structure:

- **Province/basin type** .... *Bally, 114; Klemme, III C*

Tectonic history

Structural framework of São Mateus platform, as well as Espirito Santo basin, is similar to that of other East Brazilian marginal basins whose main features reflect continental breakup events resulting from African and South American plate separation.

Regional structure

The general structure of the Precambrian basement is a monocline dipping to the east, affected by normal faults.

Local structure

The local structure is formed by a north-south anticlinal feature bounded by a master fault to the west.

Basin stratigraphy (major stratigraphic intervals from surface to deepest penetration in field):

Chronostratigraphy	Formation	Depth to Top in m
Tertiary/Quaternary	*Rio Doce*	*Surface*
Tertiary	*Caravelas*	*–650*

Albian	*Barra Nova*	*-680*
Aptian	*Mariricu*	*-1100*
Precambrian	*Basement*	*-1330*

Reservoir characteristics:

Number of reservoirs *3*
Formations .... *Mariricu, Barra Nova, and Caravelas*
Ages .... *Aptian, Albian, and Tertiary*
Depths to tops of reservoirs .... *1170 m, 990 m, and 600 m*
Gross thickness (top to bottom of producing interval) .... *110 m (Mariricu Fm. Mucuri Member)*
Net thickness—total thickness of producing zones
Average .... *16 m (Mariricu Fm. Mucuri Member)*
Maximum .... *41 m*
Lithology
Mariricu Formation Mucuri Member: coarse to conglomeratic, arkosic/subarkosic, poorly sorted sandstones
Barra Nova/Regência: oncolitic/oolitic limestones
Caravelas: oolitic limestones
Porosity type .... *Secondary intergranular porosity by calcite dissolution (Mucuri Member)*
Average porosity .... *18% (Mucuri Member)*
Average permeability .... *100 md (Mucuri Member)*

Seals:

Upper
Formation, fault, or other feature .... *Mucuri Formation/Itaúnas Member*
Lithology .... *Anhydrite*
Lateral
Formation, fault, or other feature .... *Fault*
Lithology .... *Anhydrite*

Source:

Formation and age .... *Mariricu Formation (Neocomian and Aptian)*
Lithology .... *Shale*
Average total organic carbon (TOC) *4%*
Maximum TOC *7.0%*
Kerogen type (I, II, or III) .... *I and II*
Vitrinite reflectance (maturation) .... $R_o = 0.6\text{–}0.8$
Time of hydrocarbon expulsion .... *Not available*
Present depth to top of source .... *Above 1500 m*
Thickness .... *Around 30 m*
Potential yield .... *Not available*

Appendix 2. Production Data

Field name .... *Rio Itaúnas field*

Field size:

Proved acres .... *450 ha*
Number of wells all years *79*
Current number of wells *79*
Well spacing .... *200 m*
Ultimate recoverable .... *5.4 million bbl*
Cumulative production (through 1989) .... *3.8 million bbl*
Annual production (1989) .... *0.39 million bbl*

Present decline rate *30%*
Initial decline rate *30%*
Overall decline rate .... *NA*
Annual water production .... *0.23 × 10^6 bbl*
In place, total reserves .... *26.0 million bbl*
In place, per acre-foot .... *NA*
Primary recovery .... *5.4 million bbl*
Secondary recovery .... *NA*
Enhanced recovery .... *NA*
Cumulative water production .... *0.87 × 10^6 bbl*

Drilling and casing practices:

Amount of surface casing set .... *1665 m*
Casing program .... *13⅜-in. at 25 m; 9⅝-in. at 300 m; ½-in. to the well bottom*
Drilling mud .... *Water-base mud, oil-base mud, KCl-base mud*
Bit program .... *17½-in. up to 25 m; 12¼-in. up to 300 m; 8½-in. to the well bottom*
High pressure zones .... *None*

Completion practices:

Interval(s) perforated .... *1180–1285 m*
Well treatment .... *Xylene fracturing, acidizing*

Formation evaluation:

Logging suites .... *Induction, SP, GR, density, neutron, dipmeter*
Testing practices .... *Open and cased hole drill-stem test, wireline test*
Mud logging techniques .... *Not used*

Oil characteristics:

Type .... *Paraffinic*
API gravity .... *17° API*
Base .... *NA*
Initial GOR .... *38 m^3/m^3*
Sulfur, wt% .... *NA*
Viscosity, SUS .... *45 cp*
Pour point .... *NA*
Gas-oil distillate .... *NA*

Field characteristics:

Average elevation .... *7 m*
Initial pressure .... *126 kgf/cm^2 (–1250 m)*
Present pressure .... *70 kgf/cm^2*
Pressure gradient .... *0.1 $kgf/cm^2/m$*
Temperature .... *132°F (55.6°C)*
Geothermal gradient .... *0.012°F/ft (0.022°C/m)*
Drive .... *Gas cap expansion*
Oil column thickness .... *75 m*
Oil-water contact .... *–1260 m*
Connate water .... *NA*
Water salinity, TDS .... *68.2 g/L*
Resistivity of water .... *0.065 ohm-m at 130°F (54.4°C) (60,000 ppm NaCl)*
Bulk volume water (%)

Transportation method and market for oil and gas:

Oil is transported by truck to Fazenda Cedro field processing station and from there to Regência oil terminal by oil pipeline. From there it is transported to Rio de Janeiro oil refinery. The gas is reinjected into the gas cap of the field.

Cod Field—Norwegian Sector
Central Graben, North Sea

MICHEL D'HEUR
Petrofina; Fina Exploration Norway
Brussels, Belgium

FIELD CLASSIFICATION

BASIN: North Sea
BASIN TYPE: Rift
RESERVOIR ROCK TYPE: Sandstone
RESERVOIR ENVIRONMENT OF DEPOSITION: Distal Edge of Submarine Fan System
RESERVOIR AGE: Paleocene
PETROLEUM TYPE: Gas
TRAP TYPE: Dome over Salt with Stratigraphic Components

LOCATION

The Cod gas condensate field lies in the southern area of the Norwegian sector of the North Sea, in block 7/11, 280 km southwest of the Norwegian coast (Figure 1). The field is also situated in the North Sea Central graben, 46 mi (75 km) north-northwest of the giant Ekofisk field (rank 98, after Carmalt and St. Johns, 1986) and a half mile (1 km) east of the median line between the English and Norwegian sectors. Water depth in the area is 230 ft (70 m). The reservoir is the upper Paleocene sandstone of deep marine distal turbidite origin, located at a depth of 9400 ft (2870 m) subsea. Cod is expected to produce more than 280 bcf of gas and 17 million barrels of condensate.

HISTORY

Pre-Discovery

In 1965, Norwegian blocks 7/11 and 7/8, designated by the Phillips Group as the "C area," were allocated to Phillips, Petrofina, and Agip as part of Production Licence PLO18 (Figure 2). Phillips and Agip combined their interests with the Petronord Group in January 1967, and the present participation interests in PLO18 are Phillips Petroleum Co. of Norway (Operator, 36.96%); Norske Fina (30%); Norsk Agip (13.04%); Elf Aquitaine Norge (8.1%); Norsk Hydro (6.7%); Total Marine Norsk (4.05%); Eurafrep Norge (0.45%); Coparex Norge (0.4%); and Cofranord (0.3%).

In 1965, three "en echelon" domal structures were mapped in blocks 7/8 and 7/11 in the early stage of seismic reconnaissance. In 1967, the southern feature was divided into two distinct prospects. All prospects identified by the Phillips Group in Norway received a fish name starting with the letter that designated the specific area. Since block 7/11 fell within the "C area," the name Cod was selected for the southernmost structure. The dome is well defined on seismic horizons close to the base of the Tertiary, the base of the chalk, and the top of the Permian; it broadens slightly in the area of closure at shallower depths. At the base of the Tertiary level, the vertical closure was approximately 600 ft (180 m).

Stratigraphic control of this area was very limited (as will be discussed in the section *Discovery Methods*). It was considered that the first well on Cod would be adequate to test the potential of the very thick Tertiary section as well as that of the Mesozoic. The well had therefore five possible objective zones: lower Tertiary sandstone, Upper Cretaceous carbonates, Lower Cretaceous sandstone, Jurassic sandstone, and Lower Triassic sandstones. Since seismic data indicated that the Permian evaporite thickness could exceed 5000 ft (1500 m), the Lower Permian and pre-Permian reservoirs identified in the southern North Sea were too deep to be considered as objectives in the Cod area.

Discovery

The first well, 7/11-1X, was spudded in February 1968 near the crest of the Cod dome (Figures 2 and 3A). It encountered a gas-bearing sequence of interbedded sandstones and shales of late Paleocene age between 9400 and 10,400 ft (2870–3170 m). The upper half of this interval, unofficially called the main Cod sand, contained reservoirs of good quality, totaling 170 to 230 ft (52–70 m) of net pay (depending on pay cut-off parameters used). The best zone was perforated over 75 ft (23 m) and flowed more than 40 MMcfd of gas and some condensate. Below the upper Paleocene, the well penetrated 2000 ft (610 m) of chalk, a thin interval of black Jurassic shales, and

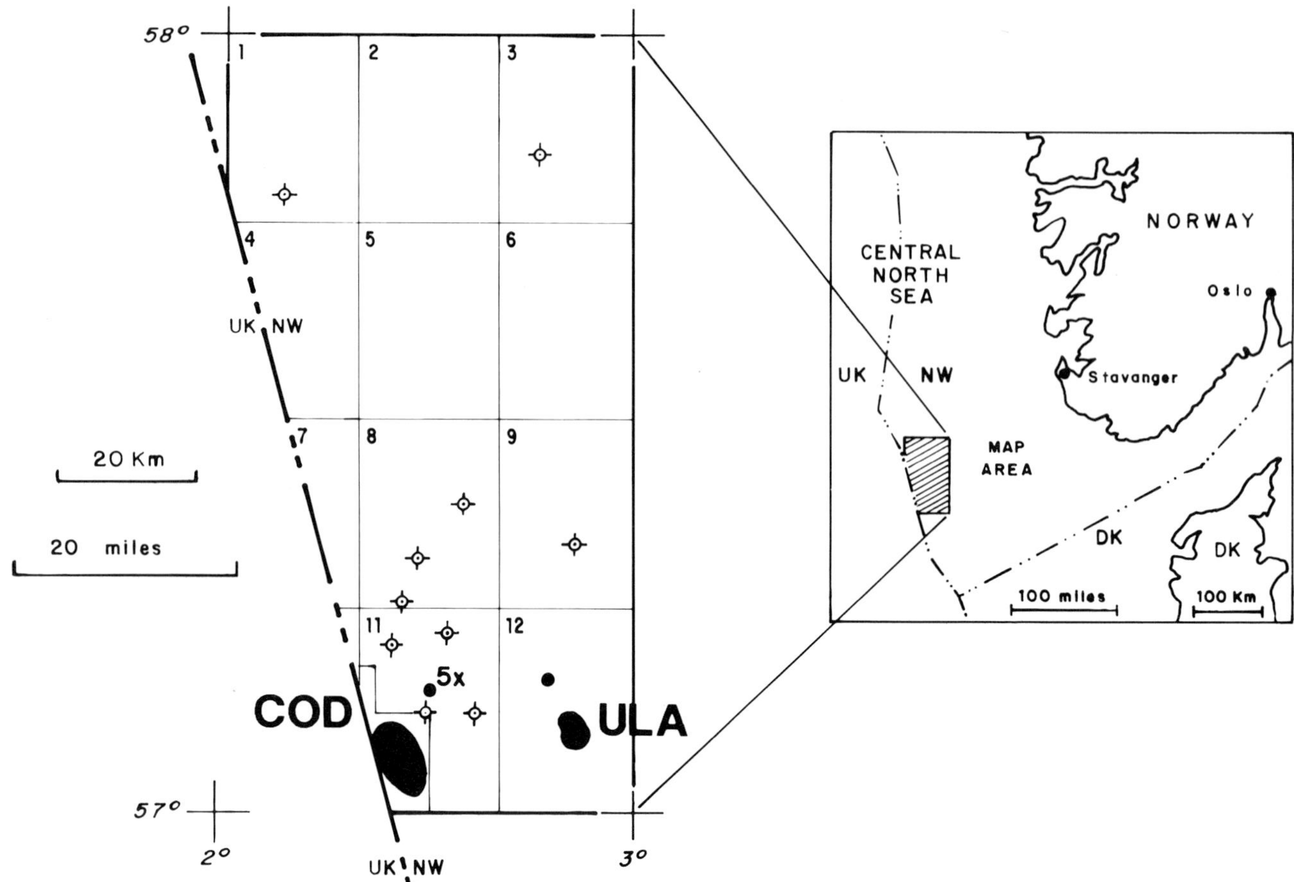

Figure 1. Location map of Cod field on the right, Norway Quadrant 7 left. UK, United Kingdom sector. N, Norwegian sector. DK, Danish sector.

white Permian anhydrite. The well bottomed in the Zechstein salt at a total depth (TD) of 13,000 ft (3965 m). It was estimated that well 7/11-1X would be capable of producing 2700 BCPD and over 40 MMcfd of gas.

The first appraisal well, 7/11-2X, was spudded in July 1968, 1.1 mi (1.8 km) west of the discovery well, on the western flank of the dome (Figure 3B). This well was also located in an area where thicker sandstones were expected. The well encountered the top of the reservoir 400 ft (122 m) deeper than in well 7/11-1X and with 170 ft (52 m) of net pay, which was 90 ft (27 m) less than prognosticated. The reservoir was gas-bearing and tested at flow rates of up to 43 MMcfd. The maximum recorded liquid flow rate was 1390 bpd of 51.6 to 61° API condensate. The average GOR was 11,900 ft^3 gas/bbl oil. Unlike the discovery well, the appraisal well had no indications of shows in the lower sandstone-shale member of the Paleocene nor in the upper layers of the chalk, where the well was drilled to TD at 11,250 ft (3431 m).

Wells 7/11-1X and 7/11-2X found no clear indication of a water table. Well 7/11-3X, spudded in October 1968, was located 1.9 mi (3 km) southeast of the discovery well to test the southeastern nose of the structure. The Paleocene sandstones were encountered between 10,050 and 10,850 ft (3065–3309 m) subsea (ss). The upper 20 ft (6 m) contained oil and water. Below 10,070 ft (3071 m) ss, the sandstone, which was of poor reservoir quality compared to that found in the previous wells, was water-bearing. The well was drilled to TD at 11,000 ft (3355 m) near the top of the lower Paleocene Danian chalk. The uppermost interval of the sandstone was tested and flowed at rates of 0.1 MMcfd, 90 BOPD, and 340 BWPD, suggesting the presence of an oil leg below the gas accumulation.

Based on the results of the three exploration wells, the productive area was reduced to 8 mi^2 (20.7 km^2), and the water level was set at -10,070 ft (-3071 m), which is 200 ft (61 m) above the spillpoint of the dome. Fluid and pressure analyses demonstrated the existence of at least five reservoir zones separated by partial or total permeability barriers. The turbiditic nature of the sands complicated the calculation of recoverable reserves, which were then estimated at 700 to 800 bcf of gas plus oil and condensate.

The fourth well on the block was drilled in July 1969 to test the main Cod sand on the satellite "Northeast Cod" prospect. Maximum structural

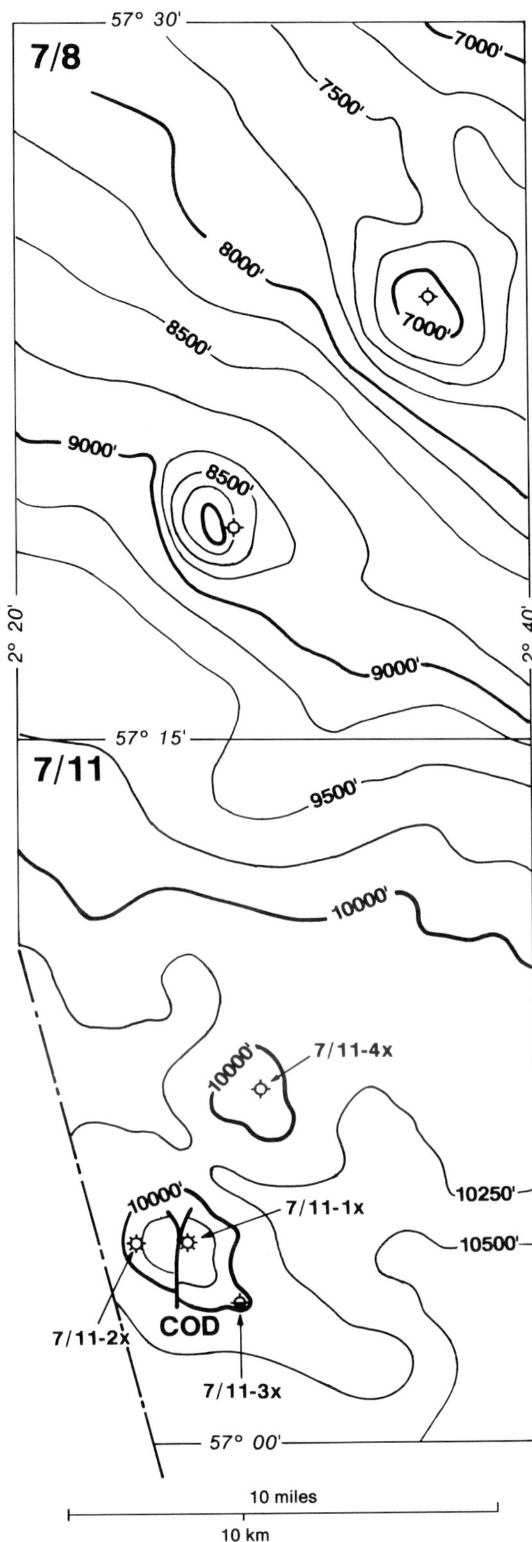

Figure 2. A 1969 depth map of the Paleocene marker showing the "en echelon" structures in blocks 7/8 and 7/11 (Phillips area C, 1969) allocated to the Phillips Group in 1965. Contour interval, 250 ft (76 m). Most of the area outside the Cod field has been relinquished in the seventies and reallocated to other companies since then. Well 7/11-4X is located on the small Northeast Cod structure.

closure at the top of the Paleocene level was 200 ft (61 m), with a closure area of 6 mi^2 (15.5 km^2). The main Cod sand or its equivalent comprised only 30 ft (9.2 m) of porous sandstone intercalated with shaly, tight sandstones and gray shales. The porous sandstones were located below the water table identified in well 7/11-3X. There was no porous interval in the upper part of the chalk layer, and since no wells in the area had proven hydrocarbons deeper in the Mesozoic, well 7/11-4X was drilled only to TD in the Upper Cretaceous.

Post-Discovery

A detailed development plan for Cod was not elaborated until the mid-seventies. In the meantime, the giant Ekofisk field and several satellites had been discovered 46 mi (75 km) to the southeast of Cod, and it was decided to develop Cod in connection with Ekofisk. A nine-slot drilling production platform was installed in 1975 on the basis of the structural map illustrated in Figure 3B. Development drilling commenced in the summer of 1976 with well A7. Despite being located only 0.6 mi (1 km) north-northeast of the discovery well, A7 encountered less than 50 ft (15.2 m) of reservoir in the main Cod sand member, and no hydrocarbons were present. This was thought to result from a drastic shaling-out of the reservoir interval northwards (Figure 4). However, well A2 drilled midway between 7/11-1 and 7/11-3 found thick water-bearing sandstone, thus demonstrating that the 7/11-3 reservoir was not part of the main Cod accumulation. Well A5, located very close to the discovery well, found a thinner reservoir interval.

To accommodate the results of these three development wells, a north-south sealing fault was arbitrarily placed as shown in Figure 3C. This interpretation led to a drastic reduction of the reserve estimate of approximately 50%. The three remaining wells of the original plan were drilled successfully during the first half of 1977. At that time, a revised development plan including two additional wells (A4 and A8) was submitted to the Norwegian authorities. All well locations are shown on a recently revised structure map in Figure 5 where a 1988 interpretation of sandstone distribution also is shown.

Well A4 was located in the small gas area of the eastern compartment, updip of the dry hole A2. It has indications of gas down to 9650 ft (2943 m) ss, while well A2 has water-bearing sandstones up to 9750 ft (2974 m) ss, indicating a gas-water contact at a depth of 9700 ft (2958 m) ss ± 50 ft (15.2 m). The second additional well, A8, drilled west of well A9 (Figure 3D), which had indications of a thickening of the reservoir to the northwest, encountered the thickest pay section in the field (230 ft; 70 m). A ninth well, A6 (also referred to as 7/11-7), drilled in 1983 to evaluate the Mesozoic section on the eastern flank of the structure following discoveries in Upper Jurassic sandstones at Ula and Northeast Cod, found no indications of hydrocarbon.

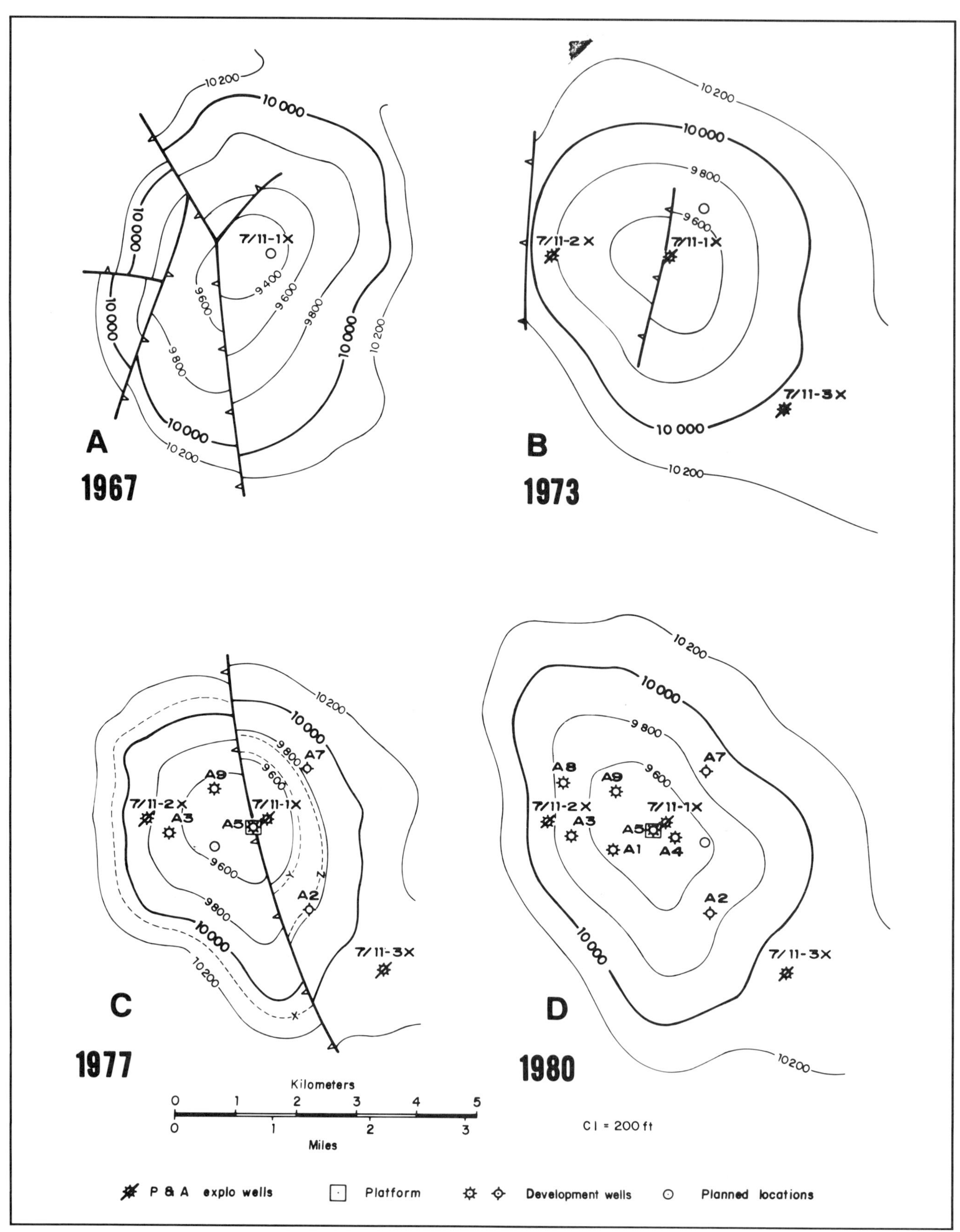

Figure 3. Successive structural interpretations of the Cod structure, top of the Forties Formation. (A) Before exploration drilling. (B) Before development drilling. (C) After drilling of two dry development wells. (D) Based on additional seismic data. The 1988 version is given in Figure 5.

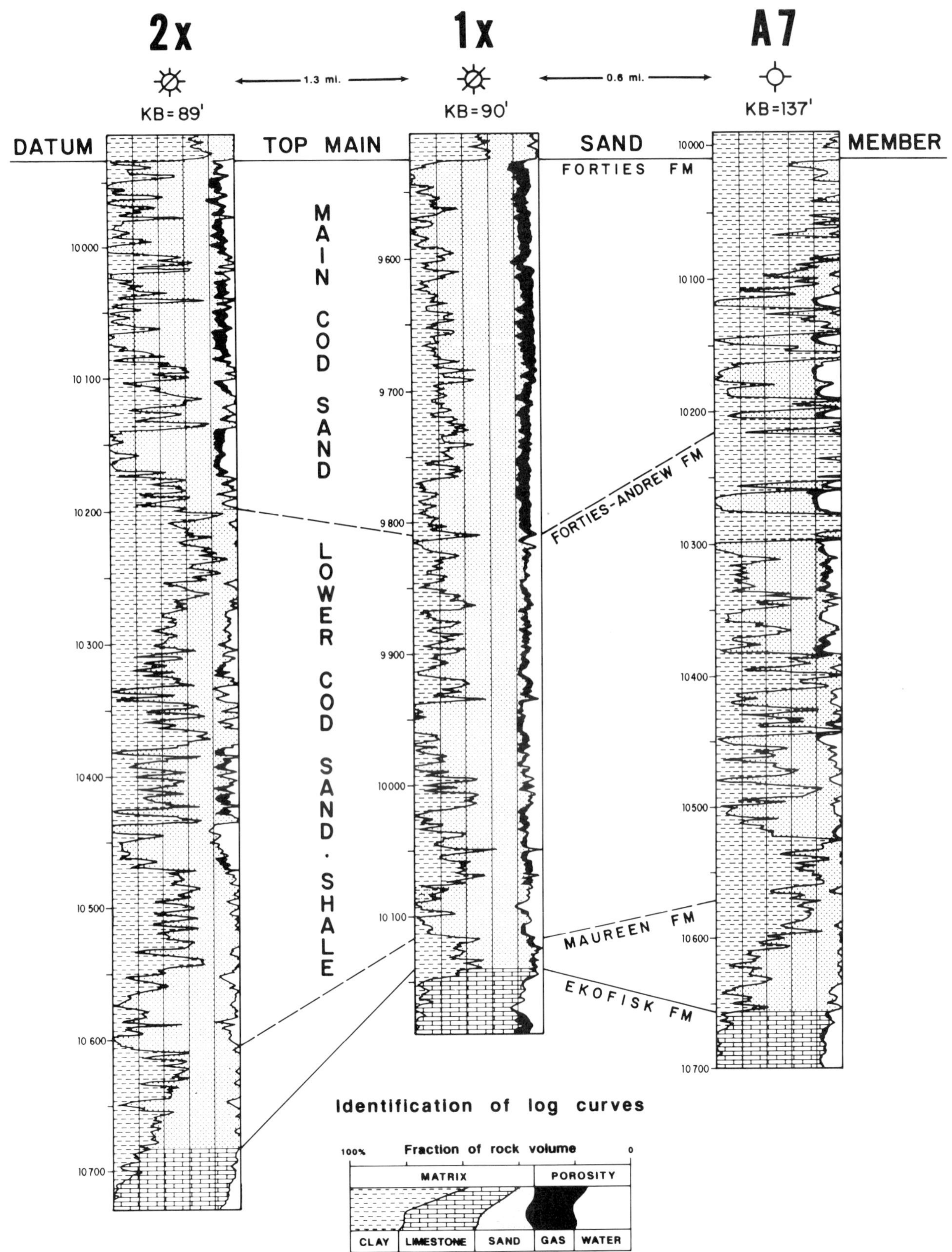

Figure 4. Lithology, porosity, and hydrocarbon saturation logs of the Paleocene sandstone interval of the discovery well (7/11-1X), the first appraisal well (7/11-2X), and the first development well A7. Note the drastic shaling out of the reservoir (main Cod sand) between wells 1X and A7, only 0.6 mi (1 km) apart.

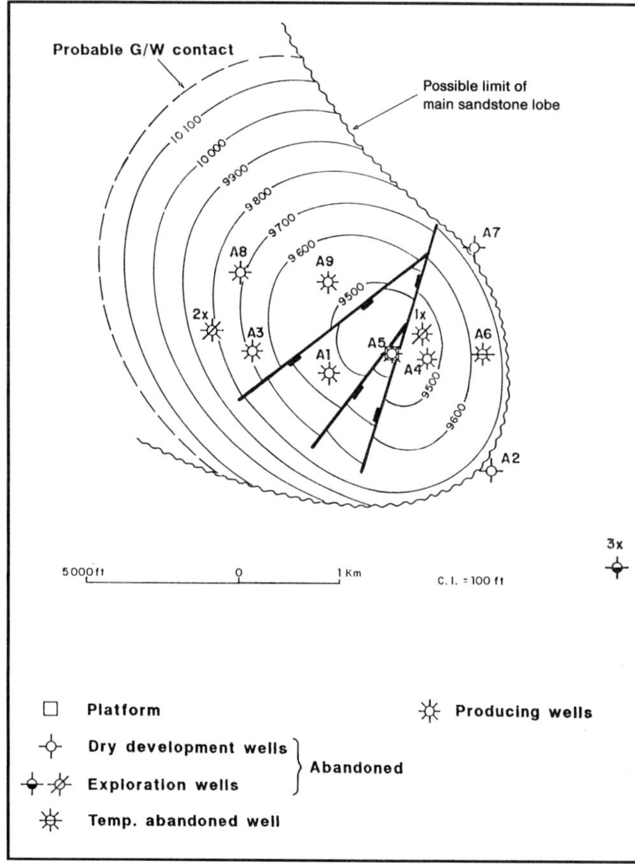

Figure 5. 1988 structural interpretation of the Cod structure at the top of the Forties Formation reservoir. Depths are subsea. Additional minor faults have not been represented.

The Cod field production facilities were linked to Ekofisk and production began in December 1977. Peak production of 115 MMcfd of gas was attained in 1980. The production rate in early 1988 was about 50 MMcfd of gas and 1400 BCPD. Cumulative production to 1990 is approximately 250 bcf gas and 16 MMbbl condensate.

DISCOVERY METHOD

Then

Exploration drilling in the North Sea began in the southern area after the discovery of the giant Gröningen field (ranked 23) in the Netherlands coastal onshore area in 1959. Huge amounts of gas were encountered in Lower Permian sandstones overlying rich gas-prone source rocks of the Carboniferous Coal Measures. Since the coal beds were also known in England, geological interpolations were rapidly made into areas offshore between these two countries. After governments had allocated offshore licenses, exploration commenced in the English and Dutch sectors in 1964 and 1967, respectively. Six gas fields totaling 23 tcf of gas were discovered in the U.K. sector in 1965 and 1966. Following such success, exploration moved northwards in 1967 to the central North Sea where much less information was available. The Coal Measures were not documented in the area, and in addition, the Lower Permian reservoir was thought to be too deep to be an attractive objective. Since the mid-sixties, however, geophysical exploration in the central North Sea had identified a thick Tertiary basin underlain by a deep Mesozoic section that clearly contained mobile halite.

Halite was in many circumstances the cap rock of the Permian reservoir in the Southern basin, but should source rock and reservoir be present in the Mesozoic to the north, then halokinesis could create structural traps. Seismic lines shot in 1965 and 1966 showed three large domes in blocks 7/8 and 7/11 (Phillips area C) in the central area of the Tertiary basin. Stratigraphic control was, however, very limited. Results of well 16/11-1, down to 5000 ft (1525 m) ss (Figure 6), just drilled by the Phillips Group 70 mi (110 km) north in Norwegian water, and of the Tenneco/Hamilton well 30/18-1, down to 9500 ft (2898 m) ss, 50 miles (80 km) to the south (Figure 6), provided the guidelines for assessment of the Tertiary section. The Tertiary of both wells was marine and shaly with possible source rock in dark gray-black shales of Eocene and Paleocene age. In addition, oil stains were observed in sandy Paleocene intervals of well 30/18-1.

It was then believed that Paleocene sandstones were derived from the mid-North Sea high, a west–east-trending paleohigh, just south of well 30/18-1. Since the sandstones were not present in well 16/11-1, they were believed to pinch out northwards, possibly north of area C. The Paleocene sandstone and other possible Tertiary sandstones became the primary objectives in the area.

Further south, in the Danish sector, two wells were believed to have tested oil in the lower Tertiary, possibly in chalk reservoirs. The Norwegian well 25/11-1, drilled in 1966, 145 mi (230 km) north of Cod, tested oil from a Paleocene–Eocene sandstone at 5500 ft (1678 m) ss. Thin porous but water-bearing zones were also identified in the Cretaceous chalk section of well 16/11-1. Finally, possible Lower Cretaceous to Triassic sandstones known in the Southern basin but as yet undrilled in the central North Sea were additional objectives in area C. As previously mentioned, Lower Permian and older rocks, probably present at depth greater than 18,000 ft (5490 m) ss, were too deep to be considered viable objectives at this location.

While geological hypotheses were being developed, new seismic mapping based on a 1967 2 km × 2 km (1.2 × 1.2 mi) grid helped to improve the structural definition. Based on a new seismic line and interpretation, the largest and southernmost dome was subdivided into two domes named Cod and Northeast Cod, separated by a saddle. A seismic line through

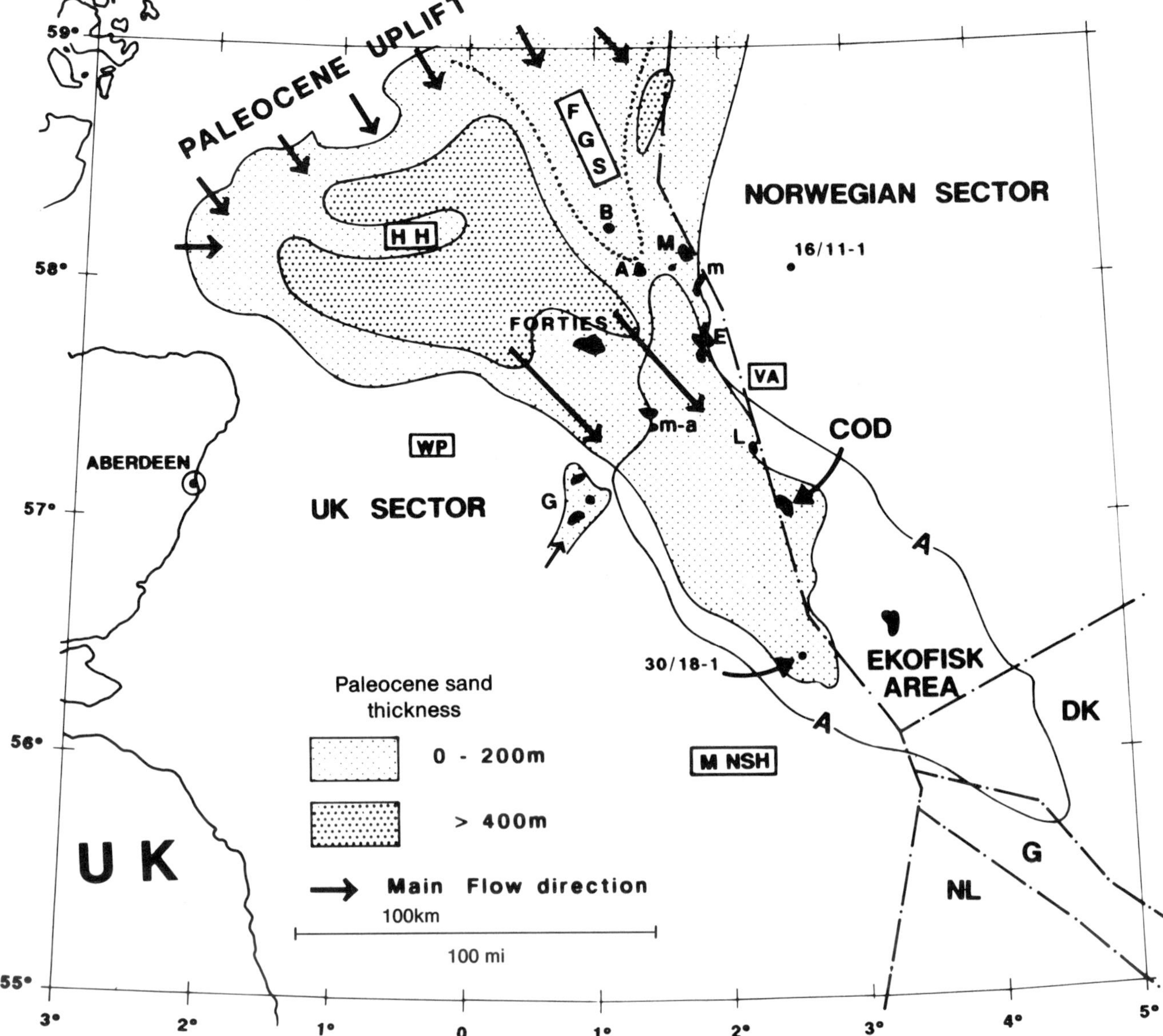

Figure 6. Distribution of the Paleocene sandstone in the Central trough and adjacent areas to the north. Structural highs controlling sand distribution are the Fladen Ground spur (FGS), the Hallibut horst (HH), the western platform (WP), the Vestland arch (VA), and the mid-North Sea high (MNSH). The most significant hydrocarbon accumulations in the Paleocene sandstones are Forties, Maureen (M); Montrose and Arbroath (m-a); Cod, Erskine (E); Andrew (A); Balmoral (B); Gannet (G); and Lomond (L). Contour A approximately coincides with isobath -10,000 ft (-3050 m) of the base of the Paleocene sandstone.

the Cod structure is shown in Figure 7A and a geoseismic section in Figure 7B. Permian salt was identified as responsible for having domed the overlying strata, creating closures up to 4000 ft (1220 m) below sea level within the Tertiary section. The dome broadens slightly in the area of closure from 7 mi^2 (18 km^2) at the deep (top salt) horizon to 12 mi^2 (31 km^2) at the base of the Tertiary seismic marker whereas vertical closure decreases from 1100 ft to 600 ft (336 m to 183 m). Radial normal faults were identified in various horizons from lower Mesozoic to Tertiary sections. After the discovery of Cod, seismic lines totaling 52 km (32 mi) were shot, in 1969, and the field was then covered by a 1 km × 1 km (0.6 × 0.6 mi) grid until development drilling started in 1976.

Now

Based on dipmeter analyses and on very limited data available in 1968, it was believed that the source of the Paleocene sands was located south of Cod, possibly in the mid-North Sea high area. Detailed mineralogical analyses of the appraisal wells and comparison with rocks from Scotland led in 1969 to a new geological model with the source of these sands

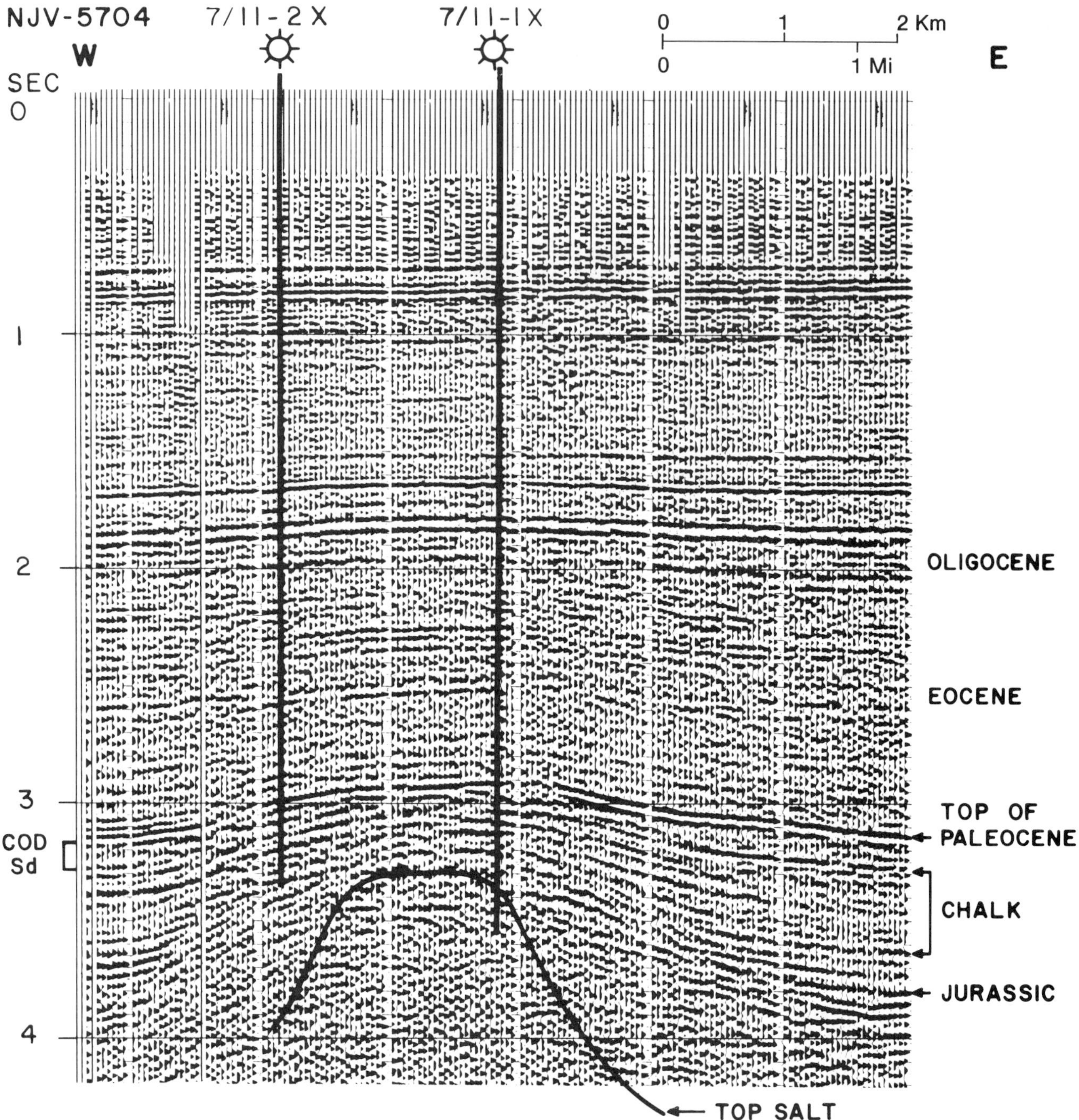

Figure 7A. West-east seismic line. (From D'Heur, 1987, with the permission of the Norwegian Petroleum Society.)

correctly located to the northwest. It was, however, inferred from the model that the thickest sand section would be present in the deepest part of the trough, which was not found to be the case.

The depositional setting of the Paleocene is now well defined (see section on *Structure*). Cod appears to be located at the southeastern edge of the upper Paleocene submarine fan complex. The 1965–1967 seismic data were adequate for exploration of the basal Tertiary strata, and no significant improvement could be expected at the Cod location. However, the problematic results of the first development wells led to further seismic acquisition, totaling 146 km (90 mi) at a spacing of 700 m (2250 ft).

The new seismic data were used in an attempt to determine continuity of the sandstone bodies, with questionable results. This seismic stratigraphy study is summarized by Kessler et al. (1980). Despite the thinness (0.2 sec) of the total Paleocene sandstone interval, a number of "pod-like" features were

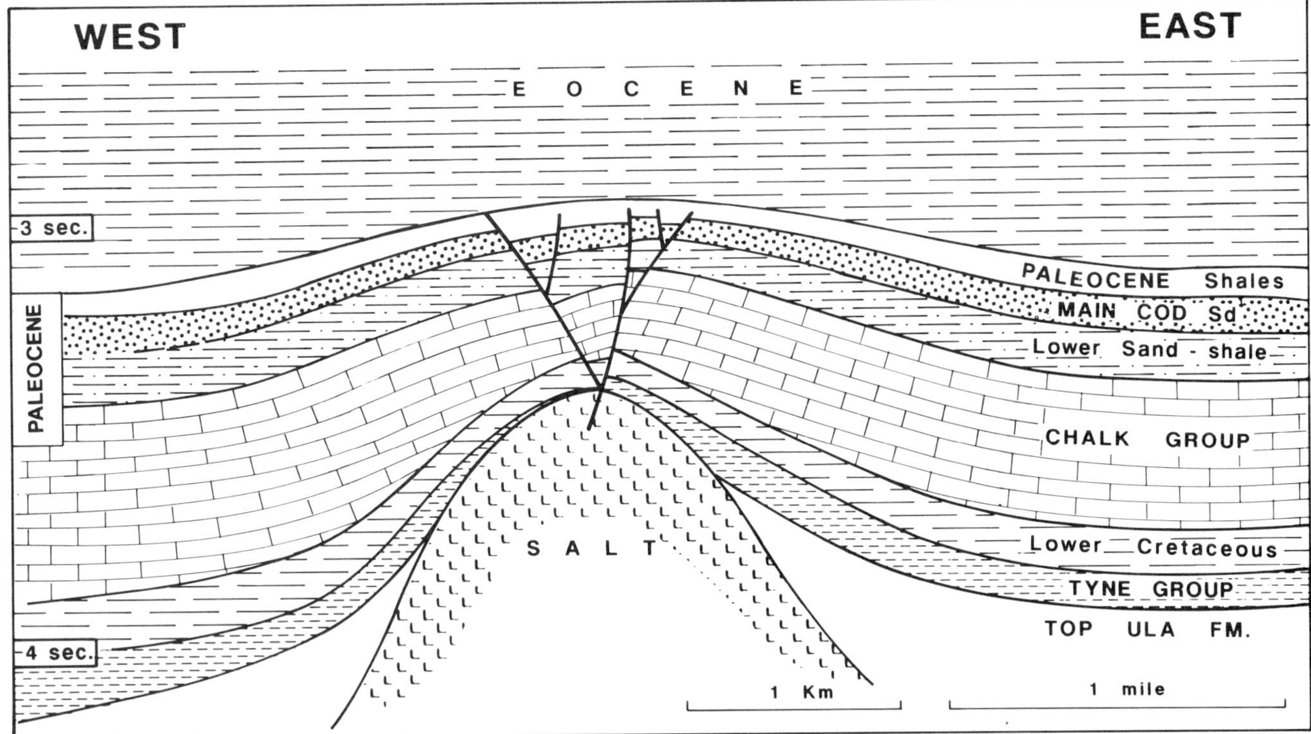

Figure 7B. West-east geoseismic section across the Cod structure. The upper Eocene to recent sequence above 2.5 sec has not been represented.

observed. They consist of zones where concave and convex or flat reflectors converge. An illustration of such a pod-like interpretation based on seismic, sedimentology, and pressure evaluation is given in Figure 8.

In 1982, following the discovery of oil 3 mi (5 km) northeast of Cod by a group led by Norsk Hydro, the development well A6, also referred to as 7/11-7X, was deepened to investigate the Mesozoic section. This well found a thin Jurassic sandstone to be water-bearing. In 1983, a new seismic survey of 320 km (199 mi) was shot in common with Norsk Hydro, which helped improve definition of Paleocene and Jurassic intervals in the area.

STRUCTURE

Tectonic History

The Cod field lies near the center of the Tertiary basin, where the thickness of the Cenozoic exceeds 11,000 ft (3350 m). The Tertiary basin, better termed the Central trough, overlies the older Central graben initiated in Early Permian time. During the Late Permian, the area was part of the Northern Permian basin where thick evaporites were deposited. The Permian Zechstein salt is the oldest rock penetrated below the Cod field.

Subsidence continued during the Triassic with thick accumulations of varicolored shallow marine claystone and shale and fluvial siltstone and sandstone. Part of the Triassic and all Lower and Middle Jurassic deposits were removed by erosion during several phases of Kimmerian movements. This was followed by a wide marine transgression during Late Jurassic time. By the end of the Jurassic, the transgression had covered the entire area and the organic-rich black shales of the Mandal Formation (Kimmeridge Clay equivalent) were deposited. During Late Jurassic and Early Cretaceous times, Cod was located between the Vestland arch to the northeast and the Central graben that had developed to the west. Early Cretaceous time was characterized by low-energy environments with deposition of marine shales, marls, and limestones.

By mid-Cretaceous time, the Central graben complex was buried and the area began to subside more regularly and regionally to form the wide Central trough where thick chalk accumulated during Late Cretaceous and early Paleocene times.

Open marine conditions prevailed during the Paleocene, and after chalk deposition, uplift, tilting, and erosion of the East Shetland Platform resulted in enormous quantities of sand being shed into the Central trough. The Forties Formation, which contains the Cod field reservoirs, represents a sequence of deep-water submarine fan sands that were deposited in the northwest-southeast direction

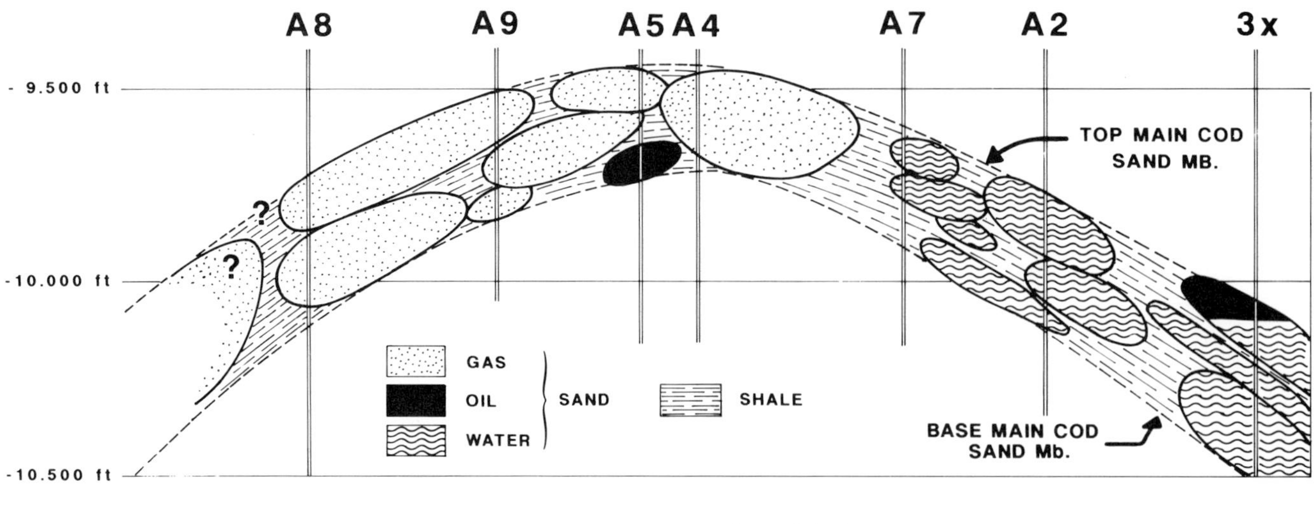

Figure 8. A combined sedimentologic/seismic/hydrodynamic interpretation of the main Cod sand member, mainly after Kessler et al. (1980), along a generally northwest-southeast section across the field. Location of wells can be seen in Figure 5.

of the axis of the trough. Subsidence persisted throughout the Tertiary with accumulation of clay, claystone, shale, silt, and sand providing 11,000 ft (3350 m) of cover.

Regional Structure

Cod is located more than 100 mi (160 km) south-southeast of the primary source of the Paleocene sands (Figure 6); consequently, the area received only a limited amount of sand. Along the axis of the trough, the sandstones extend southwards to well 30/18-1 where they were first discovered in 1967, as previously discussed.

Because it is located east of the axis of the trough, Cod contains less sand than other accumulations identified in its proximity to the west and northwest (Erskine and Lomond). The Montrose and Arbroath fields (totaling 200 MMbbl of recoverable oil) and the giant Forties field (2200 MMbbl of recoverable oil) produce from the same Paleocene formation. The Andrew field contains oil and gas in older upper Paleocene sandstones, and the Maureen field reservoir is mainly of early Paleocene age. The Gannet complex will produce from the Forties Formation equivalent, but here the sands were derived from a local supply area on the western flank of the Central trough.

Thickness of the total Paleocene sandstone is illustrated in Figure 6. The sandstone pinches out less than 10 mi (16 km) east of Cod.

Local Structure

Cod is a nearly circular dome with a 2 mi (3 km) radius resulting from halokinesis of the Permian salt, which probably started moving when loaded by the thick Triassic deposits. Halokinesis occurred in several pulses and a slight paleorelief existed during Paleocene time as suggested by thickness variations of the Paleocene interval. The trap attained its present shape during early Miocene time. The reservoir culminates at 9400 ft (2870 m) ss and the vertical closure exceeds 600 ft (180 m).

Several seismic interpretations show a fault cutting the anticline centrally in a north-northeastward direction. This fault is scissored and breaks into paired radial faults on the north and south of the dome. The fault that affects the seismic horizons from below the base of the chalk to above the top of the Paleocene has displacements averaging 70 ft (21 m).

STRATIGRAPHY

The Upper Permian Zechstein salt (Figure 9) is the oldest rock penetrated in the Cod field. In well A6, the salt is overlain by 150 ft (46 m) of varicolored shallow marine claystone, shale and fluvial siltstone, and sandstone of the Smithbank Formation (Triassic). The Smithbank Formation is in fact locally absent at the crest of the structure but thickens flankwards; the A6 well is nearly at midstructural elevation.

Lower and Middle Jurassic deposits are absent due to erosion and nondeposition. Upper Jurassic shales and marine sandstone of Kimmeridgian age rest on the Triassic rocks. These sandstones are equivalent to the upper unit of the Ula Formation (Vollset and Doré, 1984), which contain oil in Northeast Cod and which are expected to produce more than 200 MMbbl at the Ula field, 15 mi (24 km) to the east (Figure 1).

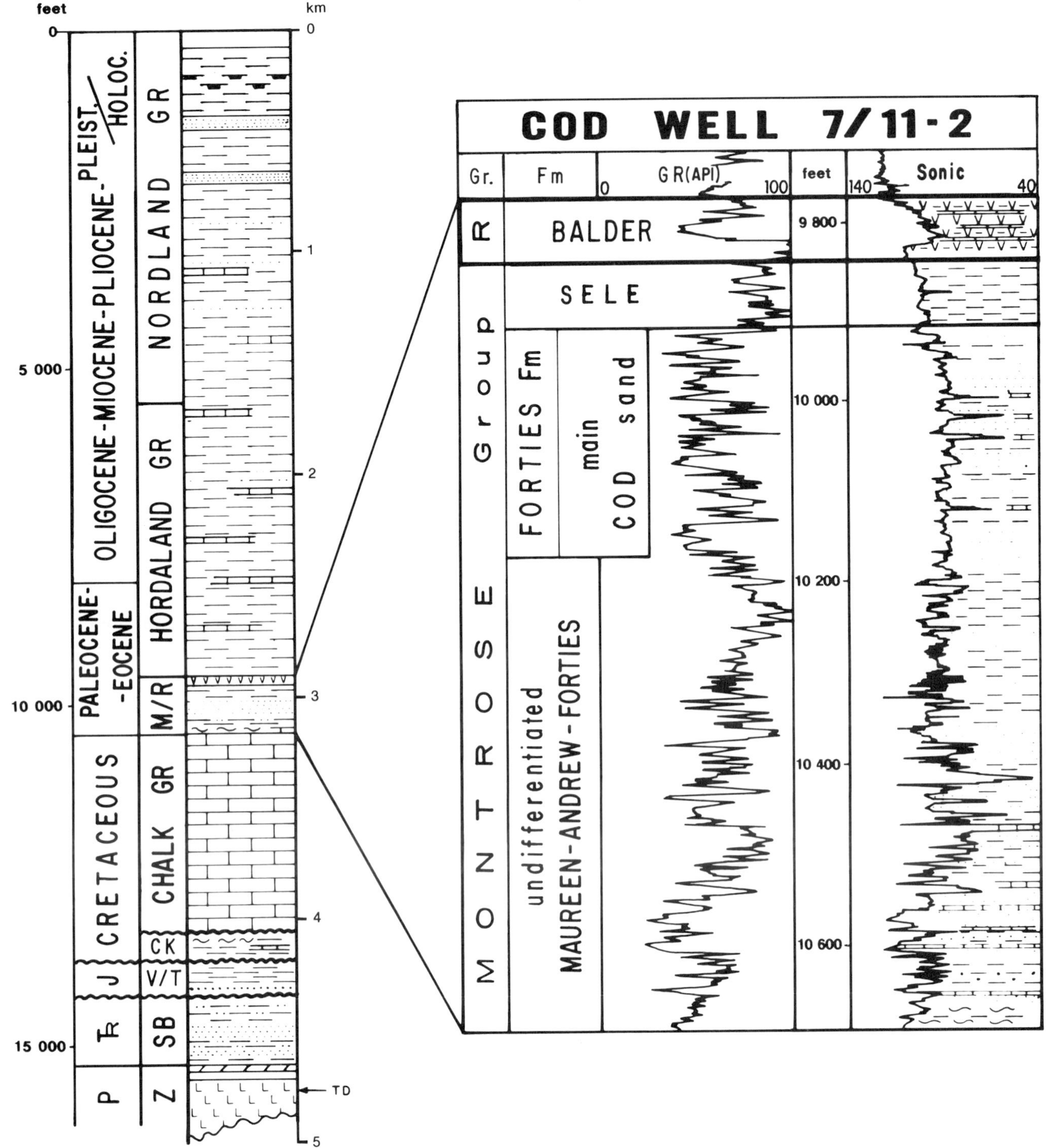

Figure 9. Stratigraphy of the Cod field. P, Permian; Z, Zechstein; TR, Triassic; J, Jurassic; SB, Smith Bank Formation; V/T, transition Vestland Formation to Tyne Formation; CK, Cromer Knoll Group; M/R, transition Montrose Group to Rogaland Group.

Below the Cod field, the Jurassic sandstones are water-bearing. They are overlain by 400 ft (122 m) of dark gray to black shales of the Mandal Formation, which is a lateral equivalent of the Upper Jurassic Kimmeridge Clay Formation. It is the source rock for hydrocarbons of the Cod field and all other fields in the area. In well A6, the Lower Cretaceous deposits consist of 300 ft (92 m) of calcareous shales and limestones of the Cromer Knoll Group.

The Upper Cretaceous deposits consist of gray to white chalk. Together with chalk of early Paleocene Danian age, they form the 3000 ft (915 m) thick Chalk

Group. The Danian carbonates of the Ekofisk Formation pass progressively into marls and shales of the Maureen Formation (Figure 9), which locally contain slumps and reworked material of Maastrichtian and Danian chalks.

Up-section, the Paleocene sediments become more and more sandy in the Andrew Formation and lower part of the Forties Formation. The upper 250 ft (80 m) of the Forties Formation contain massive sandstones intercalated with gray shales; this member is also called the main Cod sand. This reservoir is overlain by 50 ft (15 m) of gray shales of the Sele Formation, which terminate the Montrose Group.

The Rogaland Group comprises the typical ash beds of the Balder Formation. They were initially called upper Cod sand and were supposed to have produced gas in the discovery well 7/11-1X; but further analyses indicated that the gas had channeled from the main Cod sandstone behind the casing.

Shale, claystone, and clay of the Hordaland and Nordland groups, 9500 ft (2900 m) thick, cover the field. These groups, which also contain limestone stringers and rare silty and sandy intervals, conform to the general stratigraphy described by Deegan and Scull (1977).

TRAP

Trap Type

Cod is a domal structure with a vertical closure of 600 ft (185 m). The reservoir culminates at 9400 ft (2070 m) subsea. The trap consists of combined structural and stratigraphic closure of channel sandstone bodies, probably cut by several faults. The upper seal consists of Tertiary claystone and shale. The lateral seal results from complex relationships between the discontinuous sand bodies and their shaly lateral equivalents, from faults and/or truncation, and elsewhere from normal shaling-out. Although faults seem pretty obvious on the seismic lines, several "unfaulted" interpretations have been considered. The diagrammatical section of Figure 8 illustrates the discontinuity of sandstone bodies without the need of faults; this figure is appropriate to explain the difference in fluids (gas, oil, water) and pressure across the field. As shown by Kessler et al. (1980), such interpretation may be supported by seismic observations of a number of "pod-like" structures. These consist of zones where concave and convex reflectors seem to converge at the limit of the sandstone bodies. In the eastern part of the field, the gas-water contacts in the various wells occur over a depth range of 570 ft (175 m). To the west, where the sandstones are thick and massive, no clear indication of a water table has been found in the main Cod sand.

Reservoirs*

The reservoirs of the Cod field consist of sandstones of late Paleocene age, belonging to the Montrose Group (Figure 9). However, since Cod is located near the transition area between the coarse-clastic Montrose Group and its shale equivalent (Rogaland Group, Lista Formation), a precise equivalence of the Cod sandstone with previously defined formations is not yet certain. According to Deegan and Scull (1977, figure 33), the Andrew Formation extends farther to the south than the Forties Formation. For that reason, it is proposed that the Cod reservoir (the main Cod sand) belongs to the Forties Formation, whereas the lower sand-shale sequence is classified as undifferentiated Maureen-Andrew-Forties Formation.

Within the field itself, correlation of the main Cod sand is uncertain; problems of reservoir continuity were identified during the early phase of development drilling. Lenticular sandstone bodies, thinning and shaling out toward the northeast, an increase in calcite cement southeastward, and a sealing fault, all are features identified early in the appraisal of the field.

Detailed sedimentological studies were undertaken, using core material and log responses. Kessler et al. (1980) suggested that the sandstone interval was deposited in at least six major episodes of submarine channel and fan lobe progradation. The sequences range in thickness from 50 to 120 ft (15–37 m). Individual sandstone beds within a sequence range in thickness from a few inches to 20 ft (6 m). Progradational sequences often immediately overlie each other, suggesting catastrophic abandonment of a channel or a fan lobe system and then almost immediate reoccupation and progradation by another similar system just above it. Other sequences are indicative of gradual channel and fan lobe system abandonment (Kessler et al., 1980).

Kessler et al. (1980) described thin, homogeneous beds, graded beds with some associated planar bedding and ripples, and occasionally 1.5 to 6 ft (0.5–1.8 m) thick zones of dish structures (concave-upward, water-release features). These sedimentary structures are consistent with various types of sediment gravity flows, especially grain flows. Structures for a cored interval of well 7/11-3X are illustrated in Figure 10.

The facies identified by Kessler et al. (1980) for the same well are shown in Figure 11 from upper to lower units:

1. Gradual fan lobe abandonment (A).
2. Fan lobe progradation (P).
3. Fan lobe or channel abandonment (A).
4. Fan lobe or channel progradation (P).
5. Outer fan plain, thin-bedded sandstone (FP).
6. Fan lobe or channel abandonment (A).

*Most of this section has been reproduced from D'Heur (1987) with permission from the Norwegian Petroleum Society.

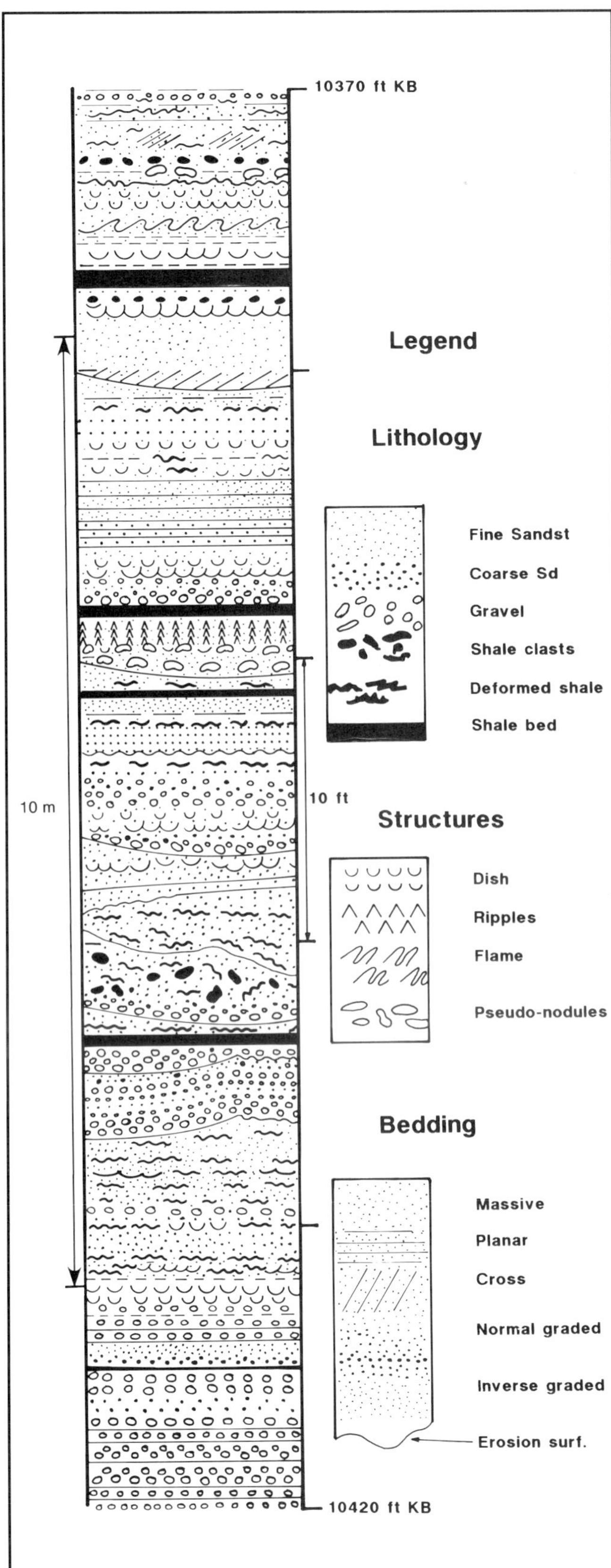

Figure 10. Lithostructural example (well 7/11-3X, 50 ft of core 3) for part of the main Cod sand member, after Kessler et al. (1980). (Reproduced with the permission of the Norwegian Petroleum Society.)

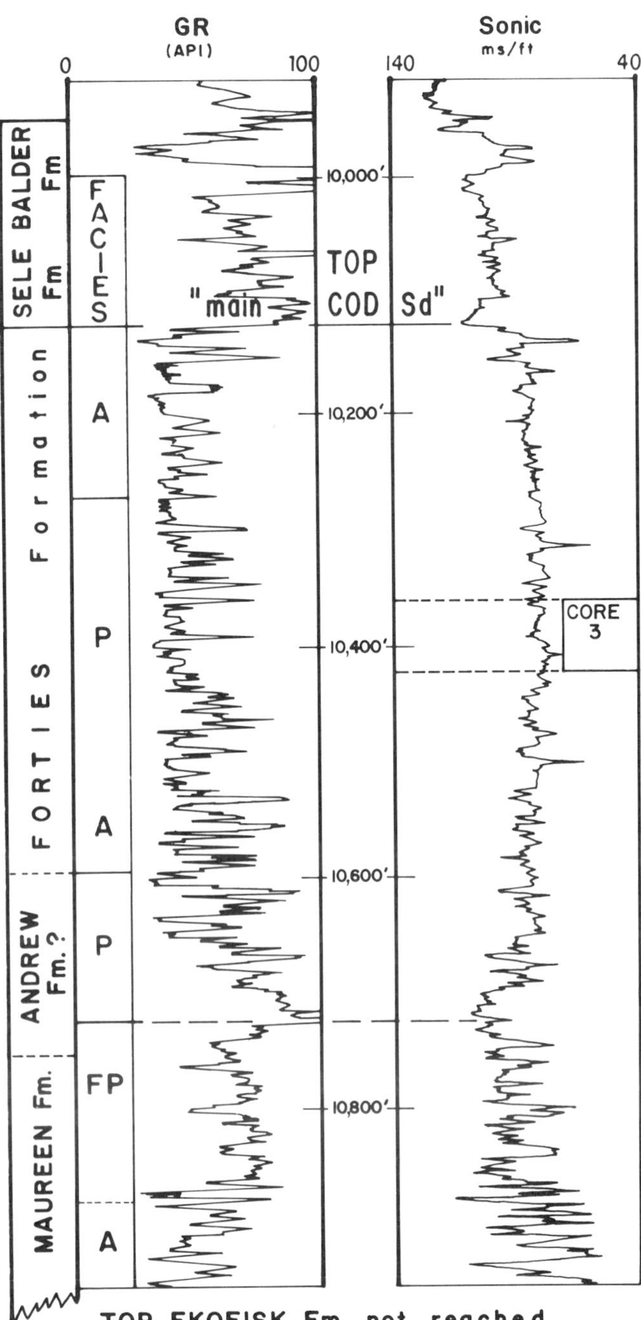

Figure 11. Sandstone facies, well 7/11-3, after Kessler et al. (1980). P, progradation; A, abandonment; FP, outer fan plain. (Reproduced with the permission of the Norwegian Petroleum Society.)

This sequence, observed in the southeasternmost well of the field, suggests a more distal depositional setting than those observed in wells to the northwest. Based on such a description, on seismic observation, and on reservoir pressure data, Kessler et al. (1980) proposed a complex sand lobe distribution (Figure 12).

The porosity and saturation logs of the main Cod sand are shown in Figures 13 and 14 for several wells. Within the correlation of the top and base of the main

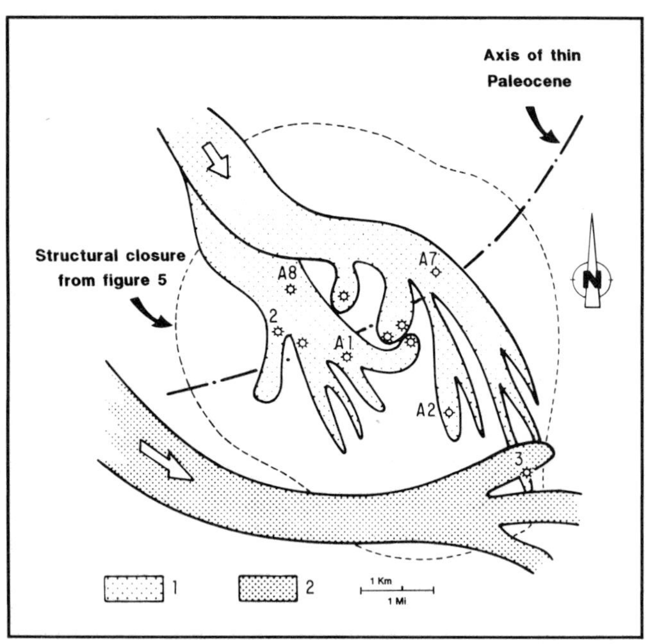

Figure 12. Hypothetical lobe distribution, modified after Kessler et al. (1980). 1, main Cod system; 2, earlier abandoned system. (Reproduced with the permission of the Norwegian Petroleum Society.)

Cod sand member, several sandstone bodies are probably discontinuous as shown in Figure 8. As regards hydrocarbon distribution, it was suggested as early as 1969 that the turbidite nature of the sandstone was responsible for the five or more distinct accumulations identified by reservoir pressure data. The main Cod sand flowed at rates of up to 43 MMcfd of gas and 1390 BCPD (0.76 g/cc). The presence of oil with a gravity of 0.80 g/cc was established in 1968 in well 7/11-3. In well A5, 0.86 g/cc gravity oil flowed from a 30 ft (9 m) interval at the base of the main Cod sand. This oil zone extends up to a depth of 9580 ft (2922 m) whereas the gas zone in well 7/11-2 extends down to a depth of 10,100 ft (3080 m).

When the reservoir was found water-bearing at -9760 ft (-2977 m) in well A2, the difference in gas/water elevation was accounted for by a major normal fault (Figure 3C, 1977). It also became obvious that 7/11-3 belonged to another small accumulation. Hydrodynamic studies in 1978 suggested that at least nine distinct accumulations exist at Cod. Recent interpretations, however, suggest that this model may not be entirely appropriate since pressure data now suggest communication between sandstone bodies that were originally considered to be separate.

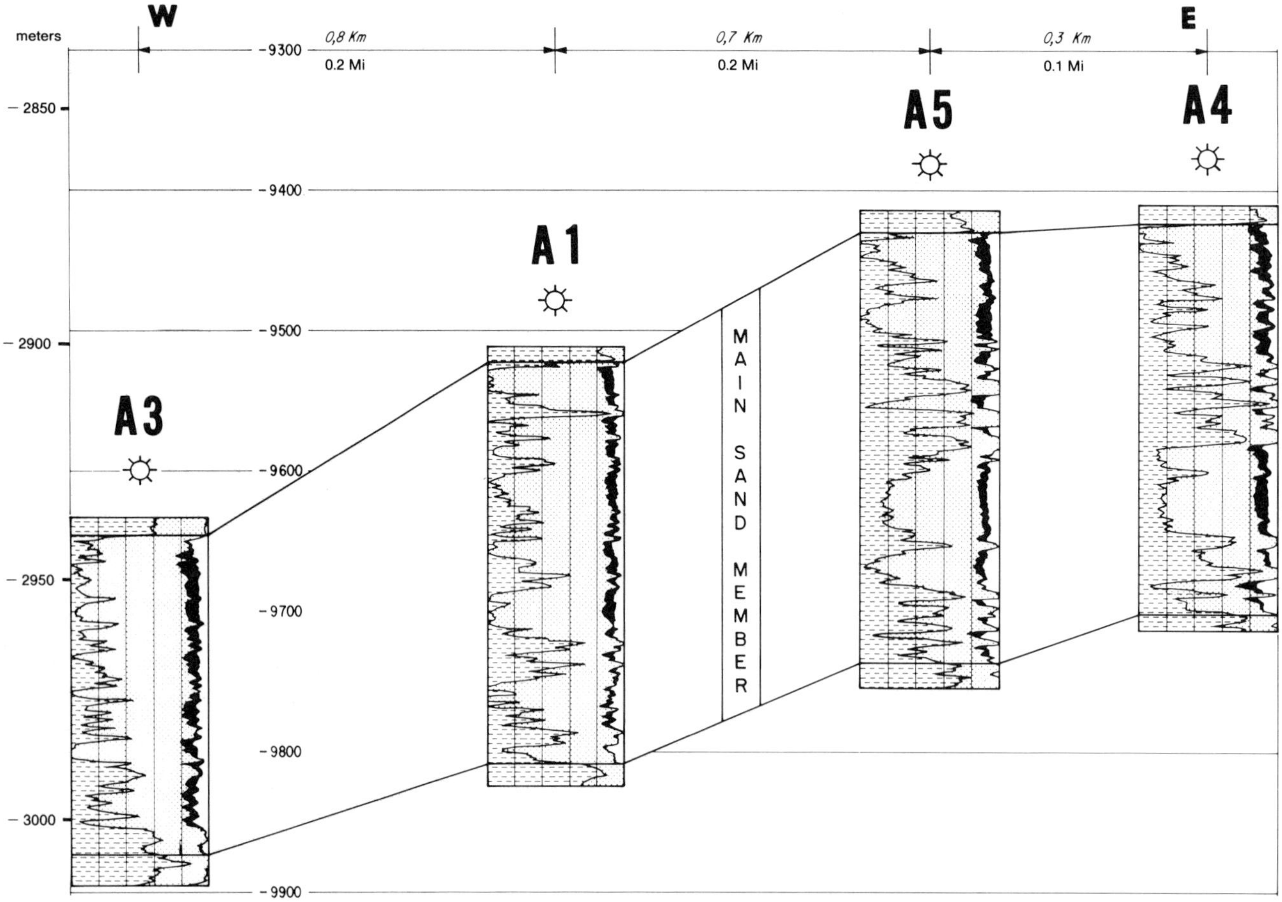

Figure 13. Lithology, porosity, and hydrocarbon saturation in the main Cod sand in the development wells A3, A1, A5, and A4. Legend in Figure 4. Location of wells in Figure 5.

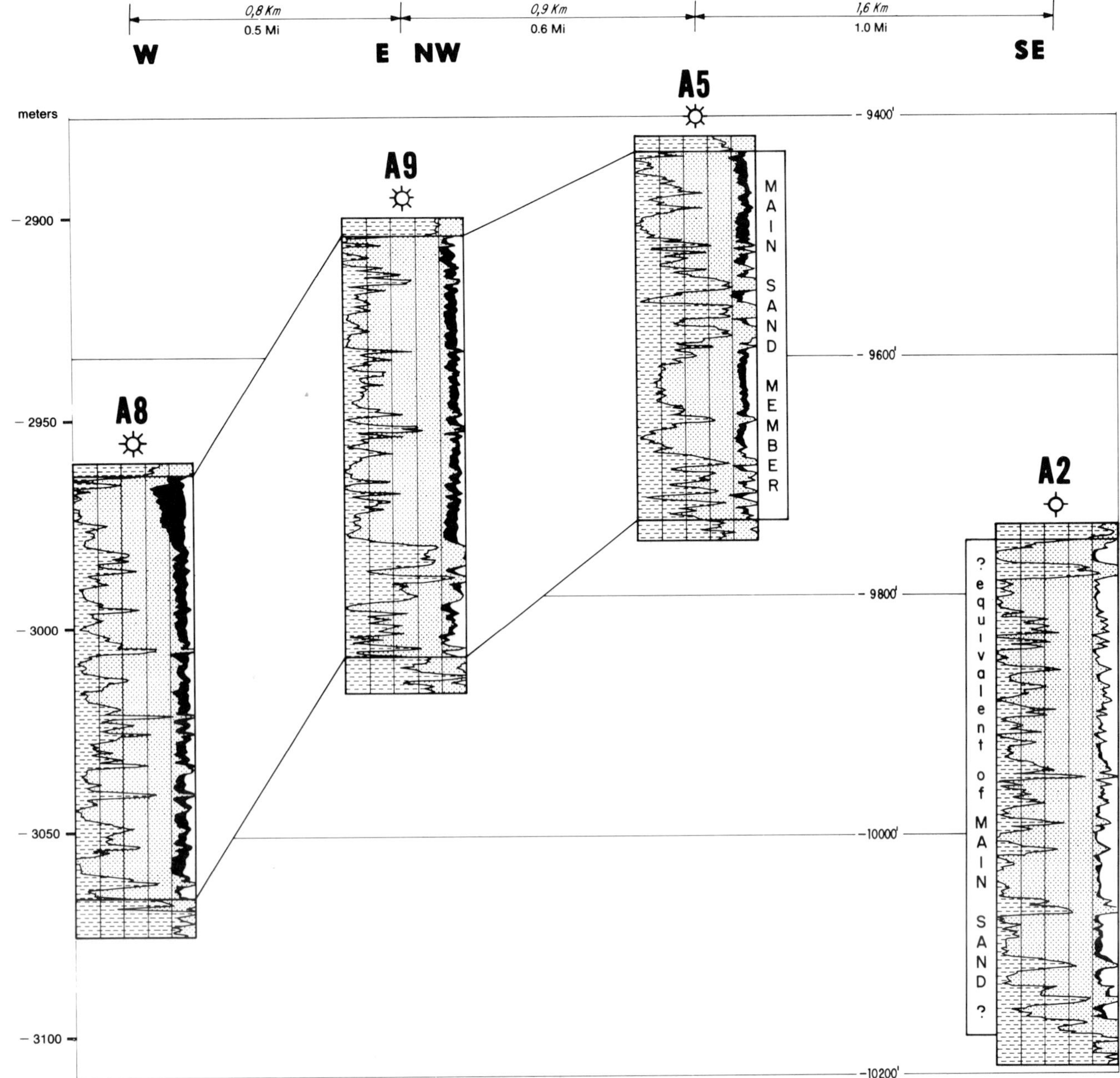

Figure 14. Lithology, porosity, and hydrocarbon saturation in the main Cod sand in the development wells A8, A9, A5, and A2. Legend in Figure 4. Location of wells in Figure 5.

Development Considerations

Successive interpretations of distribution of the net pay are shown in Figure 15 together with the original and final development plan. The initial plan was based on the assumption that the trap was purely of a structural type, and the nine-slot platform was set above the apex of the structure. The plan was modified when it was understood that most of the gas-bearing sandstone bodies were limited to the western half of the structure. Details on development drilling history are given in the *Post-Discovery* section. The production history is given in Figure 16. The original recoverable reserves of the field are estimated at 280 bcf of gas and 17 MM STB of condensate. The peak production was reached in 1980 with 120 MMcfd producing from six wells. Cod is linked to the Ekofisk Centre by a 16-in. oil and gas pipeline.

Faults

Faults affecting the Cod structure have been mapped since the mid-sixties. Their throw is generally moderate to small. Depending on preconceived ideas, workers have established highly faulted or totally unfaulted structural interpretations of the field. In fact, faults do exist but their interpretation

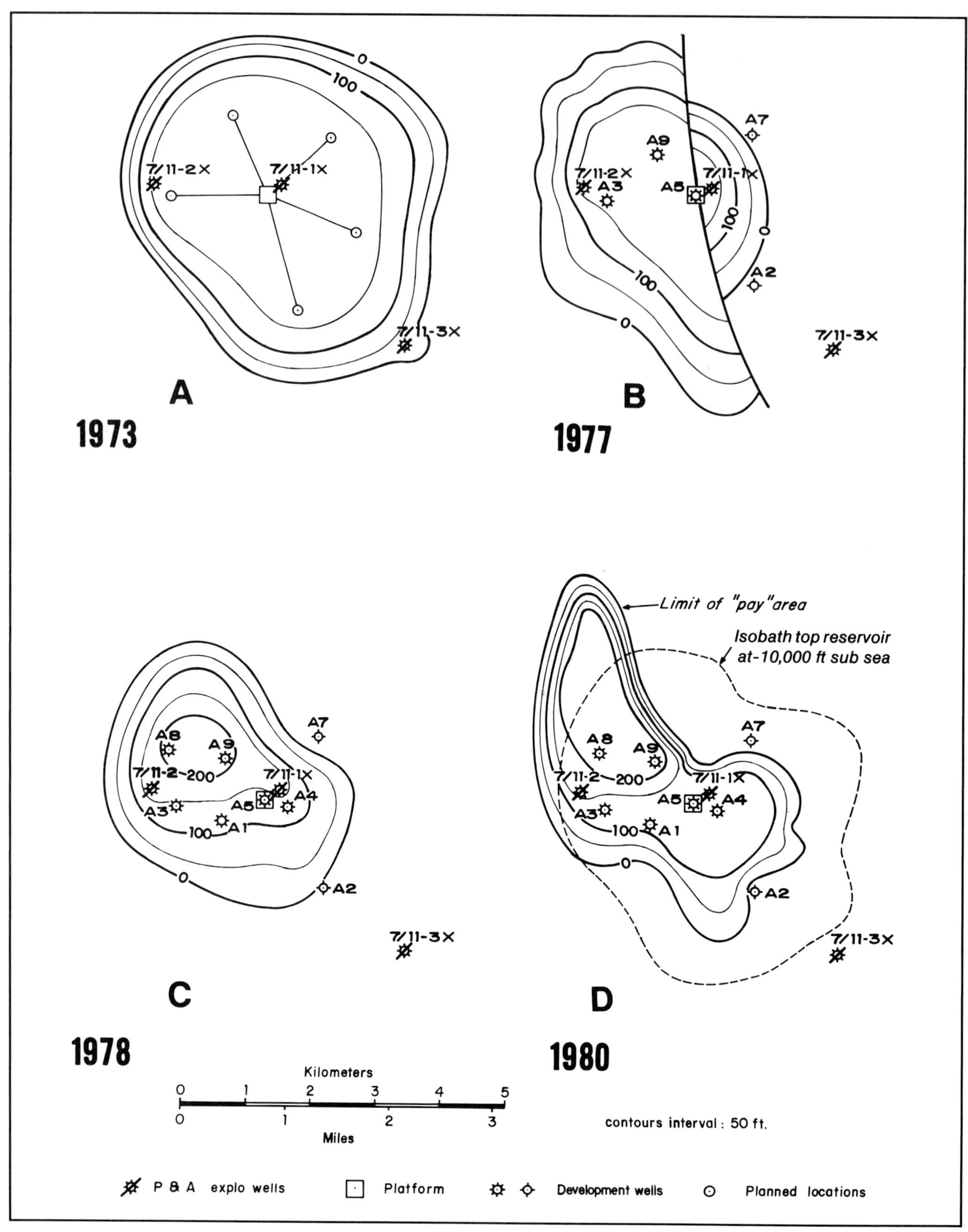

Figure 15. Successive interpretations of net pay thickness between 1973 and 1980. (A) Used for deciding on preliminary development plan, based on the interpretation that Cod was only a structural trap. (B) Taking into account the results of two dry development wells. (C) Assuming sand lobes discontinuity. (D) Based on revised geological model and seismic interpretation in 1980. Note that the limit of the main Cod sand distribution has been slightly modified since then (Figure 5).

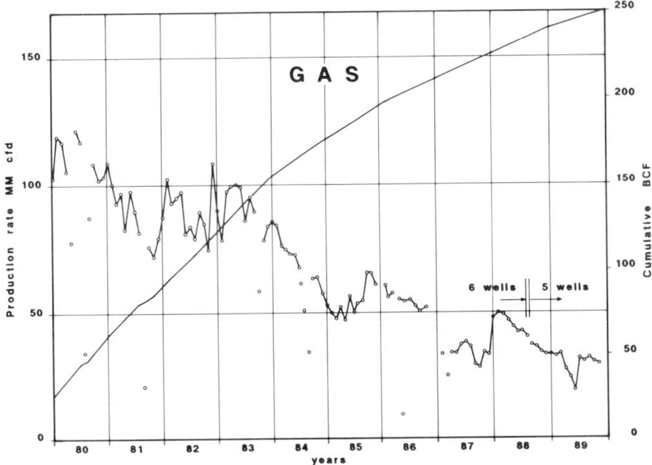

Figure 16. Gas production history and cumulative production of gas since 1980.

is complicated by rapid variations in lithology, porosity, and fluids within, above, and below the main Cod sand. The latest structural interpretation is shown in Figure 5. It is now felt that Cod is characterized by a complex system of several faulted sandstone lobes.

Source

The Upper Jurassic shale of the Mandal Formation, equivalent to the Kimmeridge Clay Formation, is the source rock for the hydrocarbons of the Cod field. Brownish-black and gray-black shales belonging to the Mandal Formation have been penetrated between 13,630 ft (4157 m) and 14,020 ft (4276 m) ss in well 7/11-7X (A6). These source rocks have a high potential for generation of wet gas, condensate, and light oil. The upper 80 ft (25 m) have an average TOC of 6.5%. The kerogen content is mainly of type II and has passed the peak of oil generation. The same interval is at a depth greater than 16,000 ft (4880 m) to the west of Cod.

EXPLORATION CONCEPTS

At the time Cod was discovered, the Paleocene sandstone play was unknown. At present the Paleocene fan complex is well defined and distinct episodes of sand influx have been clearly identified. Long-distance correlations have been established based on regional seismic and detailed age dating. The Upper Jurassic source rock is present in the whole area, below the structures or in their vicinity. The source rock has passed the peak of oil production everywhere below the Paleocene sandstone in the trough.

Structural traps result from several factors, mainly halokinesis of the Permian salt and differential compaction over buried paleorelief (ridges, horsts, or volcanoes). In the 1970s a number of fields were found in upper Paleocene structural traps, the largest being the Forties complex with more than 2200 MMbbl of recoverable oil. It is probable that no more Paleocene sandstone structural traps (four dip closure) remained undrilled in the Central trough. During the 1980s exploration of the Paleocene sandstone moved to a second phase with the objective of investigating stratigraphic traps. The eastern limit of the fan system was the most favorable area for such a purpose because the sandstones gently pinch out on the slope of a northwest-southeast-trending paleoridge. Several oil and gas accumulations have been found recently along this ridge called the Vestland arch.

In the 1990s a further exploration could be the search for individual channels, mainly in the Maureen Formation. The concept is similar to that of the new Eocene play in the central and north-central area of the North Sea.

Lessons learned from the Cod field are numerous since almost nothing was known of the Paleocene sandstone distribution at the time of the Cod discovery. Main points were the identification of the Upper Jurassic as the source rock and of the complex lobe distribution of the reservoir which was at first considered a blanket sand. The complex sandstone distribution complicated the evaluation and development of the fields in the North Sea but has created the opportunity for further, more sophisticated exploration on the margin of the fan system. It is also worth noting that no major Paleocene fields have yet been found at locations where good and thick older or younger reservoirs and traps existed simultaneously. The discovery of this kind of field could not be enhanced by any of the new techniques available today, since Cod is such an obvious structural feature. However, with the availability of 3D seismic, advances in Tertiary biostratigraphy and seismic stratigraphy, the field would probably have been better understood earlier in the phase of development drilling.

ACKNOWLEDGMENTS

The author thanks Petrofina, Fina Exploration Norway, Phillips, Agip, Elf, Norsk Hydro, and Total for their permission to publish this article. Most of the sections on trap, reservoirs, and faults and several figures have been reproduced from D'Heur (1987) with the permission of the Norwegian Petroleum Society (NPF). The opinions expressed in this paper do not necessarily represent those of any of the above-mentioned companies.

Thanks also to Gordon McLanachan for comments on the manuscript and to Dominique Moens and Joao de Lucena for their help.

REFERENCES CITED

Carmalt, S. W., and B. St. Johns, 1986, Giant oil and gas fields, *in* Future Petroleum Provinces in the World: American Association of Petroleum Geologists Memoir 40.

Deegan, C. E., and B. J. Scull, 1977, A standard lithographic nomenclature for the Central and Northern North Sea: Rep. Inst. Geol. Sci. UK, v. 77, n. 25.

D'Heur, M., 1987, Cod, *in* A. M. Spencer et al., eds., Geology of the Norwegian oil and gas fields: Norwegian Petroleum Society, Graham and Trotman.

Kessler, L. G., R. D. Zang, J. A. Englehorn, and J. D. Eger, 1980, Stratigraphy and sedimentology of a Paleocene submarine fan complex, Cod field, Norwegian North Sea, *in* Sedimentation of the North Sea reservoir rocks: Norsk Petroleumsforening, Article VIII, 19 p.

Vollset, J., and A. G. Doré, 1984, A revised Triassic and Jurassic lithostratigraphic nomenclature for the Norwegian North Sea: Norwegian Petroleum Directorate Bulletin, v. 3.

Appendix 1. Field Description

Field name .... *Cod field*

Ultimate recoverable reserves .... *280 bcf gas + 17 MM STB condensate*

Field location:

- **Country** .... *Norway*
- **State** .... *Continental Shelf (North Sea)*
- **Basin/Province** .... *Central trough*

Field discovery:

- **Year first pay discovered** .... *Upper Paleocene Forties Formation main Cod sandstone 1968*

Discovery well name and general location:

- **First pay** .... *7/11-1X, 285 km SW of Stavanger (Norway)*

Discovery well operator .... *Phillips Petroleum*

IP:

- **First pay** .... *43 MM SCFD + 1390 BCPD*

All other zones with shows of oil and gas in the field:

Age	Formation	Type of Show
Early Paleocene	*Ekofisk (tight)*	*Gas*

Geologic concept leading to discovery and method or methods used to delineate prospect

Well 7/11-1X was one of the first wells to test an anticlinal trap of halokinetic origin in the central North Sea where the lower Tertiary and Mesozoic series were poorly known but supposed to contain reservoirs and source rock. The structure was well defined by seismic reflection.

Structure:

Province/basin type .... *Cratonic basin located over earlier rift graben system; Bally 1211, Klemme IIB*

Tectonic history

During pre to mid-Cretaceous time the area belonged to the "Central graben" rift complex where mobile halite and thick organic-rich sediments existed. After Kimmerian tectonic phases, the area subsided more widely and regularly and received turbidite sands during the Paleocene.

Regional structure

Being located toward the flank of the "Central trough" and far from the source of Paleocene sand, the Cod area received a relatively small quantity of sand which forms the reservoir rock.

Local structure

Nearly circular dome with a 2 mi radius and a 600 ft structural closure at reservoir level.

Trap:

Trap type(s)

Combined structural and stratigraphic trap caused by halokinetic movements and pinch-out of channel sandstone bodies that are probably cut by sealing faults.

Basin stratigraphy (major stratigraphic intervals from surface to deepest penetration in field):

Chronostratigraphy	Formation	Depth to Top in ft (m)
Paleocene	*Balder*	*9500 (2900)*
Upper Cretaceous	*Tor*	*10,500 (3200)*
Lower Cretaceous	*Cromer Knoll*	*13,400 (4100)*
Upper Jurassic	*Mandal*	*13,700 (4200)*
Triassic	*Smithbank*	*14,200 (4350)*
Permian	*Zechstein*	*15,200 (4650)*

COD

Reservoir characteristics:

- **Number of reservoirs** *1*
- **Formations** .... *Forties Formation, also named main Cod sand*
- **Ages** .... *Late Paleocene*
- **Depths to tops of reservoirs** .... *9420 ft (2873 m) ss (apex)*
- **Gross thickness (top to bottom of producing interval)** .... *330 ft (100 m)*
- **Net thickness—total thickness of producing zones**
  - **Average** .... *130 ft (40 m)*
  - **Maximum** .... *230 ft (70 m)*
- **Lithology** .... *Deep marine distal turbidite sandstone interbedded with shale*
- **Porosity type** .... *Intergranular*
- **Average porosity** *17%*
- **Average permeability** .... *50 md*

Seals:

- **Upper**
  - **Formation, fault, or other feature** .... *Upper Paleocene, Sele Formation*
  - **Lithology** .... *Shale*
- **Lateral**
  - **Formation, fault, or other feature** .... *Forties Formation*
  - **Lithology** .... *Shale*

Source:

- **Formation and age** .... *Mandal Formation, Jurassic*
- **Lithology** .... *Shale*
- **Average total organic carbon (TOC)** .... *7% (wt)*
- **Maximum TOC** *12%*
- **Kerogen type (I, II, or III)** .... *Mainly II (sapropelic)*
- **Vitrinite reflectance (maturation)** .... $R_o = 1.1$
- **Time of hydrocarbon expulsion** .... *Eocene–Oligocene*
- **Present depth to top of source** .... *13,700 ft (4180 m)*
- **Thickness** .... *80 ft (25 m) ("hot" shales)*
- **Potential yield** .... *NA*

Appendix 2. Production Data

Field name .... *Cod field*

Field size:

- **Proved acres** .... *3200 (1296 ha)*
- **Number of wells all years** *12*
- **Current number of wells** *6*
- **Well spacing** .... *3000 ft (915 m)*
- **Ultimate recoverable** .... *280 bcf gas + 17 million bbl condensate*
- **Cumulative production** .... *240 bcf gas + 16 million bbl condensate (1988)*
- **Annual production** .... *18 bcf (1987)*
- **Present decline rate** .... *28% (1986)*
  - **Initial decline rate** *14%*
  - **Overall decline rate** *64%*
- **Annual water production** .... *NA*
- **In place, total reserves** .... *400 bcf gas*

In place, per acre foot .... *750,000 ft^3 gas*
Primary recovery .... *280 bcf gas + 17 million bbl condensate*
Secondary recovery .... *NA*
Enhanced recovery .... *NA*
Cumulative water production .... *NA*

Drilling and casing practices:

Amount of surface casing set .... *2800 ft (854 m)*

Casing program

(Vertical depth from KB to casing shoe) 30-in. at 600 ft (183 m); 20-in. at 1200 ft (366 m); 13⅜-in. at 2800 ft (854 m); 9⅝-in. at 9500 ft (2900 m); 7-in. at 10,500 ft (3200 m)

Drilling mud

1600–4000 ft (480–1220 m), sea water/attapulgite; 4000–9500 ft (1220–2900 m), sea water + polymer; 9500–10,500 ft (2900–3200 m), KCl water/polymer

Bit program .... *NA*
High pressure zones .... *Oligocene-Eocene, 4000–8000 ft (1220–2440 m)*

Completion practices:

Interval(s) perforated .... *All pay sandstone, 2 or 6 shots/ft*
Well treatment .... *None*

Formation evaluation:

Logging suites .... *2800–9500 ft (855–2900 m), GR/CAL/IES; 9500 ft (2900 m) to TD, GR/FDC/CNL/IES/microlog/microlaterolog/CAL*
Testing practices .... *No open hole tests; occasional selective tests where needed; multirate survey*

Mud logging techniques

Total Concept Mud Logging Unit from 2800 ft (855 m) to TD; sample frequency: 30 ft (9 m) from 1200 ft (366 m) to 9500 ft (2900 m) and 10 ft (3 m) from 9500 ft (2900 m) to TD

Oil characteristics:

API gravity .... *Gas, 0.68° (relative to air); oil, 54° API*
Initial GOR .... *13,000 SCF/bbl (stock tank)*
Sulfur, wt% *0*
Viscosity, SUS .... *Gas, 0.0294 cp*
Pour point .... *NA*
Gas-oil distillate .... *NA*

Field characteristics:

Average elevation .... *9620 ft (2934 m)*
Initial pressure .... *5380 psi (37,100 kPa)*
Present pressure .... *2100 psi (14,500 kPa)*
Pressure gradient .... *0.14 psi/ft (3.17 kPa/m)*
Temperature .... *265°F (129.4°C)*
Geothermal gradient .... *1.9°F/100 ft (0.035°C/m)*
Drive .... *Natural gas depletion*
Oil column thickness .... *600 ft (183 m)*
Oil-water contact .... *Below -10,070 ft (-3070 m)*
Connate water *29%*
Water salinity, TDS .... *30,000 mg/L*
Resistivity of water .... *0.06 at 250°F (or at 121°C)*
Bulk volume water (%) *44%*

Transportation method and market for oil and gas:

16-in. oil and gas line, 47 mi (75 km) to the Ekofisk Center; from there, 34-in. oil line, 220 mi (355 km) to Teeside, England; 36-in. gas line, 274 mi (440 km) to Emden, Germany

COD

Maureen Field—U.K.
Central Graben, North Sea

RICHARD C. LAMB
Fina Exploration Ltd.
Epsom, United Kingdom

PETER M. McGAUGHRIN
Phillips Petroleum (UK) Ltd.
Woking, United Kingdom

MICHEL D'HEUR
Petrofina S.A.
Brussels, Belgium

PETER L. CUTTS
Hamilton Brothers Oil and Gas Ltd.
London, United Kingdom

FIELD CLASSIFICATION

BASIN: North Sea South Viking/ Witch Ground graben
BASIN TYPE: Rift
RESERVOIR ROCK TYPE: Sandstone
RESERVOIR ENVIRONMENT OF DEPOSITION: Submarine Fan
RESERVOIR AGE: Paleocene
PETROLEUM TYPE: Oil
TRAP TYPE: Salt Induced Anticline

LOCATION

The Maureen oil field is situated (Figure 1) in the UK sector of the North Sea, approximately 160 mi (260 km) east-northeast of Aberdeen and 37 mi (60 km) northeast of the Forties field, which is ranked among the hundred largest oil fields in the world (Carmalt and St. John, 1986). Both the Maureen and Forties fields produce oil from Paleocene sandstones. The Maureen field was discovered in 1972 in Block 16/29, which was granted to the Phillips-operated UK Group in 1970 in the Third UK Round of Offshore Licensing. The field is operated by Phillips Petroleum Co. UK Ltd (33.78%) on behalf of its partners Fina Exploration Ltd. (28.96%), Agip UK Ltd. (17.26%), Gas Council (Exploration) Ltd. (11.50%), and Ultramar Exploration Ltd. (8.50%).

The field lies in a water depth of 325 ft (99 m) close to the northern limit of the North Sea Central graben at its intersection with the South Viking trough and the Witch Ground graben (Figure 1). The ultimate recovery of the field is estimated to be 210 MMSTBO (28 million tonnes) using natural water drive enhanced by water injection. Production from the Maureen field began in mid-1983 and by the beginning of 1990 the field had produced approximately 155 million bbl of oil. The field produced at a peak of nearly 100,000 BOPD, but this has subsequently declined to just under 50,000 BOPD. The facilities comprise a single triangular gravity base platform, making it the world's first steel platform to combine flotation, ballasting, and oil storage. It has a capacity for storing 650,000 bbl of oil, which is offloaded to a tanker via a 1.5 mi, 24-in. pipeline and a tanker loading column.

HISTORY

Pre-Discovery

The Maureen field sits in a thick sedimentary basin between the Fennoscandian shield (covering Norway, Sweden, etc.) to the east and the Shetland platform to the west. Within this area a post-Cretaceous sag basin overlies an intensely faulted pre-Cretaceous graben complex.

Exploration in this area began in the late 1960s after drilling had proved successful in the southern area of the North Sea in the mid-1960s where large quantities of gas had been discovered in Permian–Triassic sediments. Farther north, initially small oil

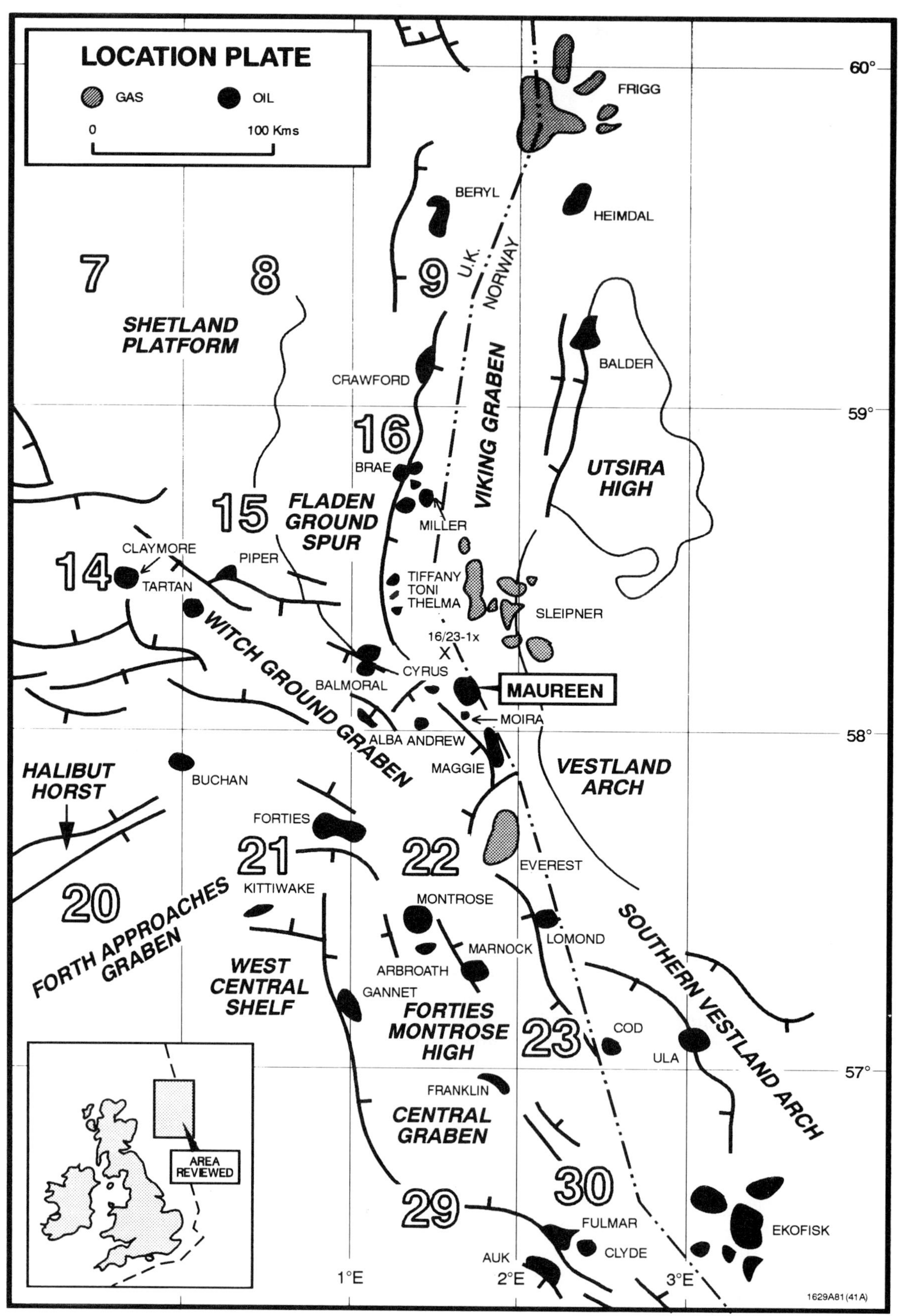

Figure 1. Structural elements map of the central North Sea and location of the main oil and gas fields.

and gas discoveries were made in lower Tertiary reservoirs. The most significant oil discovery at that time was that of the Ekofisk field (Figure 1) in the Cretaceous Chalk of the Norwegian Sector by Phillips. In 1970, the large Forties oil field was discovered by B.P. in Paleocene sandstones, and in 1971 Shell discovered the Brent field in Middle Jurassic sandstones. These three fields together held more than 6 billion bbl of oil. This encouraged other companies to explore for oil accumulations in the northern North Sea, and numerous discoveries were made in the ensuing years. Block 16/29 was known at this time to lie on a productive Paleocene sand belt that had an established extent of 250 mi (400 km) from the Cod field northwestward to B.P.'s Forties field and then northeastward to the Frigg field. The Maureen field was discovered by the Phillips Group in 1972.

As is usual for Phillips-operated areas, license blocks or groups of blocks in the UK sector are designated by an alphabetic letter and prospects within these licenses are assigned a girl's name starting with this letter. Blocks 16/24 and 16/29 were designated as area "M," within which the largest prospective structure was called Maureen.

The Maureen structure is a well-defined dome at Tertiary and Cretaceous levels, overlying complex, faulted Mesozoic strata. The first seismic survey over the block was acquired in 1970 and comprised ten east-west and north-south lines totaling 84 mi (135 km).

The original mapping of the structure (Figure 2A) showed the areal closure to be 6850 ac (2770 ha) with a vertical relief of 600 ft (180 m) at the top of the Paleocene. The expected minimum gross sand section was anticipated to be about 400 ft (120 m) and 250 ft (75 m) net, based on the section drilled in the Conoco 16/23-1 well immediately to the north. Sandstone porosity in the 16/23-1 well averaged 26%.

During November 1972, the first exploration well was drilled on block 16/29 using the semi-submersible rig, "Ocean Traveller." The well was located near the apex of the Maureen anticline (Figure 2A) and was designed to test the primary objectives of the Eocene and Paleocene sandstones. Secondary objectives were the Lower Cretaceous, Jurassic, and Triassic sandstones with a total expected Mesozoic net pay thickness of 250 ft (75 m).

Discovery

Well 16/29-1x found no Eocene sandstone but encountered 90 ft (27 m) of net oil-bearing lower Paleocene sandstone between 8310 and 8460 ft (2535 and 2580 m) (KB) in a simple, four-way, dip-closed trap. The well also penetrated 500 ft (150 m) of Cretaceous Chalk overlying a 4000 ft (1220 m) thick sequence of Lower Cretaceous marls, Upper Jurassic shales, and Middle Jurassic sandstone with interbedded coal, before reaching a total depth of 13,000 ft (3965 m) in Permian halite (Figure 3).

The Paleocene sandstone was perforated in four zones. The bottom 31 ft zone produced 2300 BOPD and 612 MCFGPD through a 1-in. choke. The upper three zones flowed oil at a rate of 3588 BOPD and 872 MCFGPD through a 1-in. choke. No gas cap was penetrated by the well.

Post-Discovery

In-fill seismic totaling 83 mi (132 km) was acquired in 1973 which resulted in a 1.6 km (1 mi) grid coverage. The revised mapping after the discovery well resulted in a 4672 ac (1892 ha) closure with 600 ft (180 m) of vertical relief. Since no indication of an oil-water contact (OWC) had been found in the discovery well, the first appraisal well, 16/29-2x, was drilled to evaluate the downdip oil limit on the western flank of the field, 4.2 mi (2.6 km) to the northwest (Figure 2B).

The appraisal well (16/29-2x) encountered 212 ft (65 m) of net porous sandstone, 82 ft (25 m) of which was above an OWC (100% water) at 8683 ft (2648.3 m) subsea. This indicated an oil column of 473 ft (144.3 m) between the two wells. Seismic mapping at this time indicated the maximum oil column to be about 600 ft (183 m). Figure 4 shows the log and core characteristics of this thickened sand section. The well was deepened to a total depth of 11,533 ft (3517.6 m) through Upper Cretaceous Chalk, Lower Cretaceous marls and shales, Upper Jurassic shale, Middle Jurassic coal and shale, and oil-stained Triassic sandstones to the total depth in Triassic siltstones and shales. The best Triassic zone tested 38° API oil at a maximum flow rate of 750 BOPD from a 190 ft (58 m) perforated interval. The primary-objective Paleocene sandstone flowed at a rate of 10,830 BOPD and 3 MMSCFD from a 75 ft (22.8 m) perforated interval. Well 16/29-2x was extensively cored and the data demonstrated an average net pay porosity of 22% and an average permeability of 500 md.

A second appraisal well, 16/29-3x, was drilled during the summer of 1974 to evaluate the northern extension of the field at a step-out location 2.5 mi (4 km) north of the discovery well (Figure 2B). The well encountered the top of the reservoir close to the original OWC seen in well 16/29-2x. This defined the northern limit of the field. The well reached a total depth of 11,500 ft (3508 m) in Triassic red shales. Additional seismic data were acquired in the following years, but no additional wells were drilled in the field until 1978.

Initially the field was considered to be marginally economic by North Sea standards. As a result of this and the fact that development options were sensitive to reserve estimation, several Maureen field interpretations were undertaken between 1974 and 1978. A major concern at the time was the very complex results obtained from development drilling in the Cod gas field. The Cod field, located about 80 mi (130 km) south of Maureen (Figure 1), is in a similar geological

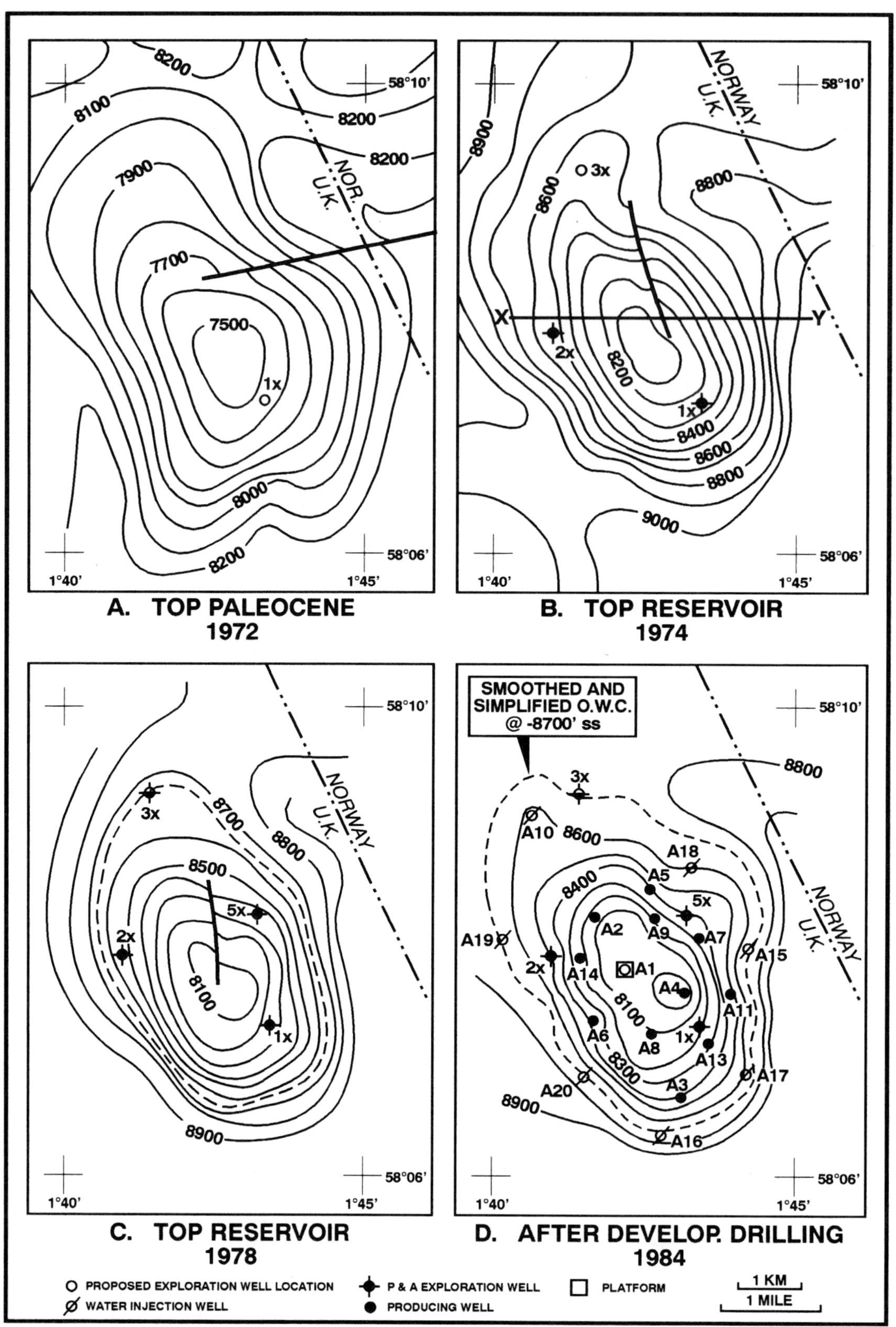

Figure 2. Successive structural interpretations of the Maureen anticline at various phases of exploration, appraisal, and development drilling. The seismic grids available in 1972 and 1974 are shown in Figure 2A and 2B by the fine straight lines. Seismic line X-Y shown in Figure 7. The seismic coverage shot in the 1980s is shown in Figure 6. Figure 2C shows the top reservoir structure map pre-development drilling, while Figure 2D shows it post-development drilling. Wells designated with an *x* are exploration/appraisal wells.

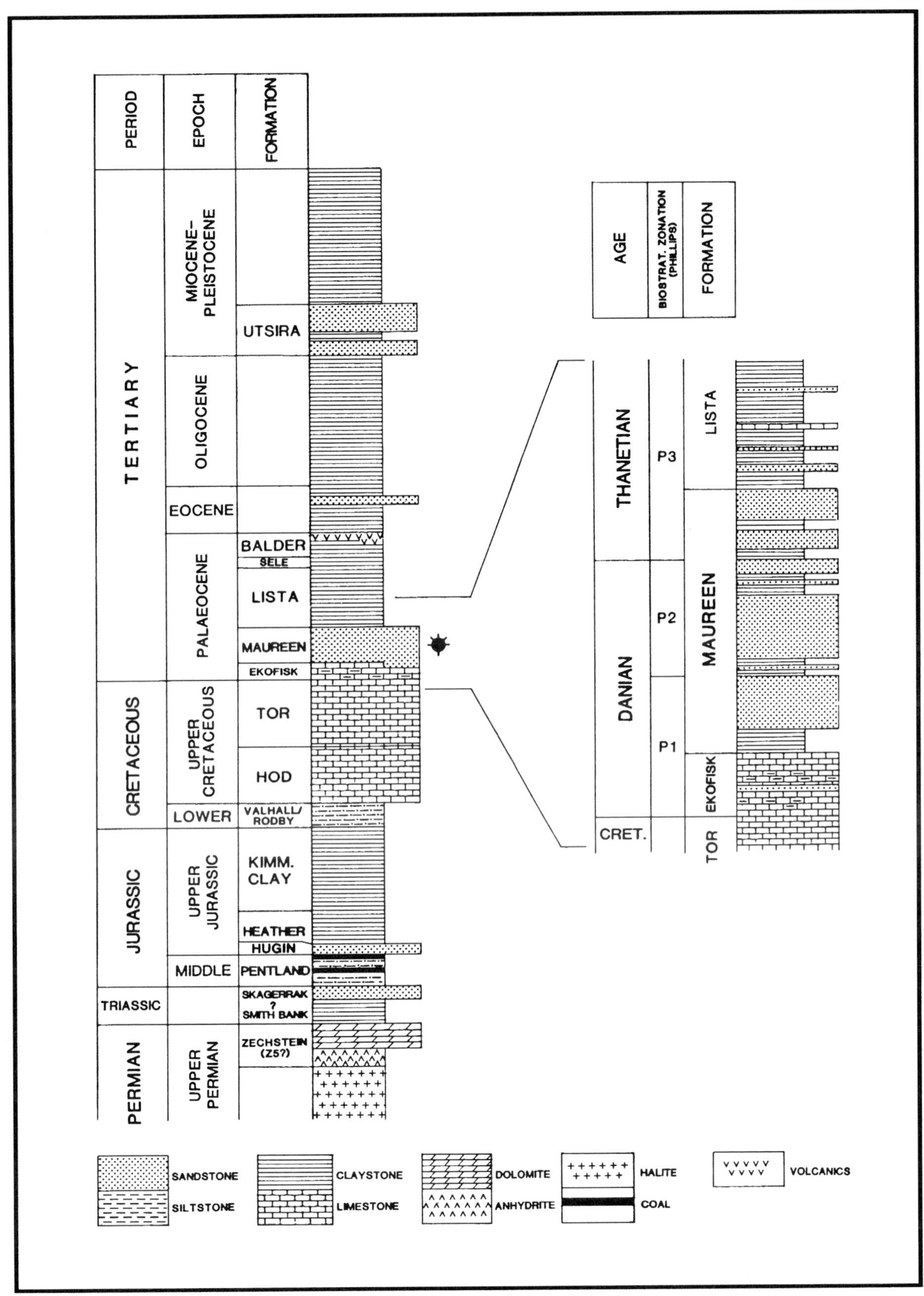

Figure 3. Maureen area stratigraphy. (Reproduced by permission of the Geological Society from "The Maureen Field, Block 16/29a, UK North Sea" by Cutts, 1991 in Geological Society Memoir 14, UK Oil and Gas Fields, 25 year commemorative volume, 1991.) (Symbol adjacent to Maureen Formation indicates oil with associated gas.)

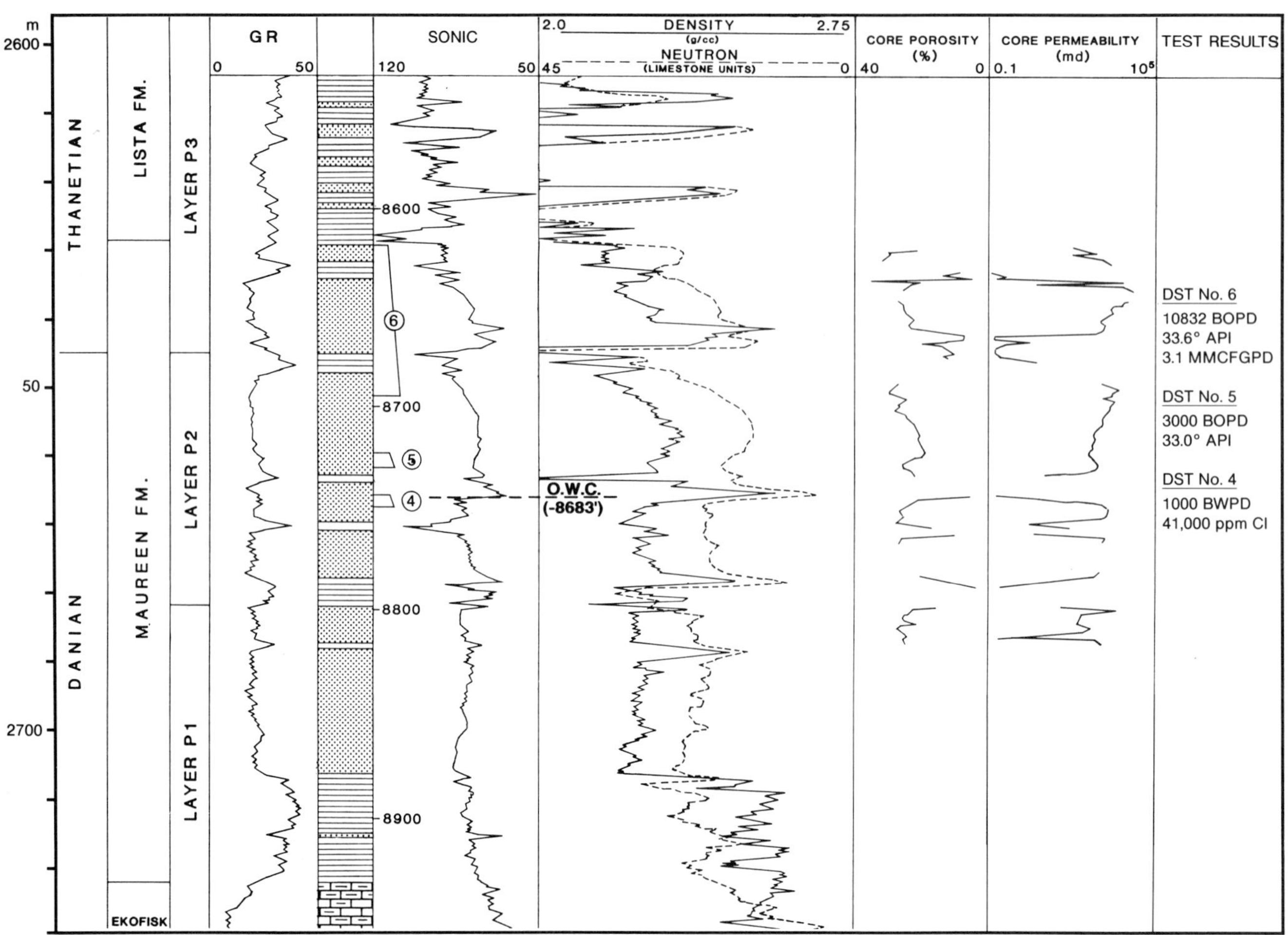

Figure 4. Well summary for 16/29-2x. (Modified from Cutts, 1991 [see Figure 3].)

situation and appeared to have a very complex Paleocene sandstone and hydrocarbon distribution. Even with the information from three exploration-appraisal wells on a relatively simple structure, the first development well on Cod was a dry hole (D'Heur, this volume), and the reserves estimate had to be downgraded significantly after drilling. Several studies on the Maureen recoverable reserves were undertaken at this time which were based on the three exploration-appraisal wells (Figure 5). The main uncertainty was the amount of sand thinning on the crest of the structure and on the flank to the east.

In order to reduce this uncertainty in the Maureen field, a fourth appraisal well, 16/29-5x (16/29-4x was located on a separate structure to the southwest), was drilled in 1978 to give additional confirmation of the OWC, reservoir quality, and sand thickness in the northeast quadrant of the field (Figure 2C). It encountered 129 ft (39.3 m) of net oil-bearing sandstone that tested at a flowrate of 5330 BOPD. This confirmed the 1975 net pay interpretation shown in Figure 5B.

Immediately following this, Annex B field development plan (Phillips Petroleum, 1978), based on a relatively conservative estimate of 289 MMBO in place (125 MMBO recoverable), was submitted to the U.K. Department of Energy in 1978. The uncertainty as to the degree of reservoir thinning at the crest had not yet been resolved and a down-side case with less than 200 MMBO in place was still possible (Figure 5C).

Pre-drilling of the development wells (prior to installation of all production facilities) from a 24-slot subsea template started in June 1979. The first production well (16/29a-A1, vertical) proved 150 ft (46 m) of net pay in the central area of the field, confirming again the 1975 interpretation (Figure 5B). The other wells were deviated from the template and comprised an initial, regular 12-spot production-well spacing pattern aligned northwest-southeast, which was modified slightly during the development drilling phase to optimize drainage based on updated mapping. This procedure also allowed a number of wells to be drilled to optimum pre-Cretaceous targets. By June 1983, the Maureen production platform had

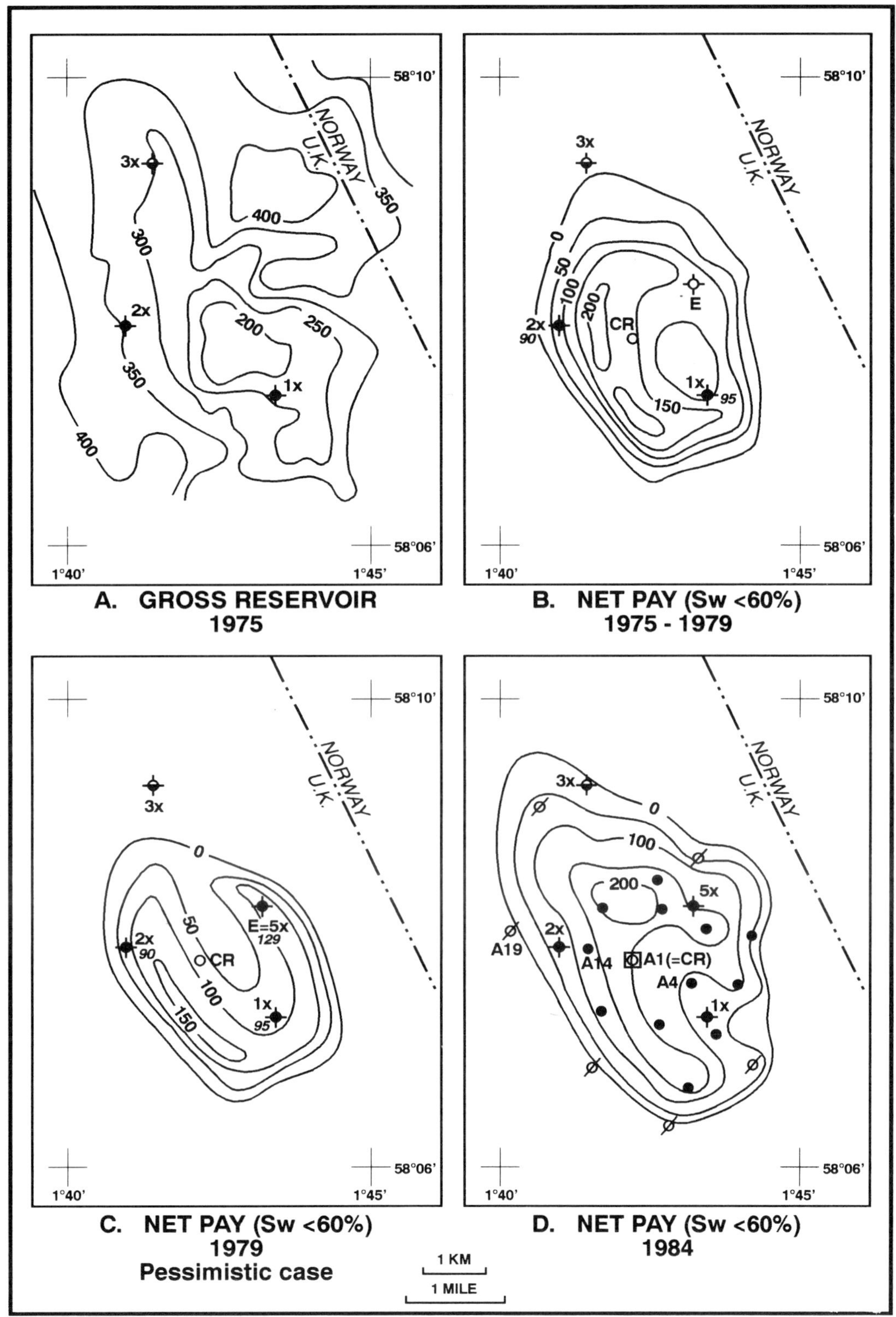

Figure 5. Successive interpretations of the reservoir thickness at various phases of exploration, appraisal, and development drilling. (Well symbols shown on Figure 2.) (A) A crestal thinning had already been recognized by seismic interpretation in 1975. OWC, -8683 ft (-2648.3 m). (B) The 1975 net pay thickness map left room for different interpretations in the crestal (CR) and eastern (well 5x) parts of the field, where it was even possible to extrapolate complete absence of the reservoir. A slightly less pessimistic assumption is shown in (C). However, after completion of the development drilling, the 1984 net pay thickness map (D) appeared to be very similar to the 1975 version.

been installed over the template and the field put on production in September 1983. Water injection began in February 1984 into the first of seven peripheral injection wells, and the injection rate was increased as additional injection wells were completed. Peak production of 100,000 BOPD was achieved in 1984.

In 1985, a revision to the Annex B was presented to the Department of Energy with a revised ultimate recovery of 185 MMSTB (Phillips Petroleum, 1985). This was based on the interpretation of 370 km (230 mi) of 1981 vintage seismic data (Figure 6), control data from 23 wells, detailed biostratigraphic analyses of sidewall and conventional cores, and a better reservoir performance than initially predicted. By the end of 1989, approximately 155 MMSTB of oil had been produced.

The oil is initially stored in three integral, platform-storage tanks before being offloaded via a remote tanker mooring-loading system, 2 km from the platform.

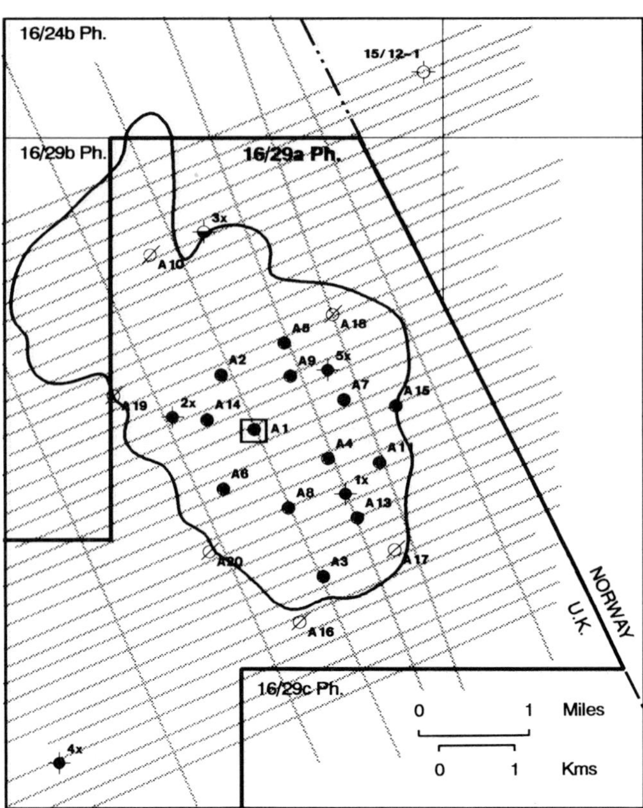

Figure 6. Seismic database from 1985 showing traverse lines.

DISCOVERY METHOD

Then (1970s)

The Maureen prospect had been clearly identified on the seismic lines available in 1970 and 1973 (Figure 7). Exploration in this part of the North Sea had only recently started, and therefore data from only a small number of wells were available to identify seismic horizons and the specific reservoir objectives of the first exploration well on block 16/29. The closest relevant well at this time was the Conoco-Phillips 16/23-1, about 9.3 mi (15 km) to the northwest (Figure 1), which had encountered 420 ft (128 m) net of thick and uniform Eocene sands and 258 ft (70 m) net of Paleocene sand with good shows of gas. Both sections were untested because of hole problems. In addition, the 16/23-1 well drilled sands of reservoir potential in the Lower Cretaceous, Jurassic, and Triassic. However, because of the simple nature of the Maureen structure and the identification from available well control of Tertiary sandstones as the primary objective, the fundamentals of the trap were more or less understood prior to drilling.

Now (Late 1980s)

It is unlikely that the discovery of the field would have been significantly enhanced by any of the new exploration methods available today since it is such an obvious structural feature in the block, nor would the development of the field be significantly different. However, with the availability of 3D seismic, advances in Tertiary biostratigraphy, and current knowledge of the natural water drive mechanism, it is possible that the balance of production wells to injectors would have been modified somewhat.

STRUCTURAL AND DEPOSITIONAL HISTORY

Tectonic History

The Maureen field is located on the Maureen shelf close to the northern limit of the North Sea Central graben at its intersection with the South Viking graben and the Witch Ground graben (Figure 1).

The oldest known sediments penetrated beneath the field are Permian in age (Figure 3), although Devonian sediments are known to occur in the area to the south and west. No Carboniferous rocks have been identified in the immediate area, suggesting that this area was emergent during Carboniferous times.

The pre-Tertiary tectonic development of this part of the basin can briefly be summarized as follows.

1. The development of a Permian evaporite basin with thick salt development to the south of Maureen.
2. Late Permian rifting within a north-south extensional regime. This produced a series of

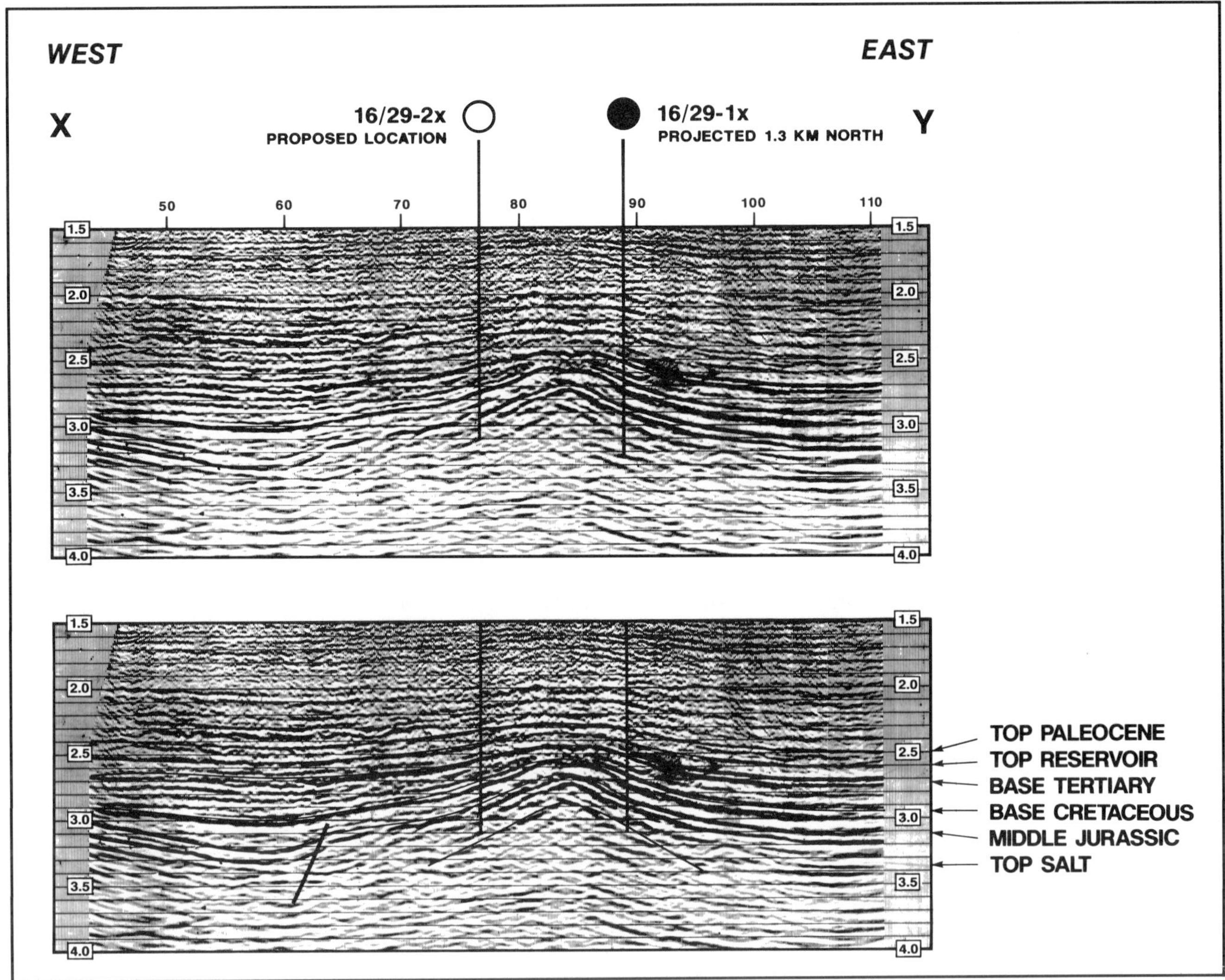

Figure 7. East-west seismic line across the field. Upper, uninterpreted. Lower interpretation prior to drilling 16/29-2x well. (See Figure 2B for line of section. Length about 3 mi or 4.9 km. Source Phillips Petroleum Co., 1973.)

Paleozoic horsts and grabens, including the initial establishment of the Central and Viking grabens. Thick continental sediments built up in these grabens.

3. Late Triassic/Early Jurassic uplift, with significant erosion creating a major unconformity.
4. The accumulation of Middle Jurassic volcanics in the Central and Witch Ground grabens.
5. The development of Early Jurassic half grabens with the deposition of shallow marine sediments.
6. Intense Late Jurassic faulting in the Viking graben, with the deposition of a deep water clastic sequence.
7. The continuation of basin development during the Cretaceous, with the accumulation of deep water carbonates.
8. The uplift and erosion of the basin margins during the early Tertiary as the basin margins were uplifted. The eroded material was transported and deposited into the continuously subsiding basin to the south and east.

The early Tertiary depositional history is described below in more detail since this time includes the deposition of the primary reservoir in the Maureen field.

Chalk sedimentation continued briefly during the early Danian before significant change in depositional conditions occurred. From the early Paleocene onwards, subsidence accelerated over a wide area. Regional tilting related to the early Tertiary opening of the Atlantic resulted in the downwarping of the Central and Viking grabens and the Outer Moray Firth basin, together with the uplift of the Scottish Highlands and the Shetlands to the northwest. Erosion of these source areas led to a massive influx of coarse clastic material that prograded and thinned eastwards and southwards onto a shallow marine shelf before being reworked. The shelf break coincided roughly with the western margin of the South Viking graben, off which the sediments passed laterally downslope into an extensive submarine fan complex dominated by sediment gravity flows (Knox et al., 1981; Stewart, 1987). Slumping and reworking of Cretaceous carbonates during the latest Mesozoic and earliest Cenozoic times is thought to have been an important depositional process close to active intrabasinal and basin margin highs. In the Norwegian part of the Central graben, 120 mi (195 km) southeast of Maureen, for example, the allochthonous chalks of the Tor and Ekofisk formations contain major oil and gas fields (D'Heur, 1986).

On the Maureen shelf, the Ekofisk Formation comprises less reworked chalk but contains minor local sandstones, probably related to the early movement on the Fladen Ground spur. Major Paleocene submarine fans were initiated as a result of the large influx of clastic material into the Central and Viking grabens (Figure 8; and in Stewart, 1987; Morton, 1979).

The oldest fan systems are Danian in age, comprising mixed carbonate and clastic deposits and are attributable to the Maureen fan. During the Danian, thick coarse clastic sediments were transported into the Central and Viking grabens, the latter acting as a sand fairway. The subsequent Andrew fan was deposited in a similar manner but with an axis offset to the west of the Maureen fan. The youngest Paleocene fan comprises the Forties Formation and is represented by submarine fan channels and lobes developed in the Forties field (Carman and Young, 1981), the Montrose and Arbroath fields (Fowler, 1975), and extending southeastwards into the Cod field (D'Heur, 1987). A later localized submarine fan was located in the Gannet field (Armstrong et al., 1987). The major fan systems in the Central graben are interpreted as lowstand submarine fans and the thick basinal claystones as deposition during intra-Paleocene sea level highstands.

The later larger clastic fan systems that have sediment thicknesses in excess of 2000 ft (600 m) in their axial zones are not seen in the Maureen area because it is situated on the northeastern flank of the main depositional basin. In the Maureen area, sequences of up to 400 ft (120 m) of sand-rich clastics are seen in the western part of the field, but these thin to 140 ft (43 m) in the east, reflecting their lateral position on a mid- to outer-fan setting. Fine clastic deposition together with sporadic and widespread volcanic tuff followed the major fan sedimentation during the remainder of the Paleocene.

The Eocene saw continued subsidence with the accumulation of thick basinal marine shales and with scattered submarine fan development in the early

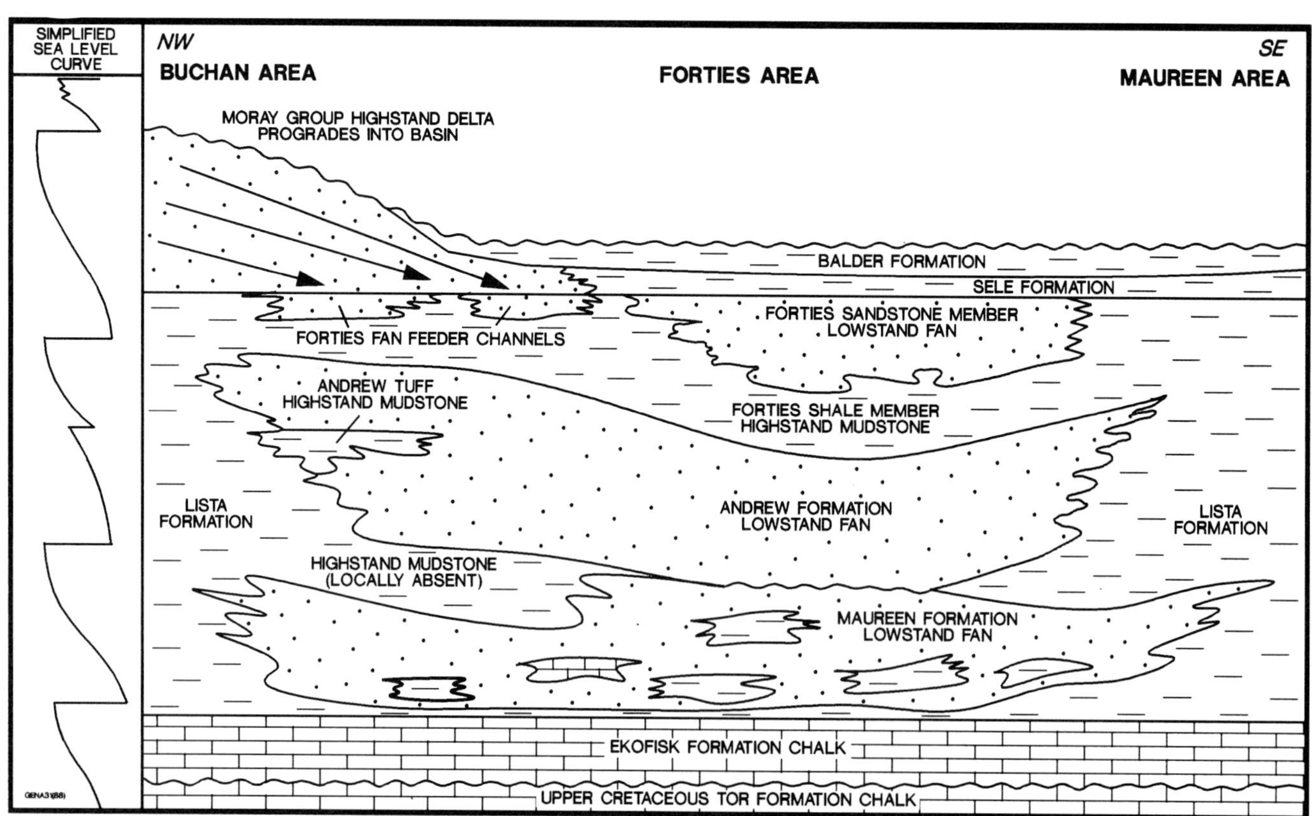

Figure 8. Paleocene depositional system summary diagram for the Northern Central graben. No scale. Schematic represents a section of about 80 to 100 mi (130-160 km). (Buchan area lies to the northwest of Forties field along the boundary of Quads 20 and 21.)

Eocene. The thick shale accumulations (Eocene and younger) were instrumental both in burying the underlying Jurassic source rocks (primarily the Kimmeridge Clay) to hydrocarbon generation depths and in acting as effective seals for the Tertiary traps.

Local Structure and Deposition

A simplified tectonic history of the Maureen structure is illustrated schematically in Figure 9. The Maureen structure is a salt cored anticline with approximately 1000 ft (300 m) of structural relief at the top of the Paleocene. Below the top of the Jurassic, faulting is both significant and complex. Halokinetic movements were initiated at least as early as the late Triassic and continued intermittently through to the Miocene. The main halokinetic events are recorded as major unconformities at the top of the Triassic, top of the Middle Jurassic, top of the Upper Jurassic, and near the top of the Cretaceous. The thinning of most of the major units over the Maureen field testifies to the long-term, periodic nature of the halokinetic uplift. The original structure of Maureen was not, however, created by the salt; during the Late Permian/Early Triassic an approximately north-south-oriented fault block existed on the northern margin of the Permian salt basin. Erosional patterns on that fault block suggest an origin of the structure that is entirely tectonic; this structure subsequently provided a focus for halokinetic movements (Figure 9).

Early Tertiary, clastic, submarine gravity-flow deposition was controlled and significantly modified by the simple anticlinal feature, with the net effect that the halokinetic structure was progressively subdued and sediment built up on its northwestern flank. The presence of the Vestland arch immediately to the east of the Maureen field also controlled sediment deposition, limiting the extent of submarine flows.

Massive sandstone units are restricted to the north and west sides of the dome while the lower energy, fine clastic deposition occurred preferentially on the eastern flanks and up onto the Vestland arch.

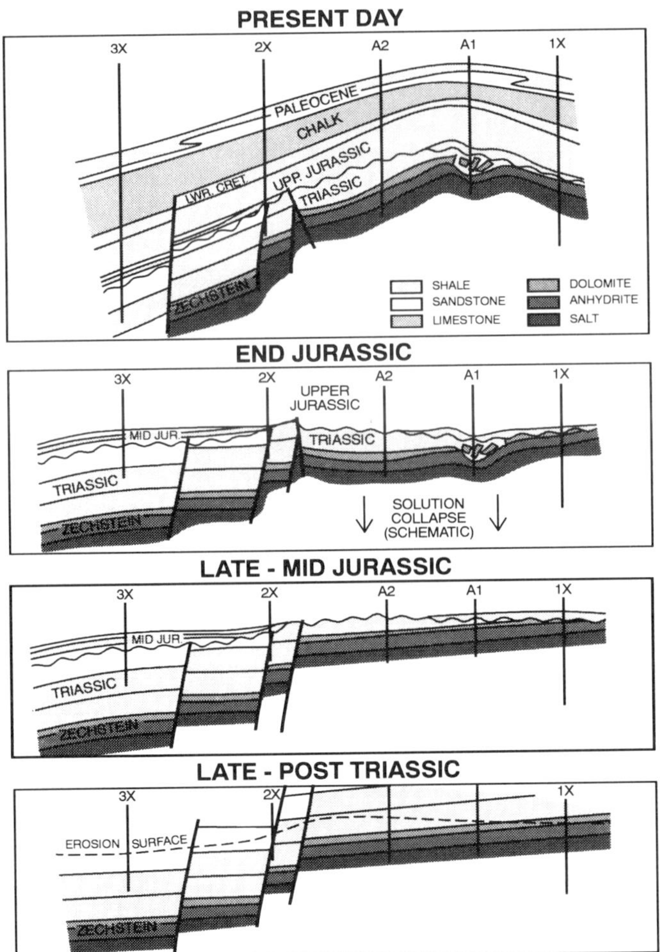

Figure 9. Simplified tectonic history of the Maureen field from the Triassic to the present day. Generally north-northwest to south-southeast section. No scale. Straight-line distance well 3x to 1x about 2.7 mi (4.4 km). See Figures 2D and 6 for well locations.

TRAP

Trap Type

The Maureen field, with its crest at about 8100 ft (2470 m) subsea, comprises a northwesterly oriented domal anticline trap (Figure 2). A fieldwide OWC, defined by the level of 100% water saturation, is taken at an average value of 8702 ft (2654 m) subsea; variations in individual well contacts of up to 50 ft (15 m) occur, but since all of the reservoir sands are in effective pressure communication, these variations are thought to result largely from well deviation survey correction errors. The field is effectively full to spillpoint with an areal closure of 4650 ac (1880 ha) and a maximum oil column of 650 ft (200 m). The reservoir is sealed above and laterally by Paleocene Lista Formation claystones; the lower reservoir limit is the underlying Danian Chalk Formation and Maureen marl unit.

Reservoir

Stratigraphy

The Paleocene reservoir of the Maureen field belongs to part of the central North Sea lower Tertiary Montrose Group (Figure 8), which itself comprises four formations—the Maureen, Andrew, Heimdal, and Forties formations. The Maureen and Andrew formations are the only sand-bearing formations present in the Maureen field. The Maureen Formation was named by Deegan and Scull (1977) after the Maureen field. Detailed micropaleontological and palynological studies, together with seismic data, have determined reservoir subdivisions (see *Geological Models* section).

Lithology

The type section of the Maureen Formation is found in the Forties oil field (well 21/10-1), some 37 mi (60 km) to the southwest. The section comprises a clastic-carbonate sequence containing a mixture of reworked chalks, mudstones, and sands with occasional conglomeratic units. The type section is not lithologically characteristic of the Maureen field reservoir section or the main body of the fan.

In six of the Maureen field wells (16/29-1x, -3x, -A6, -A10, -A13, and -A19, locations shown on Figure 5D), sandstones occur within the upper chalk sections of the Ekofisk Formation, and this section is therefore more comparable with that in the type well 21/10-1. The predominantly clastic sequence of largely Danian age above the Ekofisk Formation is referred to here as the Maureen Formation; the representative type well for the field taken by Cutts (1991) is 16/29-2x, where the Maureen Formation is both complete and largely cored (Figure 4).

The sandstones are generally of excellent reservoir quality, and neither authigenic nor detrital clays present any significant permeability problem in the oil leg. The sandstones are light gray-brown to light gray in color, fine to medium grained throughout, poorly to moderately well sorted, and subangular to well rounded. The mineralogy of the reservoir sandstones is largely consistent throughout the field and is dominated by simple strained quartz grains. The remainder of the sandstones comprise feldspar (8–12%), rock fragments (2–3%), and detrital clay matrix (1–2%). The majority of quartz grains have tangential grain contacts while the remainder have short straight or concave-convex contacts. This suggests that minor quartz dissolution has occurred and that grain welding has been minimal.

Muscovite, rare glauconite, and heavy minerals occur in trace amounts, with muscovite increasing slightly (1%) toward the base of the reservoir section. Feldspars include microcline, slightly sericitized orthoclase, plagioclase, and rare perthites. Rock fragments consist mainly of altered or partially dissolved volcanics with rare granite fragments, metaquartzite, chert, and reworked sand grains (probably Permian–Triassic[?] in age), with authigenic quartz overgrowths, some of which show signs of erosion. Heavy minerals include zircon, tourmaline, and, locally, garnet. The volcanic rock fragments often show a high degree of dissolution both internally and at their margins, together with a high degree of chlorite alteration. This may be one potential source of the chlorite seen in the sediments.

Chlorite, however, when present in substantial quantities alters the pore geometry in such a way as to increase the water saturation significantly. In fact, this increase in chlorite content, manifest as high water saturations indicated on logs of the 16/29-5x well, was interpreted by some prior to drill-stem test (DST) as an anomalously high water leg; however, water-free oil was produced by the DST of this section.

The argillaceous appearance of some of the Maureen sandstones is attributable in part to the presence of authigenic clays; X-ray diffraction analysis of the detrital clay matrix indicates a mixture of illite, mica, mixed-layer clays, chlorite, and kaolinite.

Depositional Setting

The depositional system of the Central graben during the Paleocene was characterized by the dominance of sediment gravity-flow deposits (e.g., Fowler, 1975; Walmsley, 1975; Hill and Wood, 1980; Knox et al., 1981; Carman and Young, 1981). Although the majority of the papers cited in this study are concerned with the late Paleocene, the general nature of sedimentary processes in the earlier Paleocene was similar.

The most suitable scheme for the classification of the various facies seen in the Maureen field is that of Mutti and Ricci-Lucchi (1972, 1975). Within the Maureen field, an examination of over 1200 ft (360 m) of cores has resulted in a reservoir subdivision into three simple lithofacies types.

1. *Massive sandstone facies.* This is the dominant facies of the main reservoir unit in the field, which is characteristically thick and massive and has generally blocky, low gamma ray log responses. They were deposited from fluidized flows of low transport efficiency, which resulted in thick, amalgamated clastic lobes that are massively bedded and destructured, with abundant dewatering features. Moderately sorted, predominantly fine to medium grained, with an absence of significant grading, and with subangular to rounded grains, the sandstones occur in units 3 to 50 ft (1–15 m) in thickness with sand/shale ratios generally exceeding 0.9; locally thin, laterally discontinuous claystone layers separate the sandstones. There is no evidence of fauna or bioturbation in the sandstones and sole markings are rare. Pebbles and granules of claystone are commonly dispersed throughout the sandstones, being rarely concentrated at the base of the sandstones. They are commonly associated with the mid- to outer-fan environment, where they form channels and lobes.

2. *Interbedded sandstone and claystone facies.* These facies occur as units 1 in. to 2 ft thick, with sand grain size slightly finer than in the massive sandstones. They have a wider range of textures, often distorted by compactional effects, and bed geometries more variable than in the massive sandstones; they were the result of deposition from the lower energy "tails" of large gravity flows and are normally found adjacent to mid- to outer-fan lobes and channels in the form of overbank and levee deposits.

3. *Claystone facies.* This facies comprises laterally persistent hemipelagic claystones with thin interbeds of siltstones, very fine grained sandstones, and thin limestones. Two types of claystones are present: gray-green, illitic and calcareous claystones; and black, illitic and kaolinitic, noncalcareous claystones

sometimes containing scattered dolomite nodules. Both types of claystone contain horizontally burrowing trace fossils (e.g., *Zoophycos*, *Chondrites*, *Planolites*) suggestive of relatively quiet conditions, plant fragments, and a fauna of planktonic and benthic marine foraminifera characteristic of deep water. The trace fossils are most abundant in the gray-green claystones, and each claystone type contains a different faunal assemblage. The gray-green claystones occur in the lower sections of the reservoir.

4. *Chalk melange facies.* Blocks of chalk appear in several of the wells in the field and in nearby wells. These are composed of chaotic assemblages of slumped limestones, large undeformed blocks of older chalk, and calcareous claystones, interpreted to be calcareous debris-flow deposits. They are present generally on the eastern side of the field and may have resulted from syndepositional tectonic activity and a westerly transport. They do not form widespread permeability barriers but may act as local baffles to vertical flow.

Geological Models

1. *Early interpretation (1975-1983).* Although the existence of a large Paleocene submarine fan system in the Central graben was known in broad terms, the precise age, depositional setting, and distribution of the sandstones were uncertain. Both the Maureen and Cod fields were known to be located close to the eastern and outer edges of a fan system. As was the case for Cod, a local paleo-relief seemed to have existed at the Maureen site during the earliest Paleocene, and it was logical to consider that the paleo-relief had had a strong influence on sandstone distribution. The first partnership agreed that the geological development model of Maureen was somewhat conservative and envisaged substantial thinning of the sandstones over the paleo-crest and toward the east (Figure 5A). The revised model took into account the following:

- the shape of the paleo-structure
- the general seafloor topography rising to the east
- the direction of sediment flow
- the absence of sandstone in a nearby well located in the Norwegian sector
- the problematic results of development drilling occurring at Cod at that time

At the time of Annex B submission in 1978 (Phillips Petroleum, 1978), the estimated original oil in place on a volumetric basis was 289 MMSTB of which 125 MMBO was thought to be recoverable (Figure 5). However, even at that time, more pessimistic cases (Figure 5C) were commercially viable.

2. *After completion of development drilling (1985).* Several factors led to a considerably improved geological model of the field:

- a considerably better defined Paleocene biostratigraphy (the Maureen sandstones, previously considered as upper Paleocene in age, are in fact mainly of Danian age, and as such, are older than those in the Cod, Forties, and Andrew fields)
- the data from a large number of wells presently available have led to more reliable log correlations
- a good knowledge of the paleoenvironment in and around the Maureen area from seismic and regional well control
- new seismic data and geophysical analysis within the Paleocene interval
- observations of reservoir behavior during production

Excluding a poorly defined "Ekofisk Formation" sandstone layer of relatively minor importance to production, the Maureen reservoir has been divided into three sandstone units, termed P1, P2, and P3 (Figures 10 through 13). These three units were primarily determined from detailed biostratigraphic analysis of the intervening claystones. Each sandstone unit was mapped areally using both the available well data and seismic stratigraphic techniques; their resultant distribution is shown both laterally and in cross section in Figures 10 and 13. The tops of the P1 (lowest) unit and P3 (highest) unit are generally continuous and mappable reflectors over much of the field. The reflector representing the top of the P2 unit is ambiguous and somewhat discontinuous. After over five years of production the reservoir continues to behave as a single unit, indicating that there are no major barriers to fluid flow; the discontinuous nature of the reflector in some areas of the field may indicate areas of sandstone connectivity. The lithofacies distribution across the Maureen field (Figure 10) shows sediment transport from the west and northwest, and the correlation section (Figure 13) shows the crestal thinning in the vicinity of the A1 and A8 wells. Figures 11 and 12 illustrate the correlation in more detail.

Porosity Type, Diagenesis

In general, diagenetic effects have been minimal in the reservoir and relatively high porosities (18–25%) have been preserved. Porosities have been severely reduced in thin sandstones and at sandstone/shale boundaries by extensive calcite cementation; these zones are generally thin, however, and do not constitute a serious problem in the performance of the reservoir.

The diagenetic history of the sandstones within the reservoir can be summarized as follows (starting with the earliest event).

1. *Early calcite cementation.* Calcite cementation is possibly represented by early crystal growth near contacts of sandstones with calcareous shales and marls. Calcite in the argillaceous lithologies is thought to be primary in origin.

2. *Quartz cementation.* Minor quartz overgrowths (1-2%), often incomplete, occur on quartz grains but have had little effect on the porosity and permeability of the reservoir.

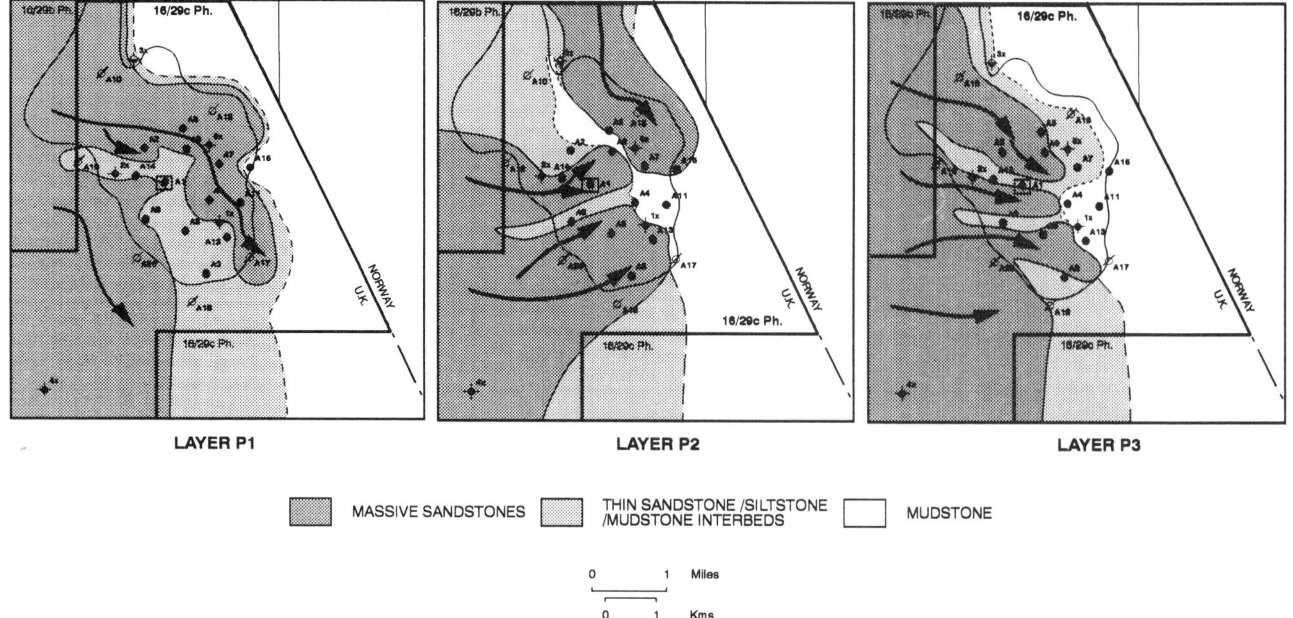

Figure 10. Three-layer reservoir model sandstone-shale distribution as perceived in 1985.

3. *Kaolinite cementation.* Authigenic kaolinite is present (2–3%) in most of the sandstones, partially infilling scattered pore spaces. SEM analyses indicate that it occurs as booklets of hexagonal platelets which contain moderate microporosity, have little influence on porosity, but marginally reduce the permeability.

4. *Chlorite cementation.* Chlorite occurs in minor amounts (3–7%) as small vermicular crystals rimming grains and partially occluding pore spaces. It is most abundant in the more argillaceous sandstones (4%), and its habit causes a noticeable effect on the permeability by sealing many of the narrow pore throats. Chlorite and kaolinite exhibit a very complex interrelationship but are rarely found within the same pore space.

5. *Late calcite cementation.* This cementation occurs near the sandstone/shale contacts but is very rare in the massive reservoir sandstones (Figure 15).

6. *Grain dissolution/welding.* Grain dissolution and welding occur to a minor degree in most sandstones but appear to be more severe in the more argillaceous sandstones. This does not significantly effect the porosity/permeability relationships of the sandstones.

Rare poikilitic, coarsely crystalline barite has been found in the thin sandstones of well 16/29-2x where it can be seen to cement thin zones associated with high-angle hairline fractures.

Fractures

Both massive and thin sandstones show abundant high-angle and oblique, hairline fracturing. All the fractures in the thin sandstones are tightly cemented, mainly with coarsely crystalline calcite; some of the fractures are cemented by secondary silica and kaolinite. The evidence from the cores indicates that the fractures were early and that they were probably cemented before oil migration occurred. They have little effect on the overall reservoir performance, and the absence of significant open fractures is confirmed by DST analyses.

Porosity/Permeability

Core data available from the oil-bearing zone of seven wells (2x, 3x, 5x, A1, A4, A8, and A10) indicate a similar trend. Figure 14 shows a clear relationship between porosities in the range 15 to 32% and corresponding permeabilities of 30 to 3000 md; the relationship is poorer in the lower quality reservoirs. The core permeabilities in the aquifer are lower, generally less than 100 md. Well tests indicate an even bigger ratio of oil leg to water leg permeabilities, i.e., at about 10:1. Detailed thin section examination of the sandstones illustrates the relationships between porosity, permeability, and cements. Figure 15 shows the detrimental effect of carbonate cement on both porosity and permeability but the relatively minor effect of secondary quartz. The proportionately higher permeability of the massive sandstones appears to be a function primarily of larger grain size and relative lack of brown clay matrix. The cause of the lower permeabilities in the water leg is largely the result of chlorite growth and more carbonate cementation. The vertical sandstone porosity distribution is shown for several wells in Figures 11 and 12.

Pay Zone Thickness

The oil-water contact for the field, as defined as the level of 100% water saturation as determined from logs and RFT data, was established in 1985 by data from five wells. Thicker in the west, the gross pay

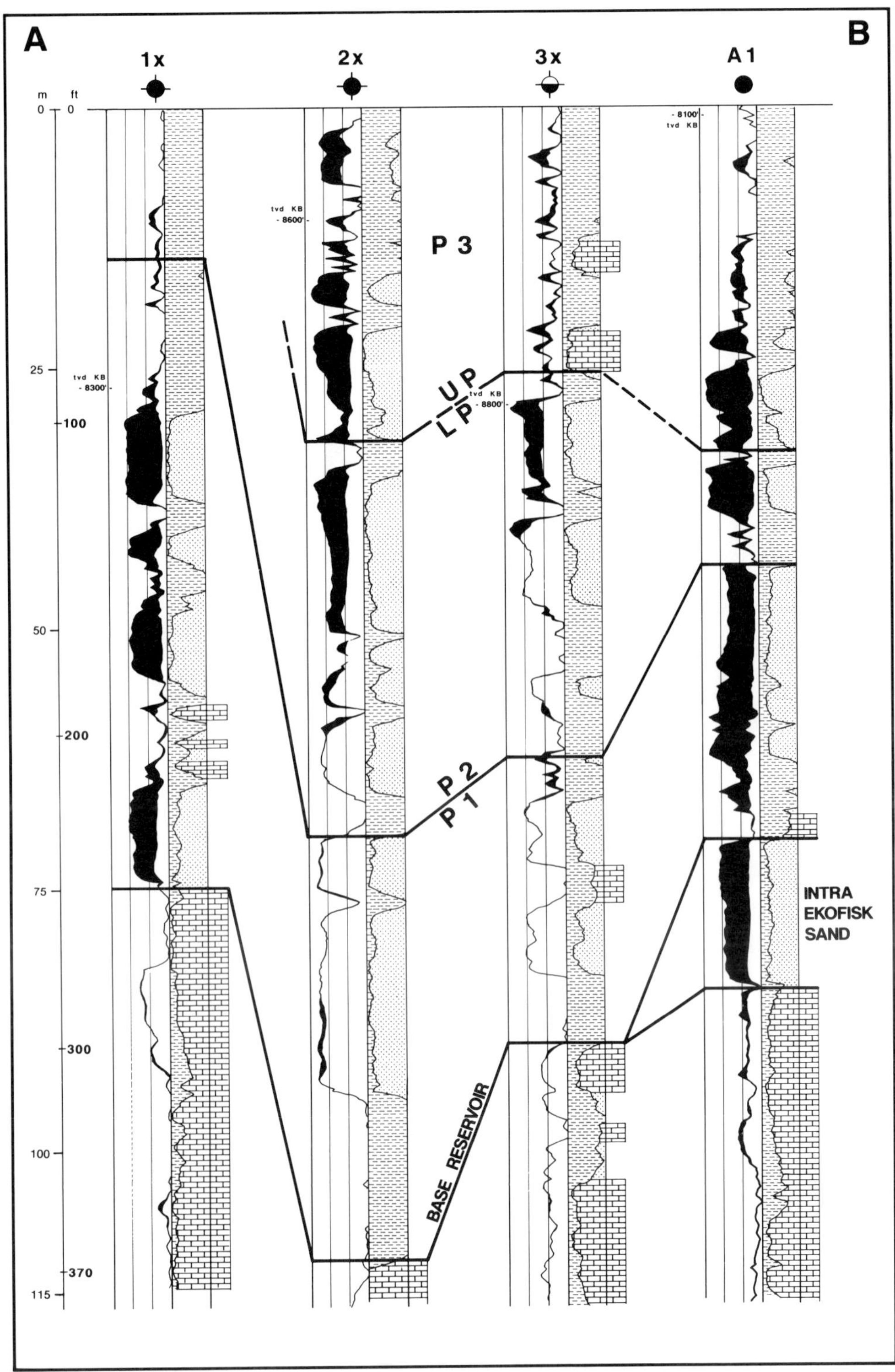

Figure 11. Interpreted lithology, porosity, and hydrocarbon saturation logs in the Paleocene reservoir section of three exploration wells and the first development well 16/29a-A1 to illustrate the difficulty in correlating the sandstone reservoir layers. The intra-Danian P1–P2 correlation and the upper-lower Paleocene boundary (UP–LP) are based on biostratigraphy. (See Figure 12 for line of section and log legend.) In the discovery well -1x, all the sandstone bodies belong to the P1 stratigraphic interval, which also contains reworked chalk beds. Sandstone rests directly on "in situ" Danian chalk in wells 1x and A1, whereas chalk progressively passes into marls and shales in wells 2x and 3x. Upper Paleocene sandstones overlay the main reservoir of early Paleocene age in wells 2x and A1. (tvd KB, true vertical depth below kelly bushing.)

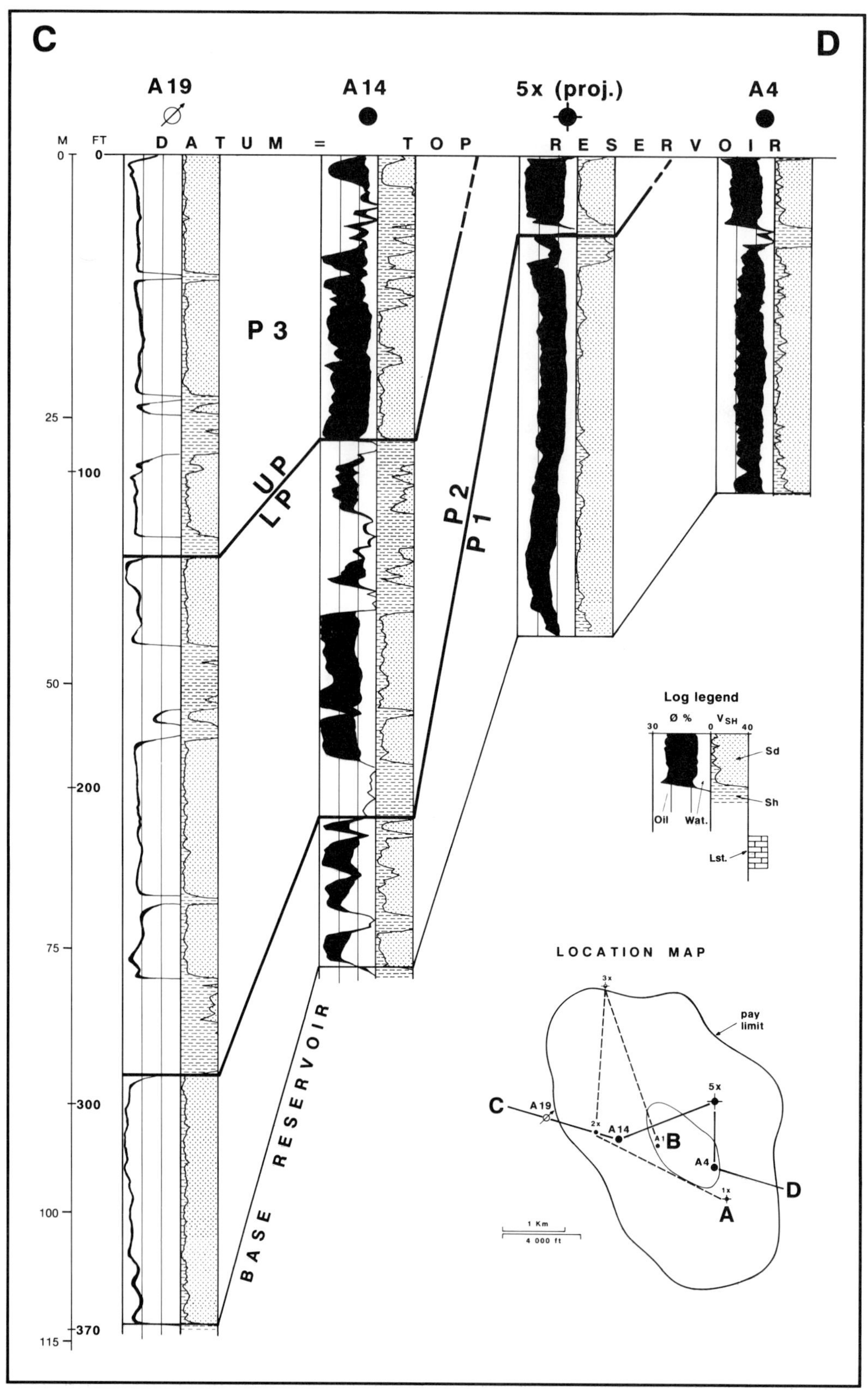

Figure 12. Interpreted lithology, porosity, and saturation logs of wells located along a west-east section across the Maureen field to illustrate thickness and age variation of the Paleocene reservoir. The lower sandstone (P1) is present over most of the field area but thins regionally to the east, outside the Maureen field. Younger sands have been deposited to the west and north. (Solid circles, producing wells; circle with arrow, injection; wells with *x* are exploratory/appraisal.)

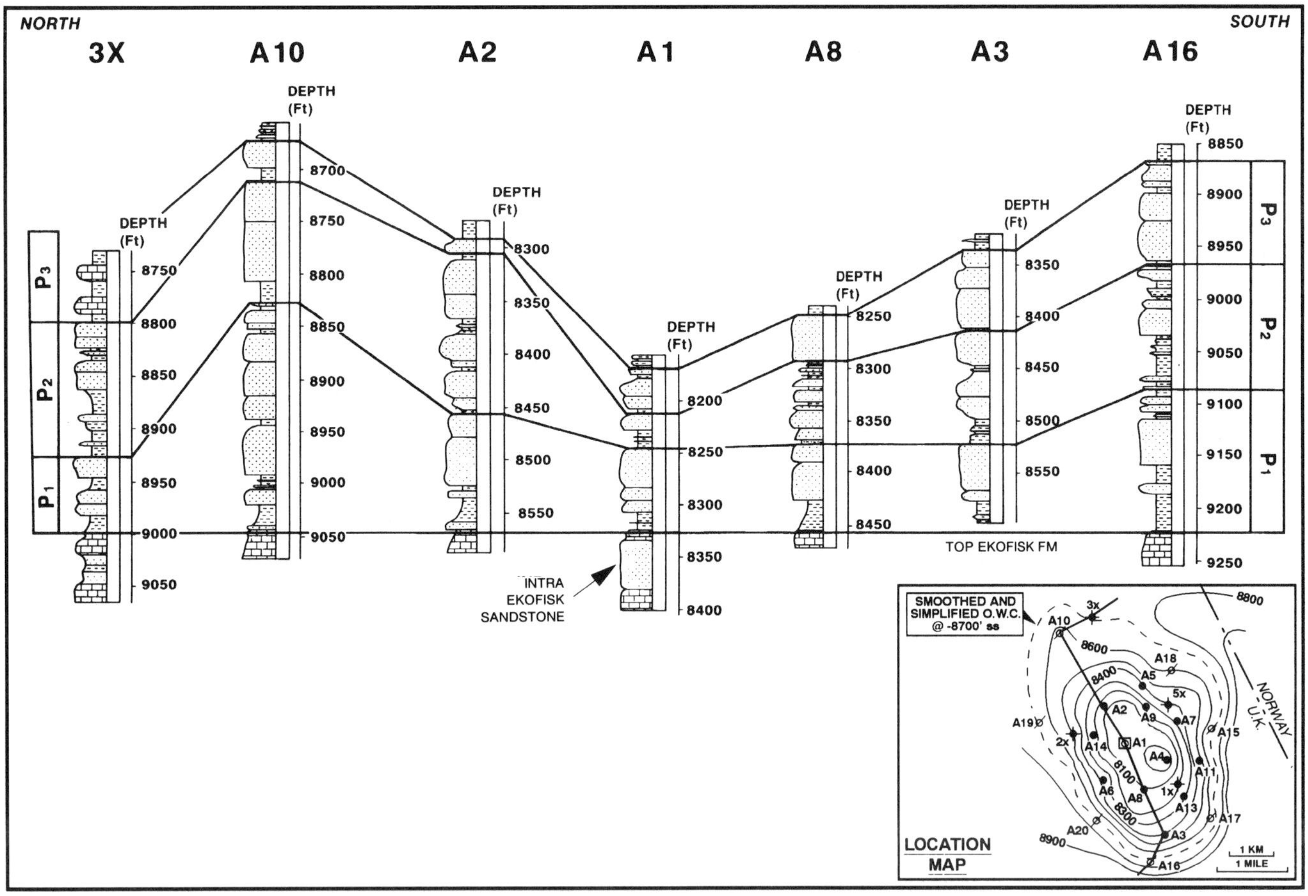

Figure 13. Well correlation as perceived in 1985. Datum top of Ekofisk formation. (See Figure 2 for well symbols.)

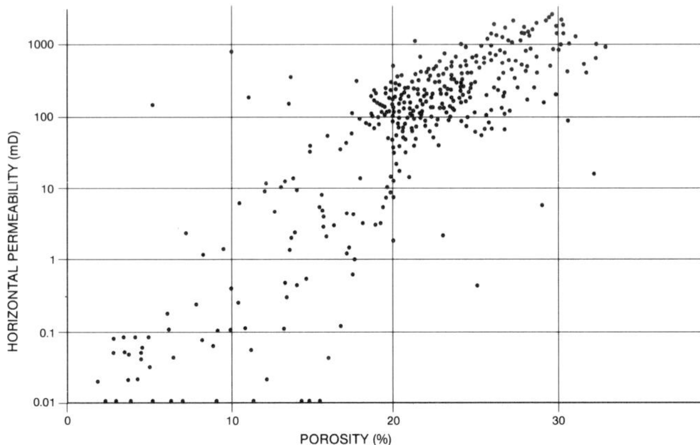

Figure 14. Core porosity/permeability data from lower Paleocene sandstones of the Maureen field.

of the field varies by nearly 400 ft (122 m). To the east, the gross pay thins as a result of the overall thinning of the depositional package. As a result of the two controls on net pay distribution (i.e., the gross sand thinning to the east and the domal shape of the top of the reservoir), the net pay distribution (Figure 5D) can be seen to consist of a structurally controlled thinning away from the field crest (wells 1x and A4) combined with the easterly sedimentary thinning. These variations in thickness also were realized at an early stage of field exploration (Figure 5B). As currently defined, the P1 unit (Figures 3, 10, and 13) contains about 25% of the field's reserves, the P2 unit about 40%, and the P3 unit around 35%.

Hydrocarbon Characteristics

The Maureen crude oil is mature paraffinic-naphthenic (undersaturated) 36° API. The reservoir temperature is 243°F (117°C), with an initial pressure of 3792 psia (267 kg/cm^2) at 8300 ft (2530 m) subsea and an initial gas/oil ratio of 392.7 scf/STB (69.2 m^3/m^3). An analysis of the crude oil is shown in the whole oil and C_{15+} saturates chromatograms (Figure 16 and Table 1) and is comparable with other North Sea crude oil sourced from the Kimmeridge Clay. It has a wax content of 14.2% and a low sulfur content of 0.35%. Carbon isotope compositions (°/oo PDB) have been measured (Figure 17 and Table 1) and they also show similar typical North Sea oil characteristics.

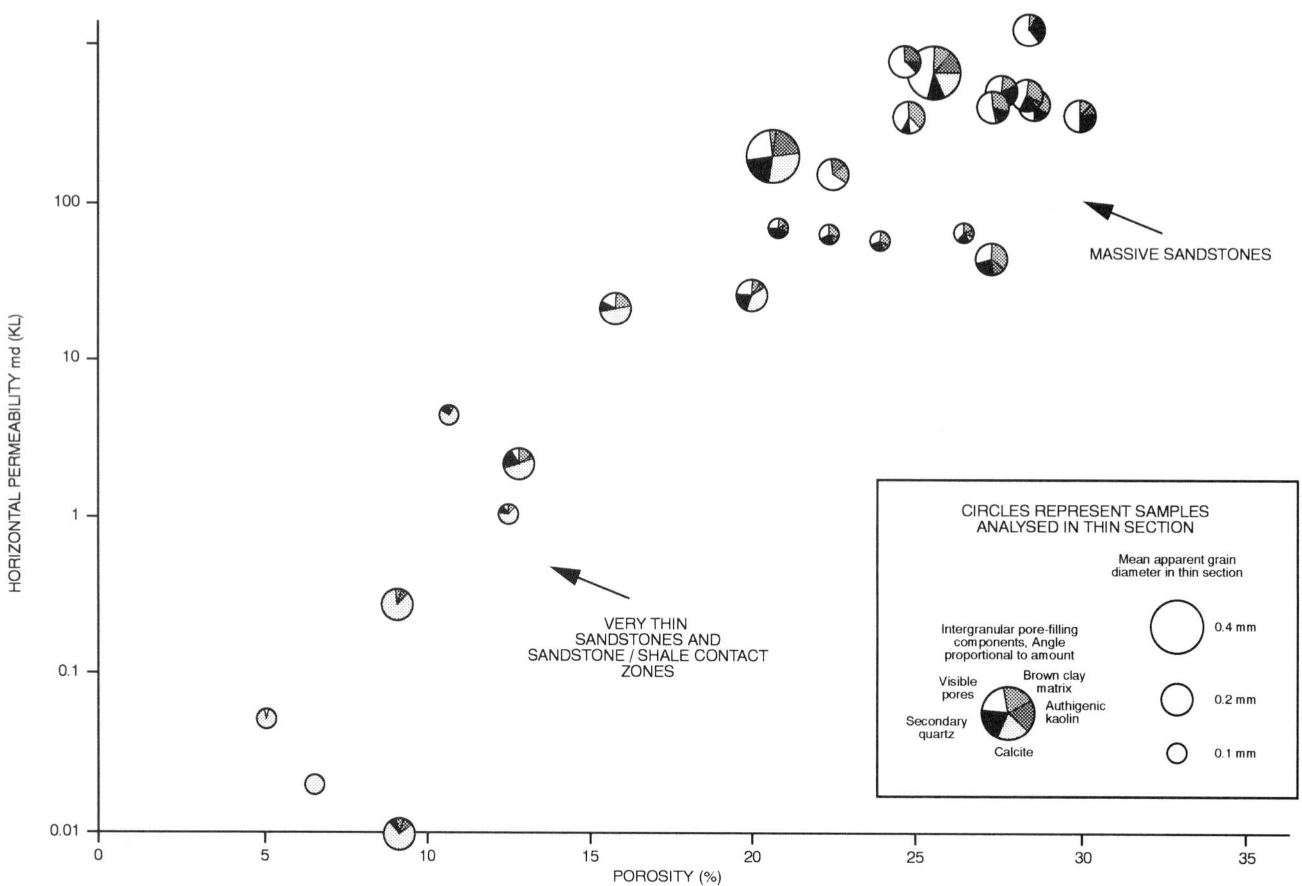

Figure 15. Relationships between permeability, porosity, grain size, and pore-filling components in lower Paleocene sandstones. (KL, Klinkenberg permeability.)

Fluid Flow Characteristics

The Maureen field came on stream in September 1983, producing via two of its 12 predrilled production wells. The remaining wells were all brought on stream by April 1984. Over this period the reservoir pressure declined by about 1000 psi (70 kg/cm^2) from initial conditions. Peripheral water injection commenced in February 1984 via two of its seven predrilled injection wells. By May 1984, six injection wells were in operation, with one well being used as an observation well. By March 1985 the reservoir pressure had declined by a further 400 psi (28 kg/cm^2). Water production from the field was first seen in April 1984 in well 16/29a-A14.

In terms of fluid flow through the reservoir, most of the injected water enters the field from the west (Figure 12) with production being mainly from the uppermost section in the central and eastern portions of the field. Water injection supplements a significant natural water drive which, because of the greater aquifer thickness to the west, has a similar pattern of support. The amount of natural pressure support from the aquifer is continuously changing. Initially most of the pressure support was from injected water, but once the aquifer started to contribute, its contribution to pressure support continuously increased. Production data indicate that all the sandstones in the reservoir are in communication, albeit with the probable existence of local shale barriers.

Development Considerations

During the first phase of development, all wells were predrilled (12 producers and seven water injectors) from a 24-slot template and completed/tied back to the single production platform that was installed in June 1983. Peripheral water injection was required to enhance the natural water drive of the aquifer and to maintain reservoir pressure above bubble point. Artificial gas lift has been employed in the field in selected wells where the water cut has risen in order to optimize levels of oil production. The maximum gas lift capacity is 10 MMSCFD. Oil production commenced in September 1983 and peaked at 29.3 MMBO per year in 1984 (Figure 18). The average daily field production in 1989 was about 50,000 BOPD. Water injection was started in February 1984, and most of the water is injected into the western half of the field. The average daily rate

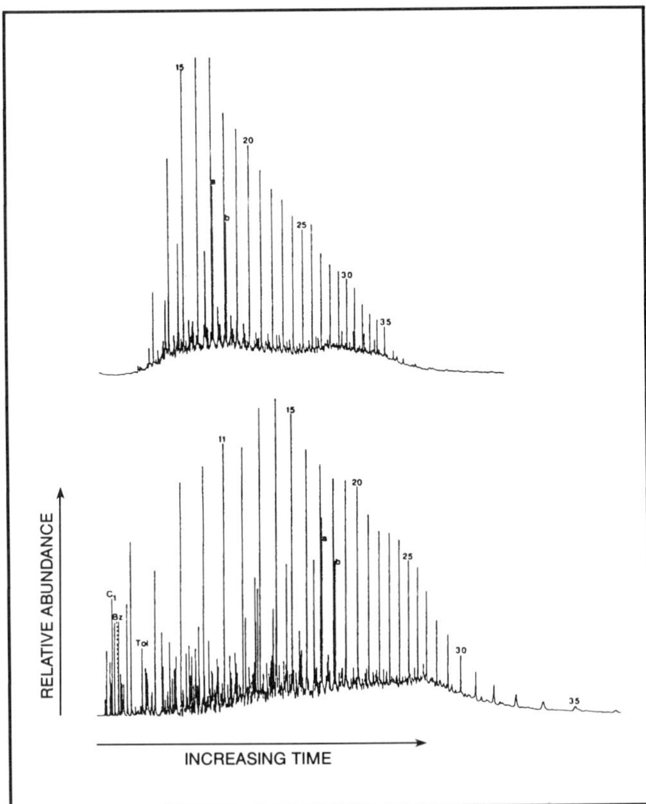

Figure 16. Maureen Paleocene reservoired oil-gas chromatograms. Analyses show that the crude oil is comparable to other Kimmeridge Clay sourced oils in the North Sea. It is highly paraffinic, non-waxy, and has a pristane/phytane ratio of 1.26 (see Table 1).

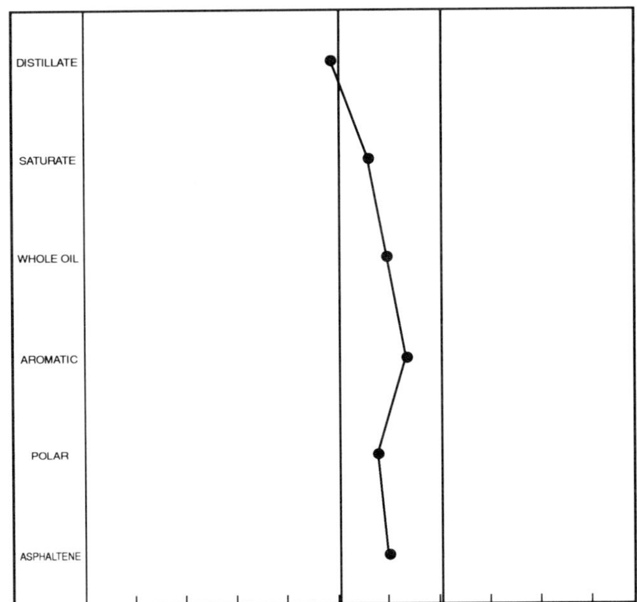

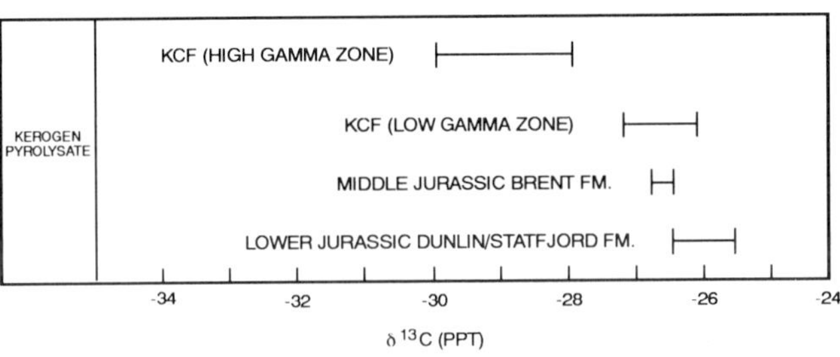

Figure 17. Maureen oil carbon isotope composition in relation to typical North Sea kerogen pyrolysate carbon isotopic signatures (Burwood et al., 1990; Bailey et al., 1990). The data show that the Maureen oil appears to have been sourced from the "hot shales" of the Kimmeridge Clay Formation (KCF [high gamma zone]).

Table 1. Geochemical characteristics of produced Maureen oil.

		Carbon isotope data ($^{\circ}/_{\circ\circ}$ PDB)	
API gravity	36°	Whole oil	-29.07
Paraffin (%)	57.39	Saturates	-29.37
Naphthene (%)	37.33	Aromatics	-28.53
Isoprenoid (%)	5.28	NSO	-29.14
Pristane/phytane	1.26	Asphaltenes	-28.99
Pristane/nC_{17}	0.52	Distillate	-30.11
Asphaltene wt%	1.56		
Sulfur (%)	0.35		
Wax (%)	14.2		

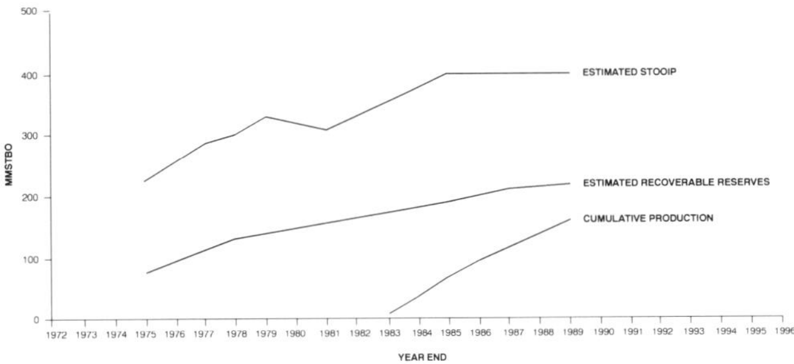

Figure 18. Oil in-place, reserves, and cumulative oil production with time. Cumulative oil production has now surpassed the original estimate of recoverable reserves, which itself has been revised upwards over the years as a result of a better understanding of the field characteristics. (MMSTBO = millions stock tank bbl oil. STOOIP = stock tank [bbl] original oil in place.)

of water injection in December 1989 was 60,000 BWPD. Associated gas is flared from the platform.

The oil-in-place estimates have varied widely since the first well was drilled on the field (Figure 18). Unlike many fields during appraisal and development drilling, the Maureen field has consistently increased its oil-in-place and reserves estimates with increased information. The oil-in-place estimate currently stands at 393 MMSTBO with an estimated primary plus secondary recovery of 210 MMSTBO (54%). Cumulative field production to the end of 1989 was approximately 155 MMSTBO, which equates to about 66% of the estimated recoverable reserves. The field is currently on decline and economic limit is estimated to be toward the end of the century.

If the field were discovered today, it probably still would be developed in essentially the same way as it was in 1983. The reservoir still would be kept flowing above the bubble point. From a development standpoint, it is likely that only one well location would be changed; to optimize recovery, the A15W would have been moved from the eastern extremity of the field to a more crestal position between wells A7 and A4. The field, in general, has performed extremely well over the past seven years and substantial changes in development would not be made.

Source

The source rock for the majority of the oil found in the North Sea, including the Maureen field, is the Kimmeridge Clay, and geochemical analyses have shown a good correlation with its hydrocarbon extracts. The shale typically has total organic carbon contents reaching 15%+ and averaging 8% (Cayley, 1987) comprizing amorphous marine kerogen. The main source area for the Maureen oil is thought to lie about 9 mi (15 km) to the southwest of the field in the Outer Witch Ground graben. Early maturity of these source rocks was reached in the mid- to Late Cretaceous, with peak oil generation occurring during the early Tertiary (Figure 19). As a result of the richness and thickness of the source rock, when mature the yield is such that it exceeds trap capacity.

The migration of the generated hydrocarbons away from the source kitchen was initially via deep basement-rooted, graben-bounding faults (Cayley, 1987). These allowed access into the relatively undeformed and laterally continuous, normally pressured aquifer of the Tertiary clastic sequence through which the oil was able to migrate for distances of up to 12 mi (20 km) to the north.

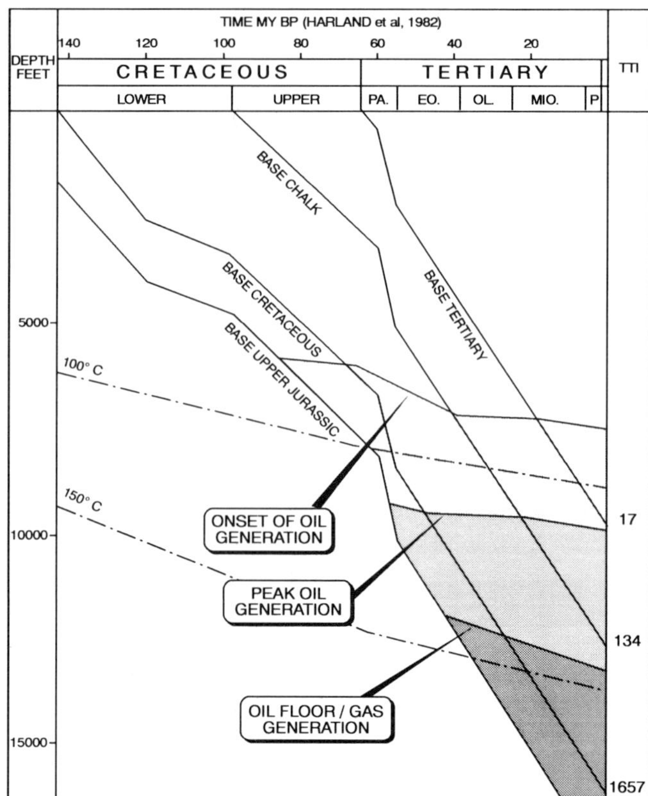

Figure 19. Burial history plot for the source area in the vicinity of the Maureen field.

EXPLORATION CONCEPTS

By the time the Maureen field discovery well had been drilled in 1972, several lower Paleocene fields such as Forties, Montrose, and Cod had been discovered. The Paleocene fan complex was still not fully understood, based as it was on regional well correlations, early seismic data, and existing age dating, but the fundamental regional and local structural concepts were clear. The uplift and erosion of the East Shetlands platform in the early Tertiary resulted in vast quantities of sand being shed eastward and southeastward into the developing basins. These were broadly coincident with the underlying Viking and Central grabens. The sands were transported over long distances by submarine fan processes deeper into the basins. Subsequent burial and a lack of significant post-depositional faulting allowed generated hydrocarbons to move east and northeast along an almost uninterrupted migration route. Closure along the route, whether by dip closure over paleo-horsts, by halokinetic effects, or by stratigraphic traps, was required to entrap these migrating hydrocarbons. The major discoveries such as the Forties and Montrose fields were made by drilling on dip closures over paleo-highs. Although dip closures caused by Permian salt movements were noted, these tended to be small and generally noncommercial; there were exceptions including the Maureen and Cod fields.

By the late 1980s, the majority of the larger dip-closed traps had been drilled and many smaller dip-closed Paleocene traps were drilled in the vicinity of existing oil installations. In addition to discoveries in dip-closure traps, investigations along the basin margins and around the intra-basin highs, drilling for updip stratigraphic pinchout has led to several discoveries, including the Maggie and Everest fields to the south of Maureen.

In the early 1970s, the limited knowledge of the detailed Paleocene stratigraphy in the Central North Sea resulted in difficulties in understanding the relationships between the various fan systems.

Detailed correlations of the reservoir sequence between fields and of reservoir layering within the Maureen field area were, therefore, problematical and questionable.

The lessons learned from the Maureen field are similar to those outlined by D'Heur (this volume) for the Cod field. These have mainly been the understanding of the complexity of the sandstone lobe distribution, of the bathymetric control of deposition, and of reservoir stacking geometries. This greater understanding has led to a better comprehension of the oil distribution, which, in the case of the Maureen field, has allowed the original oil-in-place estimate to be refined through time and ultimately to be confirmed by material balance calculations. The need for detailed biostratigraphic dating to allow a confident correlation of the sandstones and shales is required to construct a reliable predictive model if development is to maximize oil recovery from a field in this depositional environment, especially if it is of a relatively small size.

ACKNOWLEDGMENTS

The authors would like to thank Fina Exploration Ltd, Phillips Petroleum UK, Agip UK Ltd, Gas Council Exploration UK Ltd, and Ultramar Exploration UK Ltd for their permission to publish this field study. The opinions expressed are not necessarily those of all of the co-venturers. The authors would also like to extend their appreciation to the AAPG reviewers for their constructive criticisms, which have been helpful in improving the quality of this paper.

REFERENCES

Armstrong, L. A., A. Tenhave, and H. D. Johnson, 1987, The geology of the Gannet fields, central North Sea, U.K. Sector, *in* J. Brooks and K. W. Glennie, eds., Petroleum geology of north-west Europe: London, Graham and Trotman, p. 533–548.

Carmalt, S. W., and B. St. John, 1986, Giant oil and gas fields, *in* Future petroleum provinces of the world: AAPG Memoir 40, p. 11–54.

Carman, G. J., and R. Young, 1981, Reservoir geology of the Forties oilfield, *in* L. V. Illing and G. D. Hobson, eds., Petroleum geology of the continental shelf of north-west Europe: London, Heyden, p. 371–379.

Cayley, G. T., 1987, Hydrocarbon migration in the Central North Sea, *in* J. Brooks and K. W. Glennie, eds., Petroleum geology of north-west Europe: London, Graham and Trotman, p. 549–555.

Cutts, P. L., 1991, The Maureen field, Block 16/29a, UK North Sea, *in* United Kingdom oil and gas fields, 25 Years Commemorative Volume: Geological Society Memoir 14.

D'Heur, M., 1986, The Norwegian chalk fields, *in* Habitat of hydrocarbons on the Norwegian continental shelf: Norwegian Petroleum Society, London, Graham and Trotman, p. 77–89.

D'Heur, M., 1987, Cod, *in* A. M. Spencer, et al., eds., Geology of the Norwegian oil and gas fields: London, Graham and Trotman, p. 51–62.

D'Heur, M., this volume. Cod field.

Deegan, C. E., and B. J. Scull, 1977, A standard litho-stratigraphic nomenclature for the Central and Northern North Sea: Institute of Geological Sciences, Report 77/25.

Fowler, C., 1975, The geology of the Montrose field, *in* A. W. Woodland, ed., Petroleum and the continental shelf of north-west Europe. 1. Geology: Barking, Applied Science Publishers, p. 467–476.

Hill, P. J., and G. V. Wood, 1980, Geology of the Forties field, U.K. continental shelf, North Sea, *in* Giant oil and gas fields of the decade: 1968–1978, AAPG Memoir 30, p. 81–93.

Jacque, M., and J. Thouvenin, 1975, Lower Tertiary tuffs and volcanic activity in the North Sea, *in* A. W. Woodland, ed., Petroleum and the continental shelf of north-west Europe. 1. Geology: Barking, Applied Science Publishers, p. 455–466.

Knox, R. W. O'B., A. C. Morton, and R. Harland, 1981, Stratigraphical relationships of Palaeocene sands in the U.K. Sector of the Central North Sea, *in* L. V. Illing and G. D. Hobson, eds., Petroleum geology of the continental shelf of north-west Europe: London, Heyden, p. 267–281.

Morton, A. C., 1979, The provenance and distribution of the Palaeocene sands of the Central North Sea: Journal of Petroleum Geology, v. 2, pt. 1, p. 11–21.

Mutti, E., and F. Ricci-Lucchi, 1972, Le torbiditi dell'Appennino settentrionale: introduzione all' analisi di facies: Mem. Soc. Geol. It., 11, p. 161–199.

Mutti, E., and F. Ricci-Lucchi, 1975, Turbidite facies and facies associations, *in* E. Mutti, G. C. Parae, F. Ricci-Lucchi, M. Sagri, G. Zanzucchi, G. Chibaudo, and S. Iraccarino, eds., Examples of turbidite facies and facies associations from selected formations of the Northern Appenines: IX International Congress on Sedimentology, Nice, Field Trip A-11, p. 21–36.

Phillips Petroleum Company (UK) Ltd, 1978, Maureen field, UK North Sea, Annex B Plan of Development submitted to U.K. Department of Energy.

Phillips Petroleum Company (UK) Ltd, 1985, Maureen field, UK North Sea, Annex B (Revised) Plan of Development submitted to U.K. Department of Energy.

Ricci-Lucchi, F., 1975, Depositional cycles in two turbidite formations of the Northern Appennines (Italy): Journal of Sedimentary Petrology, v. 45, n. 1, p. 3–43.

Ricci-Lucchi, F., 1984, Fan sedimentation with emphasis on the North Sea Tertiary. Part 1: Joint Association for Petroleum Exploration Courses (UK), Course Notes No. 29, Geological Society of London, 73 p.

Stewart, I. J., 1987, A revised stratigraphic interpretation of the Early Palaeogene of the Central North Sea, *in* J. Brooks and K. W. Glennie, eds., Petroleum geology of north-west Europe, London, Graham and Trotman, p. 557–576.

Walmsley, P. J., 1975, The Forties field, *in* A. W. Woodland, ed., Petroleum and the continental shelf of north-west Europe. 1. Geology: Barking, Applied Science Publishers, p. 477–486.

Appendix 1. Field Description

Field name .... *Maureen field*

Ultimate recoverable reserves .... *210 MMSTBO (28 million tonnes)*

Field location:

Country .... *United Kingdom*
State .... *Continental shelf (North Sea)*
Basin/Province .... *South Viking graben/Witch Ground graben*

Field discovery:

Year first pay discovered .... *Lower Tertiary Maureen Formation February 1973*

Discovery well name and general location:

First pay .... *16/29-1x, 160 mi (260 km) NE of Aberdeen, Scotland, United Kingdom*

Discovery well operator .... *Phillips Petroleum Company UK*

IP:

First pay .... *3588 BOPD + 872 MSCFD*

All other zones with shows of oil and gas in the field:

Age	Formation	Type of Show
Middle Jurassic	*Pentland*	*Oil*
Triassic	*Skagerrak*	*Oil*
Permian	*Zechstein*	*Oil*

Geologic concept leading to discovery and method or methods used to delineate prospect
The structure was defined by seismic reflection.

Structure:

Province/basin type .... *Central/South Viking/Witch Ground graben intersection*
Regional structure .... *Graben*
Local structure .... *Salt induced anticline*

Trap:

Trap type(s) .... *Anticlinal trap with nearly 1000 ft (300 m) of structural relief*

Basin stratigraphy (major stratigraphic intervals from surface to deepest penetration in field):

Chronostratigraphy	Formation	Depth to Top in ft (m)
Paleocene	*Balder*	*7649 (2333)*
Upper Cretaceous	*Tor*	*8470 (2583)*
Lower Cretaceous	*Cromer Knoll*	*8920 (2721)*
Upper Jurassic	*Kimmeridge Clay*	*9169 (2796)*
Triassic	*Smithbank/Skagerrak*	*on flanks only*
Permian	*Zechstein*	*9682 (2953)*

Reservoir characteristics:

Number of reservoirs *1*
Formations .... *Maureen and Andrew formations*
Ages .... *Early Paleocene*
Depths to tops of reservoirs .... *7550 ft (2300 m) subsea (apex)*
Gross thickness (top to bottom of producing interval) .... *400 ft (122 m) (maximum)*
Net thickness—total thickness of producing zones
Average .... *125 ft (38 m)*
Maximum .... *350 ft (107 m)*

Lithology ... *Sandstone*
Porosity type ... *Intergranular*
Average porosity ... *18–25% range*
Average permeability ... *100–500 md (max. 3000 md)*

Seals:

Upper
Formation, fault, or other feature ... *Upper Paleocene Lista Formation*
Lithology ... *Claystone*
Lateral
Formation, fault, or other feature ... *Upper Paleocene Lista Formation*
Lithology ... *Claystone*

Source:

Formation and age ... *Upper Jurassic Kimmeridge Clay Formation*
Lithology ... *Claystone*
Average total organic carbon (TOC) ... *8%*
Maximum TOC ... *15–20%*
Kerogen type (I, II, or III) ... *II*
Vitrinite reflectance (maturation) ... *NA*
Time of hydrocarbon expulsion ... *Early Tertiary*
Present depth to top of source ... *12,000–15,000 ft (3660–4575 m)*
Thickness ... *250–300 m*
Potential yield ... *Approx. 50×10^6 m^3 (oil equivalent)/km^3 rock*

Appendix 2. Production Data

Field name ... *Maureen field*

Field size:

Proved acres ... *4650 ac (1880 ha) (PPCo)*
Number of wells all years ... *23*
Current number of wells ... *11 (6 producers, 5 water injectors)*
Well spacing ... *1 km (3280 ft) average*
Ultimate recoverable ... *210 MMSTBO (28 million tonnes)*
Cumulative production ... *155 MMSTBO to end 1989*
Annual production ... *96,000 BOPD (peak rate May 1984); 52,000 BOPD (average rate for 1989)*
Present decline rate ... *NA*
Initial decline rate ... *NA*
Overall decline rate ... *NA*
Annual water production ... *NA*
In place, total reserves ... *393 MMSTBO (PPCo)*
In place, per acre foot ... *NA*
Primary recovery ... *210 MMSTBO (28 million tonnes)*
Secondary recovery ... *NA*
Enhanced recovery ... *NA*
Cumulative water production ... *NA*

Drilling and casing practices:

Amount of surface casing set

Casing program

30-in. casing shoe at 500 ft (152 m); 20-in. casing shoe at 1200–1500 ft (360–460 m); 13⅜-in. casing shoe at 5000–6000 ft (1525–1830 m); 9⅝-in. casing shoe at 9000–12,000 ft (2740–3660 m); 7-in. liner from about 8000 ft (2440 m) (as required)

Drilling mud .... *16/29-1x, ligno-sulphonate mud; 16/29-2x to A9, seawater/KCl drispac; 16/29-A10 to A20, oil-based mud from 13⅜-in.*

Bit program .... *NA*

High pressure zones .... *None*

Completion practices:

Interval(s) perforated .... *As required in reservoir for production/water injection*

Well treatment .... *NA*

Formation evaluation:

Logging suites

Wells 1x to 5x: GR/SP/ILD/ILM/BHC/FDC/CNL/IES/LL/ML/MLL/HDT

Wells A1 to A20: DIL/SLS/GR/CAL (FDC/CNL/HDT) to 9⅝-in. shoe; DIL/SLS/GR/FDC/CNL/HDT/RFT/CST/VSP/CAL (MSFL) from 9⅝-in. to T.D.

Testing practices .... *Selective tests as required and through casing/liner*

Mud logging techniques

Total concept mud logging unit from 20-in. casing shoe; sample frequency: 30 ft (9 m) from 20-in. casing shoe; 25 ft (7.5 m) from 13⅜-in. casing shoe; 10 ft (3 m) from 9⅝-in. casing shoe

Oil characteristics:

Type .... *Paraffinic-naphthenic*

API gravity .... *36° API*

Initial GOR .... *392.7 scf/bbl (stock tank)*

Sulfur, wt% *0.35*

Viscosity, SUS .... *0.28 centipoise*

Pour point .... *−10°C (14°F)*

Gas-oil distillate .... *NA*

Bubble point .... *1786 psia (12,314 kPa)*

Field characteristics:

Average elevation .... *8300 ft (2530 m) subsea*

Initial pressure .... *3792 psia (26,146 kPa, 266.9 kg/cm^2)*

Present pressure .... *NA*

Pressure gradient .... *0.443 psi/ft (10.0 kPa/m) (aquifer)*

Temperature .... *243°F (117°C)*

Geothermal gradient .... *0.022°F/ft (0.040°C/m)*

Drive .... *Water injection supporting natural water drive*

Oil column thickness .... *650 ft (198 m)*

Oil-water contact .... *8702 ft (2654 m) subsea*

Connate water .... *NA*

Water salinity, TDS .... *60,000 ppm NaCl*

Resistivity of water .... *0.0382 ohm-m at 243°F (117°C)*

Bulk volume water (%) .... *NA*

Transportation method and market for oil and gas:

Oil via tanker to U.K.; gas flared on platform

Dan Field—Denmark
Central Graben, Danish North Sea

LARS NYDAHL JORGENSEN
Maersk Olie og Gas A/S
Copenhagen, Denmark

FIELD CLASSIFICATION

BASIN: Central Graben
BASIN TYPE: Rift
RESERVOIR ROCK TYPE: Chalk (Mainly Matrix Porosity)
RESERVOIR ENVIRONMENT OF DEPOSITION: Pelagic
RESERVOIR AGE: Cretaceous–Paleocene
PETROLEUM TYPE: Oil and Gas
TRAP TYPE: Domal Anticline

LOCATION

The Dan field was discovered in 1971 and was the first Danish oil field to be developed. It is located in the southwestern corner of the Danish North Sea (Figure 1), 127 mi (205 km) west of the port of Esbjerg and approximately 93 mi (150 km) south of the Ekofisk group of fields in the Norwegian sector of the North Sea.

The field lies in the southern part of the North Sea Central graben and is part of a trend of halokinetically induced structures called the "Southern Salt Dome Province" (Michelsen, 1982). The trap is a simple structural dome (Figure 2) with the bulk of oil and gas trapped in Maastrichtian and Danian chalk reservoirs (Figure 3).

The reservoir chalk is characterized by high porosities and at the same time very low permeabilities. Permeability enhancement from natural fracturing, as it is known from other Central graben chalk fields, appears to be limited. Hydraulic fracturing of wells has therefore been essential in the development of the Dan field. The total areal extent of the Dan field is 5950 ac (24 km^2).

Stock tank oil initially in place (STOIIP) and gas initially in place (GIIP) figures are 1922 MMSTB and 1500 bcf, respectively. Owing to the low permeabilities expected, primary recovery is, however, only 208 MMSTB and 541 bcf. The field is operated by Maersk Olie og Gas for the Danish Underground Consortium (DUC).

HISTORY

Pre-Discovery

In the early 1960s, the first seismic surveys were carried out in the Danish sector of the central part of the North Sea by the DUC. During a period of four years 8700 mi (13,998 km) of reflection seismic data were acquired. A number of structural anomalies previously identified from aeromagnetic surveys were defined as structural closures at various stratigraphic levels. In 1966 the A-1X well was drilled to test the first of these structures. The objectives of the well were "Tertiary and Mesozoic" sections.

The well was drilled with a drillship unable to operate in the harsh weather during winter months in the North Sea. The A-1 location was selected for this purpose with the expectation that a hole could be drilled to target depth before the advent of bad weather. Gas and oil shows found in A-1 were successfully tested by the twin well, the A-2, the succeeding year (Kraka field). The discovery was economically marginal, and development has only begun recently (Figure 1).

However, this discovery and promising results from successive wells led to the development of the chalk play concept. A number of wells tested the play, and with the fifteenth wildcat, the M-1X (Figure 2), the Dan field was discovered in May 1971 (Childs and Reed, 1975). Meanwhile, in the Norwegian part of the Central graben, exploration drilling had begun in block 2/4 in 1969. In 1970 the giant Ekofisk field was discovered, and in the following years, ten other fields were found in the Central graben in chalks of Upper Cretaceous and lower Paleocene age.

Discovery

Gulf Oil, acting as operator for the DUC, spudded the M-1X wildcat on the crest of the structure in March 1971 (Figure 2). The top of the chalk reservoir was encountered at 5792 ft (1766 m) subsea. A chalk section of 1016 ft (310 m) was penetrated. A 262 ft (80 m) thick gas zone was encountered above a 400 ft (122 m) oil section, in turn underlain by a 75 ft

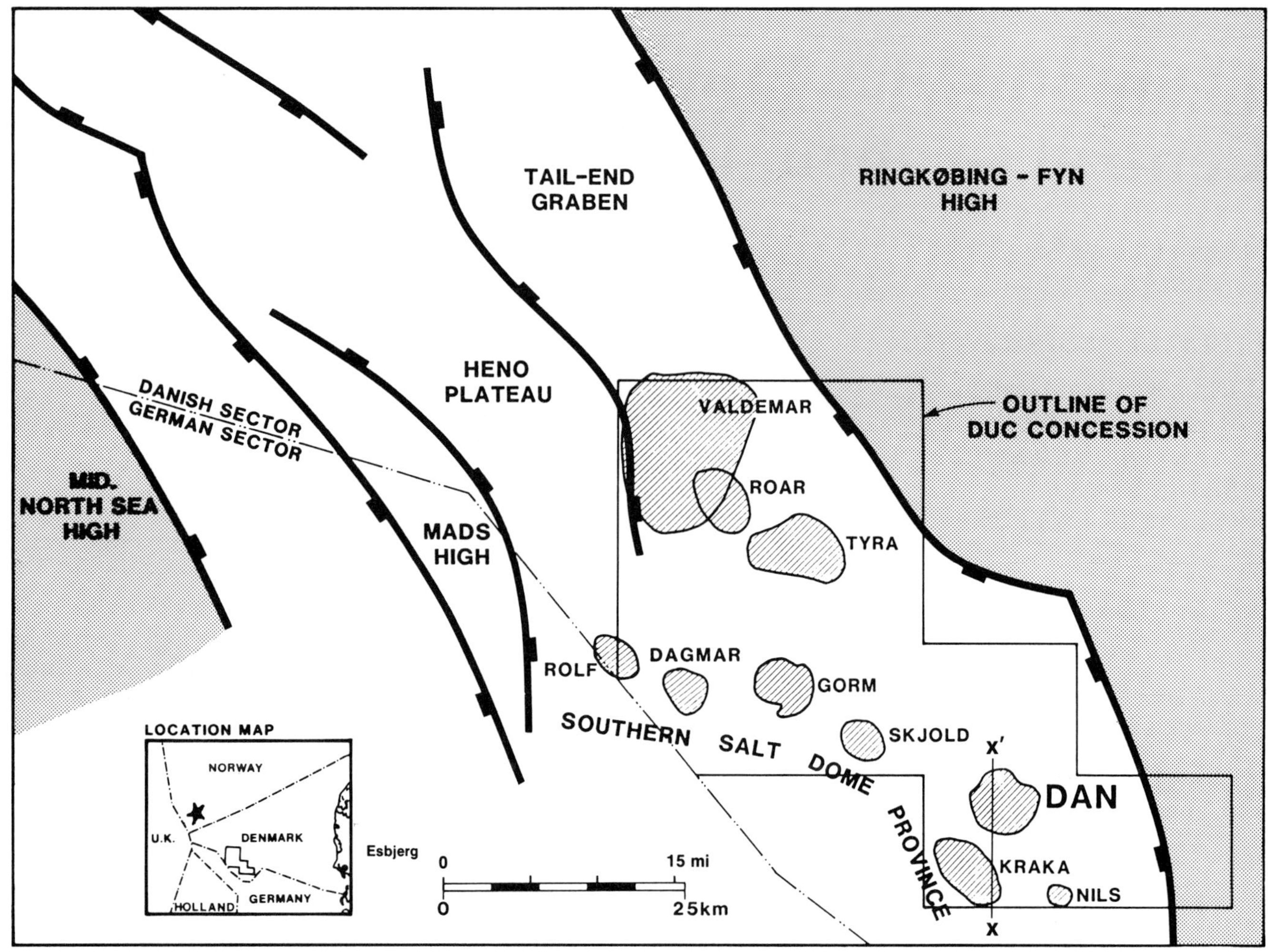

Figure 1. Location of the Dan field in relation to major structural elements in the North Sea Central graben. Star in Norwegian sector is location of Ekofisk field. Line X–X′ is approximate location of seismic line DK-83-145, Figure 4.

(23 m) oil-water transition zone. On test the well yielded a maximum flow of 3600 BOPD after acidization.

The follow-up well, M-2X, drilled in June-July 1971 northwest of M-1X, encountered 12 ft (4 m) of gas and 444 ft (135 m) of oil-bearing chalk. The top of the chalk reservoir was found at 6005 ft (1832 m) subsea, thereby establishing that the southwest-northeast-trending fault divides the field into a downthrown northwestern block (the A-block) and an upthrown southeastern block (the B-block).

Post-Discovery

The first phase of development involved five oil producers drilled on the B-block. A six-slot drilling platform (Dan-A) and a production platform were installed with production starting in July 1972. Three years later, in 1975, a second six-slot platform (Dan-D) was placed adjacent to the production platform. Three more wells were then drilled on the B-block and three wells on the downthrown A-block near the major dividing fault. In 1976 a third-six slot platform (Dan-E) was placed near the crest of the A-block and six wells were drilled on this block's western part.

Delineation drilling was concluded with the two wells M-9X (drilled 1979) and M-10X (drilled 1982–1983). These wells appraised the southern flank of the B-block and the southwestern flank of the A-block, respectively. The next phase of development, the Dan-F project, comprised 24 wells located on the A-block. Nine wells were template drilled in 1984–1985 and later tied back to a 12-slot wellhead platform, Dan FA. Three further wells were drilled from the FA jacket and 12 from the Dan FB platform. The wells came on stream in 1987.

The pay concept that governed the location of Dan-F wells was based on a mapping of Kh values (i.e., permeability × thickness). In the low permeability Dan reservoir small variations in matrix permeability strongly influence the performance of individual wells, and the Kh values therefore represent an

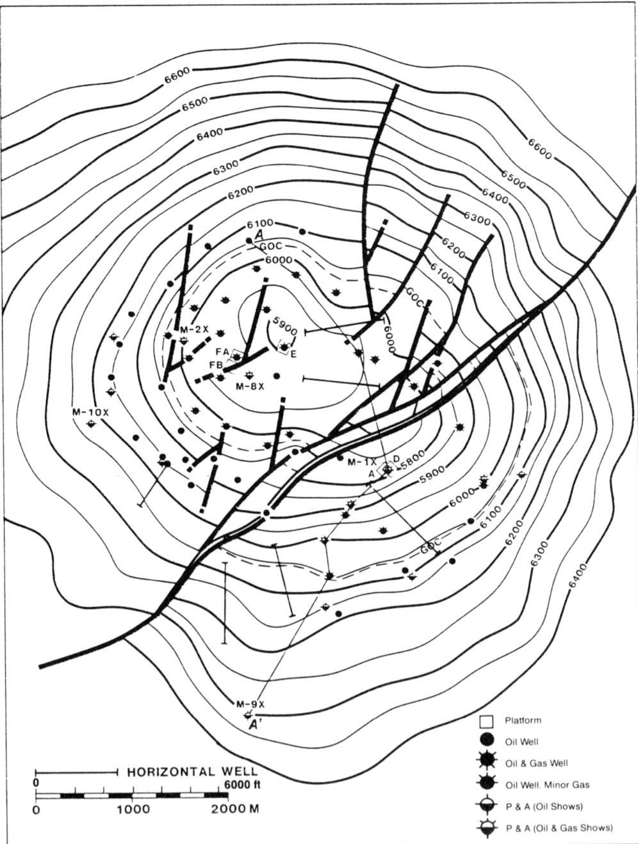

Figure 2. Top of the chalk structure map of the Dan field based upon well and seismic data. Contours are in feet; contour interval, 50 ft. Only wells and platforms mentioned in the text are named; others are indicated by symbol only. Also shown is the position of the cross section displayed in Figure 5. M-1X is the discovery well directly below the A platform.

integration of the key reservoir geological parameters.

The latest development has been the successful drilling and completion of six horizontal wells (Figure 2). The horizontal section of five of these wells exceeds 2500 ft (760 m). Oil production from the wells is more than twice that of a conventional producer, proving that horizontal drilling is a viable technique for further development of the Dan field. Future developments will likely involve infill drilling with horizontal wells and possibly implementation of a waterflood. A pilot waterflood (one injector) to evaluate the viability of a full field waterflood is currently ongoing.

To date (January 1990) 56 wells, including five re-drills, have been drilled on the Dan field. Present daily production (January 1990) amounts to 29,000 BOPD and 90 MMSCF. Cumulative production after 18 years has been 57 MMSTB and 124 bcf.

DISCOVERY METHOD

As described above, the field was discovered as a result of pursuing the chalk play that had been first indicated by the A-1/A-2 find. This play concept was to drill seismically defined structural closures at the top of the Chalk horizon for potential reservoirs in the Mesozoic and Tertiary. Several successful production tests from oil-bearing chalk were, however, required for the chalk play to gain general acceptance. It must be borne in mind that at this time chalk, tight as it is, was regarded as a rock type more suited as a seal than as a reservoir.

The discovery well, M-1X, was the fifteenth exploration well drilled by the DUC in the Danish sector. The structural closure had been defined earlier, but political considerations forced a delay in exploratory drilling. The structure is located in an area where the Danish/German national border was in dispute during the late 1960s. There was some natural reluctance to drill until the border position was determined.

The exploration philosophy would probably have been no different, had the chalk play of the Danish Central graben been explored today. The simple route followed is logical also with today's knowledge. It seems improbable that any of the tools made available over the past 20 years would have aided in determining which of the chalk structures would contain hydrocarbons and which would not.

Today, most structures at the top of the Chalk level in the Central graben have been tested, and the exploration efforts are directed toward other play types.

STRUCTURE

Regional Structure and Depositional History

The structural outline of the North Sea Central graben is indicated in Figure 1, whereas a generalized stratigraphy as compiled by Jensen et al. (1986) and Michelsen (1986) is illustrated in Figure 3. The early phases of formation of the graben took place in the Early Permian, the adjoining mid-North Sea high and the Ringkøbing-Fyn highs having been formed during the Late Carboniferous.

During the Late Permian, Zechstein evaporitic sequences accumulated in the Central graben. These sequences were particularly thick in the Northern and "Southern Salt Dome Provinces." Subsidence continued during the Triassic, and a thick section of predominantly continental clastic and evaporitic sediments was deposited.

The early Kimmerian tectonic pulse at the end of the Triassic led to a deepening of the Central graben. This was followed by a shift in the depositional regime from continental to predominantly marine or

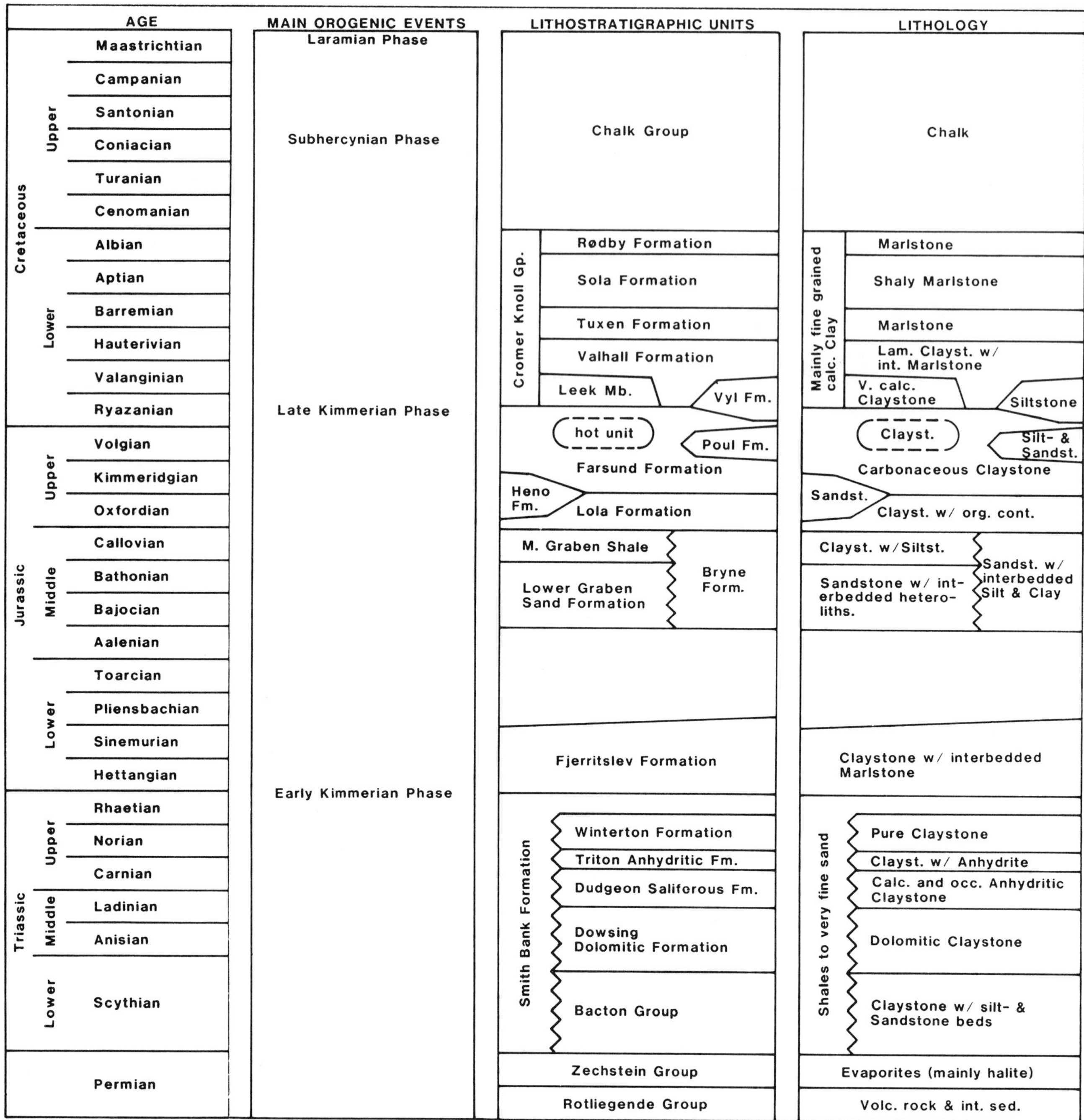

Figure 3. Generalized lithostratigraphic subdivision of the Central graben sequence with an indication of the dominant lithologies. (After Michelsen, 1986.)

transitional environments with deposition of thick sequences of shale. Due to differential movement of individual fault blocks, subsidence was highly variable across the Central graben. This gave rise to great differences in thickness of the Jurassic sequence. There are more than 10,000 ft (3000 m) of Jurassic section in the Tail End graben compared with the 4000 ft (1200 m) below the Dan field, while Jurassic deposits are locally absent on the Dogger high, 45 mi (75 km) to the northwest of the Dan field.

Permian salt, which had already been actively moving in the Triassic, was remobilized following the late Kimmerian tectonic pulse at the end of the Jurassic. This resulted in the formation of a number of salt pillows and salt diapirs in the "Northern and Southern Salt Dome Provinces."

The Early Cretaceous was characterized by a transgressional pattern that became progressively more significant with time. However, the Late Cretaceous global rise in sea level gave rise to a far more pronounced transgression. When influx of clastic material was reduced, chalk was deposited over the entire North Sea area. The Chalk group deposited during the Late Cretaceous and early Tertiary has very variable thicknesses within the "Southern Salt Dome Province," ranging from less than 500 ft (150 m) to more than 3500 ft (1050 m). This was probably mainly a result of the then existing fault block and salt dome topography.

During the Late Cretaceous, the tectonic regime changed from the previously dominant rifting to a phase of more gradual subsidence. This was coincident with a regional downwarp and the establishment of a compressional strike-slip stress regime responsible for localized inversion tectonism. These movements are believed to have been caused by the sub-Hercynian or the Laramide orogeny (Andersen et al., 1982), and they may be the cause for the renewed halokinetic activity in the "Southern Salt Dome Province."

After the Laramian tectonic event, the sedimentary pattern again changed, and fine-grained clastics became the predominant deposits throughout the North Sea basin during the Tertiary.

Local Structure

As illustrated by the seismic section (Figure 4), the Dan field dome results from uplift by a salt pillow of Triassic age, while the neighboring Kraka field to the south was formed by movement of older salt deposits (Zechstein). The Dan salt pillow experienced several growth episodes from the Late Jurassic to the late Tertiary. The major growth of the structure occurred during the Eocene, with continued limited movements during the Oligocene and early Miocene, terminating toward the end of the Miocene. The structure map (Figure 2) shows the structural configuration of the field based upon well and seismic data. The relatively simple geometry is evident from this map.

A northeast-southwest-trending normal fault system downthrown to the northwest bisects the structure. The fault has approximately 300 ft (92 m) of throw near the crest of the structure and dies out in the synclines on the northeast and southwest of the field. Smaller associated normal faults are present in the downthrown A-block. The orientation of major and associated faults conforms with fault patterns generated in model experiments of doming (Withjack and Scheiner, 1982).

At late Miocene to Pliocene time, post-hydrocarbon migration, the Dan field experienced a tilt toward the southwest, believed to result from a regional basinward downwarping.

STRATIGRAPHY

The pre-Tertiary stratigraphic succession is summarized in Figure 3. For specific depth references, please refer to Appendix 1. The deepest well in the Dan field, the M-8X, reached its total depth in salt of Triassic age. This is the salt that forms the salt pillow seen in the seismic section, Figure 4.

The salt and the overlying shales belong to the Dudgeon Saliferous Formation (Jacobsen, 1982). This formation is a claystone characterized by the associated occurrence of halite. This distinguishes the formation from the overlying Triton Anhydritic Formation, which is mainly an anhydritic claystone. The top part of the Triassic sequence is occupied by the Winterton Formation, which is a relatively pure claystone. None of the penetrated Triassic formations appear to have any reservoir or source rock potential in the Dan field area.

The Lower Jurassic Fjerritslev Formation consists of slightly calcareous, silty claystones with interbedded thin marlstone stringers. No reservoir potential has been observed in this formation in the Dan area. The formation may have a source potential although this has not been documented.

A major unconformity separates the Fjerritslev Formation from the overlying Middle Jurassic "Lower Graben Sand Formation." The formation consists of interbedded claystones and heterolithic sand-siltstones with subordinate clean sandstone interbeds. These sands form a secondary trap in the Dan field. In the M-8X well, oil and gas shows were recorded, but no flow was observed during testing. The overlying "Middle Graben Shale Formation" is similar in lithology but with a decreasing number of sandstone interbeds.

The Upper Jurassic Farsund and Lola formations are claystones with siltstone and limestone interbeds. The Farsund Formation is partly time-equivalent to the Kimmeridge Clay (Deegan and Scull, 1977). This formation is known as an excellent source rock, and indeed, geochemical investigations of samples from the M-8X well indicate a source potential of the Farsund Formation in the Dan area.

Only slightly more than 300 ft (92 m) of Lower Cretaceous sediments are present in the Dan field area. The sediments belong to the Cromer-Knoll Group and are dominated by claystones and silty claystones with thin limestone stringers.

The Upper Cretaceous and lower Paleocene Chalk group consists of white coccolith limestones with a wide variation in clay, glauconite, and chert content. The Dan reservoir is located in the top part of the Chalk group in the equivalents of the Tor and Ekofisk formations (defined in the Norwegian Central graben; Deegan and Scull, 1977). The stratigraphy of the reservoir units is discussed in detail in the *Reservoir Stratigraphy and Facies* section.

The Cenozoic Rogaland, Hordaland, and Nordland groups overlie the chalk. These lithostratigraphic units are, strictly speaking, defined outside the Danish Central graben. In the Danish Central graben only informal lithostratigraphic names have been assigned to the Cenozoic units (Kristoffersen and Bang, 1982).

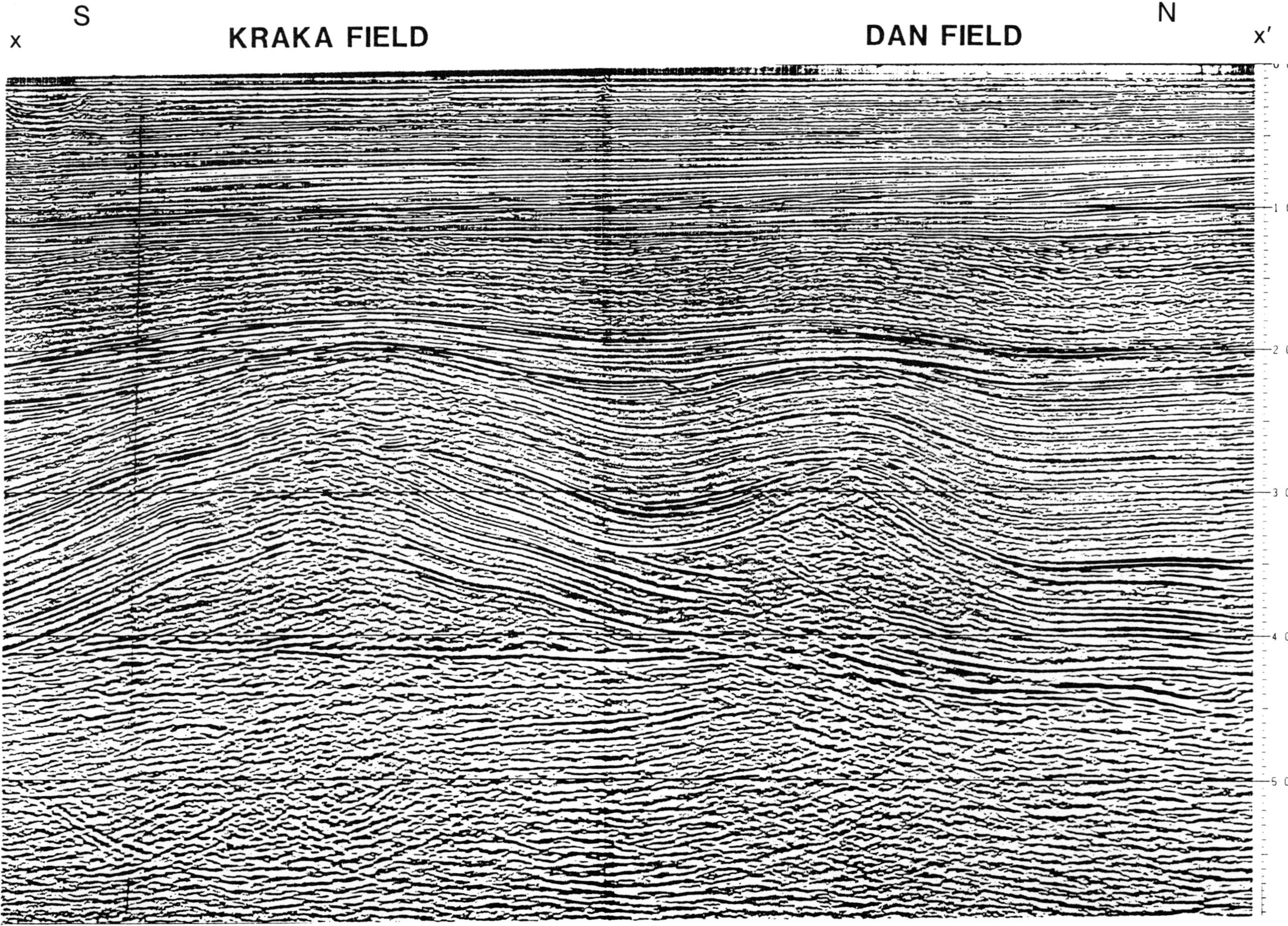

Figure 4A. Seismic line DK-83-145 running approximately north-south.

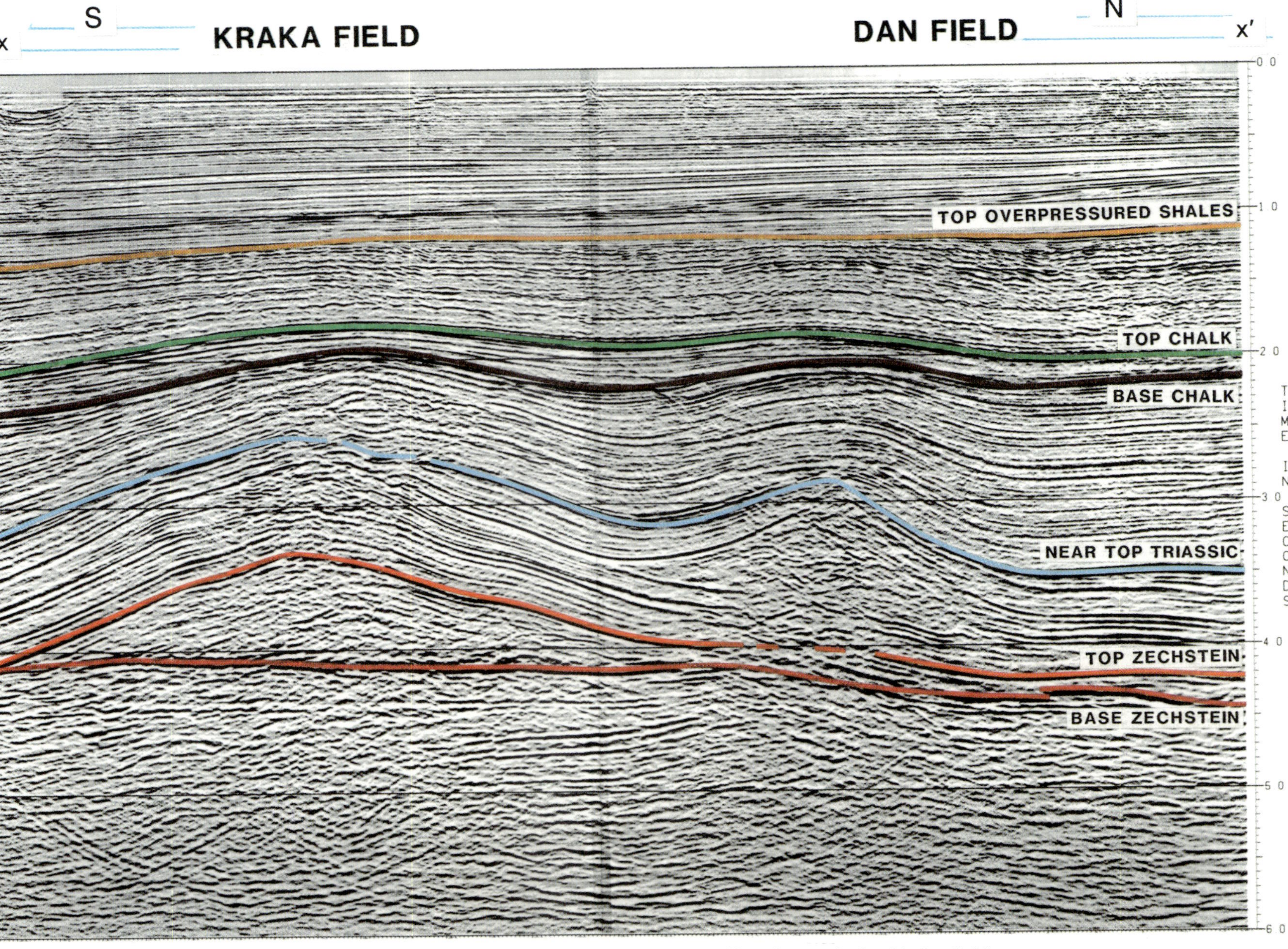

Figure 4B. Same seismic line with interpretation. Note that the salt pillow beneath the Dan field is younger (Triassic) than the salt pillow beneath the Kraka field (Upper Permian, Zechstein). See Figure 1 for location.

The Rogaland group is of Paleocene age. The basal part of the group is the "North Sea Marl," equivalent to the Maureen Formation of Deegan and Scull (1977). This marl is overlain by a noncalcareous claystone, equivalent to the Lista Formation (Deegan and Scull, 1977). The claystone is in turn overlain by a sequence of shales and claystones corresponding to the Sele and Balder formations in the Norwegian Central graben. The tuffs of the upper part of the Sele and the Balder are widely distributed and form a very important chronostratigraphic marker.

The overlying Hordaland group is of Eocene to middle Miocene age. It consists almost entirely of greenish-gray shales and clays. Thin limestone stringers occur, particularly in the Oligocene.

The middle Miocene to Pleistocene Nordland group is a series of marine clays that often contain silty and sandy interbeds. These sands become increasingly abundant toward the top of the unit. Here they often contain shell fragments and, occasionally, lignite.

TRAP

The Dan field dome is a simple structural trap (Figure 2) with the bulk of oil and gas trapped in Danian and Maastrichtian chalk. Paleocene shales of the Rogaland group equivalent form the seal across the entire dome. Sands of the Middle Jurassic "Lower Graben Sand Formation" form a secondary trap holding only insignificant amounts of hydrocarbons.

An extensive gas cap is present in the Danian reservoir in both the A-block and the B-block. In the Maastrichtian, only a very limited gas cap is present. The gas-oil contact occurs at a depth of 6030 ft (1839 m) subsea (ss) in the A-block, and 30 ft (9 m) deeper, at 6060 ft (1848 m) ss, in the B-block. This variation most likely indicates that the major northeast-southwest-running fault system is sealing.

The vertical distribution of oil is characterized by a long (400 ft [122 m] +) section of high oil saturation (water saturation, $S_w < 20\%$) underlain by a comparatively long transition zone, often in excess of 200 ft (61 m). The long transition zone is a result of the low permeability of the chalk.

Relative permeability measurements have shown that the effective oil permeability will be exceedingly low at S_w above 60%. From experience in the field it is known that oil production can be sustained only when S_w is below 50%. Therefore, the 50% S_w level is regarded as the lowest boundary for economic oil recovery.

As illustrated in Figure 5, the 50% S_w level is not constant throughout the field. The 50% S_w level occurs at a higher level in the Danian than in the Maastrichtian units. This is probably not a reflection of a complete separation of the two reservoirs (initial pressures were equal) but can be explained as a capillary effect caused by the differences in permeability.

Furthermore, the 50% S_w level is seen to be dipping gently toward the south-southwest (less than 1° dip). The same is true for other levels of equal water saturation, including the oil-water contact. Seismic sections show a tectonic tilt of late Miocene to Pliocene age of broadly the same orientation and magnitude as the dip of fluid levels. Since the mobility of oil is very limited at the high water saturations near the oil-water contact, the fluid levels may not have re-equilibrated. Hence the fluid level tilt may result, at least in part, from the tectonic tilt, but it could also be related to a regional hydrodynamic gradient.

RESERVOIR

Stratigraphy and Facies

The hydrocarbon-bearing chalks of the Dan field are of lower Paleocene Danian and Upper Cretaceous Maastrichtian age. The majority of the reservoir belongs to the informal Chalk-5 and Chalk-6 units established by the Geological Survey of Denmark (DGU) (Lieberkind et al., 1982; Nygaard et al., 1983). These units are in turn correlatable with the upper part of the Tor Formation and the Ekofisk Formation of Deegan and Scull (1977). The DUC subdivision of the Dan field chalk reservoir is based on differences in porosity as defined by the neutron-density log signature (Figure 6).

Danian Chalk (DGU Chalk Unit 6, Ekofisk Formation Equivalent)

The Danian chalk has been subdivided into two main units: an upper, very porous section (D_1) immediately below the sealing Tertiary shale, and an underlying less porous section (D_2) characterized by a high chert and clay content (Svendsen, 1979). The D_2 is further subdivided into an upper D_{2a} unit of very low porosity and a lower D_{2b} unit of comparatively higher porosity.

The *D_1 unit* is predominantly a homogeneous, beige to light gray chalk. The argillaceous content of this facies is restricted to rare wispy lamination and occasional millimeter-thick marly layers (Figure 7A). The facies is generally strongly bioturbated, leaving few indications of the original depositional processes.

In certain intervals, trace fossil species are recognizable. Within these intervals Chondrites and Zoophycos are very common. Planolites occur frequently, while Bathichnus (Nygaard, 1983) has been observed only in a very few cases. The unit is interpreted as a predominantly autochthonous pelagic deposit.

However, some intervals display millimeter-thick lamination of vague color contrast. As suggested by

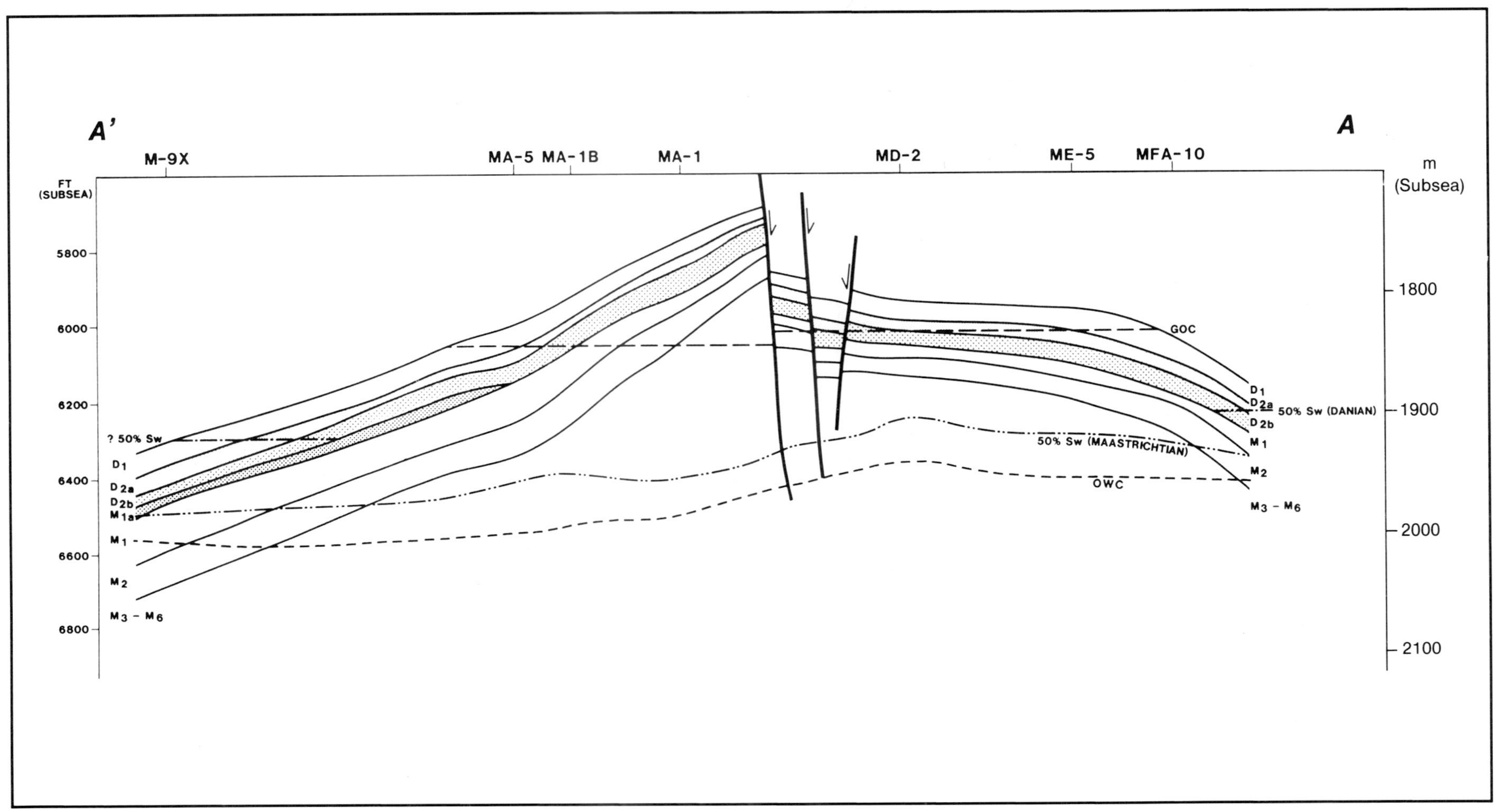

Figure 5. North-south structural cross section. Note the dipping fluid levels (50% S_w and OWC), the different position of these in the Danian compared to the Maastrichtian reservoir, and the general thickening of Maastrichtian units toward the southwest. For location and horizontal scale, see Figure 2.

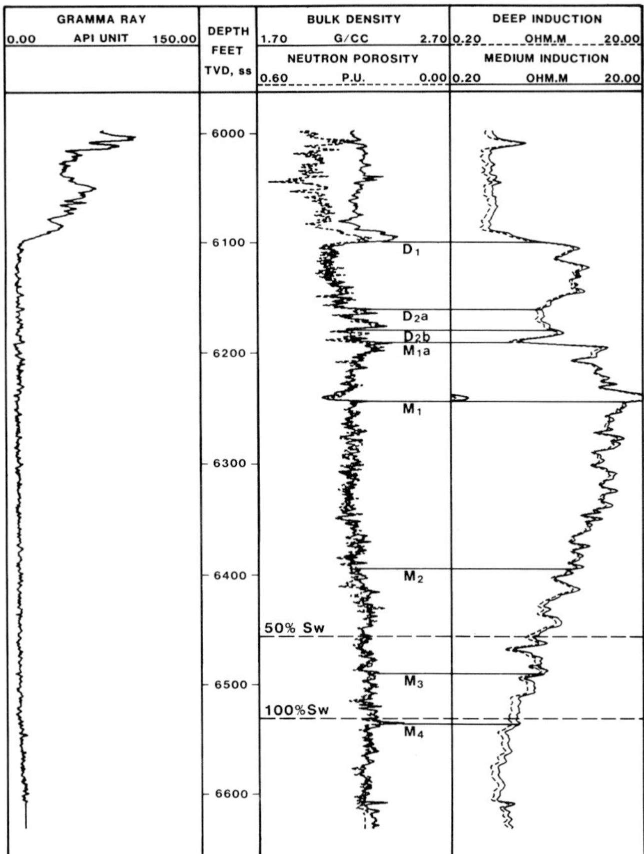

Figure 6. Type log of the Dan field (from the MFB-7 well). The definition of DUC reservoir units is indicated. The OWC (100% S_w) and the 50% S_w levels are shown. The 50% S_w can be regarded as the lower boundary for economic oil recovery.

Figure 7. Core photos of key facies of the Danian reservoir units. (A) D_1 reservoir unit; note laminations interpreted to be of allochthonous origin and the high abundance of healed hairline fissures. (B) D_{2a} reservoir unit; note alternation between a porous facies with chert nodules and a tight wispy laminated facies. This alternation constitutes the periodite facies association. Core diameter is 6 in.

Hatton (1986), these have been interpreted as a distal allochthonous facies resulting from low-density turbidites or other gravity flow generated mudclouds. Such interpretation is substantiated by the associated occurrence, albeit rare, of convoluted folds and intraclasts. Chert nodules are common. Onshore surface exposures and cores acquired in horizontal wells have shown that these nodules frequently occur in bands. Matrix silica content is relatively high, amounting to more than 5%. Stylolites occur at a frequency of 1/ft (1/30 cm). Healed hairline fractures are abundant in some intervals.

In thin section, the facies is a foraminiferal wackestone grading to mudstone with a fine micritic matrix. At a finer scale, SEM investigations show that the matrix consists of unbroken coccoliths floating in a matrix of coccolith fragments, secondary calcite rhombs, and subordinate amounts of clay minerals (Figure 9). The presence of a wide range of grain sizes together with the substantial calcite cementation are thought to be the reasons for the reduced permeability of the D_1 unit compared to the Maastrichtian reservoir units that have porosities of the same order as the D_1 unit.

The D_{2a} *unit* is generally a poorly developed reservoir. Sedimentary facies show a very characteristic variation (Figure 7B) within the unit: A slightly argillaceous facies with abundant chert nodules alternates with an argillaceous, wispy laminated facies. The latter facies appears to have extensive matrix cementation.

In this facies, a mottling is observed in certain restricted intervals. This mottling appears as elongated dark gray specks sometimes contorted in intricate patterns. The feature resembles burrows, and indeed burrow systems may be responsible for faint chemical gradients later accentuated during diagenesis to form the structure. In the nearby Tyra field (Figure 1), this structure is more strongly developed and displays patterns that have given rise to the informal name "leopard structure." A similar feature was described by Dixon (1976) from lower Paleozoic and Triassic carbonates from various Canadian localities.

Chert is abundant throughout the unit in both facies. In contrast with the D_1 unit, stylolites occur only rarely.

SEM studies of both D_{2a} facies show that the range of grain sizes is similar to that of the D_1. However, matrix clay content is higher and associated microspar cementation is much more pronounced (Figure 9). Matrix silica content often exceeds 10%. The reduced permeability of the wispy laminated facies seems to result from variations in clay content rather than from variations in disseminated silica.

The above described alternation of facies has been termed "periodite facies" (Kennedy and Garrison, 1975; Kennedy, 1987). It is thought to be characteristic of an autochthonous chalk deposit (e.g., Brewster et al., 1986), with the cyclicity possibly controlled by global climatic fluctuations (Hatton, 1986).

The D_{2b} *unit* strongly resembles the D_{2a}, displaying a similar rhythmic variation of facies. However, instead of the mottled, wispy laminated facies of D_{2a}, a much purer chalk with fair to good reservoir parameters is present. This results in an overall slightly higher porosity of the unit compared to the D_{2a}.

Although disseminated silica content is locally high, this does not seem to reduce the reservoir quality. SEM studies suggest that this is a result of the silica forming a rigid framework capable of partly withstanding porosity-reducing compaction.

Figure 8. (A) Core section of the hardground at the Maastrichtian-Danian boundary with overlying shale of Danian age. Log evaluation suggests that this hardground is only partly developed in some areas of the structure. (B) The M_1 reservoir unit showing alternation between a bioturbated, in part massive, facies and a laminated facies. Note dense network of healed hairline fractures and the high frequency of stylolites. Core diameter is 6 in.

Maastrichtian Chalk (DGU Unit 5 [Upper Part], Tor Formation Equivalent)

The Maastrichtian has been subdivided into reservoir units based mainly upon differences in porosity (Figure 6). The upper part of the Maastrichtian section is comprised of the high porosity M_{1a} and M_1 units, whereas the lower part is commonly less porous but variable enough to allow subdivision (M_2 to M_6).

A depositional unconformity at the Danian-Maastrichtian boundary is represented by a hardground occupying the uppermost 2 ft (0.5 m) of the Maastrichtian chalk (Figure 8A). Evaluation of logs, however, indicates that over parts of the field the hardground is not fully developed.

The hardground is intersected by burrows and borings; it is strongly calcite cemented and has a high glauconite content. Chert occurs disseminated in the matrix as well as in nodules of a size greater than the width of cores. In two cores the unconformity is directly overlain by a 5 cm thick shale of Danian age. The tight nature of the hardground and the associated shale suggest that this will act as a baffle to vertical fluid flow.

The M_{1a} *and* M_1 units are the most productive units of the Dan field. The M_{1a} is only present on the western part of the structure. There seems to be little lithological difference between the two units. In general, however, the M_{1a} is less porous in its upper interval and more porous in the lower part in comparison with the M_1.

The predominant lithologies of the two units are bioturbated, homogeneous chalks (with a trace fossil assemblage analogous to that of the D_1) and millimeter-scale laminated chalk, interpreted to be of allochthonous origin as in the D_1 (Figure 8B). Clay and silica contents are generally very low (less than 3 wt.%). The frequency of stylolites is notably higher than in the Danian units; i.e., 4 stylolites/ft (4/30 cm) compared to the 1/ft (1/30 cm) in the D_1.

Healed hairline fissures are, in general, very abundant. In thin section it is seen that the hairline itself is accompanied by a millimeter-wide zone of strong matrix cementation. In places these fissure zones coalesce to form centimeter-wide totally healed veins. The high residual oil content of these veins is thought to be a reflection of a reduced permeability compared to the surrounding matrix. Open extension fractures, mostly subvertical and associated with stylolite tips, have been observed in rare instances.

Based upon thin section studies, the M_{1a} and M_1 chalk is classified as a foraminiferal wackestone grading to mudstone. In SEM the M_{1a} and M_1 chalk appears distinctly purer than the Danian chalk (Figure 9). Although the porosities are lower than or equal to those of the D_1, the average coccolith and fragment size is larger in the M_{1a}/M_1 chalk. In addition, clay minerals capable of blocking pore throats are largely absent. These two factors explain the relatively higher permeabilities of the M_{1a}/M_1 chalk.

Microfaunal and nannofloral data indicate that the two units, M_1 and M_{1a}, are partly time equivalent. The fact that the M_{1a} unit only occurs in the southwestern part of the field, together with its observed predominant consistence of chalk interpreted to be of allochthonous origin, may be an indication that the M_{1a} unit could be composed of redeposited M_1 material.

The deeper Maastrichtian *M_2-M_6 units* are lithologically very similar to the M_{1a} and M_1, although progressively less porous with depth. The finely laminated facies dominates the M_2 unit, whereas in the deeper units a bioturbated, homogeneous, and strongly stylolitized chalk becomes increasingly dominant. In the M_3-M_6 units these two facies alternate on a foot by foot scale.

Argillaceous seams, chert bands, and chert nodules are notably more abundant in the M_2 and lower units than in the higher units. Healed hairline fissures are common in the M_2 unit but become less frequent and eventually disappear in the lowermost units. Calcispheres have been found to be a significant grain type in the M_2 and lower units, and the rock is in general developed as a calcisphere-foraminiferal wackestone or mudstone.

SEM investigations show that the M_2 and lower units appear similar to and only slightly less pure than the M_{1a} and M_1 (Figure 9). A higher degree of calcite cementation and of coccolith fragmentation appears to be the reason for the declining porosity and permeability with depth.

Porosity Distribution

The Dan field chalk is only affected by a limited degree of fracturing, so the vast majority of the porosity is contained in the matrix.

Similarly the bulk of permeability comes from the matrix, although the few fractures present must cause a permeability enhancement. The exact nature of this enhancement is not yet fully understood. The degree of matrix porosity variation within the individual reservoir units is illustrated on the reservoir type log (Figure 6). Average porosities and permeabilities can be summarized as follows.

Unit	Porosity Range (%)	Average Porosity (%)	Average Permeability (md)
D_1	30-37	34	2.05
D_{2a}	16-32	25	0.40
D_{2b}	21-32	27	0.50
M_{1a}	24-41	29	1.75
M_1	25-39	30	2.30
M_2-M_5	20-32	22	1.45

Despite the limited lateral facies variation, porosities do show a significant change within the individual reservoir units going from crest to flank. In Figure 10 a map displaying the variation within the M_1 unit is presented as an example. A roughly concentric pattern is seen with crestal porosities in excess of 38% and decreasing to 25% in downflank areas.

Such distribution is characteristic of all reservoir units of the Dan field and has also been observed in other fields in the Danish sector of the North Sea—e.g., Gorm (Hurst, 1983)—and in a number of fields in the Ekofisk area (D'Heur, 1984). The trend is explained by the reduction of chemical cementation and compaction brought about by the invasion of the chalk by hydrocarbons.

Thickness Distribution

There is some irregularity in the distribution of the Danian units, although over large areas the total Danian isochore is fairly constant. Pronounced variations in the D_2 thickness are compensated by opposite variations in the D_1. This pattern suggests that irregularities in the D_{2a} and D_{2b} have controlled the deposition of the D_1. Core observations and nannofloral stratigraphic data suggesting hiatuses in the D_2 indicate that the irregularities could be erosional features.

The Maastrichtian units show a different pattern (see cross section, Figure 5). There is a distinct tendency for the Maastrichtian units to thicken toward the southwest. For the M_{1a} and M_1 units, this is partly ascribable to a greater truncation at the Danian-Maastrichtian unconformity toward the northeast. Toward the southwest the general thickening trend can be explained by larger amounts of allochthonous material being deposited away from the inferred Maastrichtian paleocrest on the northeast side of the structure. Core observations are in line with this interpretation.

In a number of Norwegian chalk fields (notably the West Ekofisk), the effect of increased preservation of porosity toward the crest of the fields has been found to be reflected also in the isochore (vertical thickness) of individual reservoir units, creating a concentric pattern on isochore maps (D'Heur, 1984, 1986). D'Heur found that when mapping the solid thickness (i.e., excluding porosity) the concentric pattern disappeared. A similar pattern has not been found in the Dan field, where reservoir unit thickness

Figure 9. Scanning electron micrographs of key facies. (D_1) The D_1 facies consisting of unbroken coccoliths floating in a matrix of coccolith fragments, secondary calcite rhombs, and minor amounts of clay. (D_{2a}) The wispy laminated facies of D_{2a}. Matrix clay content and calcite cementation is more pronounced than in the D_1 facies. (M_1) The M_1 reservoir unit; note distinctly purer appearance than the Danian units. (M_2) The M_2 reservoir unit; this facies is similar to and only slightly less pure than the overlying M_1.

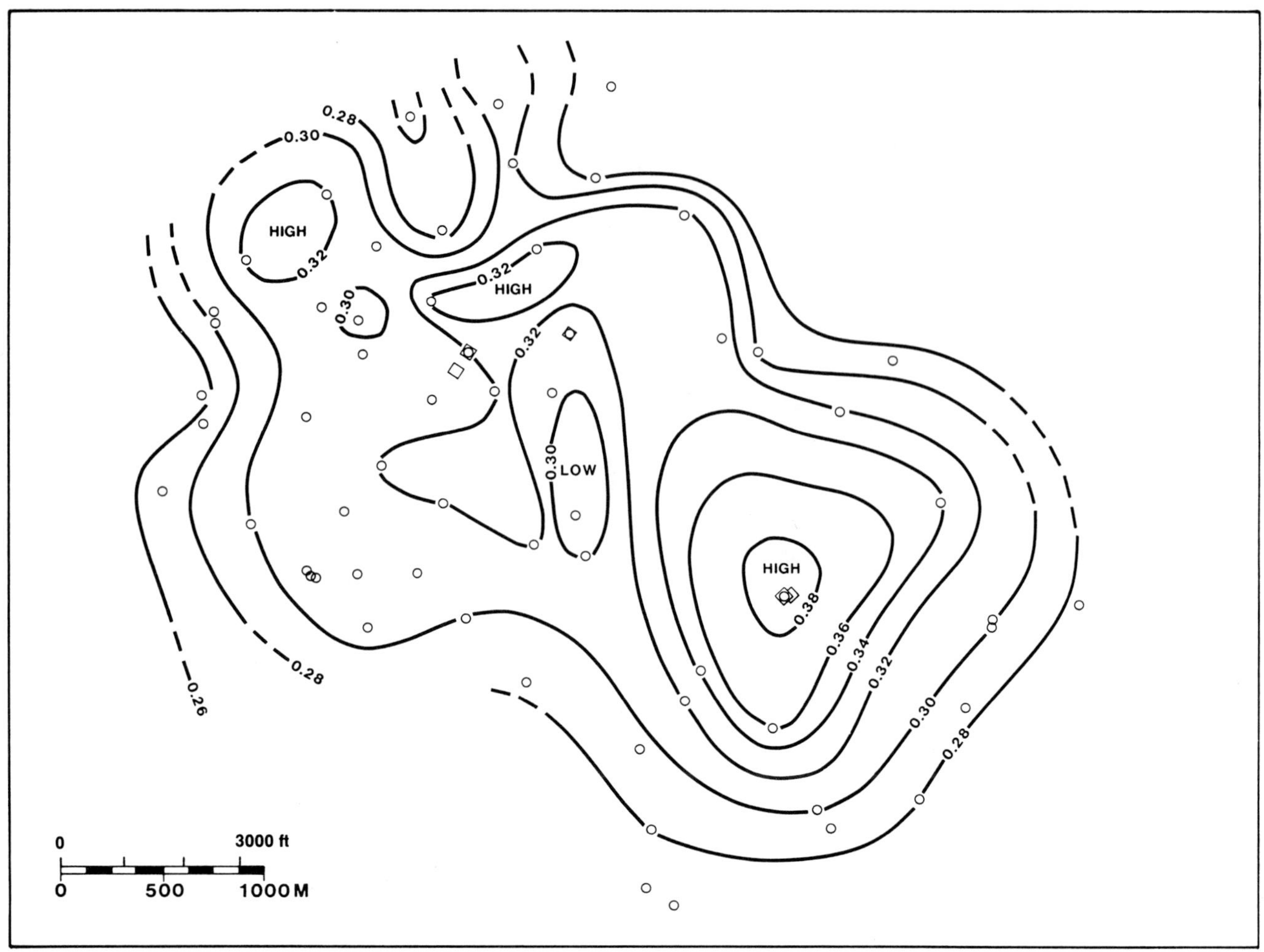

Figure 10. Isoporosity map displaying the lateral variation of porosity within the M_1 reservoir unit. A clear reduction in porosity from crest to flank is seen. Contour interval, 0.02 (2%).

variations seem to reflect mainly the original depositional thickness variations.

Fractures

Fractures occur much less frequently in the chalk of the Dan field than in most other Central graben chalk fields described in the literature (i.e., Brewster et al., 1986; Hurst, 1983). In the Dan field, two main types of fractures have been recognized: healed hairline fractures and open stylolite associated extension fractures. Tectonic fractures have been reported only rarely. This contrasts with observations from other Central graben fields where tectonic fractures are common; i.e., in the Norwegian fields in the Ekofisk area (Brewster et al., 1986) and in the Danish fields, Gorm (Hurst, 1983) and Skjold (Jensenius, 1987).

The hairline fractures often form anastomosing networks. They are completely healed by calcitic cement and appear to have lower porosity and permeability than the surrounding matrix, and hence do not enhance reservoir permeability. These fractures are thought to have been formed relatively early after deposition of the chalk.

The stylolite-associated extension fractures are interpreted to be open in the reservoir. They have been observed in only two cores and in only limited numbers. Nelson (1981) described these fractures from the Norwegian Central graben and interpreted them as tension gashes. They are generally less than 3 cm long and less than 0.5 mm wide; they are subvertical and in most instances propagate from the tips of stylolites.

Watts (1983) described similar fractures from the Albuskjell field and considered these to be formed during or after the formation of stylolites. It has been shown that a reduced vertical (i.e., overburden) stress is a prerequisite for extension rather than shear fractures to form. The development of high pore pressures in the chalk (3800 psi or 262 bar at 6000 ft [1830 m], subsea) therefore has been essential for the late formation of extension fractures.

The exact orientation of the extension fractures has not been possible to map with the limited available data. An anisotropic stress regime may have existed at the time of formation of the stylolites. This could give rise to a common orientation of the extension fractures. An anisotropic stress regime may still prevail, which may have a severe influence on the propagation of artificially induced fractures.

The present stress regime is therefore the subject of extensive investigations. These investigations include borehole breakout analyses of dipmeter records and anaelastic strain recovery measurements on core samples. Preliminary data suggest that the stress regime may be understood in terms of a fairly simple model of stress heterogeneity. In this model, the maximum horizontal compressive stress direction is radial on the flanks of the dome. On the crest, the stress direction may be related to the regional stress pattern. This model is in agreement with the results of Withjack and Scheiner (1982).

The limited occurrence of open fractures would indicate that their contribution to permeability is limited. This is in line with the observation that there is a very good match between core-measured matrix permeabilities and the permeabilities derived from analyses of well tests.

POTENTIAL VERTICAL PERMEABILITY BARRIERS AND IMPEDANCE TO FLOW

A number of lithological anisotropies may cause restrictions on vertical flow although none of them is thought to be completely effective as a barrier.

1. In the Danian D_1 and D_2 units, chert occurs in a number of fashions ranging from isolated nodules to laterally continuous, although nodular, beds.
2. In the D_{2a}, chert occurs disseminated in the chalk, often in association with the mottled "patterned chalk" described previously. The overall permeability of this facies is usually very low (as low as 0.02 md).
3. In the Danian units, the wispy laminations sometimes coalesce to form argillaceous bands a few centimeters thick. The lateral continuity of these bands is not known, but the zones of general high clay content where these bands form can in some areas be correlated over several hundreds of meters.
4. At several stratigraphic levels, horizons of strong matrix cementation are observed. Some of these may be hardgrounds that reflect periods of nondeposition. Of these, however, only the hardground at the Maastrichtian–Danian boundary is sufficiently developed to form a regional impedance to flow.
5. Stylolites occur almost exclusively in the Maastrichtian units. They may reduce vertical permeability, but there are indications that they act as conduits to lateral flow; e.g., in fields where dolomites occur in juxtaposition with the chalk, minerals derived from the dolomitic sequences have been precipitated in the chalk preferentially along the stylolite sutures. This phenomenon has been taken as an indication that significant fluid flow occurred along the stylolites.

The first four of these lithological heterogeneities are associated with the lower part of the Danian sequence. While any one of them, taken singly, may not constitute a barrier, their combined effect possibly could severely hamper vertical flow.

RFT measurements from a number of wells indeed show a pressure differential across the D_2 resulting from a higher depletion in the Maastrichtian M_1/M_{1a} reservoir units compared to the less permeable Danian D_1. The pressure differential is a strong indication that the D_2 acts as a partial barrier, at least.

HYDROCARBON CHARACTERISTICS

Oil produced from Dan is naphthenic, 30° API, with 0.64 wt. % sulfur. The viscosity is 8.24 (centistokes) at 104°F (40°C).

PVT results indicate that an initial solution GOR of about 640 scf/stb is required to saturate the oil at the initial reservoir pressure of 3820 psia (263 bar) at the Danian GOC.

Fluid and gas properties at a reference elevation of 6060 ft (1848 m) (ss) can be summarized as follows:

Initial pressure:	3820 psia (26,339 kPa)
Initial temperature:	167° F (75° C)
Oil formation volume factor:	1.3 rb/stb
Solution GOR at 3820 psia:	640 scf/stb
Tank oil gravity:	30.4° API
Initial gas formation volume factor:	0.736 rb/Mscf

SOURCE ROCKS

The Late Jurassic Kimmeridge Clay is regarded as the most important source rock in the North Sea. A review of the known occurrences of this and other potential source rocks in the Danish North Sea is given by Lindgren et al. (1982) and by Østfeldt (1987). Geochemical investigations of the Kimmeridge Clay equivalent from the Dan field M-8X well revealed that early mature, dominantly vitrinitic source rocks are indeed present beneath the Dan field (see Appendix 1).

Oil produced at Dan has characteristics similar to a number of other oils in the "Southern Salt Dome Province." The dominant kerogen facies in this area are, however, in general relatively unfavorable for good oil generation (kerogen type III). For this reason it can be speculated whether the Dan field and neighboring fields have been sourced from areas outside the province.

A source rock development of the Kimmeridge Clay superior to that observed at Dan is known to occur farther northwest in the Central graben across the Norwegian-Danish border (Robertson Research, 1983) and in the Tail End graben to the north (Damtoft et al., 1987). Although potential migration routes are not known, it is certainly possible, from a structural point of view, that the Dan field could have been sourced from this area to the north.

The migration is believed to have commenced in the late Eocene and continued during the Oligocene into the Miocene—broadly simultaneous with the main growth of the structure. This is obviously a favorable situation in that any fracture system created by the structural growth will have provided paths for the migrating oil.

EXPLORATION AND DEVELOPMENT CONCEPTS

In the mid-1960s chalk was not regarded as a potential reservoir. The A-1/A-2 discovery in 1966-1967 was therefore met with some skepticism, and encouragement to pursue the chalk play concept was only achieved after promising results in several wells that had been drilled with combined Tertiary-Mesozoic objectives. All wells in this early phase of exploration of the Danish North Sea were drilled on seismically mapped structural closures. The discovery well of Dan, M-1X, was drilled with the specific objective of evaluating the Danian and Upper Cretaceous chalk section.

Production from the Dan field commenced in July 1972, only 15 months after the discovery in April 1971. A very rapid decline rate of the first wells was soon observed. The reservoir characteristics causing this decline were not fully understood at the time. With more experience gained during the 1970s, it was gradually realized that natural fracturing of a chalk reservoir has an extreme impact on well productivity. The Dan field is only affected by natural fracturing to a limited degree. This contrasts with the giant Ekofisk field in the Norwegian Central graben and with a number of fields in the Danish sector; e.g., Skjold, Rolf, Dagmar, and Nils.

Hydraulic fracturing of oil-producing wells compensates for the lack of natural fractures in Dan, and the advances in this technology have played an important role in the development of the field. Today, artificial fracturing is carried out in all wells drilled on the Dan field and is considered an important prerequisite for economic oil recovery from the field. Massive propped hydraulic fractures with up to 2.1 million lb (950,000 kg) of sand proppant have successfully been placed in the reservoir.

Matrix permeabilities are generally low, so the relative variation of this parameter across the field becomes very important. The mapping of Kh values (permeability × thickness calculated from core and log data) to characterize the lateral variation in productivity was introduced prior to the Dan F drilling campaign in 1984-1985. This proved to be a powerful tool for well planning, and significant effort is continuously directed toward updating and refining the Kh map.

The recent introduction of horizontal producers with massively propped hydraulic fractures has proven to be a cost effective technique for the continued development of the Dan field, and it is beyond doubt that the bulk of wells to be drilled in the future in the Dan field will be horizontal wells.

The Dan field has provided the DUC with very important development experience. This together with the advances of technology during the past two decades would probably lead to a different approach should a chalk field with limited natural fracturing be discovered today in conditions similar to those of the Dan field. A thorough appraisal as in Dan would certainly still be required, not only to delineate the reservoir, but also to assess variations in natural fracturing and variations in matrix permeability.

As opposed to the existing Dan development, the field development of today would be characterized by a predominance of horizontal producers that would all be hydraulically fractured.

ACKNOWLEDGMENTS

Shell Olie og Gasudvinding Danmark, B.V., Texaco Denmark Inc., and Maersk Olie og Gas A/S kindly permitted publication of the paper.

The field description represents an integration of recent studies with numerous earlier unpublished works, notably those of N. Svendsen and N. C. Munksgaard.

M. R. Yusas and M. C. Doyle of Maersk Oil and E. Nygaard of the Geological Survey of Denmark kindly reviewed the manuscript and offered valuable criticism and suggestions for improvement.

REFERENCES CITED

Andersen, C., J. C. Olsen, O. Michelsen, and E. Nygaard, 1982, Structural outline and development, *in* O. Michelsen, ed., Geology of the Danish Central Graben: Geological Survey of Denmark, Series B, n. 8, p. 9-26.

Brewster, J., J. Dangerfield, and H. Farrell, 1986, The geology and geophysics of the Ekofisk field waterflood: Marine and Petroleum Geology, v. 3.

Childs, F. B., and P. E. L. Reed, 1975, Geology of the Dan field and the Danish North Sea: Geological Survey of Denmark, Series III, n. 43.

Damtoft, K., C. Andersen, and E. Thomsen, 1987, Prospectivity and hydrocarbon plays of the Danish Central Trough, *in* J. Brooks and K. Glennie, Petroleum Geology of north-west Europe: London, Graham and Trotman, p. 403-417.

Deegan, C. E., and B. J. Scull, 1977, A proposed standard lithostratigraphic nomenclature for the central and northern

North Sea: Rep. Inst. Geol. Sci., n. 77/25.
D'Heur, M., 1984, Porosity and hydrocarbon distribution in the North Sea chalk reservoirs: Marine and Petroleum Geology, v. 1, p. 211-238.
D'Heur, M., 1986, The Norwegian chalk fields: habitat of hydrocarbons on the Norwegian continental shelf: Norwegian Petroleum Society (Graham and Trotman), p. 77-89.
Dixon, J., 1976, Patterned carbonate—a diagenetic feature: Bull. Can. Petr. Geol., v. 24, n. 3, p. 450-456.
Hatton, I. R., 1986, Geometry of allochthonous chalk group members, Central Trough, North Sea: Marine and Petroleum Geology, v. 3, p. 79-98.
Hurst, C., 1983, Petroleum geology of the Gorm field, Danish North Sea: Geologie en Mijnbouw, n. 62, p. 157-168.
Jacobsen, F., 1982, Triassic, *in* O. Michelsen, ed., Geology of the Danish Central Graben: Geological Survey of Denmark, Series B, n. 8, p. 32-37.
Jensen, T. F., L. Holm, N. Frandsen, and O. Michelsen, 1986, Jurassic-Lower Cretaceous lithostratigraphic nomenclature for the Danish Central Trough: Geological Survey of Denmark, Series A, n. 12.
Jensenius, J., 1987, High-temperature diagenesis in shallow chalk reservoir, Skjold oil field, Danish North Sea: evidence from fluid inclusions and oxygen isotopes: AAPG Bulletin, v. 71, n. 11, p. 1378-1386.
Kennedy, W. J., 1987, Sedimentology of the Late Cretaceous and Early Paleocene Chalk Group in the North Sea basin and on its margins: JAPEC Course Notes No. 54, Geological Society, London.
Kennedy, W. J., and R. E. Garrison, 1975, Morphology and genesis of nodular chalks and hardgrounds in the Upper Cretaceous of southern England: Sedimentology, n. 22, p. 311-386.
Kristoffersen, F. N., and I. Bang, 1982, Cenozoic excl. Danian limestone, *in* O. Michelsen, ed., Geology of the Danish Central Graben: Geological Survey of Denmark, Series B, n. 8, p. 62-71.
Lieberkind, K., I. Bang, N. Mikkelsen, and E. Nygaard, 1982, Late Cretaceous and Danian limestone, *in* O. Michelsen, ed., Geology of the Danish Central Graben: Geological Survey of Denmark, Series B, n. 8, p. 49-62.
Lindgren, H., E. Thomsen, and P. Wrang, 1982, Source rocks, *in* O. Michelsen, ed., Geology of the Danish Central Graben: Geological Survey of Denmark, Series B, n. 8, p. 73-86.
Michelsen, O., 1982, Geology of the Danish Central Graben: Geological Survey of Denmark, Series B, n. 8.
Michelsen, O., 1986, The Danish pre-Tertiary lithostratigraphy. Status of the lithostratigraphic nomenclature, onshore and offshore Denmark: Geological Survey of Denmark, Internal Report 12.
Nelson, R. A., 1981, Significance of fracture sets associated with stylolite zones: AAPG Bulletin, v. 65, p. 2417-2425.
Nygaard, E., 1983, Bathichnus and its significance in the fossil association of Upper Cretaceous chalk, Mors, Denmark: Danm. Geol. Unders., Årbog 1982, p. 107-137.
Nygaard, E., K. Lieberkind, and P. Frykman, 1983, Sedimentology and reservoir parameters of the Chalk Group in the Danish Central Graben: Geologie en Mijnbouw, v. 62, p. 177-190.
Robertson Research, 1983, The Danish North Sea area: the stratigraphy and petroleum geochemistry of the Jurassic to Tertiary sediments: Unpublished report.
Svendsen, N., 1979, The Tertiary/Cretaceous chalk in the Dan field of the Danish North Sea, *in* T. Birkelund, and R. G. Bromley, Cretaceous-Tertiary Boundary Events Symposium, v. II, University of Copenhagen.
Watts, N. L., 1983, Microfractures in chalks of Albuskjell field, Norwegian Sector North Sea: possible origin and distribution: AAPG Bulletin, v. 67, p. 201-234.
Withjack, M. O., and C. Scheiner, 1982, Fault patterns associated with domes—an experimental and analytical study: AAPG Bulletin, v. 66, n. 3, p. 302-316.
Østfeldt, P., 1987, Oil-source rock correlation in the Danish North Sea, *in* J. Brooks, and K. Glennie, Petroleum geology of northwest Europe: London, Graham and Trotman, p. 419-429.

SUGGESTED READING

Hancock, J. M., 1976, The petrology of the chalk: Proc. Geol. Assoc., v. 86, n. 4, p. 499-535.
Hardman, R. F. P., 1982, Chalk reservoirs of the North Sea: Bulletin of the Geological Society of Denmark, v. 30, p. 119-137.
Kennedy, W. J., 1983, Sedimentology of Late Cretaceous-Palaeocene reservoirs, North Sea Central Graben: Proceedings of 3rd Conference on Petroleum Geology of North-West Europe.
Scholle, P. A., 1977, Chalk diagenesis and its relation to petroleum exploration: oil from chalks, a modern miracle?: AAPG Bulletin, v. 61, p. 982-1009.
Ziegler, P. A., 1982, Geological atlas of western and central Europe: Shell Internationale Petroleum Maatschappij, B.V.

Appendix 1. Field Description

Field name .... *Dan field*

Field location:

Country .... *Denmark, Danish North Sea*

Basin/Province .... *North Sea Central graben*

Field discovery:

Year first pay discovered .... *Maastrichtian and Danian chalk reservoir 1971*

Discovery well name and general location:

First pay .... *M-1X, 205 km west of port of Esbjerg*

Discovery well operator .... *Gulf Oil on behalf of the DUC (Danish Underground Consortium)*

IP:

First pay .... *13,700 BOPD (avg. first year)*

All other zones with shows of oil and gas in the field:

Age	Formation	Type of Show
Middle Jurassic	*Lower Graben Sand Formation*	*Weak gas and oil*

Geologic concept leading to discovery and method or methods used to delineate prospect

Structure at Top Chalk level identified seismically. Chalk play concept developed from earlier discoveries in vicinity.

Structure:

Province/basin type .... *Bally 1211; Klemme IIIa*

Tectonic history

The Central graben of the North Sea was initiated during the Early Permian. Evaporitic sequences accumulated in the Late Permian and Triassic were activated in several phases resulting in the formation of a number of salt pillows and salt diapirs folding the overlying Jurassic through Tertiary sediments.

Regional structure

The field is a symmetrical dome resulting from halokinetic uplift by a salt pillow of Triassic age.

Local structure

Symmetrical dome with dips generally less than 5°, bisected by a major northeast-southwest-trending fault downthrown to the northwest.

Trap:

Trap type(s) .... *Simple structural arch with reservoirs in Danian and Maastrichtian chalk*

Basin stratigraphy (major stratigraphic intervals from surface to deepest penetration in field):

Chronostratigraphy	Formation	Depth to Top in ft (m)
Pleistocene-middle Miocene	*Nordland group*	
Middle Miocene-lower Eocene	*Hordaland group*	*4260 (1299)*
Upper Paleocene	*Rogaland group*	*5890 (1796)*
Lower Paleocene-Upper Cretaceous	*Chalk group*	*6013 (1834)*
Lower Cretaceous	*Cromer-Knoll Group*	*7190 (2193)*
Upper Jurassic	*Farsund Formation*	*7514 (2292)*
	Lola Formation	*8551 (2608)*
Middle Jurassic	*Middle Graben Shale Fm.*	*10,143 (3094)*
	Lower Graben Sand Fm.	*10,260 (3129)*
Lower Jurassic	*Fjerritslev Formation*	*10,467 (3192)*
Upper Triassic	*Winterton Formation*	*10,771 (3285)*
	Triton Anhydrite Formation	*10,793 (3292)*
Middle Triassic	*Dudgeon Saliferous Fm.*	*11,559 (3525)*

Reservoir characteristics:

- **Number of reservoirs** *1*
- **Formations** .... *Ekofisk and Tor formations equivalent chalk*
- **Ages** .... *Upper Cretaceous Maastrichtian and lower Paleocene Danian*
- **Depths to tops of reservoirs** .... *5700 ft (1738 m) on crest of field*
- **Gross thickness (top to bottom of producing interval)** .... *640 ft (195 m)*
- **Net thickness—total thickness of producing zones**
  - **Average** .... *375 ft (114 m)*
  - **Maximum** .... *550 ft (168 m)*
- **Lithology**

  Chalk of both allochthonous and pelagic origin with low matrix permeability; generally low clay content; frequent occurrence of chert
- **Porosity type** .... *Interparticle matrix porosity with minor interparticle (foram chambers)*
- **Average porosity** .... *28% (unit averages ranging from 22 to 34%)*
- **Average permeability** .... *1.75 md*

Seals:

- **Upper**
  - **Formation, fault, or other feature** .... *Shales of Paleocene Rogaland group*
  - **Lithology** .... *Shale*
- **Lateral**
  - **Formation, fault, or other feature** .... *Shales of Paleocene Rogaland group*
  - **Lithology** .... *Shale*

Source:

- **Formation and age** .... *Kimmeridge Clay equivalent beneath Dan (Farsund Formation); a richer potential source (same age) occurs in NW part of Central graben farther away*
- **Lithology** .... *Shale*
- **Average total organic carbon (TOC)** .... *<2%*
- **Maximum TOC** *8%*
- **Kerogen type (I, II, or III)** .... *III, I/II*
- **Vitrinite reflectance (maturation)** .... R_o *= 0.5–0.8%, 0.6–1.2%*
- **Time of hydrocarbon expulsion** .... *Oligocene–Miocene*
- **Present depth to top of source** .... *7400 ft (2257 m)*
- **Thickness** .... *1400 ft (427 m)*
- **Potential yield** .... *NA*

DAN

Appendix 2. Production Data

Field name .... *Dan field*

Field size:

- **Proved acres** .... *5950 ac (2410 ha)*
- **Number of wells all years** *56*
- **Current number of wells** *48*
- **Well spacing** .... *40 ac (16 ha)*
- **Ultimate recoverable** .... *208 MMSTB; 1500 bcf (primary)*
- **Cumulative production** .... *57 MMSTB; 124 bcf*
- **Annual production** .... *9.3 MMSTB; 30 bcf*
- **Present decline rate** .... *NA*
  - **Initial decline rate** .... *NA*
  - **Overall decline rate** .... *NA*

Annual water production .... *424,000 bbl*
In place, total reserves .... *1922 MMSTB; 1500 bcf*
In place, per acre foot .... *148 bbl/ac-ft*
Primary recovery .... *208 MMSTB; 541 bcf*
Secondary recovery .... *NA*
Cumulative water production .... *NA*

Drilling and casing practices:

Casing program

26-in. to 380 ft; 18⅝-in. to 1520 ft; 13⅜-in. to top of overpressured shales (approx. 4000 ft); 9⅝-in. to top of reservoir (approx. 6000 ft); 7-in. liner to TD

Drilling mud .... *NA*
Bit program .... *Varies*
High pressure zones .... *Shales overpressured below ±4000 ft (1220 m)*

Completion practices:

Interval(s) perforated .... *Multiple in Maastrichtian and Danian reservoirs*
Well treatment .... *Hydraulic fracturing*

Formation evaluation:

Logging suites .... *Nonessential section above reservoir none or only GR; reservoir: neutron, density, sonic, induction, GR, RFT*
Testing practices .... *NA*
Mud logging techniques .... *Full geological and gas monitoring to TD*

Oil characteristics:

Type .... *Naphthenic*
API gravity .... *30.4°*
Base .... *NA*
Initial GOR .... *1147 (first year production)*
Sulfur, wt% *0.64*
Viscosity, SUS .... *8.24 CSt (at 104°F, 40°C)*
Pour point .... *Low, not waxy*
Gas-oil distillate .... *NA*

Field characteristics:

Average elevation .... *-6000 ft (1830 m) (reservoir)*
Initial pressure .... *3820 psia (26,339 kPa)*
Present pressure .... *3000-3600 psi (20,685-24,822 kPa)*
Pressure gradient .... *0.33 psi/ft (approx. 7.465 kPa/m)*
Temperature .... *167°F (75°C)*
Geothermal gradient .... *2.8°F/100 ft (0.04°C/m)*
Drive .... *Depletion/solution gas*
Oil column thickness .... *NA*
Oil-water contact .... *Dipping: 6367-6592 ft (1942-2010 m)*
Connate water *20%*
Water salinity, TDS .... *70,000 ppm NaCl*
Resistivity of water .... *0.055 ohm-m*
Bulk volume water (%) .... *NA*

Transportation method and market for oil and gas:

Pipeline to Gorm field platform; from there 134 mi (217 km) pipeline to shore; then onshore pipeline to Fredericia refinery

Silver Springs/Renlim Field—Australia
Bowen Basin, Queensland

ROWAN ROBERTS
Bridge Oil Limited
Sydney, Australia

FIELD CLASSIFICATION

BASIN: Bowen
BASIN TYPE: Foreland
RESERVOIR ROCK TYPE: Sandstone
RESERVOIR ENVIRONMENT OF DEPOSITION: Fluvial
RESERVOIR AGE: Triassic
PETROLEUM TYPE: Gas
TRAP TYPE: Paleo-High Onlap
TRAP DESCRIPTION: Onlap of reservoir around an eroded basement outlier; structurally enhanced by subsequent differential compaction

LOCATION AND OVERVIEW

Silver Springs/Renlim gas field is located in southeastern Queensland, Australia (Figure 1). Gas is contained in fluvial quartz arenites of the Middle Triassic (Anisian) Showground Formation (Figure 2) on the western flank of the Upper Carboniferous to Upper Triassic Bowen basin, which is overlain in this region by the Lower Jurassic to Early Cretaceous Surat basin.

A hydrocarbon province was initially established around Roma, 71 mi (115 km) to the north-northwest of Silver Springs No. 1 (Figure 1), following the discovery of gas in a Jurassic reservoir sandstone intersected in a water bore drilled in 1900. With the 1970 Boxleigh No. 1 Showground Formation gas discovery (Figure 1) and the Silver Springs No. 1 discovery in June 1974, commercial reserves of the province were extended significantly to the south, which led to commercial development in 1978. Gas is produced for markets in Brisbane, 250 mi (400 km) east, via the Silver Springs and the Roma-Brisbane pipelines.

Estimated ultimate recoverable reserves from Silver Springs/Renlim range from 55 to 105 bcf (1.56–2.97 $\times$ 10^9 Sm^3) sales gas and 2.5 to 4.8 million bbl associated condensate. Since 1985, liquified petroleum gas (LPG) has been recovered through the Wallumbilla LPG plant, 32 mi (52 km) east of Roma, at the intersection of the Silver Springs and Brisbane pipelines (Figure 1). Estimated ultimate recovery of LPG ranges from 2.9 to 5.6 million bbl. Reserves quantification is the subject of ongoing production monitoring, simulation, and appraisal drilling. To December 1987, the field has produced 30 bcf (0.85 $\times$ 10^9 Sm^3) of sales gas (Figure 3)

Reservoir depth to the Showground Formation is approximately 6200 ft (1900 m). Gas is trapped by stratigraphic onlap to a basement paleo high, and closure has been enhanced by subsequent compactional drape and minor deformation. The top seal is a regionally pervasive brackish or marine "shale." Igneous and metamorphic "basement" is an effective basal seal.

Smaller fields with similar characteristics occur nearby. Boxleigh field 3.7 mi (6 km) to the southwest and Sirrah field 6 mi (10 km) to the southeast are being developed through facilities centered on Silver Springs/Renlim, while the Taylor No. 1 discovery (1986) approximately 8.7 mi (14 km) to the northeast is still being appraised. Oil accumulations in the same reservoir sandstone are known to occur both as minor oil legs beneath gas and as isolated occurrences in the general area. Significant exploration potential is the subject of ongoing evaluation.

HISTORY

Pre-Discovery

On 16 October 1900, the townspeople of Roma, Queensland, had their hopes for a permanent water supply set back when the Queensland Government No. 2 water bore on Hospital Hill blew out with a gas flow. The serendipity in this "disaster," however,

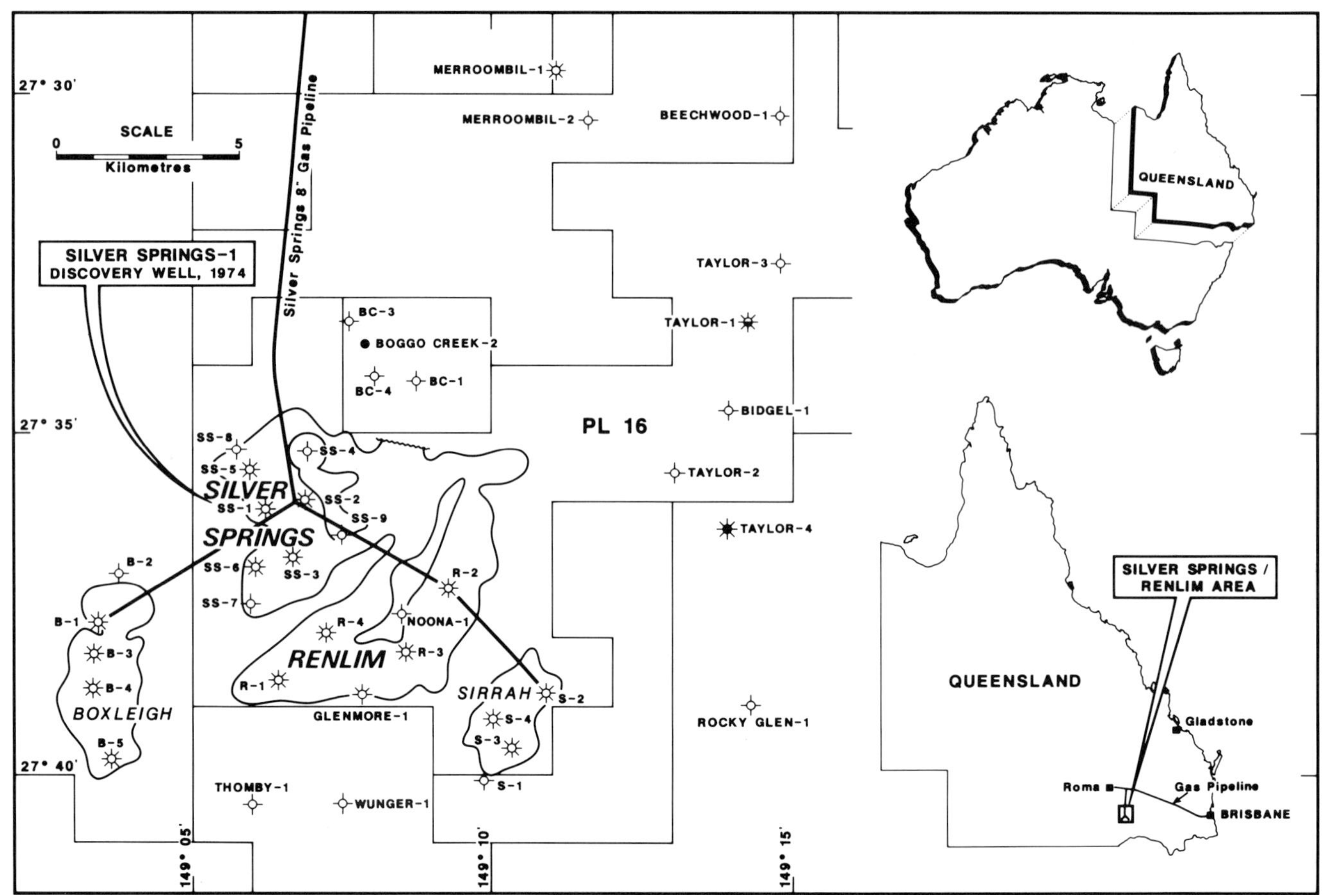

Figure 1. Location map, Silver Springs/Renlim gas field, with position of major gas gathering system indicated.

was not to yield commercial success until the 1960s, the reality of which culminated in the commissioning of the Roma to Brisbane gas pipeline in 1969 (Wilkinson, 1983, 1988).

A joint venture of Union Development Corporation and Tenneco Australia Inc. (and prior to that Union, Kerr County and Land, and Australian Oil and Gas Corporation) had explored the Silver Springs/Renlim area before Bridge's entry in 1969. In a period highlighted by the discovery of oil approximately 70 mi (110 km) to the east at Cabawin No. 1 and Moonie No. 1 in 1961, the latter being developed as Australia's first commercial oil field in 1964, and Alton oil field 28 mi (45 km) to the south in 1964, the results of drilling in the Silver Springs/Renlim area were disappointing.

Wunger No. 1 (Figure 1), approximately 5 mi (8 km) south-southeast of the subsequent Silver Springs No. 1 discovery, had flowed gas at 180 mcf (5100 Sm^3) per day and 10% oil-cut water from the Showground Formation in 1963, but minimal updip potential was recognized. More significantly, Major No. 1, 12 mi (20 km) to the west, had flowed gas at 1.8 mmcf (51,000 Sm^3) per day in 1965 from a stratigraphic equivalent of this unit. Unsuccessful follow-up drilling and lack of gas market meant that the discovery was not commercial.

Through the Union-led exploration effort, the petroleum geology of the area was rigorously interpreted and documented. Union-Tenneco had identified the Middle Triassic Showground Formation fluvial sandstone as an excellent potential reservoir on the basis of log and core data. However, they had not extended earlier successes from elsewhere in the basin. They had acquired regional airborne magnetic data and gravity surveys and conducted regional single-fold seismic surveys.

In May 1969, Bridge Oil N.L. ("Bridge"—subsequently Bridge Oil Limited) farmed into Authority to Prospect (ATP) Number 145P covering approximately 18,000 mi^2 (46,600 km^2), centered 60 mi (100 km) south-southeast of Roma. Bridge funded the drilling of Noona No. 1 (Showground absent through nondeposition) and Glenmore No. 1 (Showground wet with minor gas flow) in 1969 (locations shown in Figure 1) and initially "missed" what was to be the largest Showground Formation gas field in the Bowen basin. With a subsequent farm-in by Offshore Oil N.L. ("Offshore"—now Petroz N.L.), the Boxleigh No. 1 Showground Formation gas discovery (Figure 1)

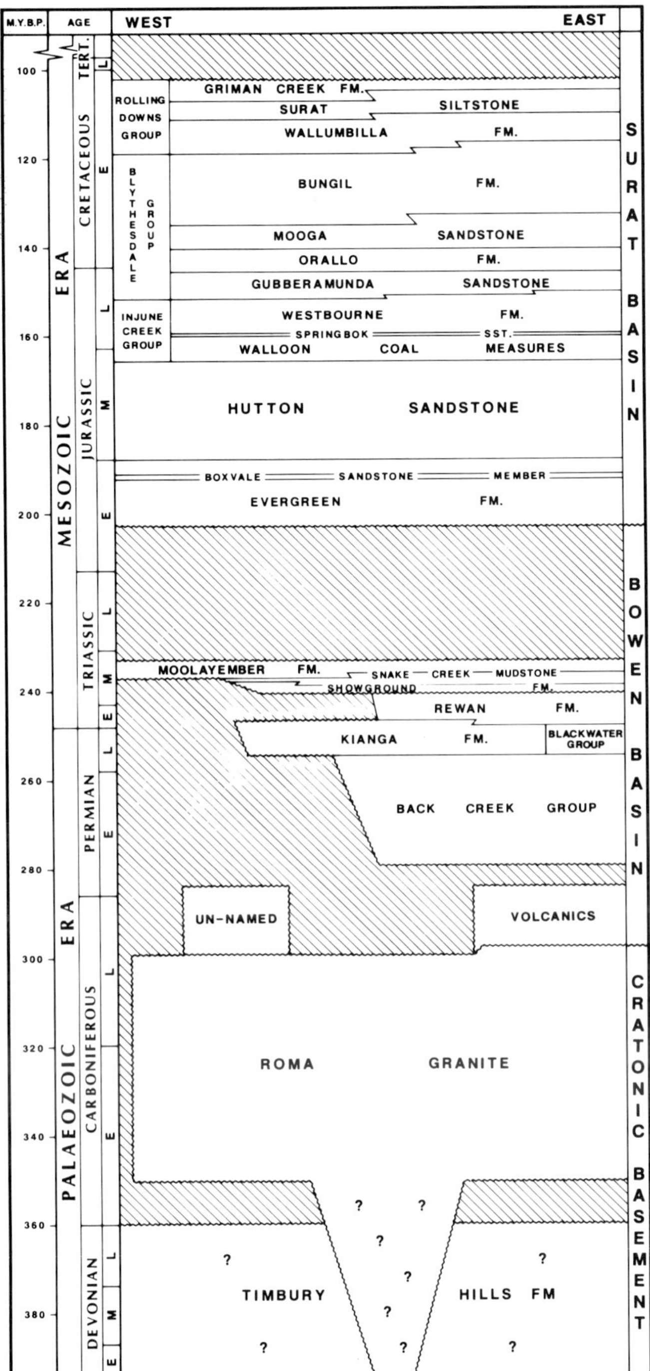

Figure 2. Stratigraphic column, Silver Springs/Renlim area. Deposition was essentially continuous within the Bowen and Surat basins. The two sequences are separated by a significant Late Triassic hiatus.

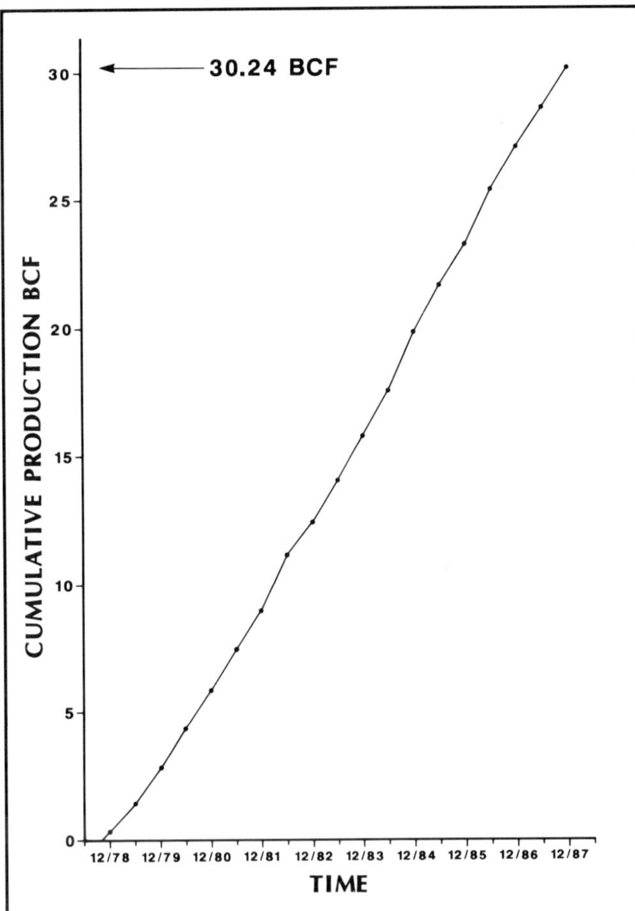

Figure 3. Cumulative sales gas production, Silver Springs/Renlim field.

was made in May 1970, but reserves of between 6 and 10 bcf (0.2 and 0.3 × 10^9 Sm^3) were inadequate to justify a 63 mi (102 km) pipeline north to the main Roma-Brisbane line.

Seismic operations in early 1974 were directed toward detailing poorly controlled structures to the northeast of Boxleigh with 400% CDP data and identified the "Boggo Creek-Noona Anticline," subsequently Renlim, and the "Silver Springs Anticline" (Bridge Oil, 1974). Despite the different names assigned to these structural features, it was clear even from initial mapping that they were part of one regional feature with independent culminations along the crest.

Discovery

Silver Springs No. 1 was drilled in June 1974 by Bridge (operator) and Offshore. The well flowed gas from the Showground Formation at 8.5 MMCFD (240,000 Sm^3/d) through a ½-in. (12.7 mm) choke (Figure 4). The company's two geologists, A. J. Kapel and D. A. Short, their consultant geophysicist, J. E. Milner, and their consultant geologist, the late C. E. B. Conybeare, were instrumental in the development of the prospect leading to discovery.

Silver Springs No. 1 was drilled with Australian federal government assistance under the Petroleum Search Subsidy Act (P.S.S.A.), 1959–1969 which was

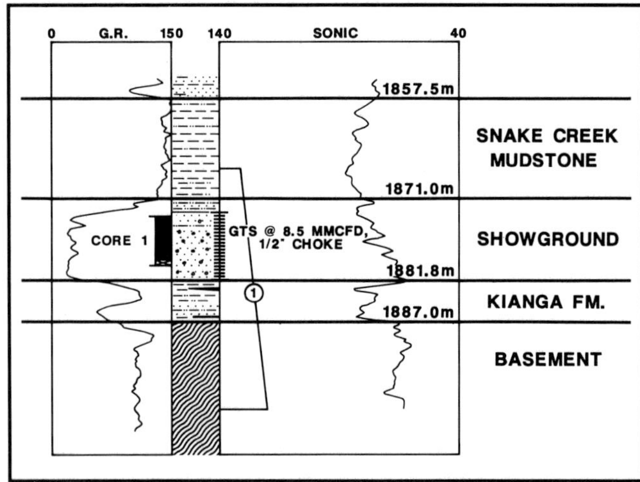

Figure 4. Silver Springs No. 1 gas discovery. Log displays "blocky" nature of G.R. curve of reservoir sand facies, with increasing porosity (increasing ΔT) from the base (which is conglomeratic) to the top (which is sandstone). Core, perforations, and test interval are indicated.

designed to stimulate exploration across the continent. The Company's P.S.S.A. submission indicates Silver Springs No. 1 was located structurally higher than Glenmore No. 1 (which may have intersected a gas-water contact) but lower than Noona No. 1 in which the reservoir was absent and also above the assumed gas-water contact from the earlier Boxleigh discovery to the west. Semiregional studies had also revealed that isopach mapping of the Evergreen Formation to basement interval (Figure 2) predicted that the reservoir would be absent if this interval were less than 650 ft (198 m) and would be present with the underlying Permian Kianga Formation if greater than 675 ft (206 m) as a result of onlap and compactional drape over the feature. The predicted thickness at Silver Springs No. 1 was 700 ft (213 m), and the actual was 698 ft with all predicted formations present. Seismic grid spacing at the time was 1 to 1.5 mi (~ 2 km) of largely 100% and some 400% data.

Post-Discovery

Silver Springs No. 2 appraisal well drilled in April 1976 defines the field's highest known gas to date, despite being programmed as a structurally lower follow-up well, downdip from the discovery. Significant statics variation in the area was the cause of this prognosis error and still remains probably the most significant limitation to structural mapping in the region, despite the apparent mapping accuracy from the discovery well. Nevertheless, after the drilling of the well, initial estimates of recoverable reserves ranged from 22 to 43 bcf (0.6–1.2 $\times$ 10^9 Sm^3) and when added to reserves estimates of between 5 and 13 bcf (0.1–0.4 $\times$ 10^9 Sm^3) from Boxleigh meant that a pipeline to join the Roma-Brisbane line (Figure 1) could be justified economically. A further successful well, Silver Springs No. 3, was drilled in late 1976. Contracts for the supply of gas to Australian Pacific Fertilizer and Allgas, both of Brisbane, were signed in 1977 and 1978, respectively. A 62 mi (102 km) long 8-in. (0.2 m) diameter gas pipeline to the Roma line was commissioned, and sales of gas commenced in October 1978.

Key pre- and post-discovery wells are highlighted in Figures 5 and 6. The culmination of the Renlim Structure was mapped in 1982. It was thought to be a separate closure, though in all 1974 and later mapping it was clearly seen to be a part of a larger feature. In January 1984, a gas-water contact was observed in Renlim No. 2 approximately 25 ft (7.5 m) above the top of porosity in Glenmore No. 1 after production from Silver Springs wells of approximately 16 bcf of gas. By March, 1987, Renlim No. 2 had watered out. The interconnection of the "Renlim" and "Silver Springs" reservoirs was established. Supported by present mapping (Figure 5) with an interpreted original gas-water contact at the top of the reservoir interval in Glenmore No. 1, this connection was not readily apparent in the mid-1980s when Renlim No. 2 (drilled in 1984) and Silver Springs No. 7 (drilled in 1981) wells indicated field limits that could have been from separate structures.

Sirrah field, 6 mi (10 km) to the southeast of Silver Springs, with sales gas reserves of approximately 10 bcf (0.3 $\times$ 10^9 Sm^3), was discovered in June 1982 soon after Renlim. With approximately 80 to 95 bbl of condensate per mmcf (4.5–5.3 $\times$ 10^{-4} kL per Sm^3) of sales gas, about twice that of Silver Springs and Renlim, this discovery contributed significantly toward a decision by Bridge and partners, Offshore and International Oil, to build a liquids and LPG recovery plant at the intersection of the Silver Springs and Roma-Brisbane pipelines. The Wallumbilla LPG plant, built subsequently in 1984, became operational in 1985 with a capacity of approximately 30 mmcf (850,000 Sm^3) per day. It significantly enhanced the value of the Silver Springs/Renlim and other Bridge-operated fields in the vicinity.

Other Showground Formation discoveries at Glen Fosslyn, Roswin (both approximately 12 mi or 20 km northeast), and Merroombil (Figure 1), though significantly smaller, will eventually be tied into existing facilities. The largest Showground Formation discovery, post-Sirrah field, is the Taylor No. 1 gas well drilled in 1986 (Figure 1). The same geological and geophysical principles as were used to successfully find gas updip from water-bearing sandstone and downdip of depositional onlap edges at Silver Springs/Renlim have been used to find gas at Taylor No. 1 in 1986, and most recently (1988) have found an exploitable oil leg below gas at Taylor No. 4, which flowed 864 bbl of oil and 1.8 mmcf (51,000 Sm^3) of gas per day from a drill-stem test.

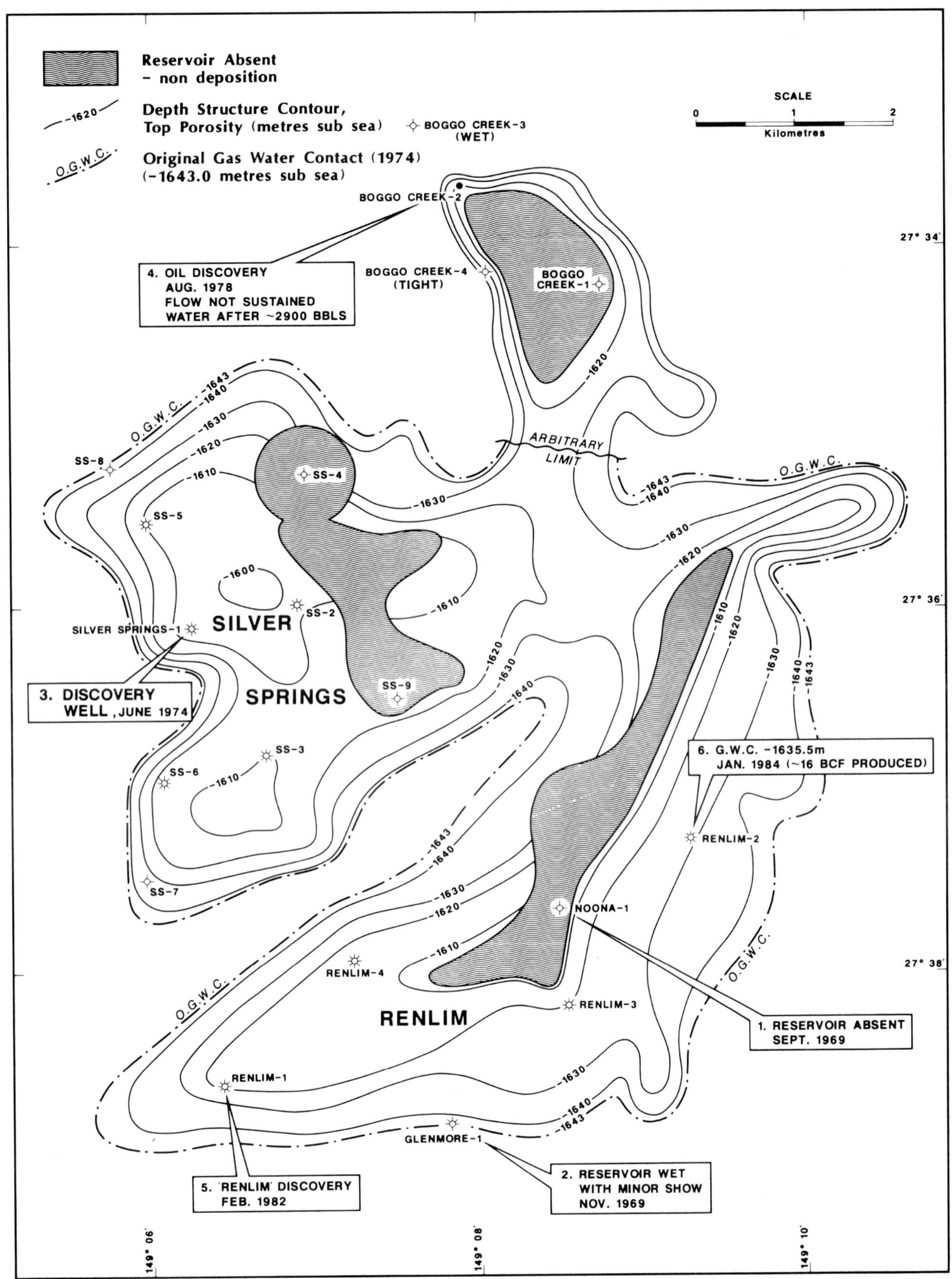

Figure 5. Structure contour map on top of porosity. Contour interval, 10 m. The sequence of drilling of key wells and their significance in defining field limits is indicated.

DISCOVERY METHOD

Serendipity must surely be a major factor in the initial discovery of hydrocarbons in the Bowen basin. No surface seeps are known in the Surat or Bowen sequences, though methane associated with Permian coals is common. The Hospital Hill water bore "blow out" near Roma in 1900 was responsible for the subsequent interest of both local and internationally based explorationists (Wilkinson, 1983). Using regional magnetics and gravity surveys, the basin geometry and major highs were recognized. Seismic, predominantly 100% data, was used to detail other large structural closures, and the rest is history.

After the region's potential was recognized, the key to finding Silver Springs/Renlim was a detailed examination of the stratigraphy, structure, and petrophysics of a potential reservoir section. The first indications that commercial hydrocarbons might be present in the local area was at Major No. 1 (1965), 12 mi (20 km) west, which flowed gas from the "Showground" but proved subcommercial. In 1970, the Boxleigh No. 1 gas discovery (Figure 1) strengthened this observation but again proved subcommercial.

On the basis of these earlier subcommercial discoveries and regional well control, Silver Springs/Renlim was predicted and subsequently shown to be located within a regional fluvial depositional axis of Showground sandstone. The coincidence of excellent reservoir-quality sandstone and significant structural closures in a position ideally situated for the trapping of hydrocarbons migrating out of the basin depocenter to the east is crucial to the overall commerciality of this area.

STRUCTURE

The Silver Springs/Renlim field is located on the western flank of the Bowen basin (Figure 7), which covers an area of approximately 77,000 mi^2 (200,000 km^2; Exon, 1976). The prospective Bowen basin is largely covered by the thick Mesozoic Surat basin sequence, and its detailed tectonic history remains obscure. The basin is a foreland basin (Figure 8) of Late Carboniferous to Late Triassic age that formed behind an arc which was migrating intermittently eastward during the Phanerozoic (Veevers, 1984). This is referred to as the Camboon arc (Murray, 1985). Hobday (1986, 1987) considered this arc system to have been of Gondwana scale, extending through eastern Australia, Antarctica, and Southern Africa. Day et al. (1978) interpreted a complex orogenic history with considerable longitudinal changes and less Paleozoic continental accretion. Harrington and Korsch (1985a and b) attempted to model the tectonic development of this and associated regions in some detail, but depend upon considerable hypothesis. Austin and Williams (1978) relate both longitudinal and accretionary phases to Gondwana rotation.

Compressional stress associated with waxing orogenesis is expressed most vividly along the Moonie-Goondiwindi reverse fault (Figure 7). Local Early to middle Permian extensional tectonics accompanied by extensive acid and basic volcanism produced north-south-trending troughs on the Western shelf (Roberts 1987). Deposition in the intracratonic Early Jurassic to middle Cretaceous Surat basin followed a hiatus and a local erosional period through the Late Triassic.

Silver Springs/Renlim field is located on the northern end of the north-plunging Wunger ridge (Figure 7), which was a positive feature throughout deposition of the Bowen sequence. This ridge has significantly influenced the location of fluvial paleodrainage axes for the Showground Formation and subsequent hydrocarbon migration.

The Silver Springs/Renlim feature itself is a pre-Permian erosional remnant of the Wunger ridge. The major southwest-plunging low that divides the "Silver Springs" and "Renlim" accumulations (Figure 5) was a significant fluvial paleodrainage axis off this high until its complete filling with and burial by Permian and Triassic sediments. The feature has the relatively flat crestal areas and steep flanks of a mesa (Figures 9 and 10).

Structural expression at higher stratigraphic levels is more subdued. Tertiary movement, as observed from seismic data, has significantly affected Lower Cretaceous sediments at less than 0.4 seconds two-way time (1800 ft/550 m; Figures 9 and 10), and Cenozoic weathering has obscured any surface expression. The present-day crest and paleo-crestal areas are not exactly coincident (Figure 5) because of this movement.

STRATIGRAPHY

The stratigraphic sequence present in the Silver Springs/Renlim area is summarized in Figure 2. Detailed stratigraphic description of this sequence has been presented by Power and Devine (1970), Gray (1972), Exon (1976), and Day et al. (1982).

The Upper Carboniferous to Upper Triassic Bowen basin sequence has been described by Dickens and Malone (1973) and Jensen (1975) from where it outcrops in the north. Butcher (1984) detailed the stratigraphy of the Triassic Showground Formation, which does not crop out in the area surrounding the Silver Springs/Renlim field, and Roberts (1987) offers minor refinement based on more recent work.

The Surat basin (Mesozoic) sequence is described by Mack (1963) and Power and Devine (1970). Golin and Smyth (1986) presented the most recent synthesis of the main Surat basin hydrocarbon-

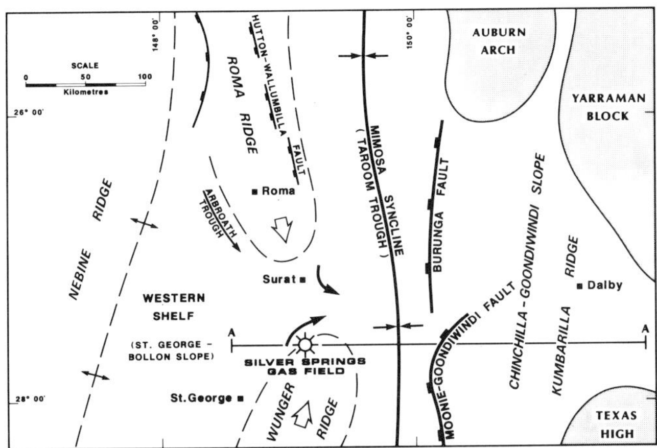

Figure 7. Tectonic elements map for the Surat and Bowen basins, modified after Power and Devine (1970). Position of cross section A to A′ (Figure 8) indicated. The Jurassic to Cretaceous Surat basin sequence overlaps the Upper Carboniferous to Upper Triassic Bowen basin sequence. Solid black arrows at Surat and Silver Springs indicate plunge of structural lows. Large faults east of Mimosa syncline are thrust faults; faults to the west are normal faults (boxes on downthrown side).

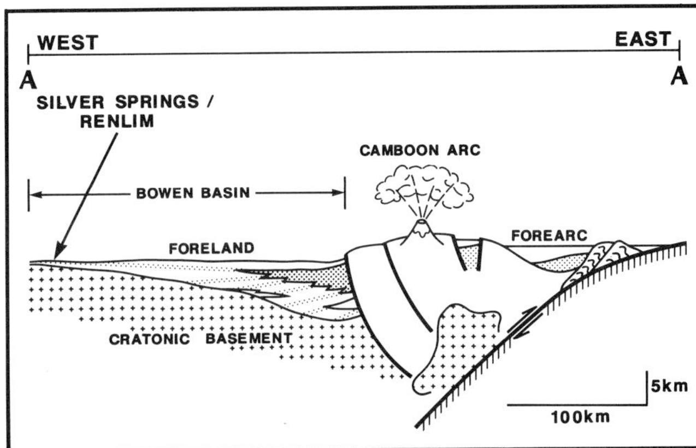

Figure 8. Generalized plate-tectonic origin model for the formation of the Bowen basin showing Silver Springs/Renlim on the western cratonic flank of a Late Carboniferous to Late Triassic foreland basin. For section location see Figure 7. Reproduced by permission of the Geological Society from Gondwana Coal Basins of Australia and South Africa: Tectonic Settings, Depositional Systems and Resources by D. K. Hobday in Geol. Soc. Special Publication No. 32, 1987.

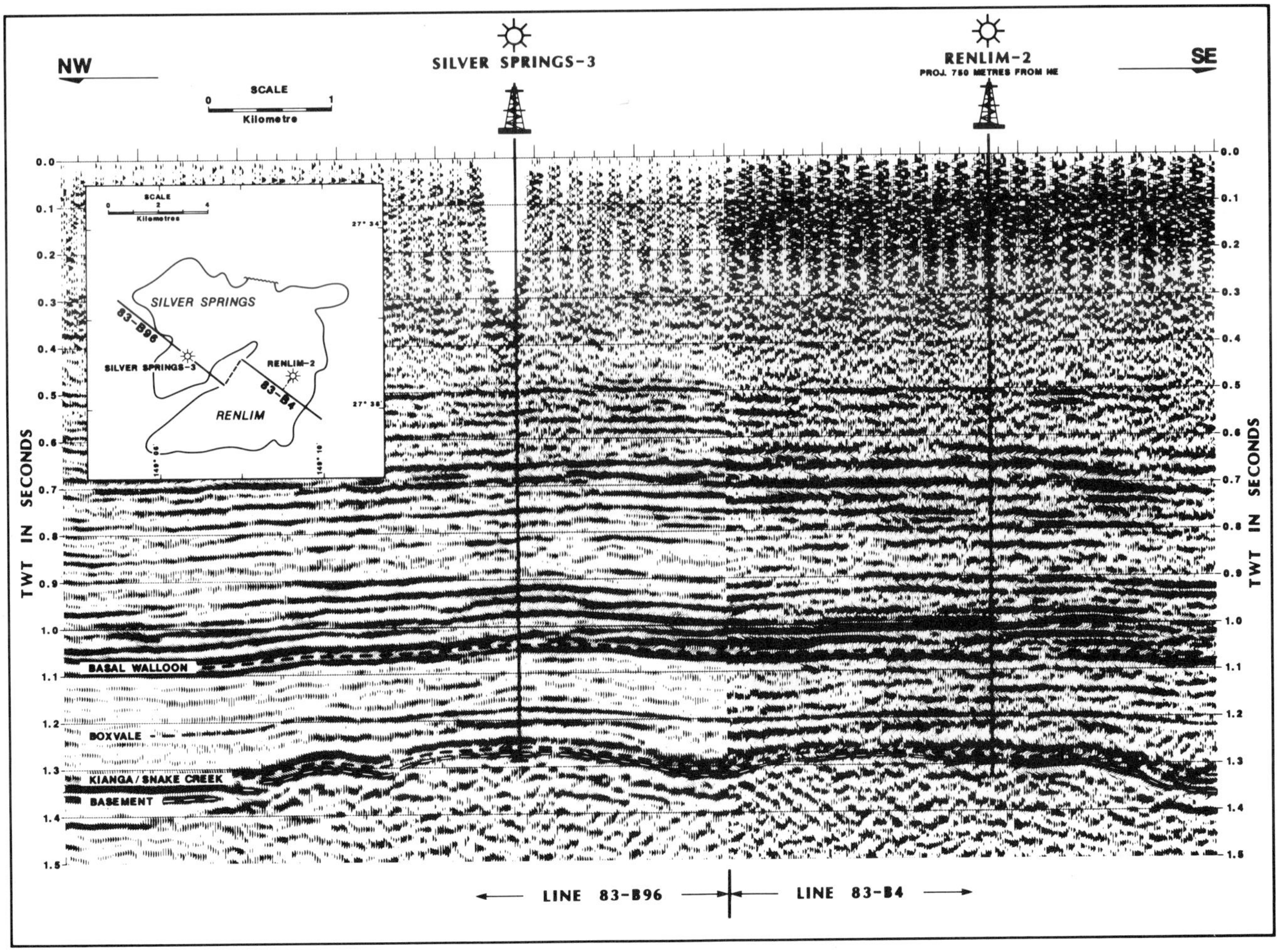

Figure 9. Composite seismic line across the Silver Springs/Renlim field shows Jurassic Evergreen (Boxvale) to basement thinning over the old high and the relatively steep flanks.

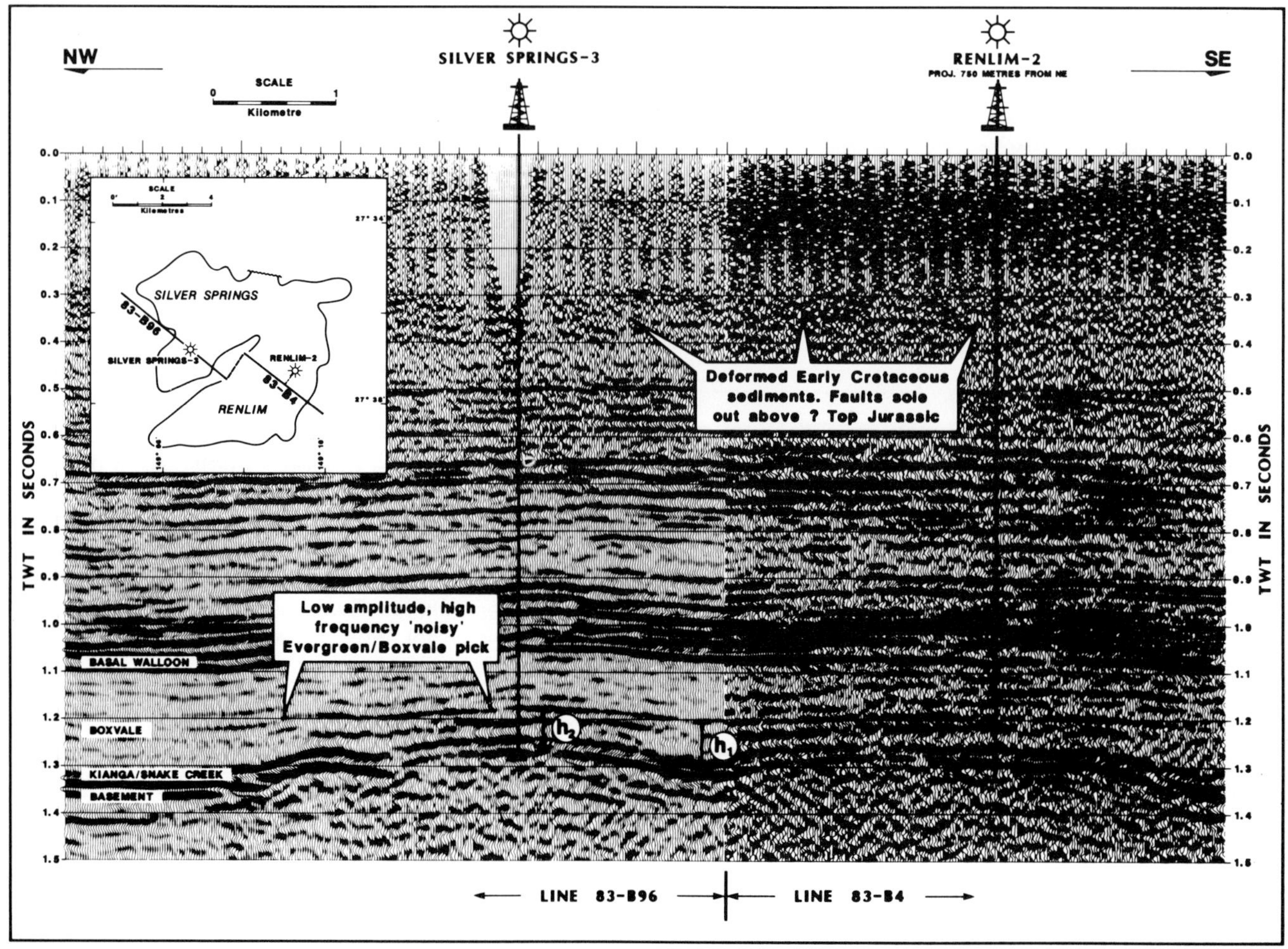

Figure 10. Composite seismic line across the Silver Springs/Renlim field. Note character changes on the Boxvale (intra-Jurassic Evergreen Formation) pick, which complicates accurate mapping; "thinning" of the Boxvale to Permian isochron (h_1 to h_2) onto the high; and Tertiary deformation of Lower Cretaceous and higher sediments above 0.4 seconds two-way time.

bearing units as they relate to the Silver Springs/Renlim areas where they are to date nonproductive.

Earliest Bowen basin deposits are Upper Carboniferous and Lower Permian acid and intermediate volcanics associated with the Camboon arc. Thicknesses may attain thousands of meters in the Taroom trough. While these have generally been considered to be economic basement, they are not hydraulic basement (Hitchen and Hays, 1971). They have been observed to contain locally mature interbedded marine source rocks (Roberts, 1987). Oil flowed to surface from volcanics interpreted to be age-equivalent to earliest Bowen basin deposits in Taylor No. 1 (Figure 1). Nomenclature adopted for these volcanics is loose, and they are known as Kuttung volcanics, Camboon andesite (volcanics), Combarngo volcanics, and Lizzie Creek volcanics in different parts of the basin.

Clastic deposition in the Bowen basin up to Upper Permian is predominantly shallow marine shale, siltstone, and lithic sandstone (Back Creek Group; Exon, 1976) in the depocenter east of Silver Springs/Renlim. This may be a response to eustatic sea level rise caused by a combination of waning continental glaciation (Crowell and Frakes, 1971) and crustal sag resulting from lithostatic loading behind the Camboon arc.

The Upper Permian Kianga Formation (Blackwater Group), a coal-bearing sequence, was deposited and effectively blanketed the Silver Springs/Renlim area. These Permian deposits, which reach a maximum thickness of approximately 15,000 ft (5000 m; Dickens and Malone, 1973) in the depocenter, are considered the most likely source of Silver Springs/Renlim gas. Conformable sedimentation continued in the basin center into the Early Triassic. The thick siliciclastic Rewan Formation was deposited to the east of the field and reached a thickness of approximately 12,000 ft (4000 m) in the east of the basin.

From the Late Permian, sedimentation was predominantly fluvial with minor marine incursions. This prolonged terrestrial sedimentation may relate to postulated erosional crustal rebound (Lambeck and Stephenson, 1986) following extensive middle Paleozoic orogenesis along the eastern margin of the accreting plate boundary.

The Rewan Formation was subsequently onlapped on the western flank of the basin by Middle Triassic Showground Formation quartz arenites and sublitharenites (Figure 11), thickening from typically 30 ft (10 m) in the Silver Springs area to over 130 ft (40 m) to the east. This formation passes eastward into the Clematis Sandstone (Gray, 1972), which reaches a maximum thickness of 1500 ft (450 m; Dickens and Malone, 1973) in the north of the basin.

Provenance of the Showground Sandstone was largely granitic, from the west, and depositional environments range from proximal low-sinuosity fluvial systems to marginal marine (Butcher, 1984). The Showground is locally diachronous, indicating a genetic relationship with the overlying transgressive Middle Triassic Snake Creek Mudstone (Roberts, 1987), and may overlie either Permian Kianga Formation or nonsedimentary basement (Figure 6). A map of percentage clean sandstone (Figure 12) indicates the position of local depositional axes. The sand body has attributes of both "sheet" and "valley" sands (Pettijohn et al., 1973) though is probably a complex of depositional environments. Facies relationships defined by Butcher (1984) confirm this observation. This structural/stratigraphic setting is considered integral to establishing commercial reservoir fairways in the region.

The overlying Snake Creek Mudstone (Figure 13) is a dark gray brackish to marine mudstone that may locally contain interbedded fluvial sandstone near its onlap edge to basement (Roberts, 1987). Snake Creek thickness in the Silver Springs area is 50 to 65 ft (15-20 m), and this formation also thickens into the basin. It is laminated, lacking evidence of bioturbation, with a moderately developed fissility, is organically rich, contains achritarchs, and is locally pyritic, indicating a euxinic depositional environment in part. Roberts (1987) divided it informally into two "members" on the basis of log character, both of which are regionally pervasive and texturally similar. The Snake Creek is diachronous with overlying Middle Triassic Moolayember fluvial and deltaic sandstones and siltstone, which, although dominantly siliciclastic, contain significant volcanic material.

Deposition is thought to have ceased locally but farther east it was continuous to Late Triassic. Deposition to the east in the Bowen basin was terminated by uplift, in the east, probably associated with further orogenic activity.

The Bowen basin was subsequently overlain by over 3300 ft (1000 m) of Lower Jurassic to middle Cretaceous Surat basin sediments. To date, commercial hydrocarbon accumulations in this younger sequence have only been found in sandstones of the Lower Jurassic Evergreen Formation to the northwest of the area shown in Figure 1 (Cosgrove and Mogg, 1985).

TRAP

The Silver Springs/Renlim gas accumulation is trapped by stratigraphic onlap of Middle Triassic Showground Formation sandstone onto and around an eroded basement outlier. The trap was structurally enhanced by subsequent differential compaction. It is sealed above by a "shale" (Figure 13) of predominantly marine origin that is regional in distribution. Basement and/or Permian Kianga Formation act as a basal seal.

The accumulation is bounded downdip by a normally pressured freshwater (approximately 7000 ppm TDS) aquifer of considerable size and with high transmissibility in the area of clean sandstone (Figure 12). All wells in this area are "feeling" pressure drawdown in the reservoir via production from Silver Springs and surrounding fields. Aquifer recharge has been and is from the northwest over geological time, but it is uncertain at present whether it can affect pressure maintenance during production.

After production of between 9 bcf (1981) and 16 bcf (1984) (0.25 and 0.45 $\times$ 10^9 Sm^3) sales gas, appraisal drilling for the lower limits of the Silver Springs/Renlim field was carried out with the drilling of Silver Springs No. 7 and Renlim No. 2 in 1981 and 1984, respectively (Figure 6). These wells were thought to have established a gas-water contact (GWC) at about -5364 ft (-1635 m) subsea (Figure 5). Renlim No. 2 was found to be "watered out" in March 1987 despite the fact that production had been from Silver Springs wells only, and it was clear then that the two accumulations were connected in the gas phase and that any GWC seen after production startup was dynamic.

A pressure/depth plot of wells drilled before production commenced (Figure 14) shows considerable scatter in the data. A gas gradient extrapolated from the top of porosity in Glenmore No. 1, which was thought to have been drilled at or very near a hydrocarbon-water contact in 1969, provides an acceptable "best fit." Further support for this interpretation is provided by a plot of reservoir pressure versus time (Figure 15), which shows a near-linear relationship to at least 1987. A virgin pressure in October 1978 of 2790 psia at -5300 ft (1615.4 m) subsea is extrapolated and observed to be only 6 psi from a gas gradient intersecting top porosity in Glenmore No. 1 (Figure 14). Considering the accuracy of Amerada gauges used and the gas show at Glenmore No. 1 when drilled, an original gas-water contact is interpreted to have been present at or near the top of porosity in this well (Figures 5 and 6). At Silver Springs/Renlim, an original gas column

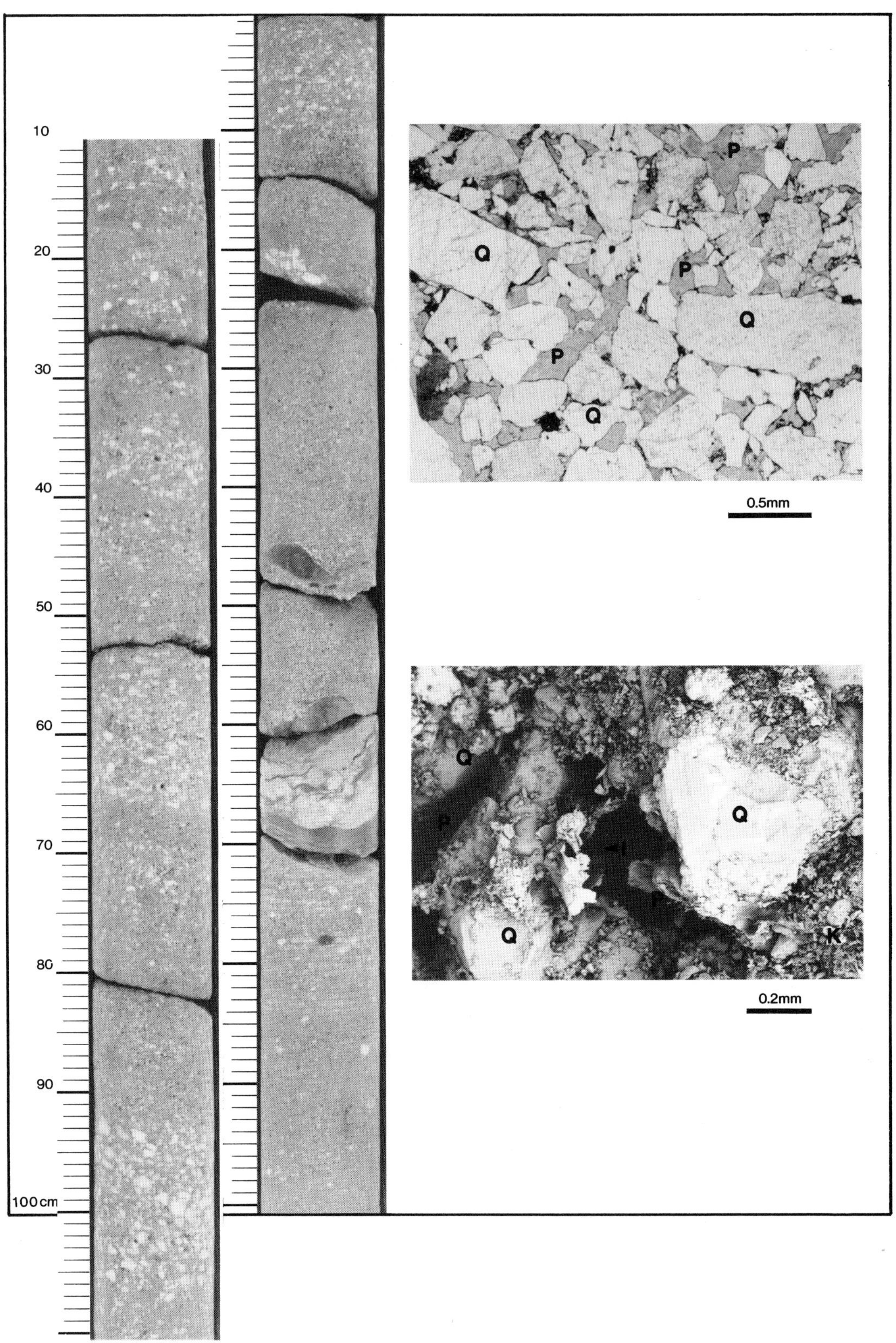

Figure 11. Typical fabric and petrology of the Middle Triassic Showground Formation shown by core photograph, thin section, and SEM. The reservoir is a medium- to coarse-grained, poorly sorted quartz arenite (sandstone and conglomerate). Primary porosity (P) and quartz (Q). Porosity has been reduced by quartz overgrowth and by formation of authigenic kaolinite (K) and illite (I). SEM micrograph sample is coated with significant drilling mud contamination.

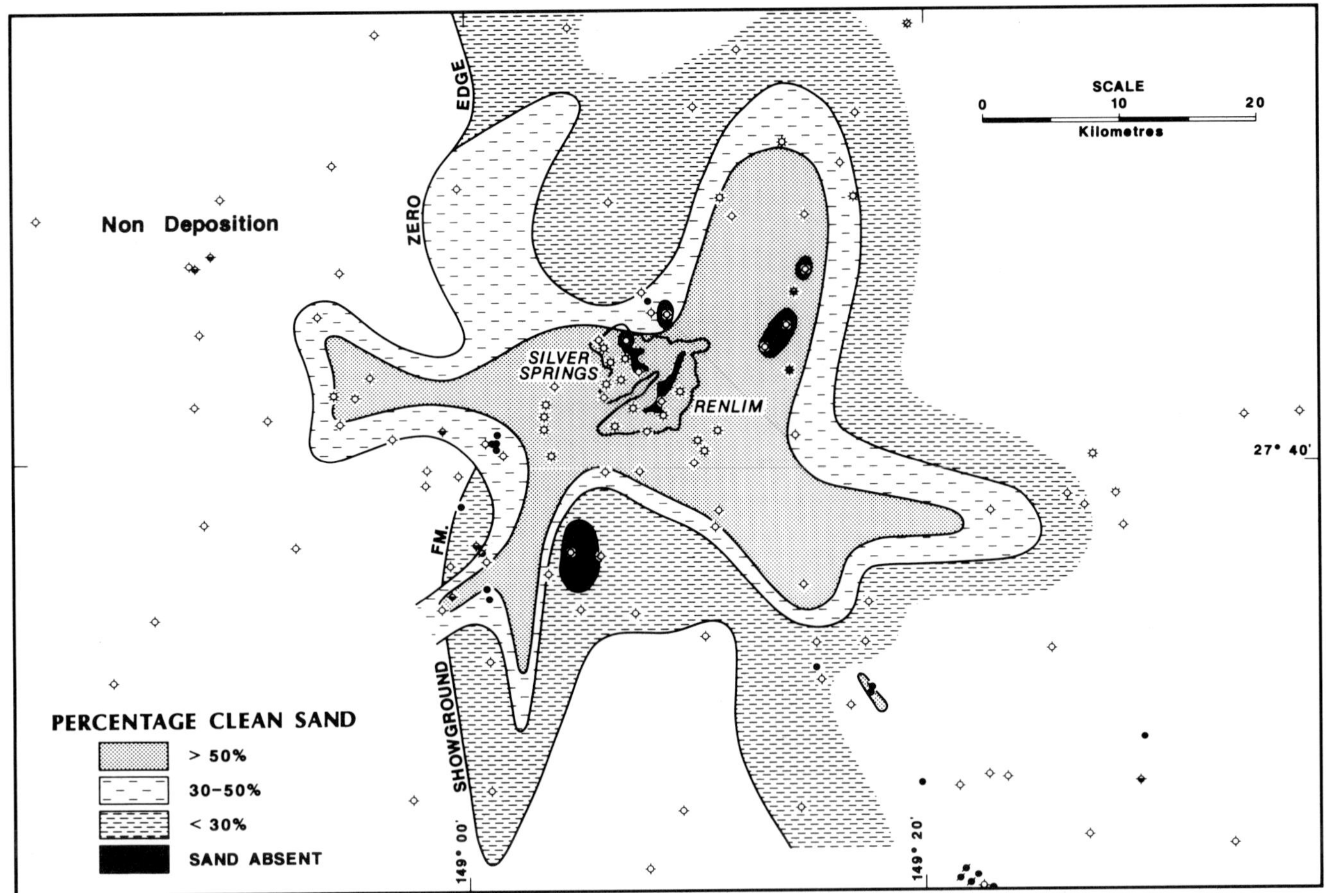

Figure 12. Percentage clean sandstone map, Showground Formation. Clean sand is determined with a gamma ray/V_{sh} cutoff. (Updated after Butcher, 1984.)

height of 131 ft (40 m) has been interpreted and from this a cap pressure on the top seal of 42 psi (290 kPa) can be calculated.

Reservoir

The Showground Formation in the Silver Springs/Renlim field is a medium- to coarse-grained, poorly sorted, quartzose sandstone and granule to pebble conglomerate (Figure 11) deposited as accretionary bed load in confined low-sinuosity drainage systems (Butcher, 1984). Clast size ranges up to rare cobbles, though are predominantly pebble size and generally clast supported. Depth to top of reservoir is typically 6200 ft (1900 m).

Showground gross thickness ranges up to 43 ft (13 m) in Renlim No. 2. On the basis of existing well control, average gross thickness is approximately 22.5 ft (6.9 m) and average net thickness is 21.5 ft (6.6 m). Non-net intervals are typically thin discontinuous mudstones and muddy sandstones.

Porosity is predominantly primary, with reduction in effective porosity caused by quartz overgrowth and the formation of authigenic kaolinite from dissolution of feldspar and secondary illite (Figure 11). The significance of observed secondary porosity is considered minor. Stylolites at grain boundaries indicate late further reduction through compaction with quartz dissolution. A plot of porosity versus permeability (Figure 16) reveals considerable scatter. However, when the data set is plotted with a grain size subdivision into "sandstone" and "conglomerate," expected lithology related populations are obvious. This plot highlights lithologic and petrophysical heterogeneity of the reservoir. A plot of core porosities at ambient versus overburden pressure (Figure 17) supports the log-derived field average porosity of 11.7%. It is apparent that porosity determined at overburden pressure is approximately 4 porosity units (P.U.) less than that determined at ambient conditions and is a significant consideration in core analysis. The existing data set is considered inadequate for a more sophisticated analysis.

Minimal special core analysis is available for Silver Springs/Renlim. Such studies, however, are part of current reservoir management and simulation efforts, which should lead to a better understanding of petrophysical parameters.

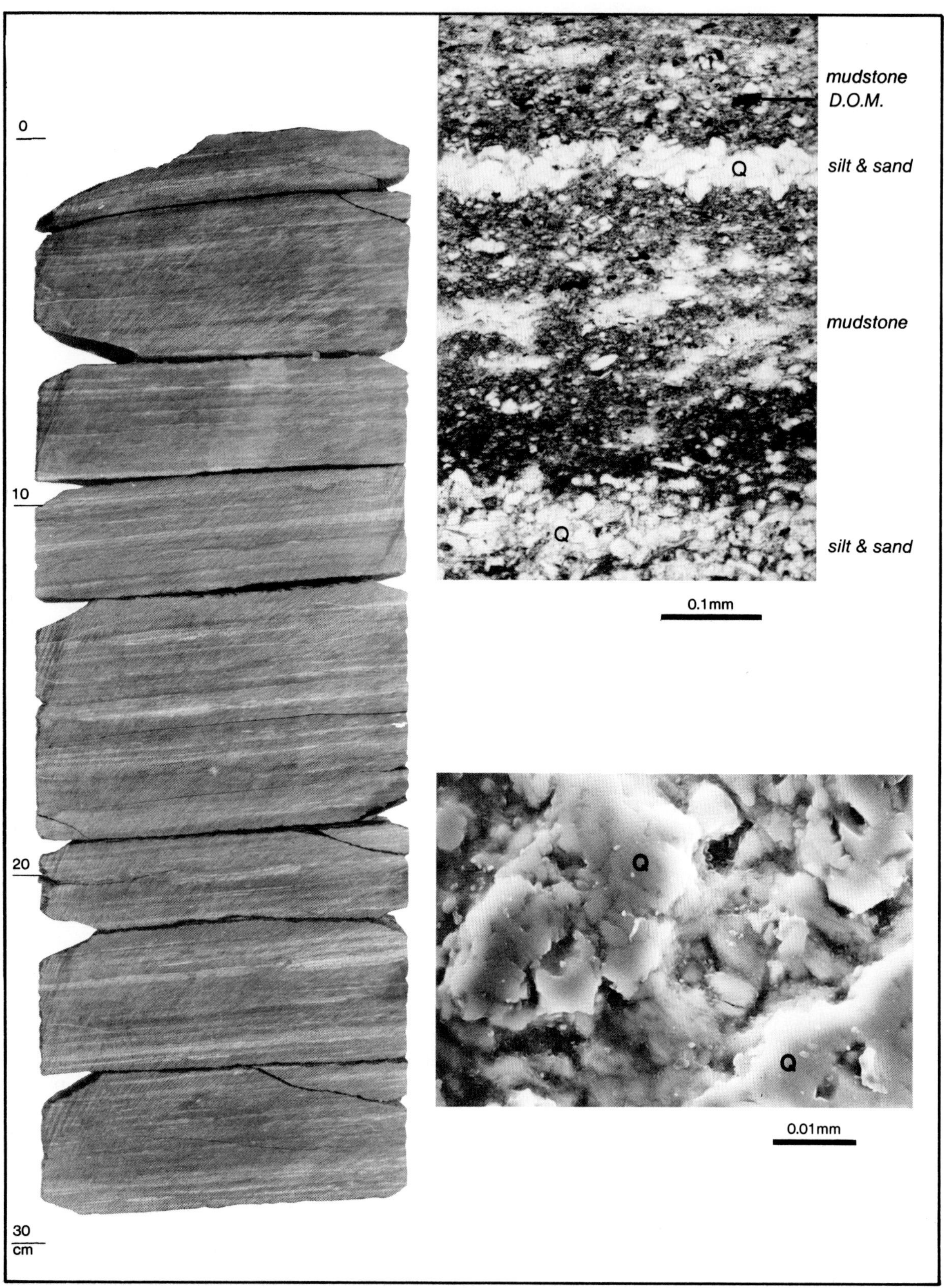

Figure 13. Typical fabric and petrology of the Middle Triassic Snake Creek Mudstone shown by core photograph, thin section, and SEM micrographs. Silt-sized quartz (Q) grains laminated with vermiculite-rich clays (C), secondary chlorite, and mica. D.O.M., dispersed organic matter. Pore size is typically less than 1μ.

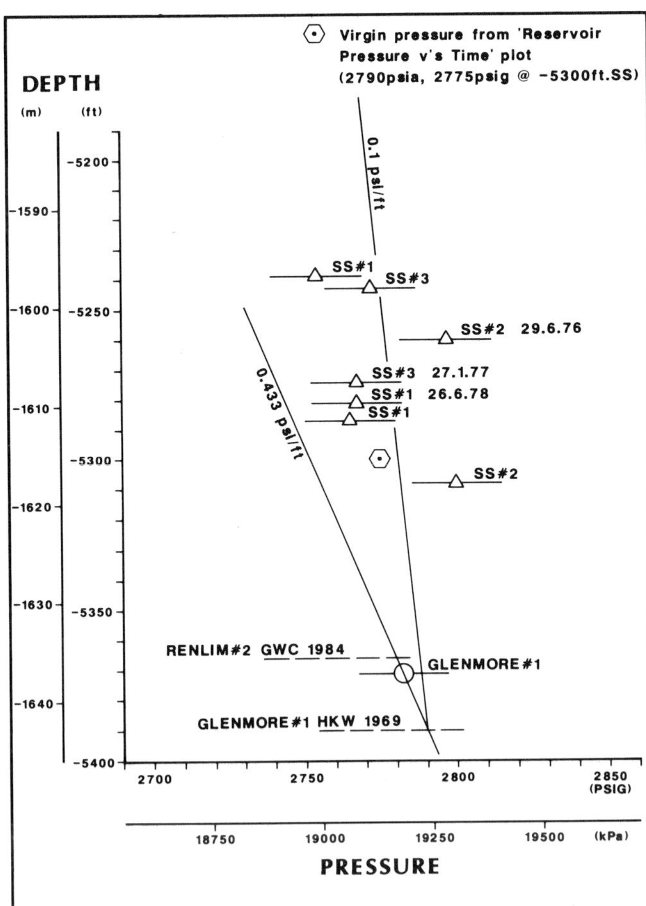

Figure 14. Pressure/depth plot of pre-production well data (Silver Springs gas phase pressures, indicated by triangles) with an interpreted gas-water contact at 5390 ft (1643 m) subsea (Glenmore No. 1 water phase; circle). An initial reservoir pressure at the fields' accepted datum of -5300 ft (-1615 m) in the gas phase extrapolated from post-production start-up data (hexagon; Figure 15) is consistent with this interpretation. Error bars for the gauges used are shown. HKW, highest known water; GWC, gas-water contact.

Permeability of the Showground is typically several darcys and averages in excess of 2 darcys. The average calculated water saturation for Silver Springs/Renlim is 31%.

Systematic development drilling of this reservoir has not been required because production to date is market constrained and significantly less than potential deliverability. Absolute open flow calculations for this reservoir are typically greater than 50 mmcf (1.4×10^6 Sm^3) of gas per day per well, while required deliverabilities from the whole field are typically 12 to 15 mmcf ($0.34–0.43 \times 10^6$ Sm^3) of gas per day. Drainage radii of approximately 1.2 mi (2 km) are considered conservative, and to date drilling has been largely motivated by reserve appraisal. Primary recovery is presently expected to be 73% of original gas-in-place. Estimated ultimate recovery may be as much as 105 bcf (2.98×10^9 Sm^3) of sales gas with associated condensate and LPG. No secondary or enhanced recovery is anticipated.

A gas composition from drill-stem test No. 1 in Silver Springs No. 1 is given in Figure 18.

Source

Gas in the Silver Springs/Renlim field has almost certainly been sourced from Permian marine siltstones and shales of the Back Creek Group and/ or Kianga Formation coal measures. Burial history modeling of an interpreted "kitchen" area, approximately 56 mi (90 km) to the northeast (Rigby and Kanstler, 1987) (Figure 19), indicates that these sediments entered the oil window in Late Triassic time, but probably did not generate significant gas until the Cretaceous. The rocks are presently mature for gas generation in the Taroom trough.

Vitrinite reflectances measured for Kianga Formation type III organic matter from wells surrounding Silver Springs/Renlim are typically 0.7% and generally support the key conclusions from the model. The potential for Gondwana type III organic matter to generate significant hydrocarbons is well documented (Smith and Cook, 1984; Smyth, 1983; Taylor et al., 1988).

Gas has been, (and probably still is), generated east of Silver Springs/Renlim at depths greater than 8000 ft (2500 m) in the Taroom trough (Figure 7). Hydrocarbons have therefore migrated distances of up to 60 mi (100 km) or more from the trough areas. Well control in the interpreted source area is poor. Philip and Gilbert (1986) support "intermediate distance" migration for western flank oils based on biomarker evidence.

Type II organic matter exists in the Back Creek Group, and it would have matured earlier; but this sequence is largely restricted to the trough areas, requiring migration over significant distances. The Snake Creek Mudstone may contain type II organic matter but is marginally mature locally; at best it may be responsible in part for sourcing small oil accumulations.

Total organic carbon (TOC) content (vol. %) in the interpreted source rocks is variable though typically 0.5% to 3.0% and locally extremely high in the Upper Permian coal measures.

Accepting that no local gas sources exist, migration from Permian source rocks would be controlled by the position of younger regional seals such as the Rewan Formation and Snake Creek Mudstone (Thomas et al., 1982; Butcher, 1984). More permeable sandstone conduits and favored directions of migration were described by Power and Devine (1970). The percentage of clean sandstone map (Figure 12) of the Silver Springs/Renlim region shows the field to be ideally located to capture westward (updip) migration of hydrocarbons from trough areas to the east.

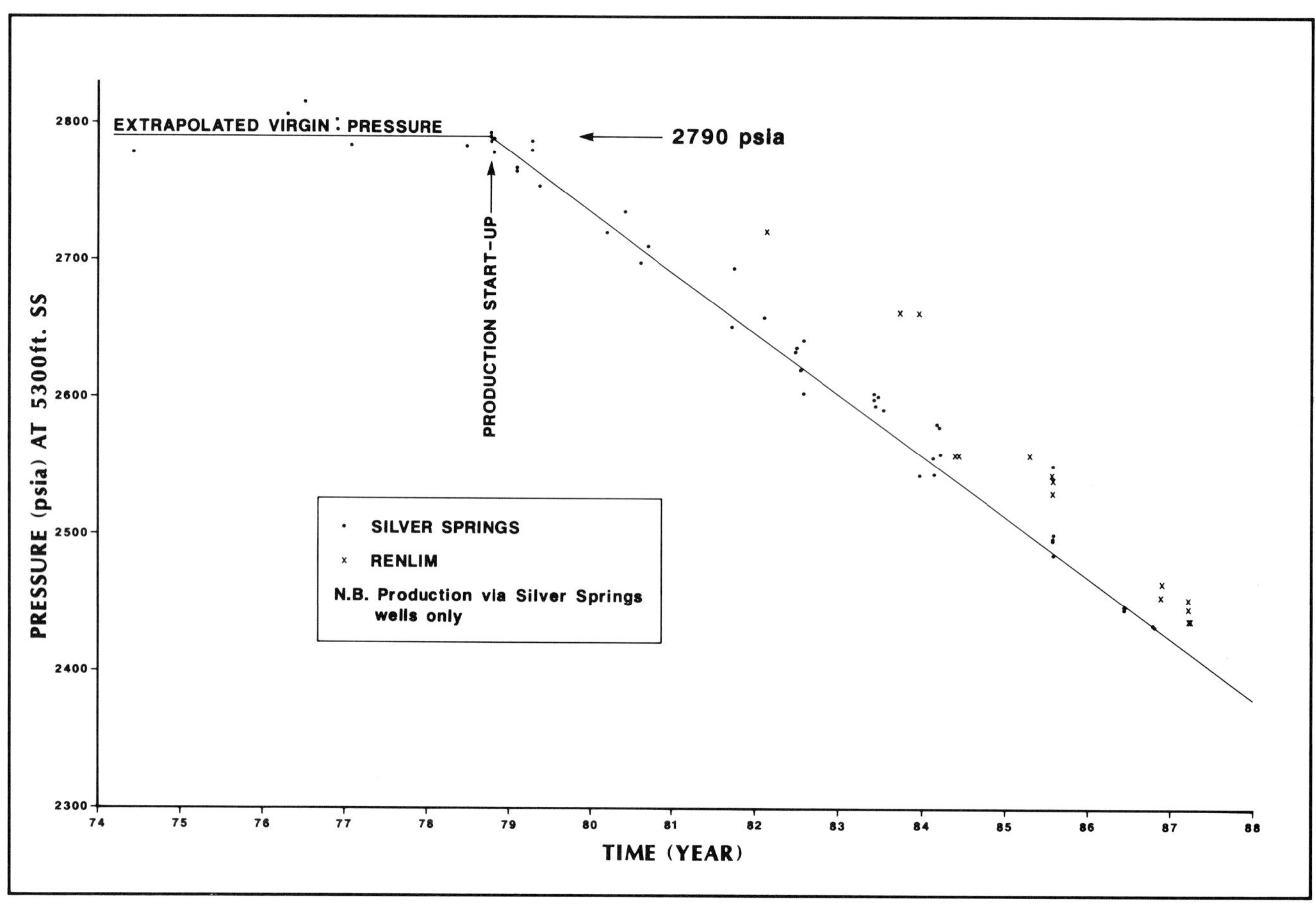

Figure 15. Reservoir pressure in gas phase (at the fields' accepted datum depth) versus time plot, showing an effectively linear decline after production start-up.

Any oil that could have been formed and migrated prior to gas entrapment has been either displaced from Silver Springs/Renlim and similarly located fields or thermally cracked in situ. The latter is discounted on the basis of observed oil in the area at greater depth combined with the observation that Silver Springs/Renlim appears to be filled to its spill point. This is consistent with observations that significant Jurassic oil and gas shows and accumulations exist west of the Snake Creek (and Showground) onlap edges and that significant oil accumulations exist in the Showground Formation in poorer quality reservoir where displacement has not occurred.

EXPLORATION AND DEVELOPMENT CONCEPTS

The Silver Springs/Renlim Showground Formation "play" or "fairway" is established by the presence of a large hydraulically connected fluvial sandstone body (Figure 12), the geometry of which is ideally suited to the capture of migrating hydrocarbons out of a deep trough region, and it is located in an area where counter regional structural dip is likely to occur (Figure 7). It is directly overlain by a thick regional seal. In Silver Springs/Renlim field the trap is in part stratigraphically controlled by reservoir onlap against a basement high, though in the nearby fields of Boxleigh and Sirrah (Figure 1) this is not a prerequisite for accumulation. This play is being pursued along the entire western flank of the Bowen basin, and it is considered likely that further undiscovered fairways of this kind remain along this trend. The ingredients that make this play successful are likely to occur wherever similar tectonostratigraphic histories exist.

Elliot and Brown (1988) have estimated that 378 bcf (10.7×10^9 Sm^3) of gas reserves have been identified on the Roma shelf and Wunger ridge with almost half of these produced to date from reservoirs within Permian, Triassic, and Jurassic formations. They also estimate that potential exists for a further 300 bcf (8.5×10^9 Sm^3) of gas to be found, "much of [which] will be found in new exploration plays." In light of recent (1988) success in the 3.6 mi (5. 8 km) Taylor No. 4 step-out well (Figure 1), drilled on a feature with 9.7 mi^2 (25 km^2) under interpreted

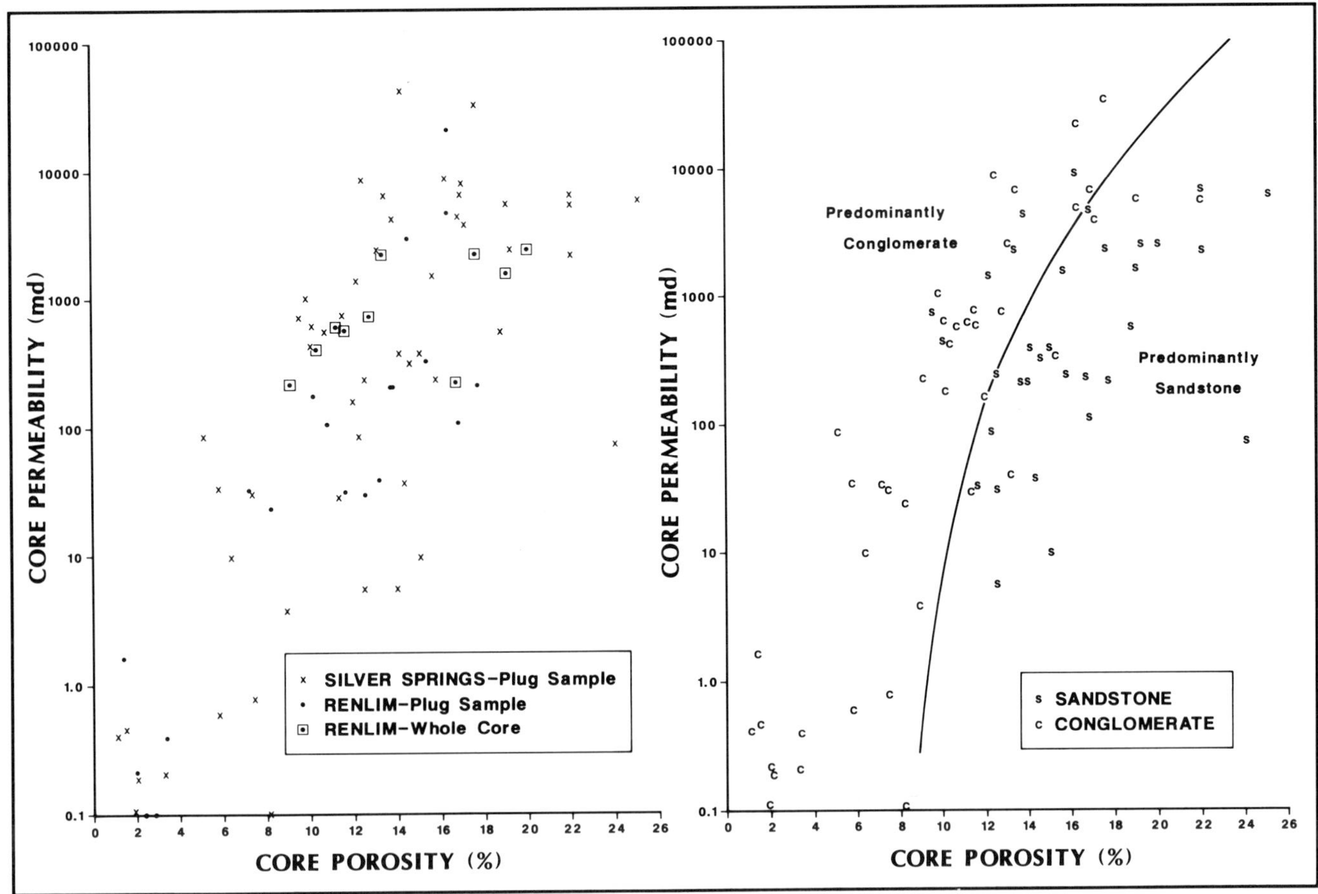

Figure 16. Porosity versus permeability plot for the Showground Formation. Considerable scatter in the raw data set (left) is explained by a grain size subdivision of the reservoir into "sandstone" and "conglomerate" (right).

closure (Roberts, 1987), the potential for this and nearby undrilled features to provide a significant proportion of this estimate from the Silver Springs/Renlim Showground Formation play-type in the Wunger ridge area alone must make this estimate of undiscovered resources conservative. Indeed, successful fracture stimulation of pre-Showground (Rewan and Kianga formations), low permeability, hydrocarbon-saturated sandstones would upgrade this estimate, the exploitation of which is price dependent.

The development of this resource has been restricted by inadequate market, with the result that appraisal drilling has been delayed after commercial feasibility was initially established. Consequently, the full potential of the Silver Springs/Renlim resources are yet to be realized.

ACKNOWLEDGMENTS

This paper is published with the permission of Bridge Oil Limited and Petroz N.L. (formerly Offshore Oil N.L.). The assistance of management and staff of Bridge Oil is appreciated.

Thanks also go to C.S.I.R.O. Australia, Divisions of Mineral Physics and Geomechanics, Sue Doyle, Macquarie University, for supplying SEM micrographs, and the Shell Company of Australia Limited for release of the burial history plot.

REFERENCES CITED

Austin, P. M., and G. Williams, 1978, Tectonic development of late PreCambrian to Mesozoic Australia through plate motions possibly influenced by the earth's rotation: Journal of the Geological Society of Australia, v. 25, n. 1, p. 1-21.

Bridge Oil N.L., 1974, Silver Springs No. 1, ATP 145P, Queensland, Application for Subsidy, Petroleum Search Subsidy Act, 1959-69. BMR, Canberra (Open File).

Butcher, P. M., 1984, The Showgrounds Formation, its setting and seal in ATP 145P, QLD: APEA Journal, v. 24, n. 1, p. 336-357.

Cosgrove, J. L., and W. G. Mogg, 1985, Recent exploration and hydrocarbon potential of the Roma Shelf, Queensland: APEA Journal, v. 25, n. 1, p. 216-234.

Crowell, J. C., and L. A. Frakes, 1971, Late Palaeozoic glaciation of Australia: Journal of the Geological Society of Australia, v. 17, n. 2, p. 115-155.

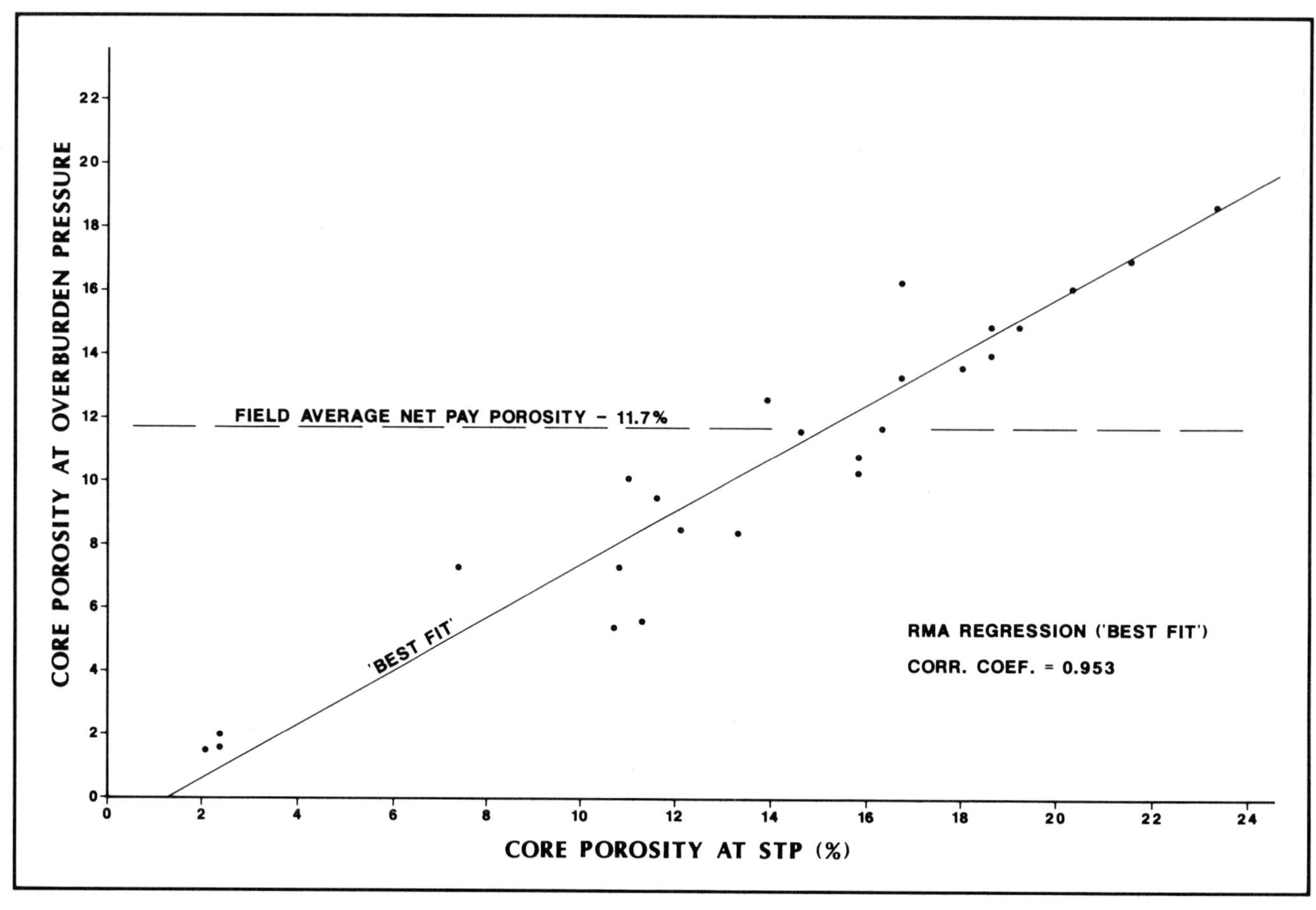

Figure 17. Core porosity at ambient conditions (standard temperature and pressure, STP; X-axis) versus at overburden pressure (Y-axis) reveals considerable porosity reduction for this analytical consideration and supports a log-derived net pay average field porosity of 11.7%. RMA, reduced major axis (best fit).

Day, R. W., C. G. Murray, and W. G. Whitaker, 1978, The eastern part of the Tasman orogenic zone: Tectonophysics, v. 48, p. 327-364.

Day, R. W., W. G. Whitaker, C. G. Murray, I. H. Wilson, and K. G. Grimes, 1975, Queensland geology, a companion volume to the 1:2,500,000 scale geological map. Geological Survey of Queensland Publication 383.

Dickens, J. M., and E. J. Malone, 1973, Geology of the Bowen basin, Queensland: BMR, Geol. Geophys. Bulletin 130. Australian Government Publication Service, Canberra.

Elliot, L. G., and R. S. Brown, 1988. The Surat and Bowen basin—a historical review, *in* Petroleum in Australia—the first century: APEA, p.15-22.

Exon, N.F., 1976, Geology of the Surat basin in Queensland: BMR Geol. Geophys. Bulletin 166, Australian Government Publication Service, Canberra.

Golin, V., and M. Smyth, 1986, Depositional environments and hydrocarbon potential of the Evergreen Formation, ATP 145P, Surat basin, Queensland: APEA Journal, v. 26, n. 1, p. 156-171.

Gray, A. R. G., 1972, Stratigraphic drilling in the Surat and Bowen basins, 1967-70: Geological Survey of Queensland Report 71.

Harrington, H. J., and R. J. Korsch, 1985a, Late Permian and Cainozoic tectonics of the New England orogen: Australian Journal of Earth Sciences, v. 32, p. 181-203.

Harrington, H. J., and R. J. Korsch, 1985b, Tectonic model for the Devonian to Middle Permian of the New England orogen: Australian Journal of Earth Sciences, v. 32, p. 163-179.

Component	Mole %
H_2S	0·0
CO_2	0·3
N_2	4·4
C_1	77·6
C_2	6·1
C_3	6·3
C_4	0·9
C_5	1·1
C_6	1·1
C_7	1·3
C_{8+}	0·9

Figure 18. Silver Springs No. 1 drill-stem test 1 composition.

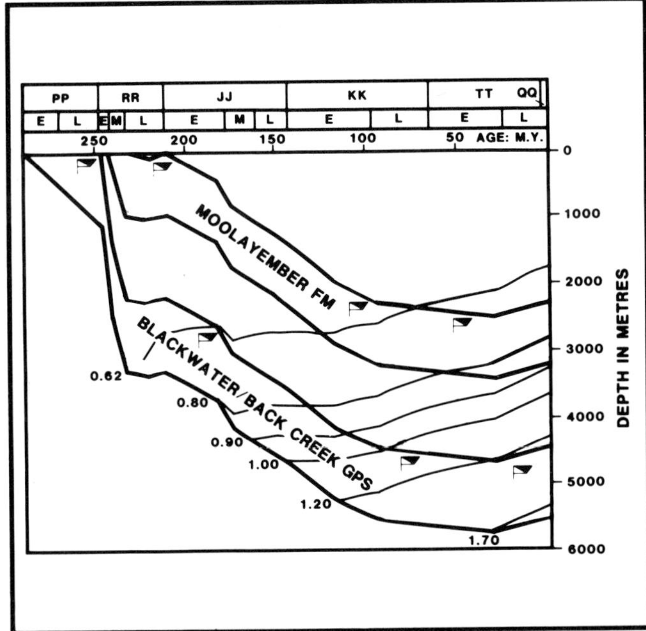

Figure 19. Burial history reconstruction and attained maturation levels (expressed in values of estimated vitrinite reflectance) from source rock ("kitchen") area in the Taroom trough (after Rigby and Kanstler, 1987). PP, Permian; RR, Triassic; JJ, Jurassic; KK, Cretaceous; TT, Tertiary; QQ, Quaternary; E, early; M, middle; and L, late. Flags are source rocks.

Hitchen, B., and J. Hays, 1971, Hydrodynamics and hydrocarbon occurrences, Surat basin, Queensland, Australia: Water Resources Research, v. 7, n. 3, p. 658-676.

Hobday, D. K., 1986, Gondwana coal basins of Australia and South Africa: tectonic setting, depositional systems and resources, *in* A. C. Scott, ed., Coal and coal-bearing strata: recent advances: Geological Society Special Publication 32, p. 219-233.

Jensen, A. R., 1975, Permo-Triassic stratigraphy and sedimentation in the Bowen basin, Queensland: BMR, Geol. Geophys. Bulletin 154, Australian Government Publication Service, Canberra.

Lambeck, K., and R. Stephenson, 1986, The Post-Paleozoic uplift history of south-eastern Australia: Australian Journal of Earth Sciences, v. 33, p. 253-270.

Mack, J. E., Jr., 1963, Reconnaissance geology of the Surat basin, Queensland and New South Wales: Australia Bur. Mineral Resources Geology and Geophysics Petroleum Search Subsidy Acts Pub. 40, 31 p.

Murray, C. G., 1985, Tectonic setting of the Bowen basin: Bowen Basin Coal Symposium, Abstract 17, Geological Society of Australia, p. 5-16.

Pettijohn, F. J., P. E. Potter, and R. Siever, 1973, Sand and sandstone: New York, Springer-Verlag, p. 214-227, p. 439-466.

Philip, R. P., and T. D. Gilbert, 1986, A geochemical investigation of oils and source rocks from the Surat basin: APEA Journal, v. 26, n. 1, p. 172-186.

Power, P. E., and S. B. Devine, 1970, Surat basin Australia—subsurface stratigraphy, history and petroleum: AAPG Bulletin, v. 54, n. 12, p. 2410-2437.

Rigby, S. M., and A. J. Kanstler, 1987, Murrilla Creek—an interesting stratigraphic play in the Surat basin: APEA Journal, v. 27, n. 1, p. 230-244.

Roberts, R., 1987, Exploration results and future activities in ATP 145P Surat basin, Queensland: Queensland 1987 Exploration and Development PESA (QLD) ODCAA-SPE Petroleum Symposium Volume, p. 82-100.

Smith, G. G., and A. C. Cook, 1984, Petroleum occurrence in the Gippsland basin and its relationship to rank and organic matter type: APEA Journal, v. 24, n. 1, p. 196-216.

Smyth, M., 1983, Nature of source material for hydrocarbons in Cooper basin, Australia: AAPG Bulletin, v. 67, n. 9, p. 1422-1428.

Taylor, G. H., Y. Liu, and M. Smyth, 1988, New light on the origin of Cooper basin oil: APEA Journal, v. 28, n. 1, p. 303-309.

Thomas, B. M., D. G. Osborne, and A. J. Wright, 1982, Hydrocarbon habitat of the Surat/Bowen basin: APEA Journal, v. 21, n. 1, p. 213-226.

Veevers, J. J., 1984, Phanerozoic earth history of Australia: Oxford, Clarendon Press, p. 235-269.

Wilkinson, R., 1988, Australian oil exploration history, *in* Petroleum in Australia—the first century: APEA, p. 15-22.

Wilkinson, R., 1983, A thirst for burning—the story of Australia's oil industry: Sydney, David Ell Press.

Appendix 1. Field Description

Field name .... *Silver Springs/Renlim*

Ultimate recoverable reserves .... *55 bcf (proven) to 105 bcf (proven plus probable plus possible) (1.559×10^9–2.975×10^9 Sm^3)*

Field location:

Country .... *Australia*

State .... *Queensland*

Basin/Province .... *Surat/Bowen*

Field discovery:

Year first pay discovered .... *Triassic Showground Formation 1974*

Discovery well name and general location:

First pay .... *Silver Springs No. 1, 115 km south-southeast of Roma*

Discovery well operator .... *Bridge Oil N.L. (now Limited)*

IP .... *8.5 MMCFD (240,000 Sm^3) (½-in. choke, DST 1), 26.1 MMCFD (740,000 Sm^3) absolute open flow*

All other zones with shows of oil and gas in the field:

Age	Formation	Type of Show
Permian (SS#5)	*Kianga*	*Fluorescence (minor)*
Mid-Triassic	*Wandoan*	*Gas (minor)*

Geologic concept leading to discovery and method or methods used to delineate prospect

Seismic and isopach mapping to predict depositional onlap of sandstone onto basement highs updip from dry (wet) wells and downdip from wells with no sandstone.

Structure:

Province/basin type .... *Bally 121; Klemme III A local/II B regional*

Tectonic history

Silver Springs/Renlim is located in a cratonic onlap area to a Permian-Triassic foreland (the Bowen) basin, the orogenic arc of which was intermittently migrating eastward through the Phanerozoic.

Regional structure

The field is on the western flank of the Bowen basin.

Local structure

The field has an amorphous form with a significant southwest-trending low dividing the "Renlim" and "Silver Springs" accumulations. The feature has a relatively flat crest with dips of 0.25–1° and steep flanks dipping up to 7° at the reservoir horizon.

Trap:

Trap type(s)

A combination of pinch-out trap with an anticlinal component due to post-depositional tilting. The paleo high as defined by sandstone absence is not coincident with the present structural crest.

Basin stratigraphy (major stratigraphic intervals from surface to deepest penetration in field):

Chronostratigraphy	Formation	Depth to Top in ft (m) subsea*
Lower Jurassic-Present	*Surat basin*	*Surface*
Middle Triassic	*Moolayember*	*4963 (1513)*
	Snake Creek	*5281 (1610)*
	Showground	*5330 (1625)*
Upper Permian	*Kianga*	*5383 (1637)*
Devonian	*Timbury Hills*	*5395 (1641)*

**From Renlim No. 2 well, as typical.*

Reservoir characteristics:

- **Number of reservoirs** *1*
- **Formations** .... *Showground Formation*
- **Ages** .... *Middle Triassic*
- **Depths to tops of reservoirs** .... *5269 ft (1606 m) SS (Silver Springs #2, highest known gas)*
- **Gross thickness (top to bottom of producing interval)** ..... *22 ft (7 m) (average of all wells with sand)*
- **Net thickness—total thickness of producing zones**
  - **Average** .... *22 ft (7 m)*
  - **Maximum** .... *29 ft (9 m)*
- **Lithology** .... *Medium- to coarse-grained poorly sorted quartzose sandstone and conglomerate*
- **Porosity type** .... *Primary porosity reduced by quartz overgrowth; minor secondary porosity*
- **Average porosity** *11.7%*
- **Average permeability** .... *2000+ md*

Seals:

- **Upper**
  - **Formation, fault, or other feature** .... *Snake Creek*
  - **Lithology** .... *Mudstone/shale*
- **Lateral**
  - **Formation, fault, or other feature** .... *Pinch-out (local only)*
  - **Lithology** .... *NA*

Source:

- **Formation and age** .... *Back Creek and/or Blackwater groups; Permian*
- **Lithology** .... *Siltstone, shale, coal (sandstone and tuff)*
- **Average total organic carbon (TOC)** *0.5-3.0%*
- **Maximum TOC** .... *Coal (~100)%*
- **Kerogen type (I, II, or III)** .... *III (II secondary by volume)*
- **Vitrinite reflectance (maturation)** .... $R_o = 0.6\text{–}1.2+\%$
- **Time of hydrocarbon expulsion** .... *Jurassic to Present*
- **Present depth to top of source** .... *8200 ft (2500 m)*
- **Thickness** .... *Unknown*
- **Potential yield** .... *Unknown*

Appendix 2. Production Data

Field name .... *Silver Springs/Renlim field*

Field size:

- **Proved acres** .... *4863 ac (1968 ha)*
- **Number of wells all years** *9*
- **Current number of wells** *8*
- **Well spacing** .... *540 ac (219 ha)*
- **Ultimate recoverable** .... *55 bcf (proven) to 105 bcf (proven plus probable plus possible)* $(1.559 \times 10^9\text{–}2.975 \times 10^9\ Sm^3)$
- **Cumulative production** .... *30 bcf* $(0.85 \times 10^9\ Sm^3)$
- **Annual production** .... *3.1 bcf (1987)* $(0.088 \times 10^9\ Sm^3)$
- **Present decline rate** .... *NA (Swing field not produced at capacity; production subject to market demand)*
  - **Initial decline rate** .... *NA*
  - **Overall decline rate** .... *NA*

SILVER SPRINGS

Annual water production .... *Nil*
In place, total reserves .... *144 bcf (proven plus probable plus possible) (4.081 × 10^9 Sm^3)*
In place, per acre foot .... *645,247 scf/ac-ft (proven plus probable plus possible) (18,285 Sm^3)*
Primary recovery .... *105 bcf (2.975 × 10^9 Sm^3)*
Secondary recovery .... *None expected*
Cumulative water production .... *Nil*

Drilling and casing practices:

Amount of surface casing set .... *671 ft (205 m)*
Casing program .... *9⅝-in. surface casing; 5½-in. production casing, 2⅜-in. tubing*
Drilling mud .... *Water (and minor control) to 1200 m; freshwater gel to TD*
Bit program .... *12¼-in. surface hole; 8½-in to TD*
High pressure zones .... *Nil*

Completion practices:

Interval(s) perforated .... *Variable; generally perforate all net sandstone*
Well treatment .... *Nil*

Formation evaluation:

Logging suites .... *DLL/MSFL/GR/CAL, CDL/CNL/GR, BHC Sonic (older wells DIL, MEL)*
Testing practices .... *Conventional off-bottom DST (recent wells no open-hole tests); production test after completion*
Mud logging techniques .... *Standard mudlogging—hot wire, gas chromatograph, rig sensors, UV box, sample description—operation from spud*

Oil characteristics:

Type .... *Condensate*
API gravity .... *55.8°*
Base .... *60°F (15.5°C)*
Initial GOR *19,700*
Sulfur, wt% .... *Nil*

Field characteristics:

Average elevation .... *902 ft (275 m) a.s.l.*
Initial pressure .... *2790 psia (19,237 kPa)*
Present pressure .... *2420 psia (16,685 kPa)*
Pressure gradient .... *0.08 psi/ft (1.81 kPa/m)*
Temperature .... *180°F (82.2°C)*
Geothermal gradient .... *0.165°F/ft (0.030°C/m)*
Drive .... *Water*
Oil column thickness .... *131 ft (40 m)*
Oil-water contact .... *–5390 ft (–1643 m) SS*
Connate water *31%*
Water salinity, TDS .... *6675 ppm (SS#7)*
Resistivity of water .... *0.87 ohm-m^2 @ 25°C (SS#7)*
Bulk volume water (%) .... *363 (porosity × Sw)*

Transportation method and market for oil and gas:

Gas pipeline: condensate; truck and pipeline

Cahoj Field—U.S.A.
Anadarko Basin, Kansas

W. LYNN WATNEY
Kansas Geological Survey
Lawrence, Kansas

BRYAN STEPHENS
Texaco, Inc.
New Orleans, Louisiana

JAN-CHUNG WONG
Kansas Geological Survey
Lawrence, Kansas

FIELD CLASSIFICATION

BASIN: Anadarko
BASIN TYPE: Foredeep
RESERVOIR ROCK TYPE: Limestone
RESERVOIR AGE: Pennsylvanian
PETROLEUM TYPE: Oil and Gas
TRAP TYPE: Anticline
RESERVOIR ENVIRONMENT OF DEPOSITION: Shallow Marine Carbonate Shelf
TRAP DESCRIPTION: Local closure on broader structure; multiple pays with variable porosity development

LOCATION

Cahoj field is located in Rawlins County in extreme northwestern Kansas, U.S.A., along the northern perimeter of the Hugoton embayment (Figure 1). The greater Cahoj field complex considered in this paper includes Cahoj discovered in 1959, Cahoj West (1962), Cahoj East (1966), Cahoj Northeast (1970), Cahoj Northwest (1972), and Cahoj South (1977). The combined area is 2980 ac (4.63 mi^2; 11.98 km^2) and over 9 million barrels of oil have been produced from this complex of fields (Beene, 1989). Sixty wells are presently producing. The field is located on the edge of one of the most productive and actively drilled areas in the United States, the Central Kansas uplift (Watney and Paul, 1983) (Figures 2 and 3).

The Missourian (Upper Pennsylvanian) Lansing and Kansas City groups, in which the main reservoir interval occurs, include a dozen carbonate units that are potential pay zones (Figures 4 and 5). The groups consisting of a similar succession of alternating carbonate rocks and shales are commonly referred to in the subsurface as the Lansing-Kansas City. Eleven of the carbonate units produce oil and gas in various places in Cahoj field. These units are also major reservoirs on the Central Kansas uplift and areas immediately west of the uplift on the northern edge of the Anadarko basin referred to as the Hugoton embayment (Figures 2 and 3). The Lansing-Kansas City comprises the main producing interval in the giant Hall-Gurney field on the Central Kansas uplift (140 million bbl cumulative production) (Figure 3). Fields producing from the Lansing-Kansas City west of the uplift, such as Cahoj field, are more typically the size of Cahoj and smaller. The Lansing-Kansas City has been found to be productive in approximately a third of all discoveries in Kansas in the decade of the 1980s. Both new fields and new pools and bypassed oil in existing fields will be future targets.

HISTORY

Pre-Discovery

Drilling in western Kansas had been well underway since the late 1920s and 1930s prior to the discovery of Cahoj in June 1959. The Fairport field in Russell County, discovered in 1923, was the first discovery on the Central Kansas uplift. Fairport field was found through the use of surface geologic mapping. Other fields discovered in the 1920s and 1930s were prospected by means of surface structure mapping, shallow core drilling, and random drilling. In the 1940s and 1950s new fields were drilled along the Central Kansas uplift and on the Cambridge arch to the north. Test drilling began in earnest west of

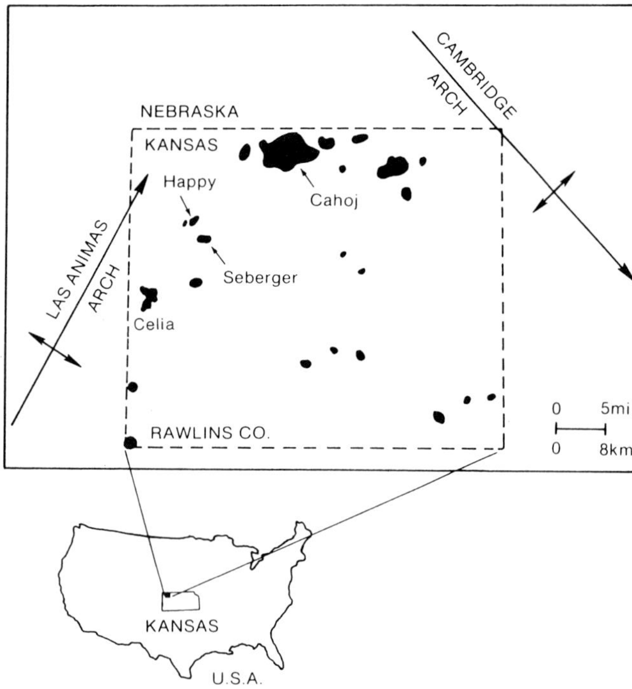

Figure 1. Location of Cahoj field on index map of Rawlins County in the state of Kansas.

the Central Kansas uplift in the early 1940s. Core drilling was used to identify structural anomalies on shallow, persistent marker beds in areas of poor surface exposure or those areas covered by the generally flat-lying Ogallala Formation. The marker beds utilized during core drilling included the Lower Permian Stone Corral Formation and base of the Upper Cretaceous Fort Hays chalk (Figure 4). Seismic reflection profiling became increasingly important in the exploration activities west of the Central Kansas uplift as this technology became available and these markers became increasingly deeper in this more western area of drilling. Seismic reflection profiling also provided a means to identify the more subtle structures that were not detected by core drilling. Specific stratigraphic targets in northwestern Kansas included the Upper Cambrian Reagan Sandstone, Cambrian-Ordovician Arbuckle Group, basal Pennsylvanian conglomerate, and the Lansing and Kansas City groups. Heightened exploration activity and numerous successes, such as Cahoj field, in the mid- to late-1950s coincided with the time (1956) in which peak production and proven reserves were established in Kansas.

Discovery

Skelly Oil Company discovered Cahoj field with the Skelly Cahoj #1, drilled in SE SE NE Sec. 17, T1S, R34W of Rawlins County in June 1959, during expanded exploration activity in western Kansas. Two small discoveries had been made prior to Cahoj in Rawlins County. Brumm field was discovered in 1956. It has produced around 48,000 bbl. Sappa Creek field, discovered in 1958, produced nearly 95,000 bbl before being abandoned in 1981 (Department of Energy, 1982).

Stanolind Oil Company had drilled a series of stratigraphic tests in western Kansas in the early 1940s. One such test by Stanolind was drilled to a depth of 4817 ft (1469 m) in 1943 on the very edge of what now is the southern productive limit of Cahoj field in Sec. 20, T1S, R34W. The well was abandoned after penetrating the top of the Precambrian. The test, which had shows of oil in cuttings, was later used to firm up a seismic prospect developed by Skelly. Skelly Oil Company shot a series of seismic reflection profiles to define the structure leading to the successful wildcat. Wilhelm field was later discovered just east of Cahoj in August of 1959 on an extension to the same structure. Ackman field in Red Willow County, Nebraska, 33 mi (53 km) northeast of Cahoj was discovered by an analogous approach by Skelly five months after the discovery of Cahoj.

Pay zones in the Cahoj field discovery well, #1 Cahoj, included the A-Zone (4052–4054 ft; 1235–1235.7 m) and the B-Zone (4065–4072 ft; 1239–1241 m) of the Upper Pennsylvanian Lansing–Kansas City. The reported initial production for the #1 Cahoj of 528 BOPD was a projected value based on limited testing.

Post-Discovery

The development drilling in Cahoj field proceeded rapidly, and, within two years of its discovery, 46 wells had been drilled. Annual oil production from the field quickly peaked in 1961 at 627,273 barrels of oil (37 BOPD per well). Production rate rapidly declined from its peak until 1969, when infill and extension drilling occurred in conjunction with a waterflood program. Three waterflooding units were started in the 1970s: Cahoj-SE unit in 1970, Cahoj unit in 1973, and Sramek unit in 1974. The initial impact of the new drilling in Cahoj was a significant increase in oil production. Production levels were maintained as a result of waterflooding operations. In fact, 1985 production was equal to that in 1968 before infill drilling and waterflooding began. In 1986, the production from the Cahoj unit waterflood (88,698 bbl), now operated by Texaco, accounted for 92% of the production from the field.

Discovery and development of new fields in the area have been occurring periodically since Cahoj was discovered. Increased drilling activity during the late 1970s and early 1980s resulted in several field discoveries from the Lansing–Kansas City groups in Hitchcock County, Nebraska, immediately to the north (Rogers et al., 1983; Prather, 1985; DuBois, 1985). Celia field, 18 mi (29 km) southwest of Cahoj field, was initially a two-well Lansing–Kansas City field, but deeper Desmoinesian Cherokee and

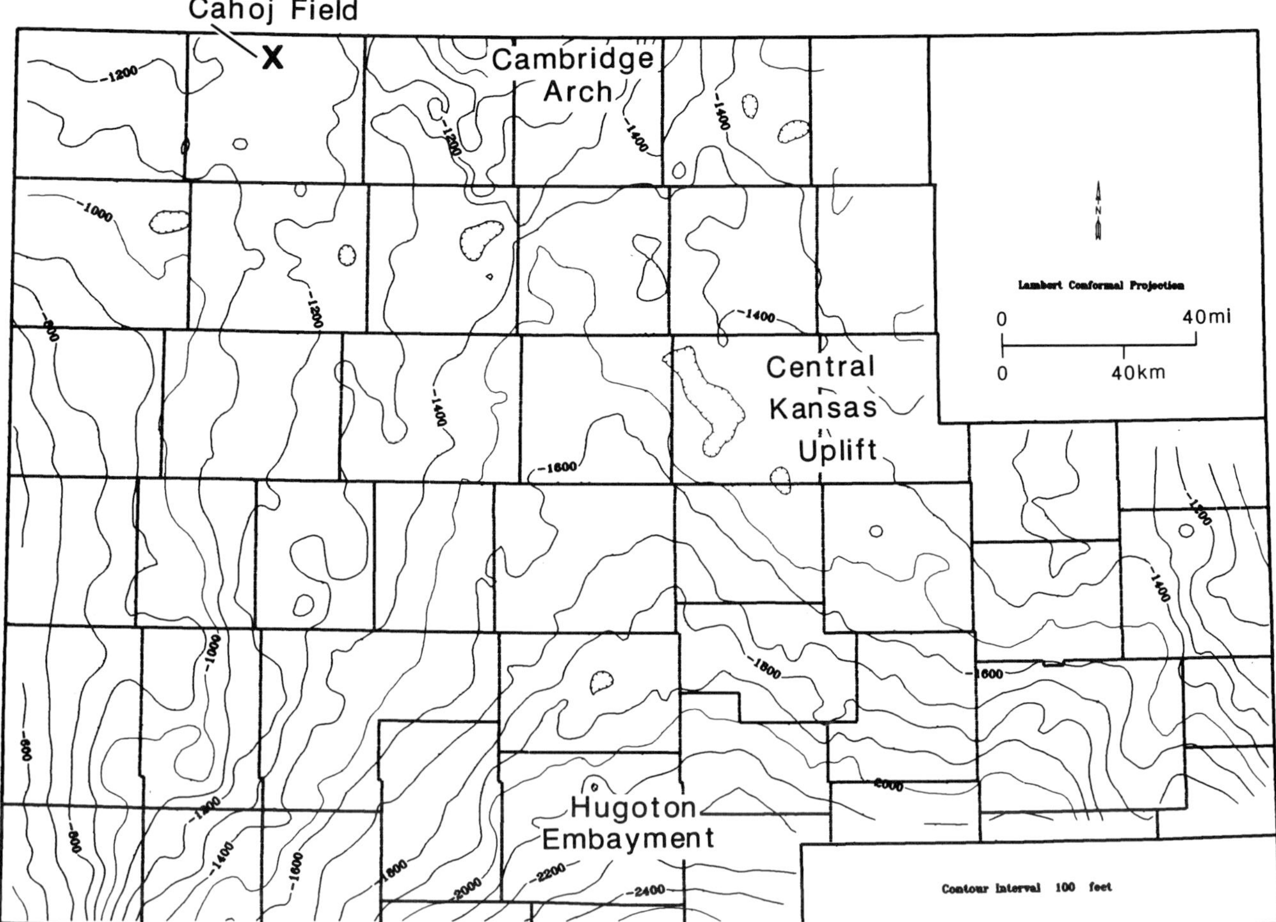

Figure 2. Structural configuration on top of the K-zone of the Lansing-Kansas City strata in the western half of Kansas (except along south border). Northern mapped area is border between Nebraska and Kansas. Western border is Colorado-Kansas line. Features identified include county boundaries (irregular rectangular grid), Cahoj field in the extreme northwestern portion of the map, and the broad structural features including Cambridge arch, Central Kansas uplift, and the Hugoton embayment.

Marmaton group carbonate reservoirs were found in it and the adjoining Celia South field (Figure 1). Both fields now cover over 1000 ac (405 ha) and contain 28 wells. Over 1.2 million barrels of oil have been produced from these two fields, primarily the result of this deeper drilling effort (~400 ft [122 m] below the base of the Lansing-Kansas City groups).

DISCOVERY METHOD

The early exploration strategy was to map and drill subsurface structural highs. This approach has stood the test of time very successfully, as it is still the primary approach taken today. Seismic reflection profiling remains an important means of identifying structures. Individual beds in the Lansing-Kansas City are too thin to be resolved individually, using conventional seismic reflection methods, and, consequently, seismic stratigraphic studies have not been utilized to date (Newell et el., 1989). Analysis of sample cuttings and wireline logs for shows of hydrocarbons and porosity in wells near a prospect well remain important criteria in lowering the risk of finding hydrocarbons in these prospects. In excess of 8% porosity is commonly used as the cutoff below which the rock is considered impermeable (Figure 6). Characterizing the nature and distribution of porosity development in these thin carbonate reservoir units through use of cuttings, cores, and wireline logs is becoming increasingly important to identifying drilling prospects in an effort to improve the success ratio.

Exploration today in the Lansing-Kansas City strata, primarily by the independent oil and gas industry, has focused on the search for more subtle occurrences of porous traps because of the advanced stage of development in many areas of the Mid-Continent. This search takes advantage of the tremendous database of subsurface information, with the availability of cutting samples and wireline logs

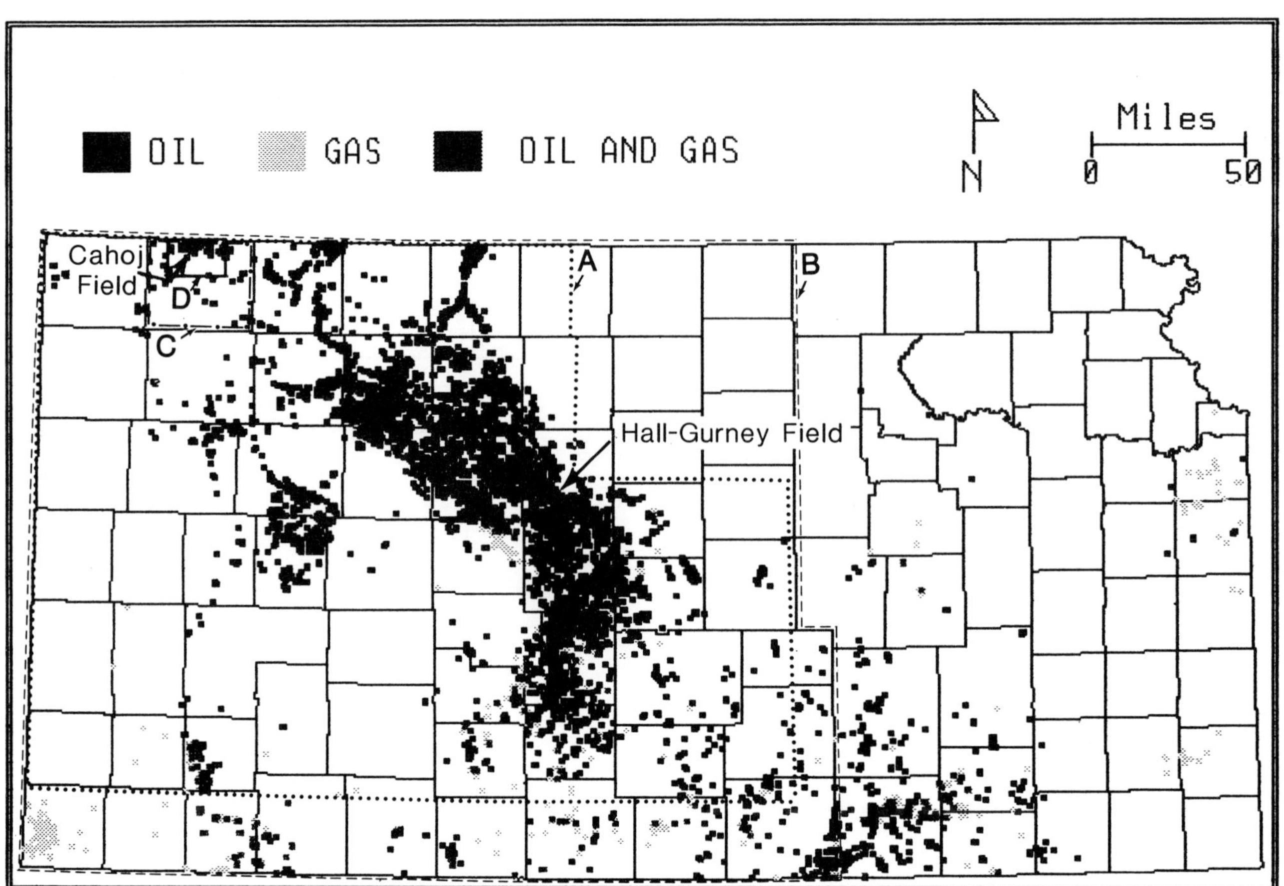

Figure 3. State of Kansas showing county boundaries. Solid squares locate wells producing from the Lansing-Kansas City groups. Cahoj field is located in Rawlins County, northwest Kansas. The Hall-Gurney, a Lansing-Kansas City giant field, is located on the Central Kansas uplift. Outlined on this map are areas covered by other maps: A—Figure 2, B—Figures 7 and 9, C—Figure 10, D—Figures 11, 13, 14, 15, 22, 23, and 24.

from many holes. The necessity to explore for more subtle indications of oil and gas accumulations has brought a need to assess the importance of the stratigraphy, lithology, and diagenesis on reservoir development in addition to structure. Accordingly, refined concepts of sedimentology and diagenesis aid in characterizing trends of porosity and qualifying interpretations of subsurface structure.

Much has yet to be learned about the extent and timing of geologic events that formed reservoir quality rock before their prediction can be made more accurate and precise. New field discoveries have tended to be progressively smaller through time in Kansas (Harbaugh and Ducastaing, 1981). These Pennsylvanian discoveries have been found on increasingly smaller structures and even structural lows in areas adjacent to the major uplifts.

Remote sensing has been reported to be useful in finding oil and gas fields in western Kansas, but details are not available. Imagery that is of sufficiently high resolution may reveal surface structures which may have undergone episodic movement which in turn may have influenced oil migration, entrapment, or porosity development or preservation.

STRUCTURE

Tectonic History

Cahoj field is situated at the northern end of the Hugoton embayment, an extension of the Anadarko basin (Figure 2). Cahoj field is located on the west flank of the Cambridge arch and northwest of the Central Kansas uplift (Figures 1 and 2). It is immediately east of the northeast-southwest structural axis of the Las Animas arch (Figure 1). The area is now part of the stable North American craton. The Central Kansas uplift has been a positive basement element since the early Paleozoic, exhibiting several periods of more pronounced tectonic activity (Merriam, 1963). Major tectonic deformation that resulted in most of the present-day structural configuration of this area occurred during the Late Mississippian and Early Pennsylvanian, including major movement on the Central Kansas uplift. Tectonic activity was pronounced throughout the Mid-Continent during early Pennsylvanian including uplift of the Amarillo-Wichita-Arbuckle Mountains

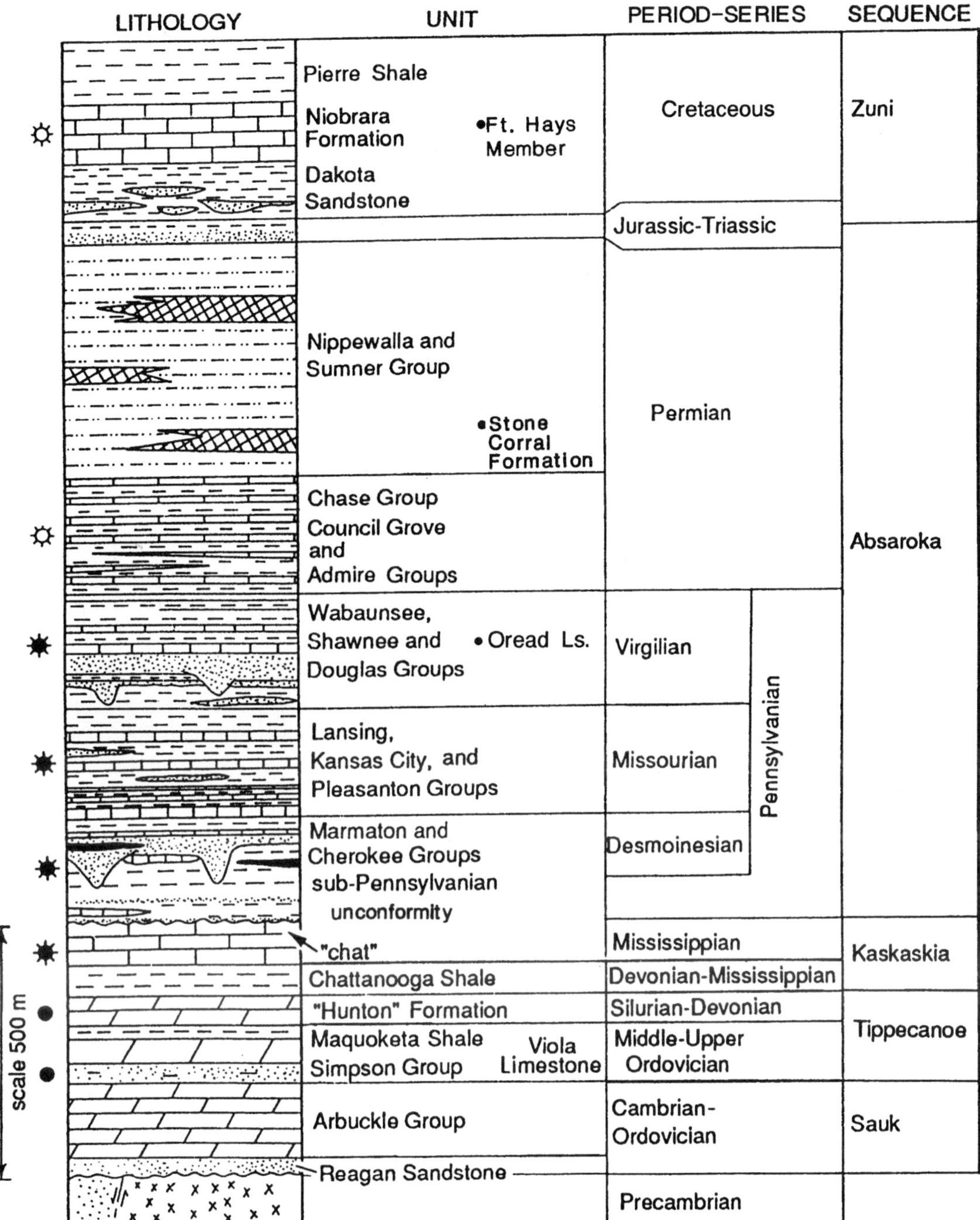

Figure 4. Stratigraphic column for Kansas extending from the Cretaceous downward to the Precambrian basement. Commonly identified stratigraphic units are labeled and their general lithologies identified using standard symbols. Symbols for oil-bearing strata (solid) and gas-bearing (radiating spokes around circle) are found along the left margin. The Lansing and Kansas City groups are referred to in the subsurface as Lansing-Kansas City. (Modified from Newell et al., 1987.)

and Ancestral Rocky Mountains, and the downwarping of the Anadarko basin (Ham and Wilson, 1967; Kluth, 1986). Deep reflection seismic profiling indicates that significant subsidence of the Anadarko basin began in Atokan time, which coincides with the northward thrusting of the Wichita Mountains for a distance of some 5 to 6 mi (8–9 km) (Brewer et al., 1983). Thrusting is attributed to plate collision along the Ouachita Mountains situated in Arkansas, approximately 300 mi (483 km) to the east of the center of the Anadarko basin. During Missourian and Virgilian time, the Anadarko basin reached its peak development; subsidence was estimated to have exceeded 0.66 ft (0.2 m)/ka (Dickinson and Yarborough, 1979). Maximum subsidence in the western Anadarko basin situated in western Oklahoma and

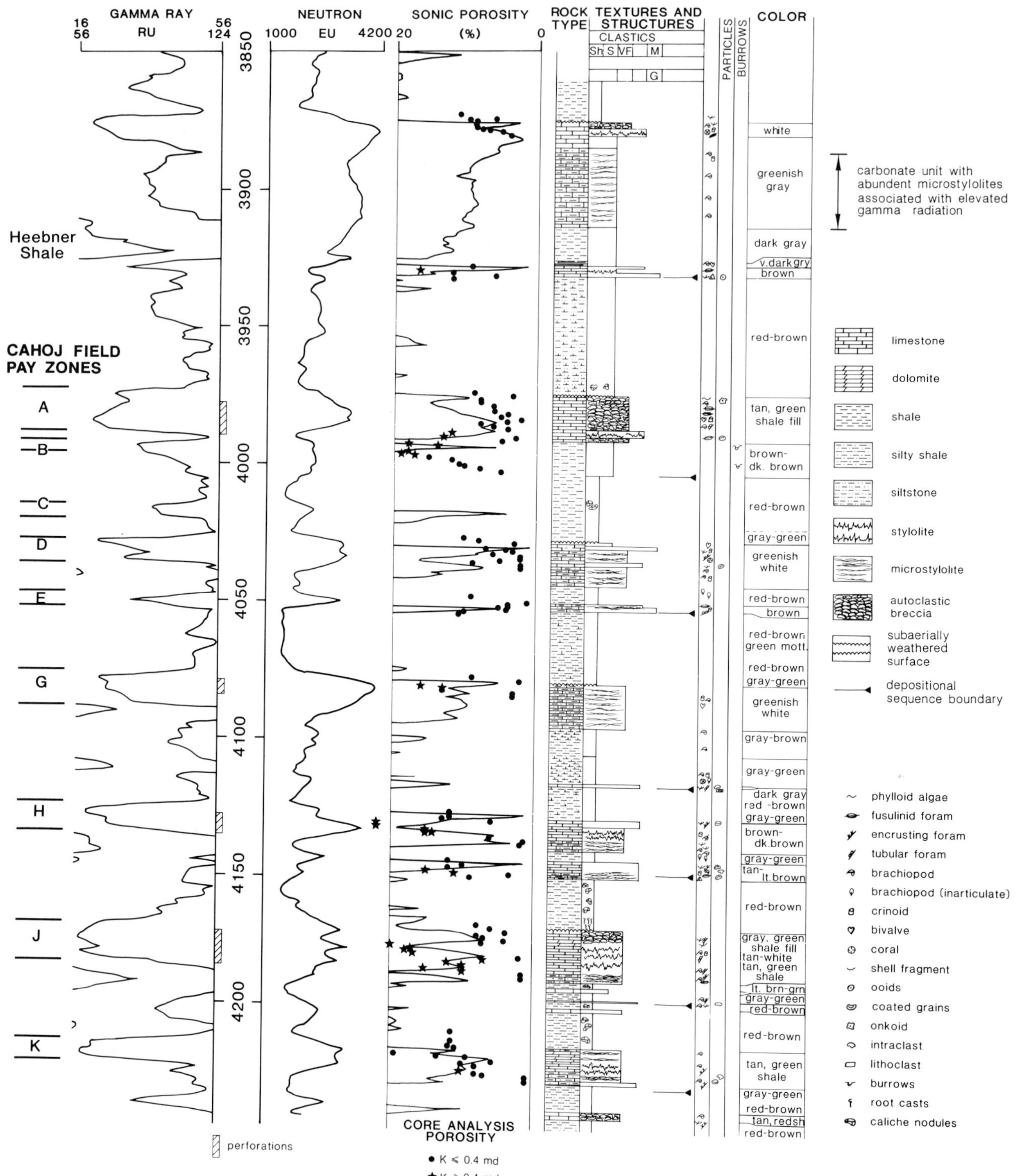

Figure 5. Graphic description of continuous core of Lansing-Kansas City strata, core analysis, and wireline log response from Skelly #1 Bartosovsky borehole located in Cahoj field in (Sec. 9, T1S, R34W). Zonal designation coincides with that used by Watney (1980). Depth provided in feet. Depositional sequences defined by hiatal bounding surfaces.

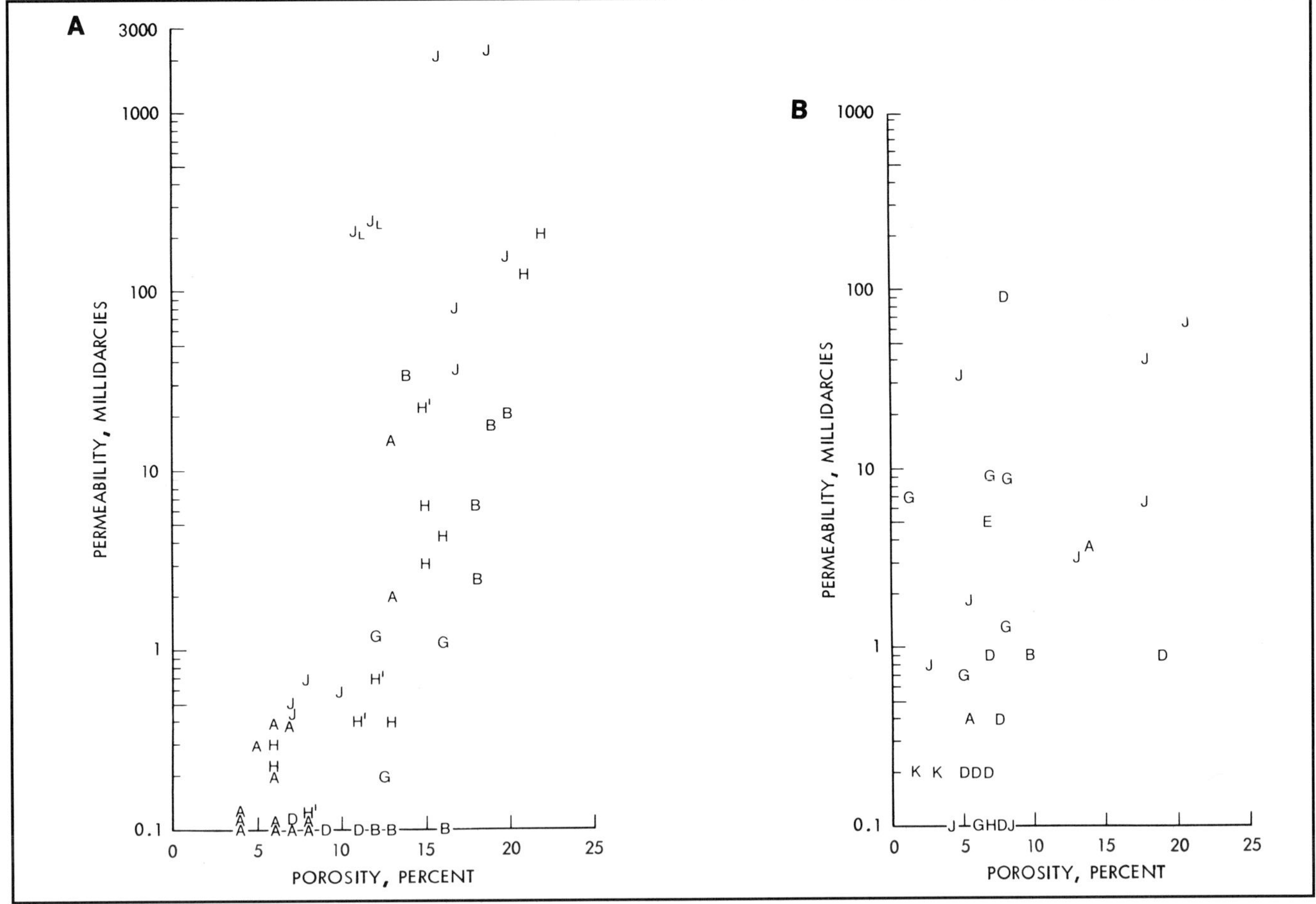

Figure 6. Semi-log plot of core-derived porosity and permeability for (A) Skelly #1 Bartosovsky (Sec. 9, T1S, R34W) in Cahoj field. This plot is compared with core analysis results of (B) Conoco Adell Unit 406 (Sec. 2, T6S, R27W) from Adell field in Sheridan County. An 8% porosity cut-off is indicated, above which pores are considered permeable. The data points are labeled according to zonal designation identified in Figure 5. (From Watney, 1980.)

the Texas Panhandle (immediately south of the Kansas shelf) is characterized by sediment-starved conditions (Galloway et al., 1977; Kumar and Slatt, 1984; Rascoe and Adler, 1983; Watney et al., 1989). Estimated relief across the shelf margin in the Anadarko basin during the Late Pennsylvanian was estimated at 1100 ft (335 m) (Kumar and Slatt, 1984).

A craton-wide regional unconformity developed at the end of the Mississippian in response to a lowering of relative sea level (base of the Absaroka sequence of Sloss, 1963). Mississippian-age strata were completely removed from the crest of the Central Kansas uplift as the result of tectonic uplift and ensuing erosion owing to concomitant sea-level fall at the end of Mississippian time. Mississippian strata were also removed prior to Pennsylvanian sedimentation from the present-day structural crest of Cahoj, indicating a similar timing of uplift that occurred considerably before Lansing–Kansas City deposition. In areas surrounding the Central Kansas uplift, Mississippian strata were also extensively weathered and partly eroded. Pennsylvanian sedimentation eventually covered this surface as sea level rose. By Upper Pennsylvanian time all of the Central Kansas uplift was covered by marine sediments, reflecting widespread marine conditions in western Kansas over a broad shelf. Subsidence accompanying the downwarping of the Anadarko basin and long-term rise of sea level acted in concert during the Pennsylvanian and Early Permian and led to thick accumulations of strata in the Mid-Continent.

Regional Structure

Depositional strike during the Pennsylvanian and Early Permian evolved from a more north-northwest trend in the Early Pennsylvanian off of the western flank of the Central Kansas uplift (Figures 7 and 8) to a more northwest-southeast trend in the same region during the time the Lansing–Kansas City was being deposited (Figure 9). Sediment accommodation during the Early and Middle Pennsylvanian reflects prominent tectonic uplift along the Central Kansas uplift and accompanying subsidence along a trough extending from southwestern Kansas to northeastern Colorado (Figures 7 and 8). Subsidence abruptly increased in extreme southwestern Kansas. Subsidence during the Upper Pennsylvanian (lower Kansas

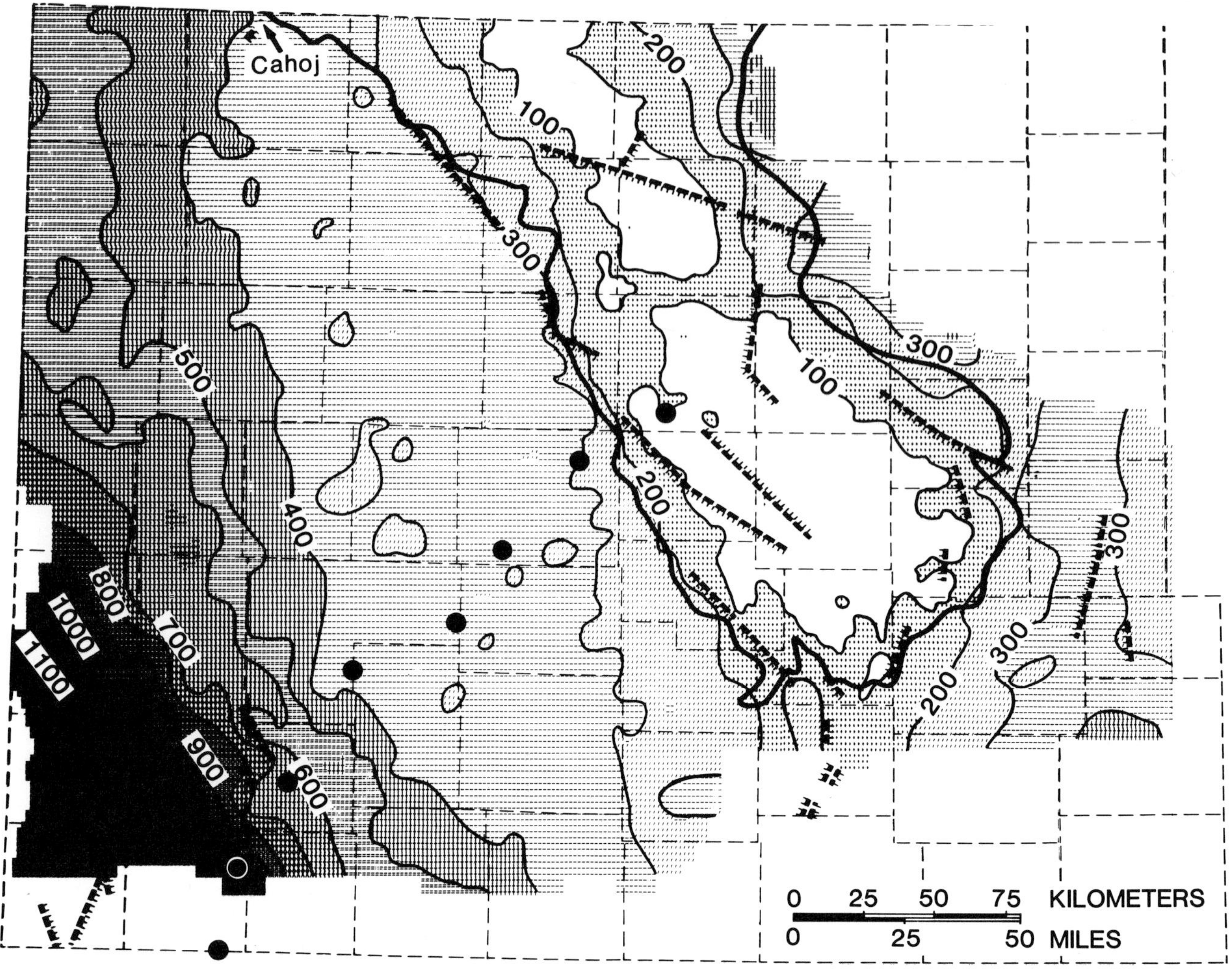

Figure 7. Map of western Kansas with contours depicting thickness of interval from base of Kansas City to base of Pennsylvanian strata. Contour interval equals 100 ft (30 m). Thinning is shown on the east side over the Central Kansas uplift. Abrupt thickening into southwestern Kansas is shown by area with diagonal lines. Thickness exceeds 1400 ft (427 m) in extreme southwestern Kansas. Series of eight dots transecting map identify wells utilized in stratigraphic cross section of Figure 8.

City Group) broadened across southern Kansas, increasing toward the Anadarko basin (Figure 9). The Central Kansas uplift apparently subsided less rapidly than the surrounding areas, suggesting that positive relief could have been maintained. Independent evidence for positive relief is inferred from regional depositional and diagenetic facies distribution in strata within the Lansing-Kansas City (Watney, 1980, 1984, 1985a, 1985b).

Cahoj field is located on a broad structural saddle that crosses northern Rawlins County as seen on the structural contour map of the K-zone, a major carbonate petroleum reservoir in western Kansas (Figure 2). The regional dip direction in the vicinity of Cahoj field varies significantly. The magnitude of the dip is low, averaging only 10 ft/mi (1.9 m/km) (Figure 2). Depositional slope in the vicinity of Cahoj field as inferred from the isopach mapping in the Lansing-Kansas City is west to southwest, closely associated with the positive Central Kansas uplift (Figure 9).

Wells producing from the Lansing-Kansas City carbonate reservoirs are concentrated along the Central Kansas uplift, rimming the east side of the Hugoton embayment (Figure 3). Production also extends down into a regional low located at the northern apex of the Hugoton embayment in Graham County (Figures 2 and 3). Additional less concentrated sites of production are spread out on the west flank of the Hugoton embayment, including its northernmost reaches in Cahoj field (Figure 2). This regional pattern of oil production may be controlled by pathways of oil migration leading from the Hugoton embayment.

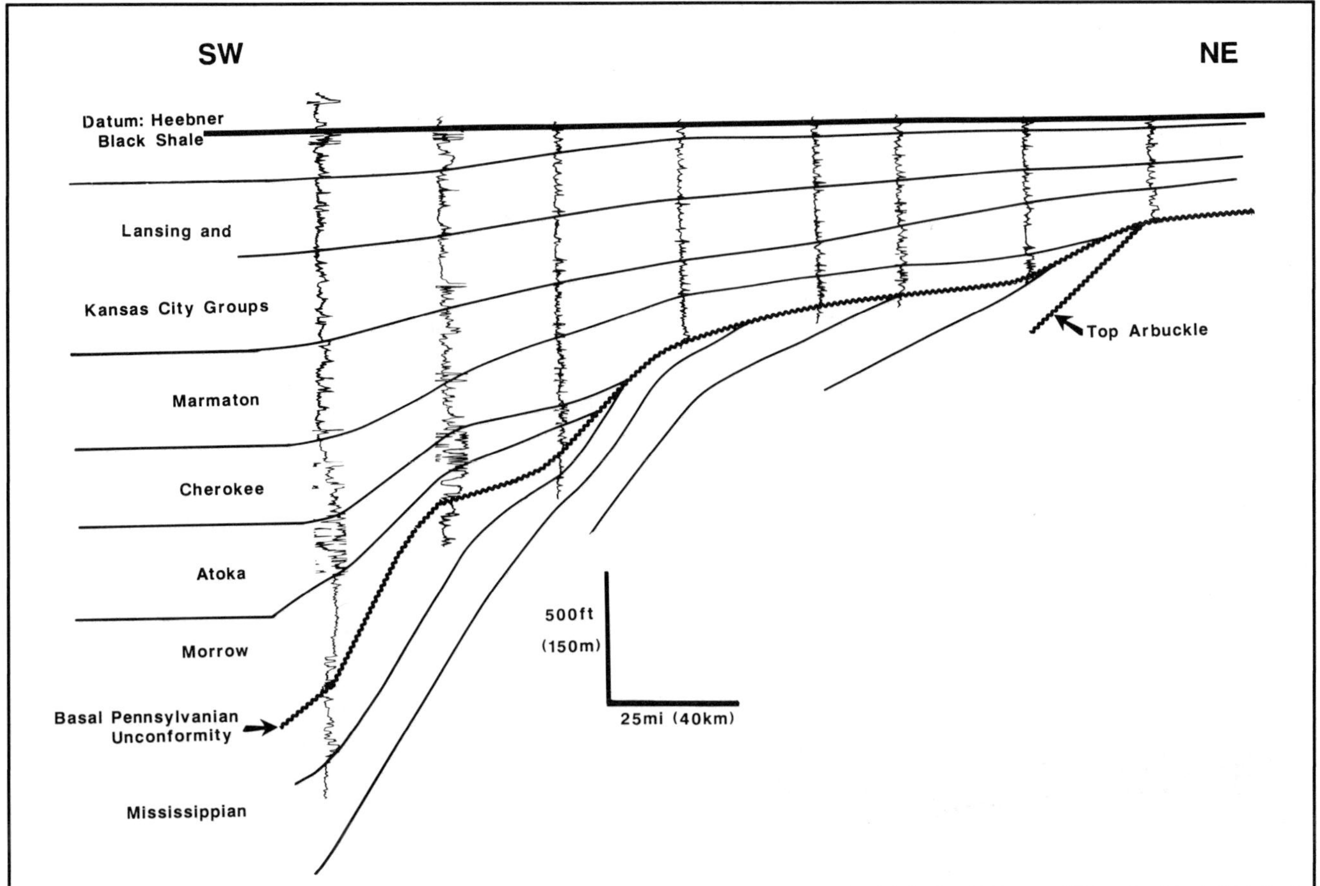

Figure 8. Northeast-southwest stratigraphic gamma-ray log cross section of Lower to Upper Pennsylvanian age strata in western Kansas. Index map is Figure 7. Datum is Virgilian-age Heebner Shale. Pennsylvanian strata thicken stepwise westward off of the Central Kansas uplift. Boundary of Central Kansas uplift is between first and second wells on right side of section. Vertical exaggeration about 260×.

Local Structure

A series of maps follows that covers six north-central townships in Rawlins County, Kansas (Figure 10). A structure map on the top of the Lansing Group (Figure 11) reveals a portion of the northeast-southwest-trending anticline on which Cahoj field resides. About 40 ft (12.2 m) of closure are indicated at this stratigraphic level in Cahoj field. Figure 12 is a structural wireline log cross section that extends over the crest of the structure and down its southeastern flank. The northern flank of the structure reaches into Nebraska. The distribution of oil is mainly controlled by structure, limited locally by porosity and permeability variation in the 11 carbonate reservoirs. An interval isopach between the base of the Kansas City Group and the base of Pennsylvanian reveals a localized thin area that is slightly offset from the area of present-day, maximum structural development (Figure 13). Thinning suggests an earlier history of relative structural uplift. The data points on which this map is based are few, offering less constraint to mapping than those that apply to the shallower intervals. Thus differences are to be expected between the maps.

Locally, Mississippian strata were removed by pre-Pennsylvanian erosion over portions of the area of maximum thinning. The present-day structural high supports the concept of earlier structural activity. Isopachs of the Lansing-Kansas City (Figure 14) and of the interval from the Lower Permian Stone Corral Formation to the top of the Lansing-Kansas City (Figure 15) indicate further thinning over the structure during those periods of deposition. Together, the isopach information suggests multiple periods of episodic movement of this structure throughout Permian-Pennsylvanian time. This attribute made some structures well suited to being detected in shallow beds through core drilling and seismic reflection.

STRATIGRAPHY

The Kansas rock chart shows various zones that produce oil and gas in Kansas (Figure 4). Rocks that serve as known sources for oil and gas in Kansas

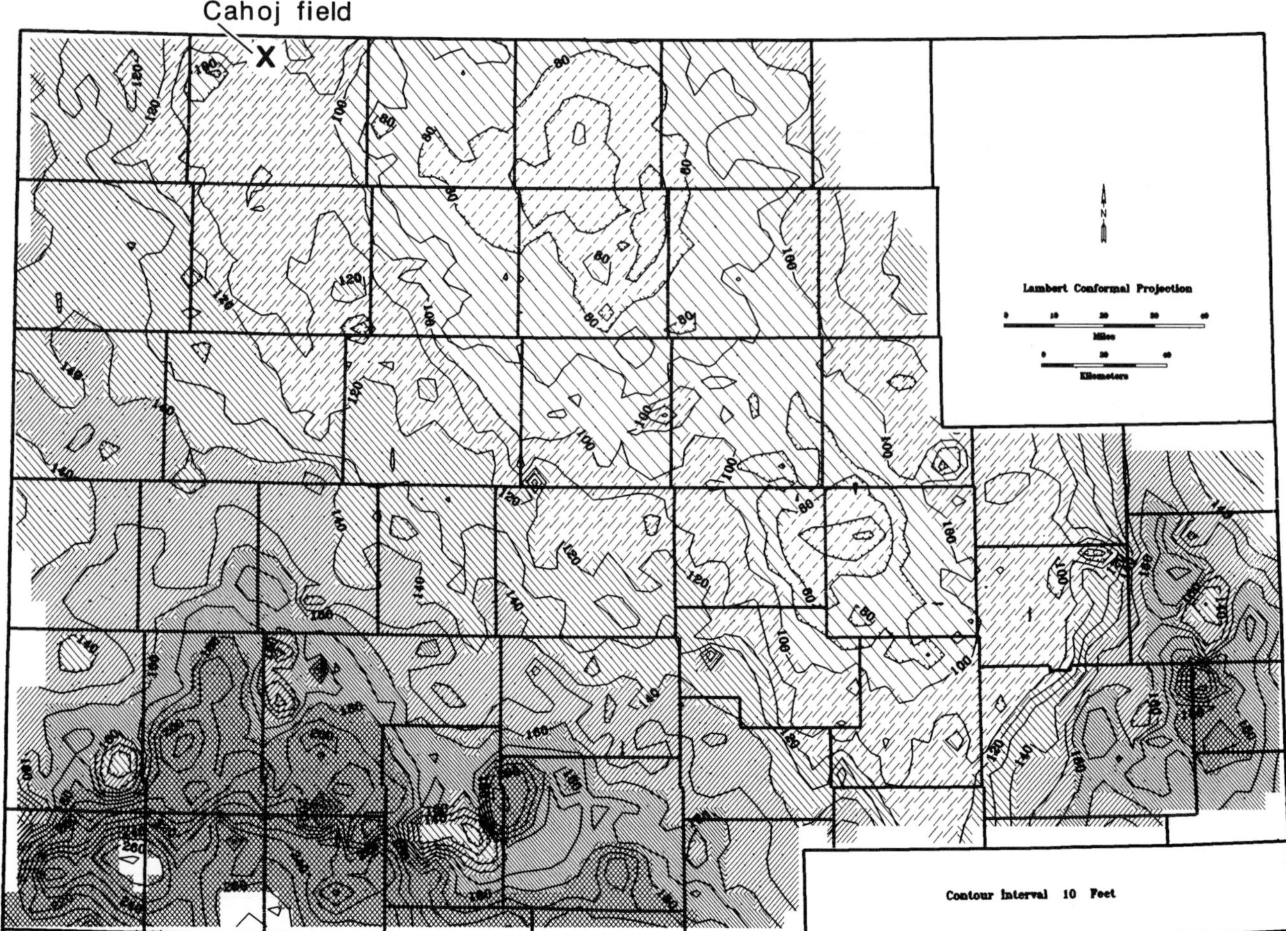

Figure 9. Isopachous map of Kansas City strata in western Kansas. Large elongate area of thinning on east side of map is the location of the Central Kansas uplift. Thickening of interval occurs westward off of the Central Kansas uplift into the Hugoton embayment.

include dark Pennsylvanian shales, Devonian-Mississippian shales (Chattanooga and Woodford), and the shales in the Simpson Group (Newell et al., 1985; Jenden et al., 1988).

The type log for Cahoj field, shown in Figure 5, is from the Skelly #1 Bartosovsky. Skelly had continuously cored the pay section of the Lansing-Kansas City, revealing a succession of alternating limestones and shales as seen in the graphic core description alongside the wireline log. Moreover, bounding surfaces, representing regional subaerial exposure, separate individual depositional sequences (Watney et al., 1988, 1989; Watney and French, 1988). Similar cyclical strata occupy much of the Middle and Upper Pennsylvanian and Lower Permian in the western Mid-Continent area.

Biostratigraphic studies indicate that individual Missourian Pennsylvanian depositional sequences are interbasinal, correlative between Mid-Continent and central Texas (eastern shelf of Midland basin) (Boardman and Heckel, 1989). Each depositional sequence on the shelf of western Kansas is characterized by four common genetic units: (1) a flooding or transgressive unit, (2) a condensed section, (3) a highstand, commonly a shallowing-upward unit, and (4) an upper shale containing characteristics of paleosols. A distinctive hiatal surface separates each sequence. The flooding and highstand units are typically carbonate rocks in this shelf setting and comprise more than 60% of the thickness of a sequence in this area. The condensed section is a marine, often a black, radioactive shale (especially south of this area toward the lower shelf). The upper shale and included paleosols consist of thin, oxidized argillaceous siltstone containing caliche. This siliciclastic unit forms the seal above each carbonate reservoir (the highstand carbonate unit). These genetic units are readily discerned on wireline logs because of lithologic contrast and sufficient thickness for resolution. The bounding hiatal surface commonly represents a distinctive dislocation of lithofacies that can be defined using wireline logs

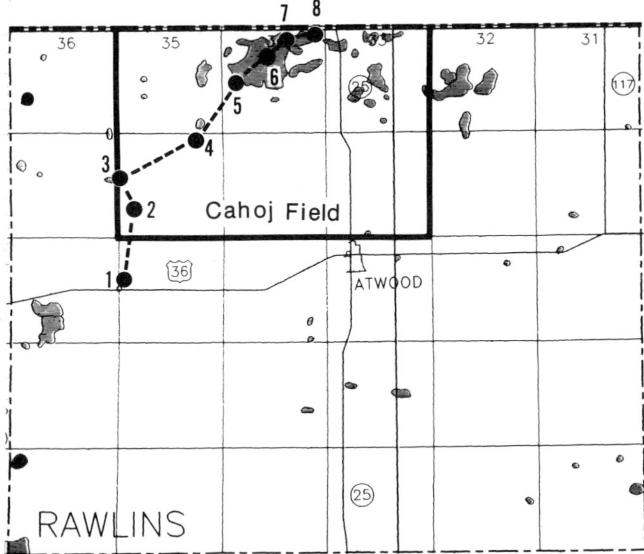

Figure 10. Index map of Rawlins County showing oil fields and wells used in stratigraphic/lithofacies cross section of Figure 16. Townships, 6 mi (9.7 km) square, are included for scale. Also six-township area located in north-central Rawlins County encompassing the Cahoj field complex is outlined. This area will be shown in subsequent maps. Northern edge of map is Nebraska-Kansas border.

(Figure 5). Each depositional sequence includes individual variability, making each vary in thickness, lithofacies, and early diagenetic overprint.

Correlations of depositional sequences, genetic units, and their included lithofacies are accomplished with the wireline logs after correlating logs to core data, which permits the mapping of stratal geometries illustrated later. Hiatuses separating depositional sequences are typically shelf-wide, with few exceptions, permitting temporal distinctions and the resultant construction of high-resolution paleogeography. Moreover, these distinctions provide detailed assessment of events responsible for sedimentation and early diagenesis, some critical to understanding reservoir development (Watney et al., 1989).

Separate depositional sequences isolate individual carbonate reservoirs as illustrated in a cross section of eight wells constructed from wireline logs, cuttings, and a core (Figure 16, modified from Watney, 1980). Sequences can be correlated shelf-wide such as the H-, I-, J-zones shown in Figure 17. Conceptual view of a single depositional sequence across the western Kansas shelf into the Anadarko basin is illustrated in Figure 18.

Each depositional sequence is identified by an informal letter nomenclature beginning with "A" at the top of the Lansing Group and extending through to "M" at the base of the Kansas City Group (Figure 5). Morgan (1952) attributed the letters to individual limestones, confirmed later by Watney (1980). The J-zone is a prominent producing reservoir limestone in Cahoj field as it is in many fields in western Kansas.

The reservoir units within each of the depositional sequences are generally the thicker, shallowing-upward (highstand) limestones that have undergone favorable depositional and diagenetic events, leading to porosity formation and preservation.

TRAP

Cahoj field is a combination structural-stratigraphic trap. Eleven zones produce from various areas in the field. Some wells on the crest of the structure have no reservoir capacity while others produce from multiple zones as commingled production. A few pay zones have been found considerably off the flanks of this structure, but the associated fields are small, ranging from 1 to 5 wells in size. A majority of the producing wells are confined in the area within 40 ft (12.2 m) of the crest of the structure, corresponding to the area of structural closure.

No oil-water contact has been recognized. "Water cut" is common to most wells in Cahoj field. A water drive is not indicated since rapid pressure decline occurs without edge wells going to water or increases in water/oil ratios in the field in general.

Erratic distribution of producing and nonproducing wells suggests that there is considerable lateral lithologic heterogeneity which limits the interconnection of porous zones.

Reservoir

Pennsylvanian reservoirs provide an estimated 23% of Kansas' original oil in place (OOIP) (Watney and Paul, 1983). Ultimate recovery from Pennsylvanian rocks in the Mid-Continent estimated at 8.84 billion barrels (1.26 billion metric tons) provides an estimate of recovery efficiency from these reservoirs. A fourth of this ultimate recovery will come from fields under 25 million barrels in size, such as Cahoj (Rascoe and Adler, 1983).

The pay zones at Cahoj field range from the Oread Limestone (Shawnee Group, Virgilian), at approximately 3850 ft (1173 m), down to the base of the Kansas City, at approximately 4300 ft (1311 m). The reservoirs are most frequently the shallowing-upward (highstand) limestones of the depositional sequences. The flooding unit (transgressive limestone) serves as a limited reservoir because of its thinness and commonly widely scattered porosity. Both limestones are identified using the informal letter nomenclature (Figure 5). Carbonate facies range from subtidal, open marine, low energy mudstones and wackestones (Dunham, 1962) in the lower portions of the shallowing-upward carbonates to high- and low-energy, intertidal and supratidal

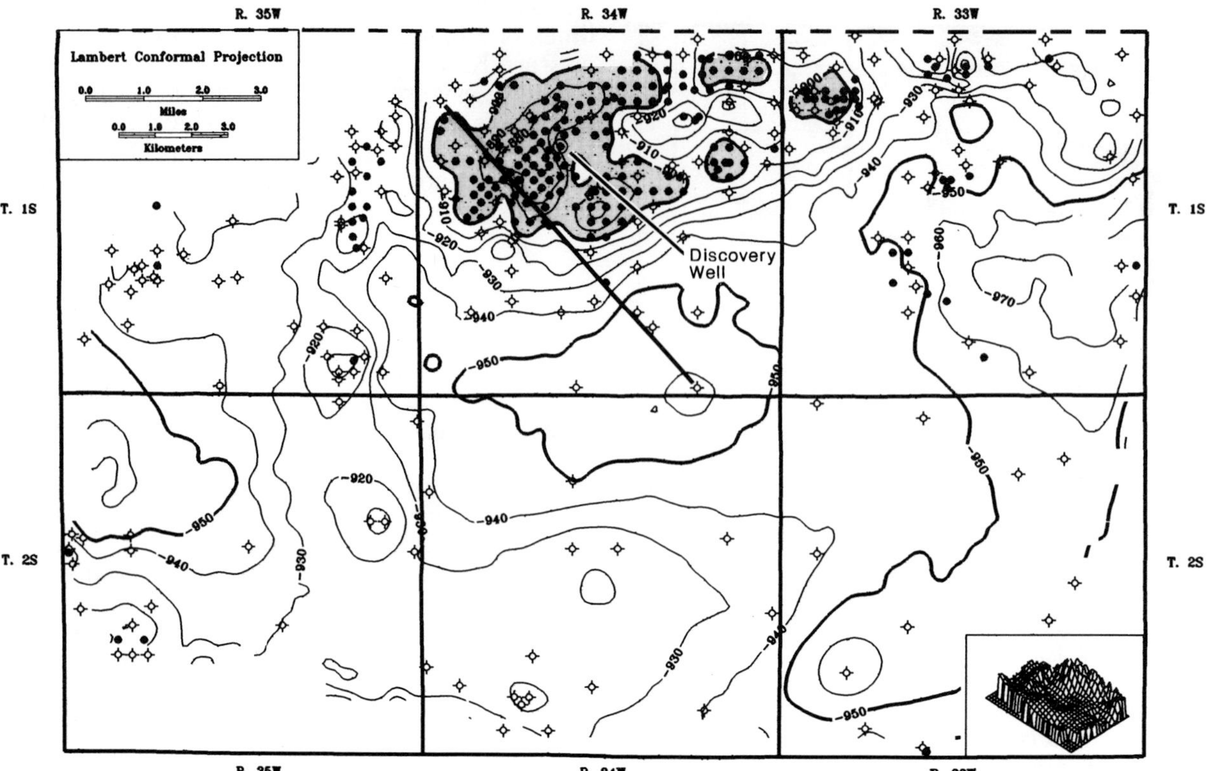

Figure 11. Structural contour map of the top of Lansing-Kansas City strata in the six township area outlined in Figure 10. Contour interval is 10 ft (3 m). Wells used in making the map are shown by their production status, including solid dots for oil wells and circles with four radiating spokes for dry holes. Discovery well (Skelly #1 Bartosovsky) also shown. Diagonal line identifies location of cross section shown in Figure 12. Maps are computer generated using SURFACE III. Inset at bottom right includes small, three-dimensional view of this structural surface looking toward the northwest.

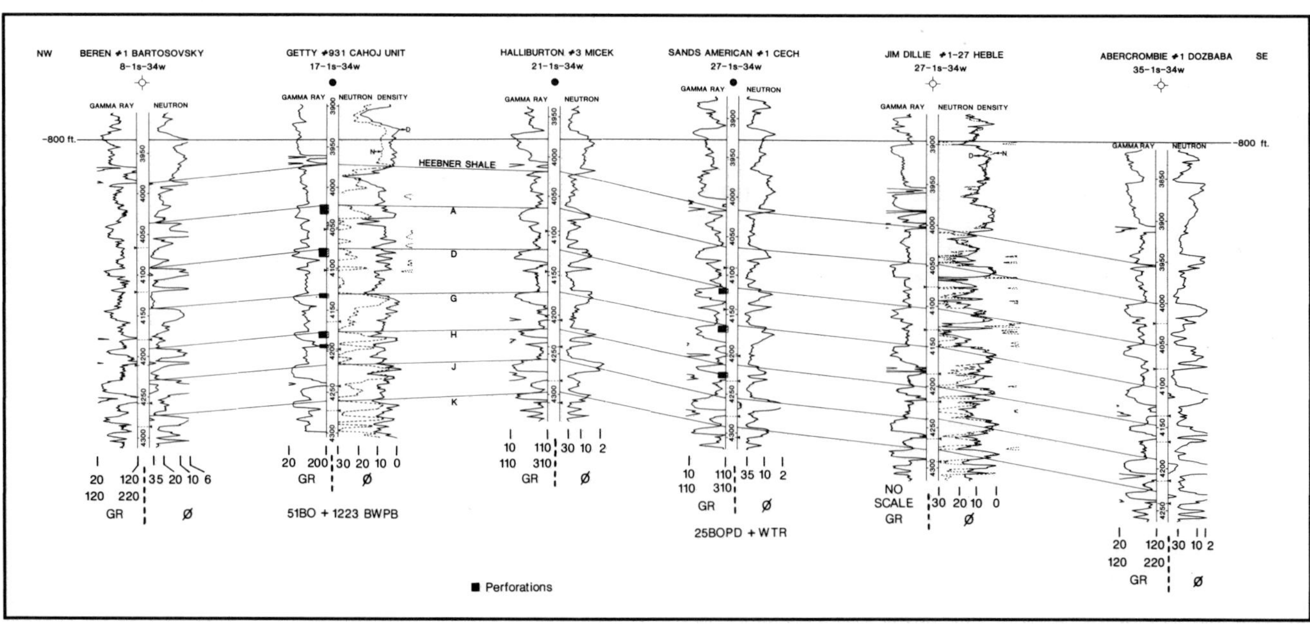

Figure 12. Structural cross section based on wells from northwest-southeast traverse across Cahoj field identified in Figure 11. Elevations in feet. Greek letter phi identifies porosity scale of porosity logs shown.

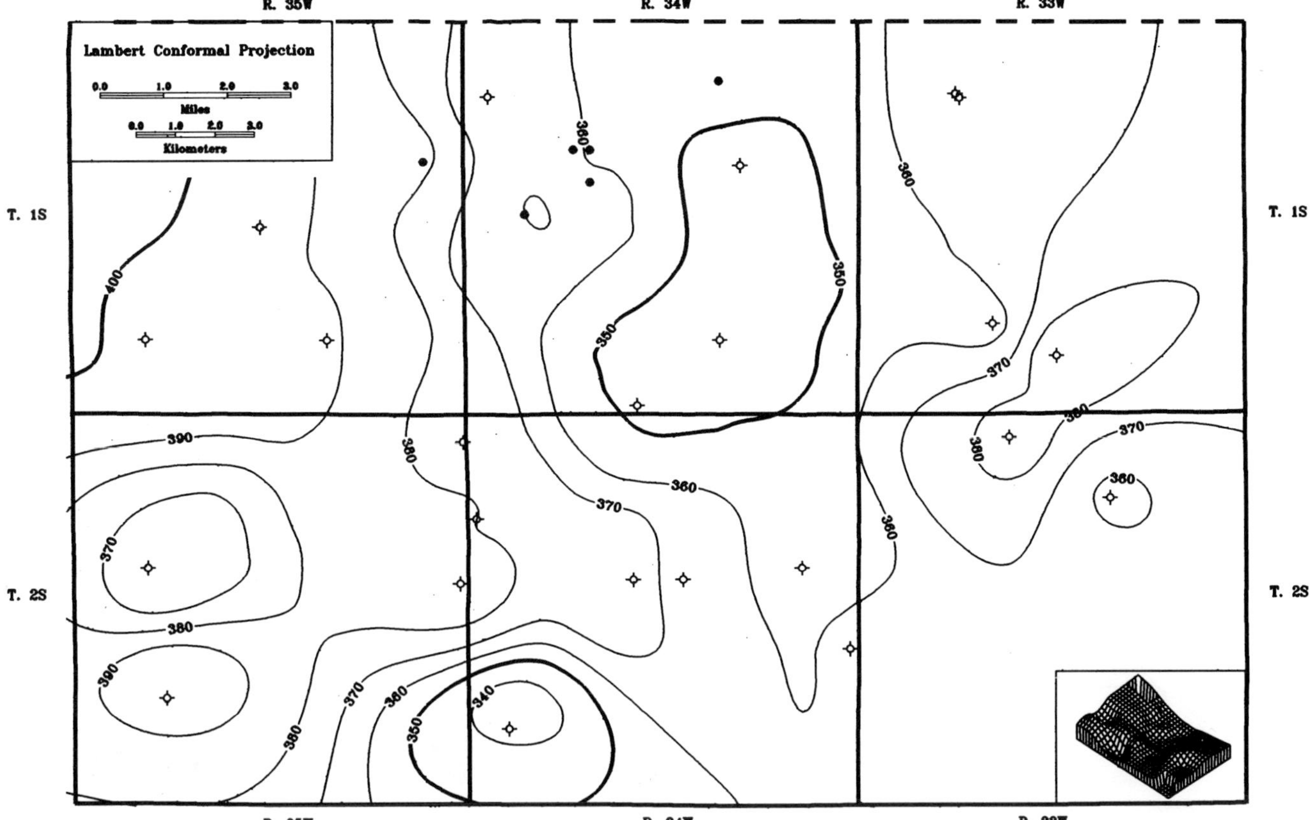

Figure 13. Isopachous map of interval from base of Pennsylvanian to base of Kansas City in six-township area outlined in Figure 10. Contour interval is 10 ft (3 m). Only wells that penetrate lower horizon are shown. Three-dimensional view in lower right box shows relative thickness of interval.

deposits nearing the top of these carbonates. Minor deepening events are evident from changes in texture and faunal content in some of the shallowing-upward carbonates, such as the J-zone.

The flooding unit occasionally consists of either sandstone or siltstone, e.g., B-zone, but typically is a carbonate unit deposited during rapid deepening. The flooding unit may be too thin to be recognized on the wireline log, e.g., at the base of the J- and K-zones, immediately above the lower sequence boundary (Figure 5). In contrast to these, the E-zone, for example, represents an unusually thick carbonate flooding unit associated with the shallowing-upward D-zone limestone.

A thin marine shale, the condensed section, overlies the flooding unit. The marine shales are commonly darker, richer in organic matter, and more radioactive farther south (Watney, 1984). High gamma radiation in these shales provides excellent markers on gamma ray logs for correlation of the sequences. Biostratigraphic techniques, such as studies of conodont and ammonoid fauna, have proven to be successful for regional correlation of individual marine shales (Boardman and Heckel, 1989).

An upper shale overlies the shallowing-upward carbonate, and, in this area, it was modified to become a paleosol, with all of the diagnostic evidence for soil (pedogenic) development: rhizoliths, ped (fracture) surfaces, and caliche nodules. Laminated subaerial crusts (calcrete), microkarst, autoclastic breccia, oxidation mottling, and fracture and fissure filling by vadose silt and red shale are common in the upper portions of underlying carbonates (Watney and Ebanks, 1978; Watney, 1980). The upper shale was deposited in a continental setting perhaps subaerially, as wind-blown silt and clay on the northern shelf prior to soil development (Prather, 1985). Other studies have also delimited the characteristics of paleosols associated with the upper shale (Schutter and Heckel, 1985; Goebel et al., 1989). Succeeding sequences above each paleosol were deposited over very well defined, subaerially exposed surfaces. At these contacts are distinct facies dislocations and lithologic contrasts. Sequence boundaries and the lithofacies on either side of them can generally be defined from well logs once a type of log response is identifiable by correlation with core information in the area (Watney, 1979; Watney et al., 1989).

Northwestern Kansas was a broad, carbonate-dominated platform that locally graded to the north into increasingly thicker shales derived from the more positive areas of the shelf. An isolated area in

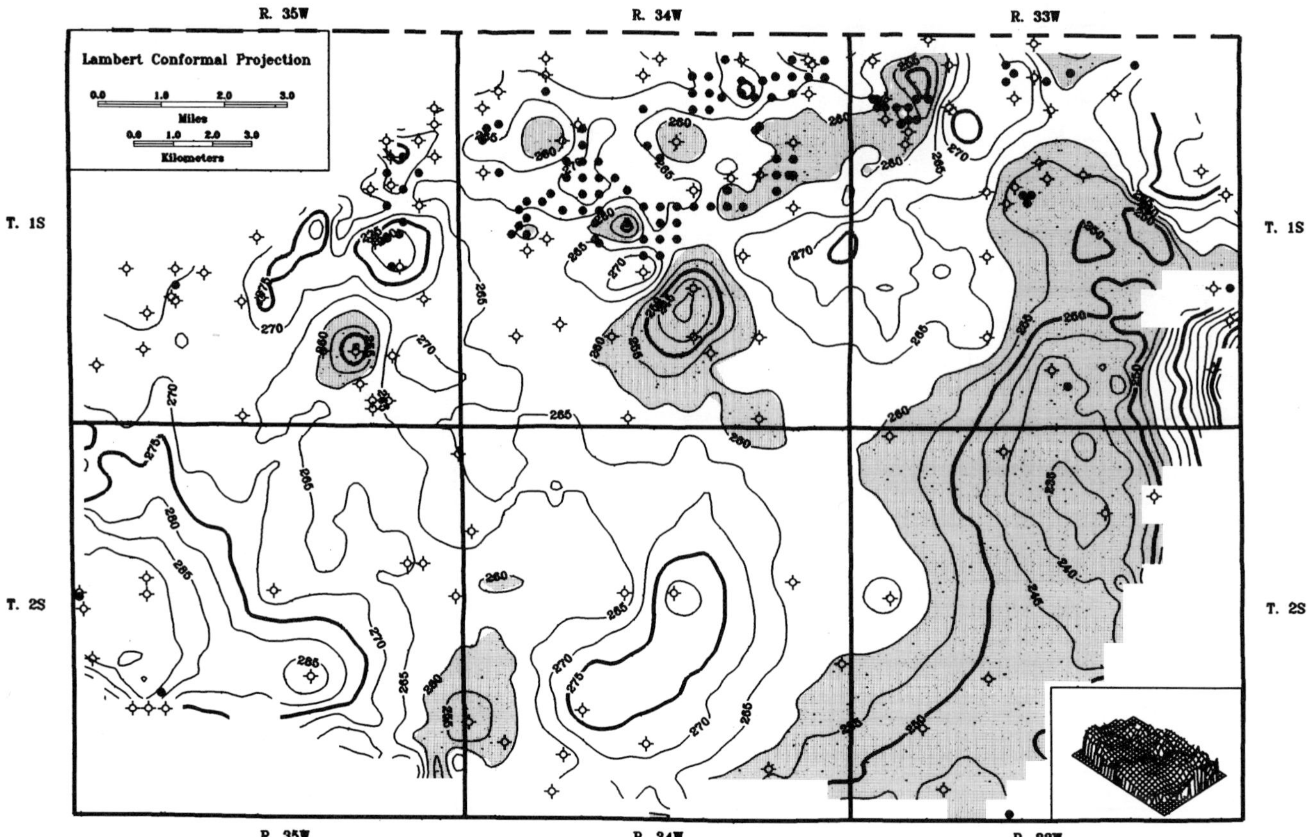

Figure 14. Isopachous map of Lansing-Kansas City interval in six-township area outlined in Figure 10. Contour interval is 5 ft (1.5 m). Stippled area represents thinner areas. Computer-generated map including three-dimensional perspective of interval thickness looking northwest shown in lower right box.

west-central Nebraska on the crest of the Chadron arch apparently was high enough to have been subaerially exposed during much of the Missourian interval (Figure 19 modified from Rascoe and Adler, 1983).

Stratigraphy is quite uniform through the immediate area of Cahoj field. However, facies change within each sequence is significant and notably affects porosity development (Figure 16 from Watney, 1980). Porosity most commonly results from early dissolution of grains, carbonate micritic matrix, or recrystallization of carbonate micrite to microgranular calcite or dolomite spar (Figures 20 and 21). Intergranular porosity in grainstones is also common. Packstone textures, where grains are in contact and intergranular spaces are filled with micrite, commonly become porous and permeable after micrite is recrystallized and some of the grains leached away.

Diagenesis is a significant component in porosity formation. Porosity is better developed in the upper portions of the shallowing-upward carbonates. This relationship appears to be related both to the presence of less argillaceous carbonate rock higher in this carbonate and proximity to early weathering and dissolution events associated with subaerial exposure and meteoric freshwater percolation associated with each sequence (forming vugs, fractures, recrystallization of micrite). Abundant evidence also exists for a late (burial) porosity development, preferentially in less argillaceous carbonate facies (Anderson, 1989). In contrast, argillaceous facies common to the lower portions of the shallowing-upward carbonate units also went through a form of dissolution but resulted in development of microstylolites (clay seams). Porosity either is not preserved or it never occurred in these shaley intervals. These intervals are recognized in log response as having very low porosity and slightly elevated natural gamma radiation (e.g., Figure 5, 3890–3910 ft; 1186–1192 m).

Discontinuous, vertical, nontectonic fractures are also present, which could locally accentuate permeability. However, fractures are not considered to be important in enhancing permeability in these reservoirs. Solution pipes (microkarst) and autoclastic brecciation are common but are typically occluded with red shale and silt shortly after and during their development.

Figure 22 is a series of maps illustrating the distribution of producing wells in the six-township area identified in Figure 10. Each map represents

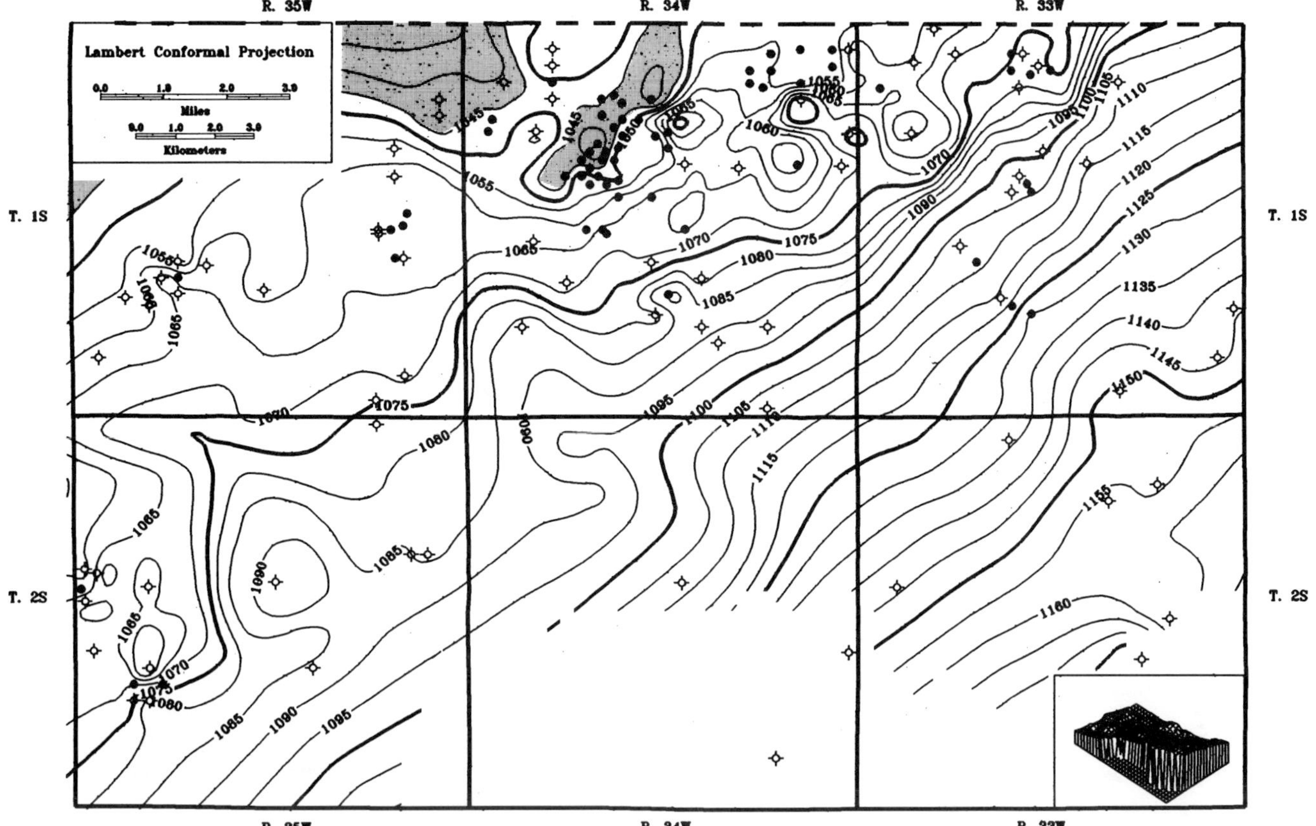

Figure 15. Isopachous map of interval from top of Lansing to top of Lower Permian Stone Corral Formation in six-township area outlined in Figure 10. Contour interval is 5 ft (1.5 m). Stippled areas represent thinner regions. Three-dimensional perspective diagram of interval thickness looking toward the northwest is shown in lower right.

one reservoir carbonate zone identified in the type log (Figure 5). The area depicted in these maps encompassing the Cahoj field complex is the same area as that of the structure map in Figure 11. Several of the maps shown in Figure 22 also include isopachous contour lines representing thickness or porosity-feet when these data are available.

Oil production from the Virgilian Oread Limestone is found on the east side of the Cahoj structure (Figure 22) on the flanks of structural closure, suggesting stratigraphic control of production from this zone. The producing wells from the A-zone (uppermost Lansing Group, Figure 22) are found on the crestal areas of the structure. Porosity in the A-zone is more widespread and structural elevation plays a more significant role in localizing oil production than in other Lansing–Kansas City zones. Porosity in the A-zone of the cored well (#1 Bartosovsky) illustrated in Figure 5 is patchy microspar in an autoclastic breccia.

The producing wells from the B-zone (Figure 22) also indicate widespread porosity over crestal areas of the Cahoj structure. The B-zone is a flooding (transgressive) sandstone of the depositional sequence of which the A-zone is also a part. The sandstone/siltstone lithology of the B-zone evidently is associated with reworking of the underlying siliciclastics. The unit appears to be sheetlike in distribution, and, therefore, pay is confined to the top of the structure.

The C-zone (Figure 22) is a zone of nodular, microcrystalline calcite in a red siltstone in the upper shale in the sequence containing the D-zone shallowing-upward carbonate unit. Carbonate clasts contain scattered ostracodes while other occurrences of carbonate are nodular and resemble caliche. Although no porosity has been seen in the cores of this unit, the porosity may be related to diagenetic recrystallization, leaching, or fracture development. It has not been established whether the interval has contributed any fluid owing to commingled production with other zones.

The D-zone is a shallowing-upward carbonate unit. The map (Figure 22) includes contours that depict southward thickening of the carbonate. The mapped area is located along the northern edge of a broad east–west-trending carbonate buildup (Figure 58 in Watney, 1980). Cuttings and cores from wells in Rawlins County indicate that this buildup is composed of mud-supported, phylloid algal-rich

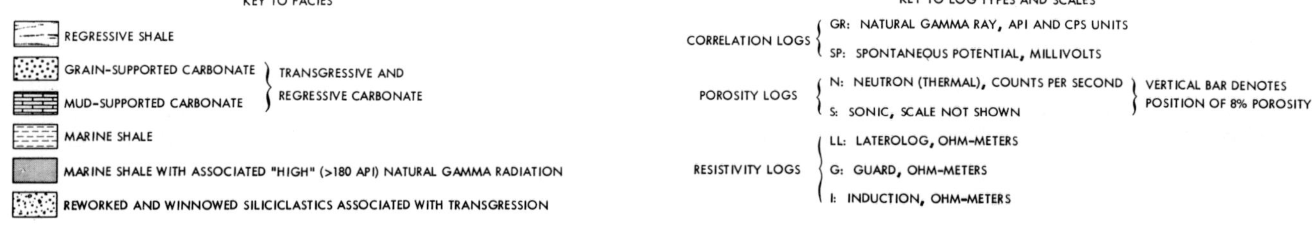

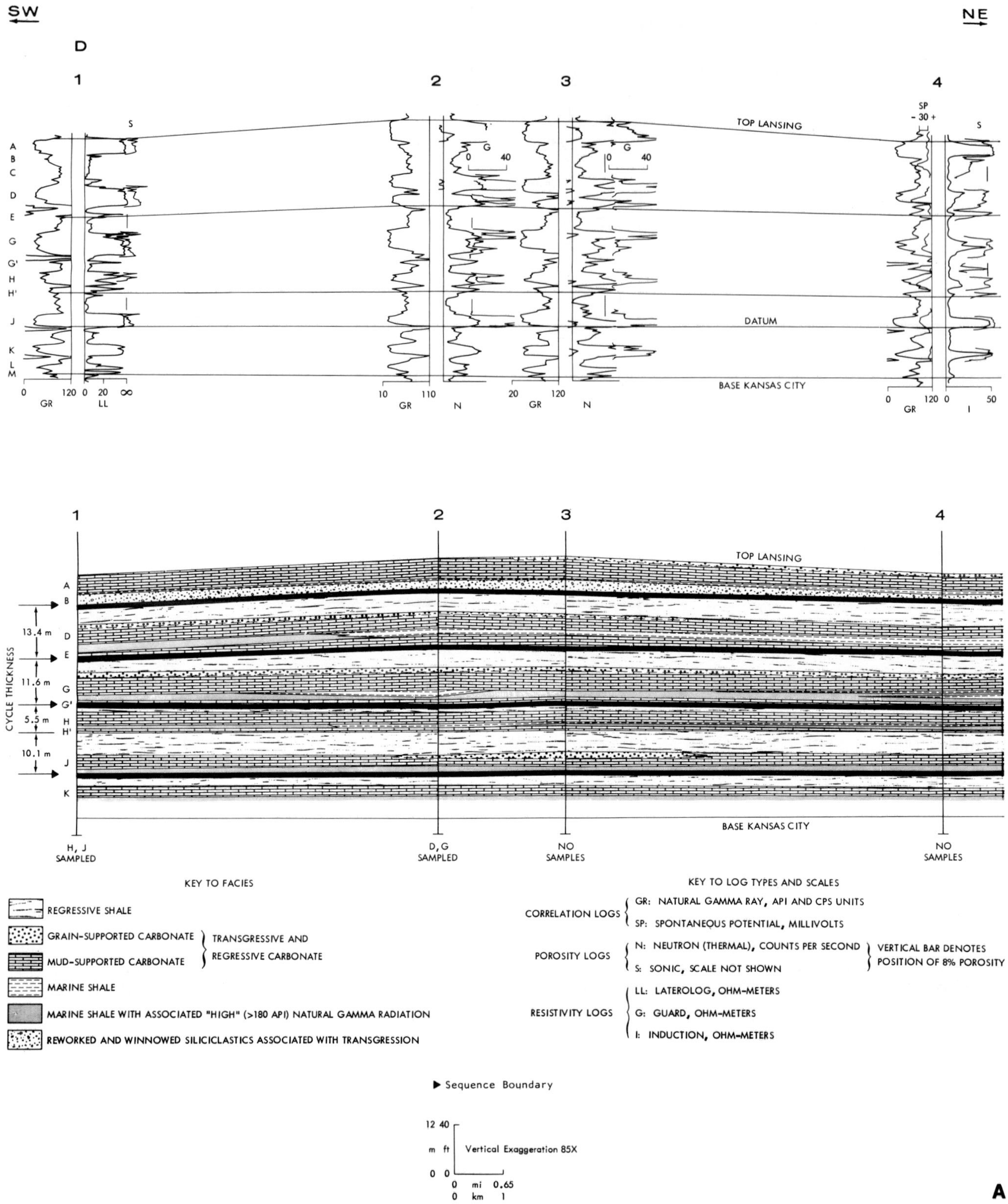

Figure 16. Matching set of well log (upper) and lithofacies (lower) stratigraphic cross sections with a datum at the base of the shallowing-upward carbonate unit of J-zone. Section extends from southwest to northeast. Index map for wells used in section is Figure 10. Shown in two sections (A) wells 1-4 and (B) wells 4-8. Note that well 4 is common to both sections. Wireline log cross section in upper portion of illustration is interpreted with use of cores and cuttings as lithofacies section shown in lower half of illustration. Vertical bars in wireline log section identify 8% porosity on neutron (N) and sonic curves (S). Stratigraphic zones are designated along margins.

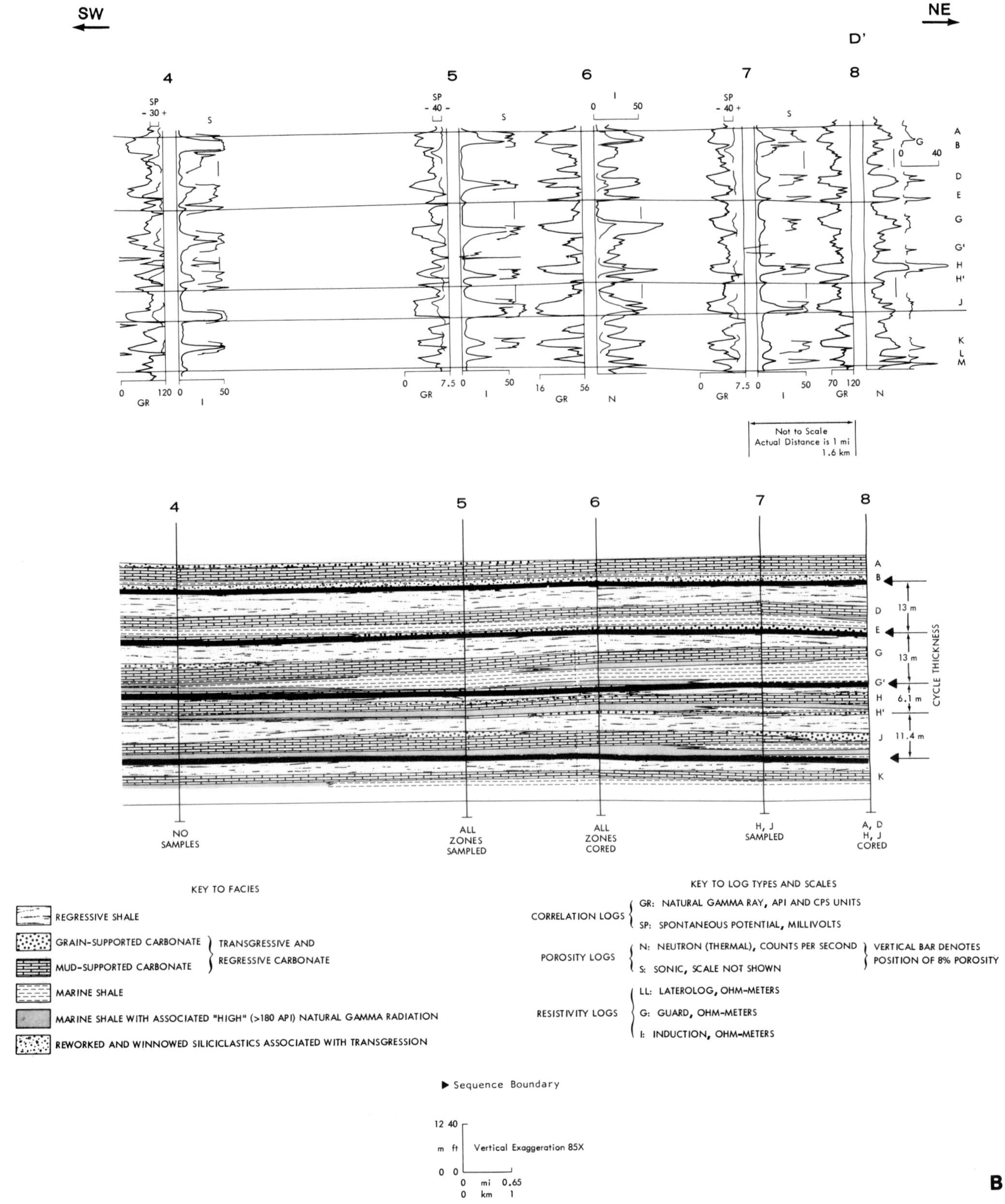

Figure 16. (Continued)

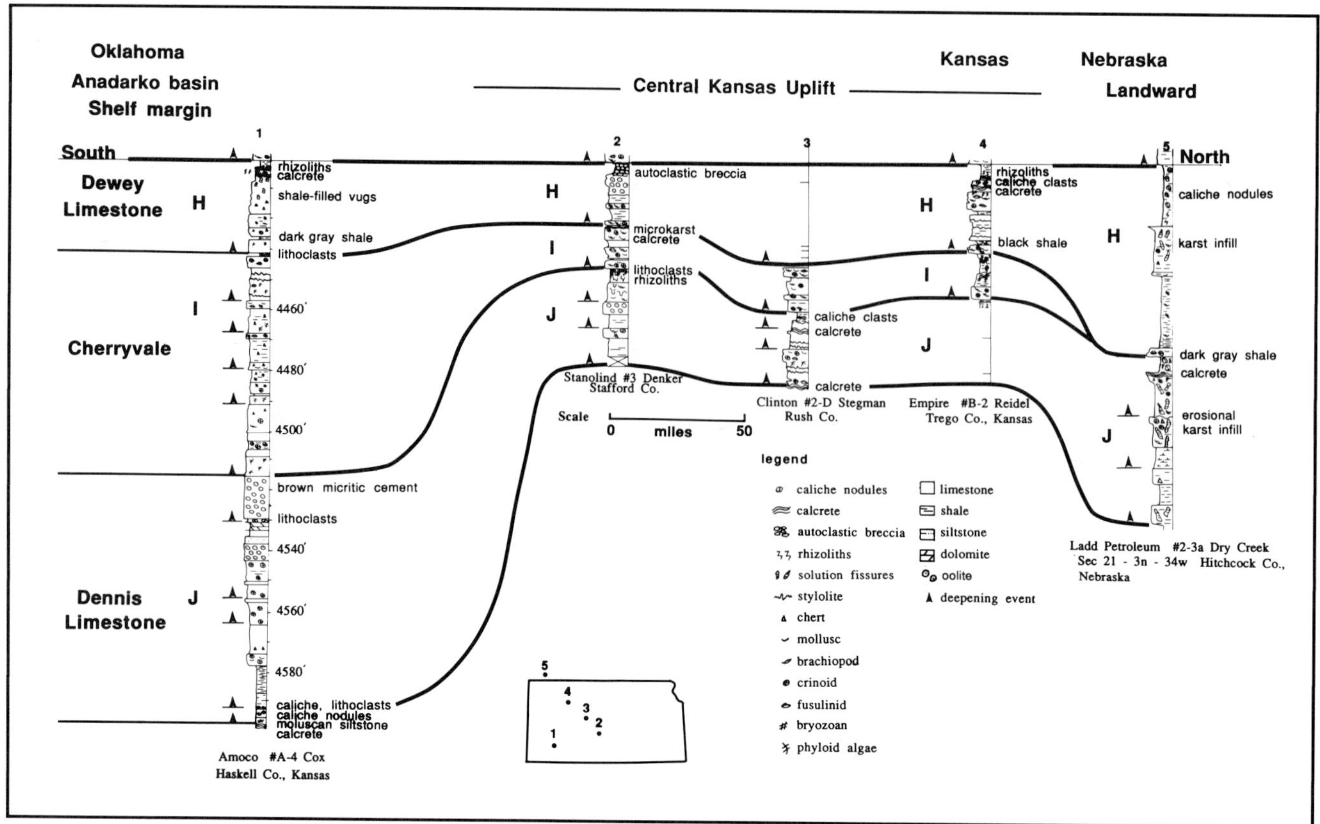

Figure 17. Regional sequence stratigraphic cross section of H-, I-, and J-zones of the Kansas City Group. Heavy lines indicate extensive subaerial surfaces that separate depositional sequences. Cross section is based on cores from wells identified in inset map of Kansas in lower part of illustration. Interval thins over Central Kansas uplift affecting wells #2, #3, and #4.

carbonate facies that are commonly capped by a bed with grainstone-packstone texture (Watney, 1980; Ebanks and Watney, 1985). The grain-supported, interparticle porosity of the upper zone and moldic porosity resulting from leached phylloid algal blades of the lower interval provide two separate reservoirs in the D-zone in the Happy field, southwest of Cahoj (Ebanks and Watney, 1985). The D-zone thins over the crest of Cahoj, and it is locally productive in a relatively thick blue-green algal-foram encrusted bioclastic grain-supported facies. Producing wells are found on the structural crest and lower flanks of Cahoj. Porosity pinch-outs are necessary in creating these flank-accumulations of oil. The lack of phylloid algae in formations on the structure and the common thinning and facies change such as to include thicker grain-supported fabrics suggest that the present-day structure was also topographically higher.

The E-zone is the transgressive carbonate unit of the sequence that contains the D-zone shallowing-upward carbonate. The E-zone is unusually thick for a transgressive unit. It thickens particularly off the Cahoj structure to form another elongate carbonate buildup (Figure 22). It again is composed of mud-supported, phylloid algal, carbonate rock. A grainstone-packstone texture is commonly present at the base of the E-zone. While the overall E-zone thins over Cahoj field, core indicates its basal grainstone thickens. The E-zone produces from this grain-supported facies in isolated porous zones along the southeast flanks of the structural crest of Cahoj.

The G-zone is a shallowing-upward carbonate unit. The thickest area of its occurrence is centered on the southwest side of the structure (Figure 22). Descriptions of cuttings indicate that this carbonate buildup is again associated with phylloid algal, mud-supported, carbonate facies. The G-zone produces oil on the crestal areas of the structure and in isolated porous carbonate lenses around the southeast, south, and west flanks of Cahoj. Not all of the wells on the crestal areas produce from the G-zone, such as the cored #1 Bartosovsky well (Figure 5). The G-zone in this well is an argillaceous wackestone with no porosity development.

The distribution of producing wells in the H-zone is very erratic, with wells located on the crests and flanks of the structure (Figure 22). The H-zone in the area of Cahoj field is thin compared to its regional

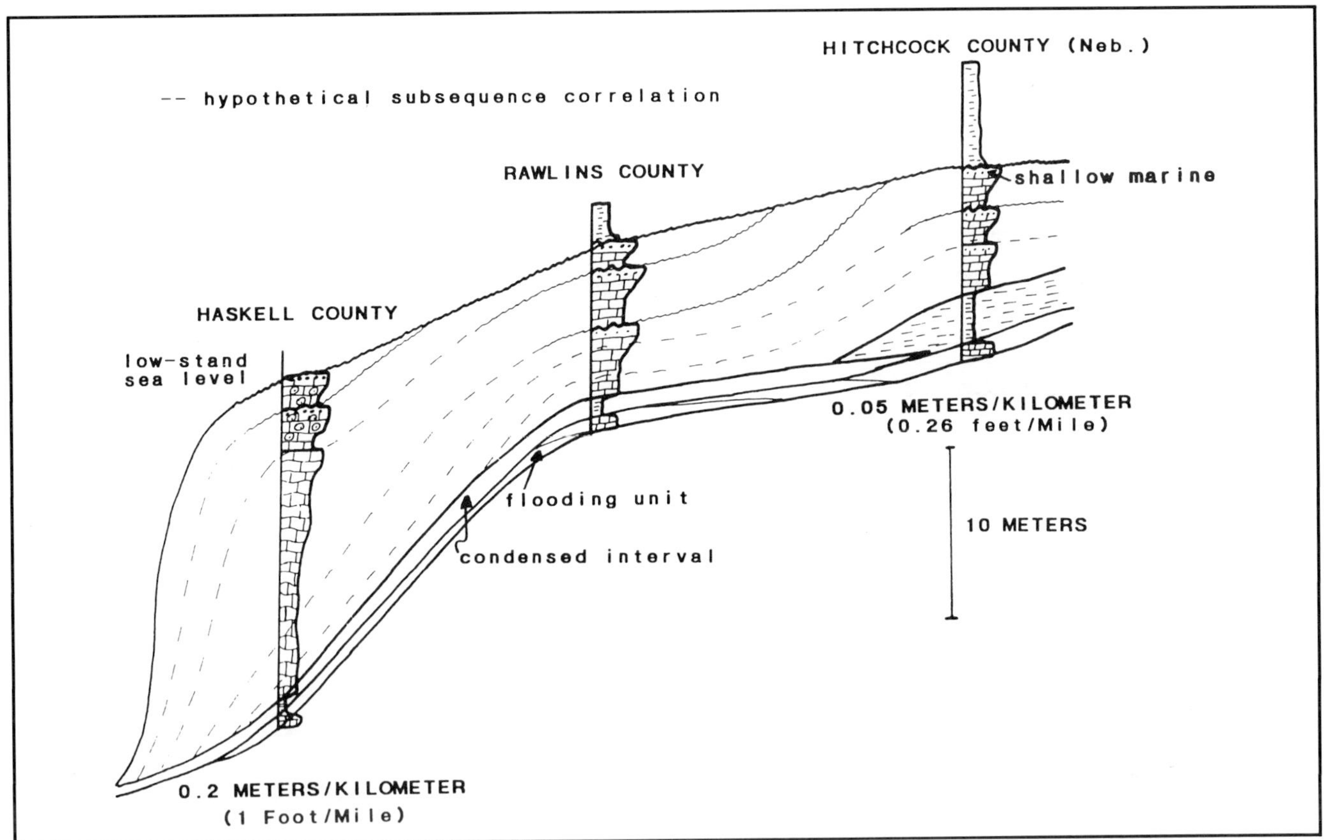

Figure 18. Conceptual regional south (left) to north (right) profile of possible geometry and internal stratigraphy of J-zone in western Kansas and southwestern Nebraska. Haskell County is in southwestern Kansas, Rawlins County in northwestern Kansas, and Hitchcock County immediately north of Rawlins County in southwestern Nebraska. Depositional sequences are characterized by thin, basal flooding units (limestone) and thin, distinctive, condensed sections (dark, commonly radioactive marine shales). These genetic units are overlain by complex shallowing-upward carbonate units and upper shales (redbeds in northern Kansas, capping the sequence and thickening northward, landward).

thickness pattern. Patches of porosity are associated with relatively thick, porous, grain-supported carbonate facies found at the top of the H-zone.

The J-zone zone is the most prolific of the field and consequently was mapped in greater detail. A map of thickness in feet multiplied by percent porosity of the upper J-zone reservoir is shown in Figure 22, using wells that produce from the J-zone. Thickness of the carbonate unit with porosity that exceeds 8% is included. Eight percent is considered the minimum value of effective porosity in these rocks, as illustrated by permeability-porosity crossplots (Figure 6). Analysis of cuttings indicates that much of the porous carbonate in the upper J-zone is associated with grain-supported textures (Watney, 1980). The upper J-zone in the #1 Bartosovsky well is nonporous wackestone that has undergone autoclastic brecciation, with shale infill between carbonate clasts. The J-zone is overlain by a prominent paleosol exhibiting caliche, rhizoliths, soil peds, and variegated coloring. Grains and micritic matrix are leached, producing much of the porosity. The main porosity trend within the J-zone extends along the axis of the Cahoj structure. Additional and apparently isolated porosity extends down its flanks. Productive wells are located mainly on highs on the structure, but some off-structure locations are also productive.

Figure 22 also illustrates a porosity-thickness map of the middle and lower portions, using the same 8% porosity cut-off. The trend is again along the axis of Cahoj, but the distribution of reservoir rock is more restricted (according to log interpretations). Porosity is related to recrystallized micrite and vugs in mud-supported carbonate as opposed to grain-support fabrics. Producing wells from the K-zone and the L/M zones (Figure 22) are restricted to the crestal areas of the Cahoj structure.

The relationships of thickness, facies, and porosity development suggest that the Cahoj structure presented some form of paleotopographic relief during deposition of the Lansing–Kansas City strata. Greater slopes along the edges of the structure and

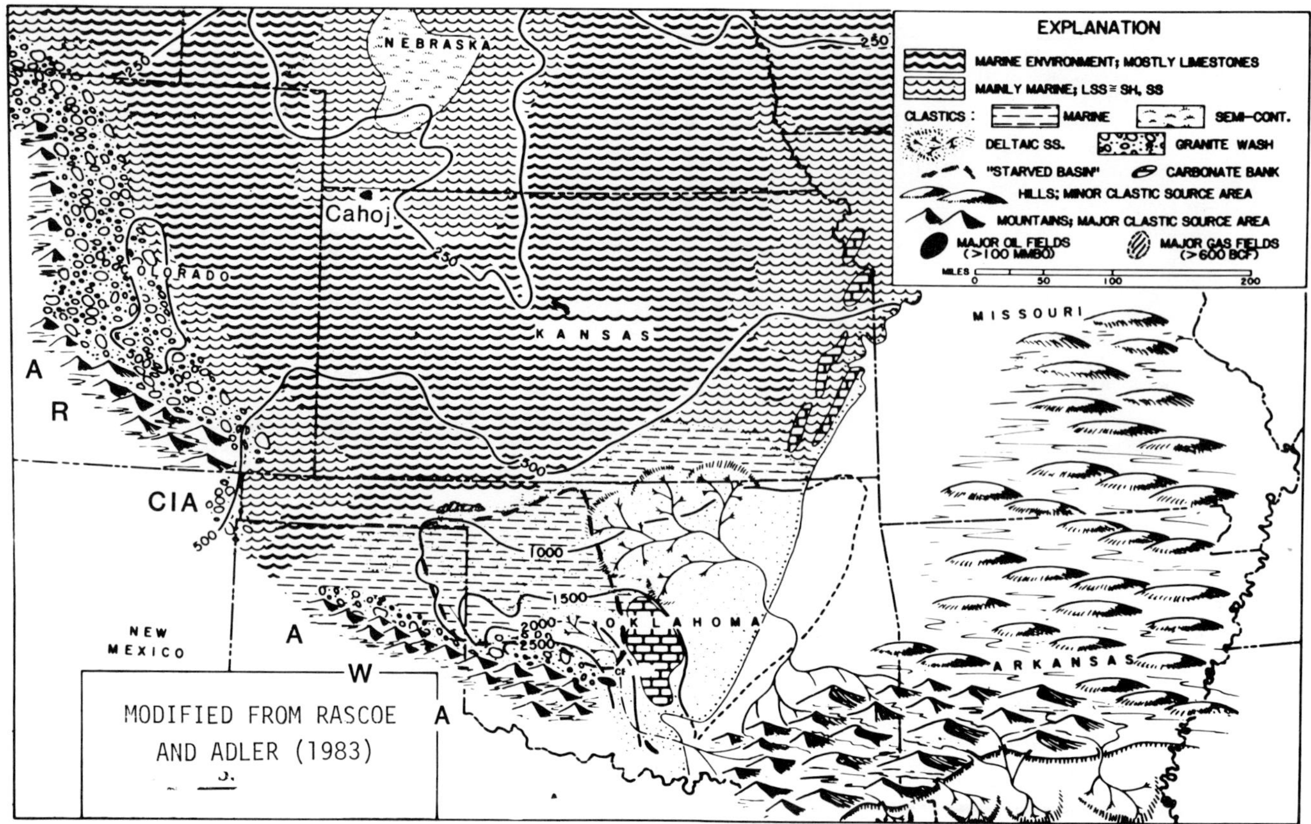

Figure 19. Map depicting a combination of paleogeography, lithofacies, and thickness of the Missourian Series of the Upper Pennsylvanian of the southern Mid-Continent U.S. (Rascoe and Adler, 1983). Interval comprised mainly of Lansing-Kansas City strata and equivalents. Location of Cahoj field indicated in northwestern part of Kansas in transitional area from mainly marine carbonate shelf to south and west and marine mixed carbonate and shale/sandstone of upper shelf bordering Chadron arch in western Nebraska. AR, Ancestral Rocky Mountains; CIA, Cimmaron arch; AWA, Amarillo-Wichita-Arbuckle uplift.

higher elevations along crestal areas apparently produced a locus for current and wave action that resulted in grain-supported depositional fabrics. Subaerial exposure and associated infiltration of meteoric water were also probably more pronounced (of longer duration and intensity) over the structure during periods of the lowering of sea level, leading to modification of porosity development. However, evidence of subaerial exposure is extensive and ubiquitous across the shelf in western Kansas. Earlier and more prolonged subaerial exposure of the "highs" may have encouraged formation or preservation of porosity.

A map of initial oil-producing potential (IP) of wells in Cahoj field and nearby area (Figure 23) indicates an apparent lateral pressure communication in the reservoirs among the wells, with greatest production potential centered on the crest of the structure. Similarly, the water production from the initial potential (IP) tests (Figure 24) that is centered on the structure has an apparent continuous, mappable pattern suggesting extensive lateral continuity of porosity and permeability in the reservoirs. This is contradictory to the more limited interconnection of flow between wells suggested by the zonal maps of producing wells (Figure 22). However, the maps of IPs include the combined production of commingled, stacked pays whose cumulative thickness of effective porosity is greatest on the crest of the structure. Thus, on a zone-by-zone basis, reservoir development is concluded to be discontinuous. Any efforts to improve recovery (waterflooding or E.O.R.) must consider the consequences of the heterogeneity when patterns of communication between wells are important.

Three waterflood projects were initiated in Cahoj field in the 1970s. Figure 25 shows total annual oil production for the field and annual production attributed to each of the three waterflood projects. The lower curve represents the change through time of the total number of producing wells for the field. The number of producing wells during the period 1975 through 1985 are not available. Unfortunately, no data are available for water production. What is evident in the annual oil production is that, soon after discovery, primary oil production peaked and

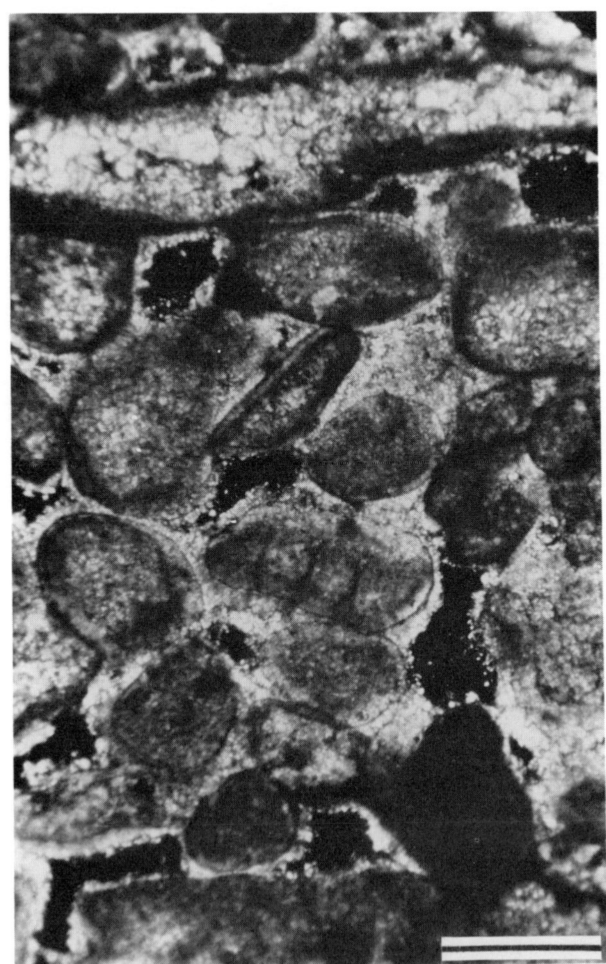

Figure 20A. Photomicrograph of top J-zone in Argosy No. 1 Wicke well located in Sec. 17, T1S, R32W at 4050-4060 ft (cuttings interval). Bioclastic grainstone with well-sorted, uncoated, micritized bioclasts. Isopachous rim of thin, finely crystalline bladed calcite cement. Later coarser calcite spar partially occludes intergranular pore space. Some porosity identified as black areas remaining between grains. Thin section stained with alizarin-red and photo taken under crossed-polars. Scale bar equals 1 mm.

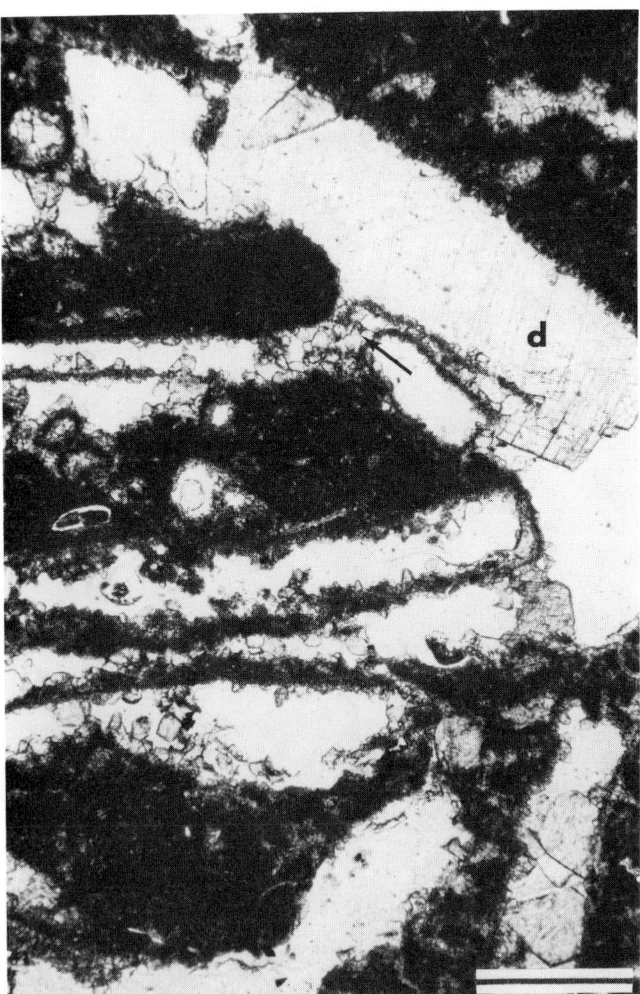

Figure 20B. Photomicrograph of middle D-zone of Murfin Drilling Company No. 1 Prentice located in Sec. 30, T2S, R35W at 4265 ft. Phylloid algal wackestone to packstone where algal blades have been dissolved. Molds of algae are partly occluded with scattered fine and coarse blocky calcite spar and baroque dolomite (d). White spaces are pores. Pores are permeable, connected by nontectonic fractures and collapsed micritic matrix. Scale bar equals 1 mm.

began a steep decline. Infill drilling coupled with waterflooding restored production to about half of the initial peak value. Production again peaked and has entered an extended phase of more gradual decline during waterflooding, apparently as reservoir pressure has been maintained. Waterflood projects, particularly the large Cahoj unit, have sustained production halting the high rate of production decline. Production has subsequently leveled off, except for the slight declines in 1985 and 1986. This was apparently due to plugging of some active wells, which was presumably related to unfavorable economics of the operations during the latest period of lowered prices.

The steep decline in primary production is attributed to a pressure depletion type of reservoir drive in the field. This is very common in Lansing-Kansas City reservoirs in the area. No gas production is associated with the field, and without significant water drive to replenish the produced fluids, the production rate has dropped off rapidly. Details of the waterflooding operation are not available, but its

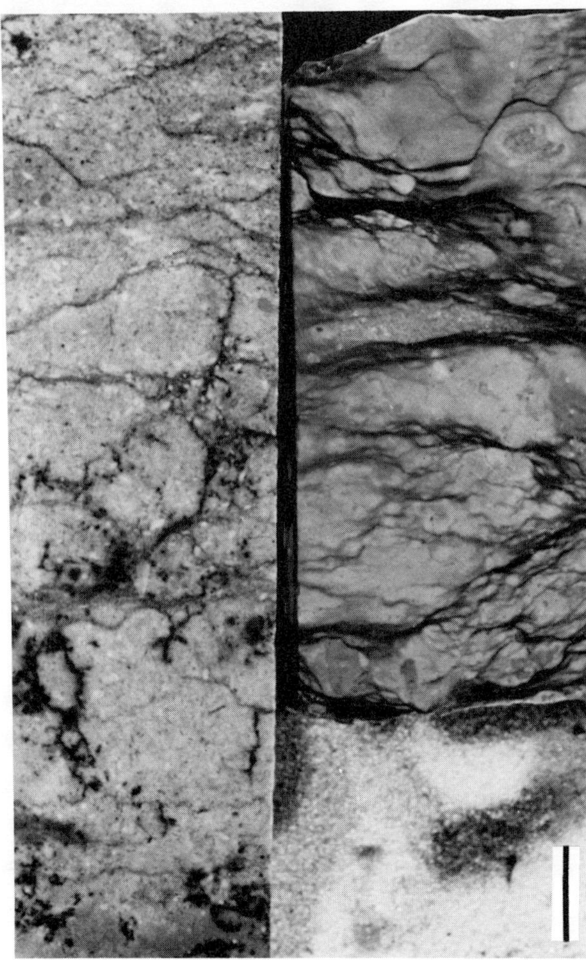

Figure 20C. Core slabs from the lower portion of the J-zone regressive carbonate from the Empire No. 10 Palmer located in Sec. 6, T1S, R33W at 4061.5 to 4062 ft. Scale on right is 2 cm. Left side of core is burrow-mottled crinoid/brachiopod wackestone with widely scattered clay seams. Dark irregular anastomosing patches and lines are areas of oil-stained microspar and nontectonic fractures. Recrystallization of micrite in these relatively clean wackestones creates excellent reservoir rocks in Cahoj field. Sample on right side of photo is nonporous wackestone that is engulfed in microstylolite swarms. This facies is typically tight and nonproductive.

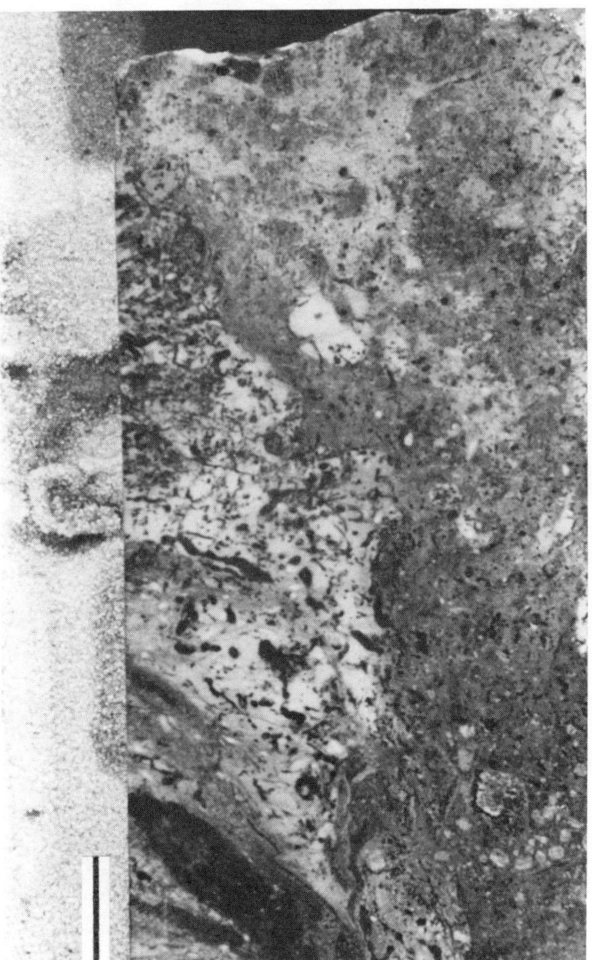

Figure 20D. Core slab of uppermost J-zone regressive carbonate in No. 10 Palmer at 4063 ft. Scale on left equals 2 cm. Left third of rock is originally deposited lithology of upper regressive carbonate (fusulinid, crinoid wackestone). This original texture has been significantly altered along what is interpreted to be a solution pipe filled with original carbonate debris and microcrystalline carbonate (caliche) derived from the subaerial exposure of the unit shortly after it was deposited.

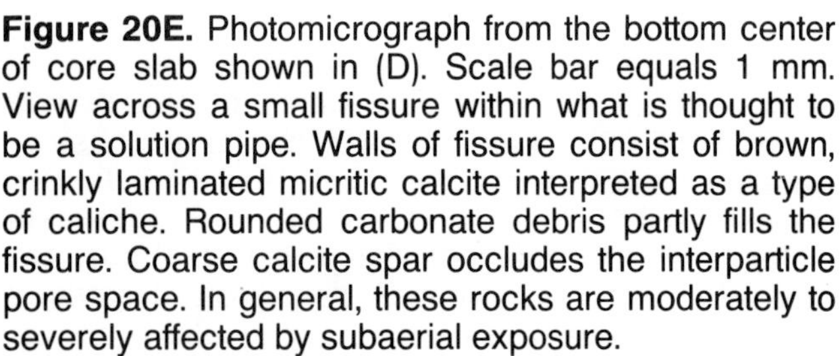

Figure 20E. Photomicrograph from the bottom center of core slab shown in (D). Scale bar equals 1 mm. View across a small fissure within what is thought to be a solution pipe. Walls of fissure consist of brown, crinkly laminated micritic calcite interpreted as a type of caliche. Rounded carbonate debris partly fills the fissure. Coarse calcite spar occludes the interparticle pore space. In general, these rocks are moderately to severely affected by subaerial exposure.

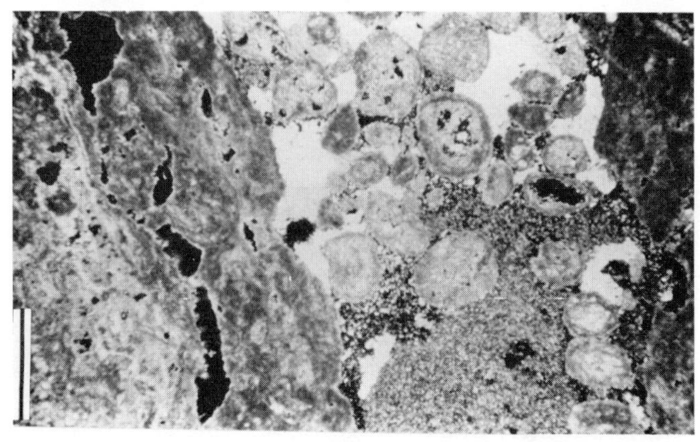

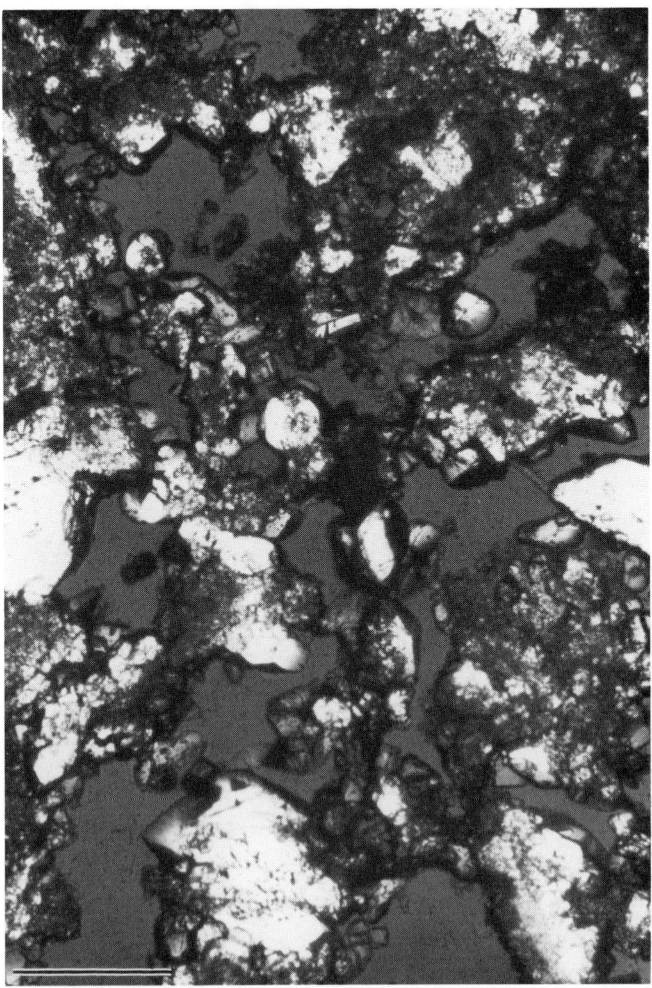

Figure 21A. Photomicrograph of upper portion of regressive carbonate of J-zone from the Skelly #1 Bartosovsky well located in SE SW SW Sec. 9, T1S, R34W at 4174.5 ft. Bivalve, encrusting foraminifera wackestone with numerous vugs and skeletal molds exhibiting good effective porosity and oil stain. This interval is the best pay zone in field. Scale bar equals 0.2 mm.

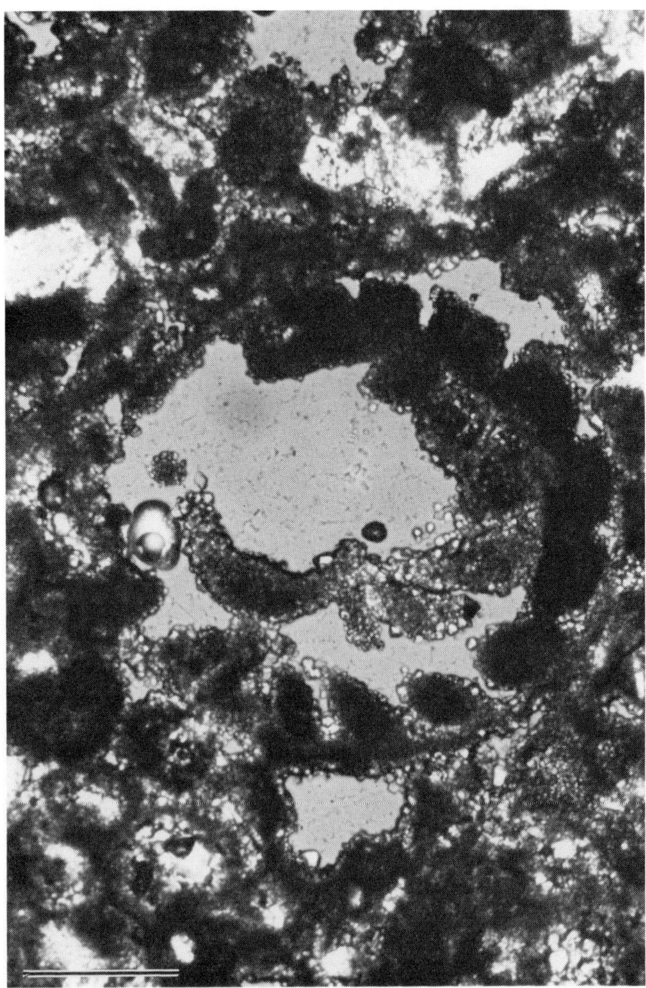

Figure 21B. Photomicrograph from sample 1.5 ft of the top of D-zone regressive carbonate from the Murfin #1 Souchek well located in C SE NE Sec. 2, T1S, R34W at 4006 ft. Peloidal bioclast packstone exhibiting dissolution of mud, peloids, and skeletal debris such as the fusulinid mold shown in the photo. Core analysis at this depth indicates porosity ranges from 10 to 21.8% and permeability 4.4 to 10.2 md. Porosity is effective (filled with injected colored epoxy). Scale bar equals 0.2 mm.

effectiveness is very evident (Figure 25). Waterflooding has sustained oil production, which now accounts for over 90% of the oil produced from the field.

Source

Although organic matter in the marine shales is of the type that can generate oil and gas, the volume of organic matter and the level of organic thermal maturity are insufficient and preclude any local generation of oil from these Pennsylvanian rocks (Figure 26) (Jerry Clayton, personal communication, 1988). The Mississippian-Devonian and Simpson Group shales are not present in this area, indicating that a certain amount of long-distance migration of hydrocarbons is necessary to explain existing accumulations.

Maximum burial in the vicinity of Cahoj field occurred during the early Tertiary (Figure 26). Maximum depth of the Lansing-Kansas City is estimated as 6000 ft (1818 m). With a geothermal gradient of 1.4°F/100 ft (25°C/km), maximum temperature reached was 153°F (67°C). Locally higher geothermal gradients estimated on the craton, such as 104°F (40°C/km), would result in a maximum temperature of 162°F (72°C). This level of heating is still insufficient to generate significant hydrocar-

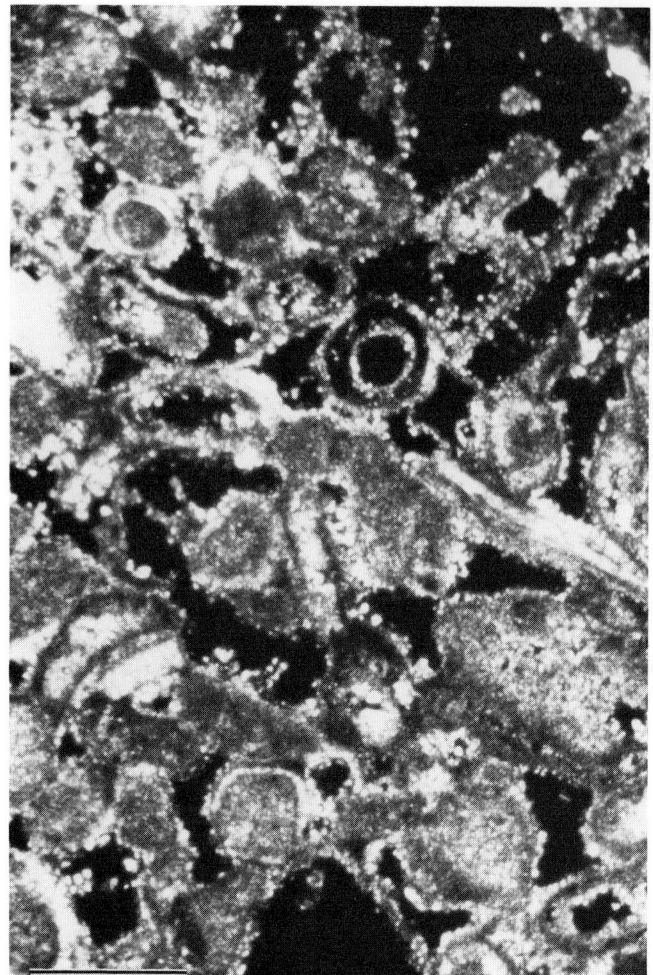

Figure 21C. Photomicrograph of upper regressive carbonate of J-zone in Murfin #1 Souchek located in C SE NE Sec. 2, T1S, R34W at 4145 ft. Micritized, rounded bioclastic, peloidal grainstone. Scattered calcspar with significant effective intergranular and vuggy porosity. Scale bar equals 0.2 mm.

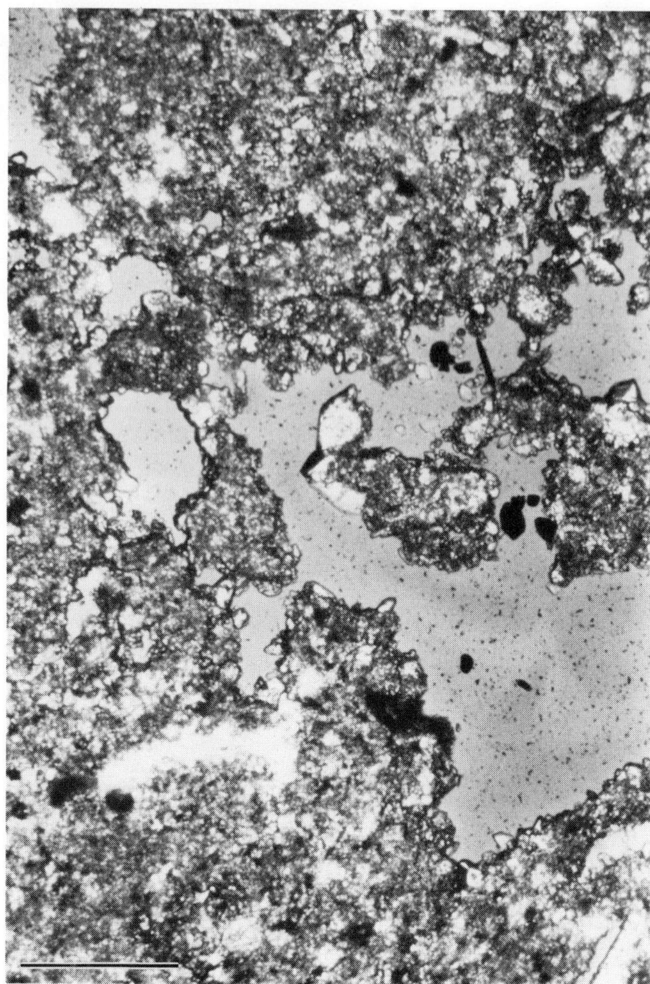

Figure 21D. Photomicrograph of middle portion of regressive carbonate of J-zone in Murfin #1 Souchek located in C SE NE Sec. 2, T1S, R34W at 4150 ft. Brachiopod wackestone exhibiting substantial effective vuggy porosity. Sixty percent of porosity in sample attributed by Anderson (1989) to late stage dissolution. Scale bar equals 0.2 mm.

bon in the immediate vicinity of Cahoj field. Walters (1958) argues that reservoirs in the Arbuckle Group on the Central Kansas uplift were probably charged by oil that migrated into the area through the Cambrian–Ordovician Arbuckle Group from the Anadarko basin during Permian time. Schmoker (1986) presents thermal maturation and burial history analysis for the Anadarko basin which indicates that generation may have begun as early as the Mississippian and extended to the present.

Recent petrographic studies of the Lansing–Kansas City in northwestern Kansas have identified secondary, oil-filled, fluid inclusions trapped in carbonate cements. The mean temperature of homogenization of these two-phase, oil-filled, fluid inclusions in the Murfin #1 Souchek (Sec. 2, T1S, R34W) is 170.4°F (76.9°C). This elevated temperature suggests that hotter fluids carrying hydrocarbons moved through these rocks (Anderson, 1989).

Geochemical data on the oils from the area and correlation to extracts from potential source rocks indicate a predominant Mississippian–Chattanooga (Woodford) shale source (J. Clayton, personal communication, 1988). Since the northern pinchout of the organic-rich facies of this potential source rock is located about 100 mi (161 km) south of Cahoj field, significant lateral migration of oil (updip to north) is inferred to have occurred. The fluid inclusion data at Cahoj support an oil phase carried by warmer waters, which were possibly related to such a long-distance, updip migration of basinal fluids. The fluid inclusions post-date burial cements and therefore suggest that oil migration was relatively late. The nature and timing of the source-carrier-trap system

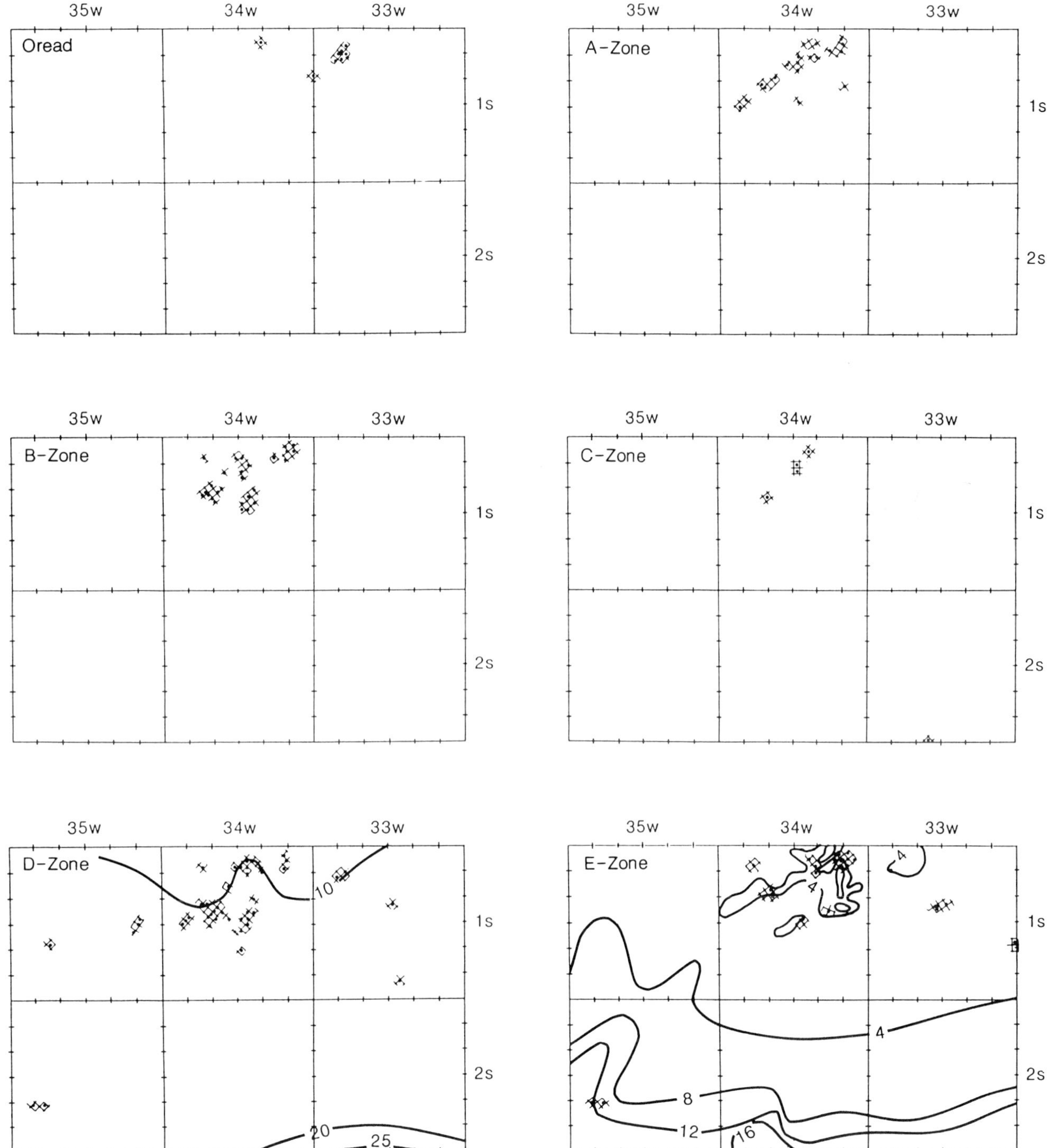

Figure 22. Zonal production maps each showing the same six-township area of the Cahoj field complex identified in Figure 10. Each township is 6 mi (9.6 km) square. Each map includes only wells perforated in the particular zone illustrated. Many wells have commingled production. Contours shown on some maps (when available) describe thickness (feet) of zone or porosity-feet of pay zone where porosity exceeds 8%. Maps include in order of presentation: Oread, A-zone, B-zone, C-zone, D-zone (with zone thickness), and E-zone (with zone thickness), G-zone (with zone thickness), H-zone (with zone thickness), upper portion of J-zone (with porosity-feet contours), lower J-zone (with porosity-feet contours), K-zone, and combined L- and M-zones. Each map discussed further in text.

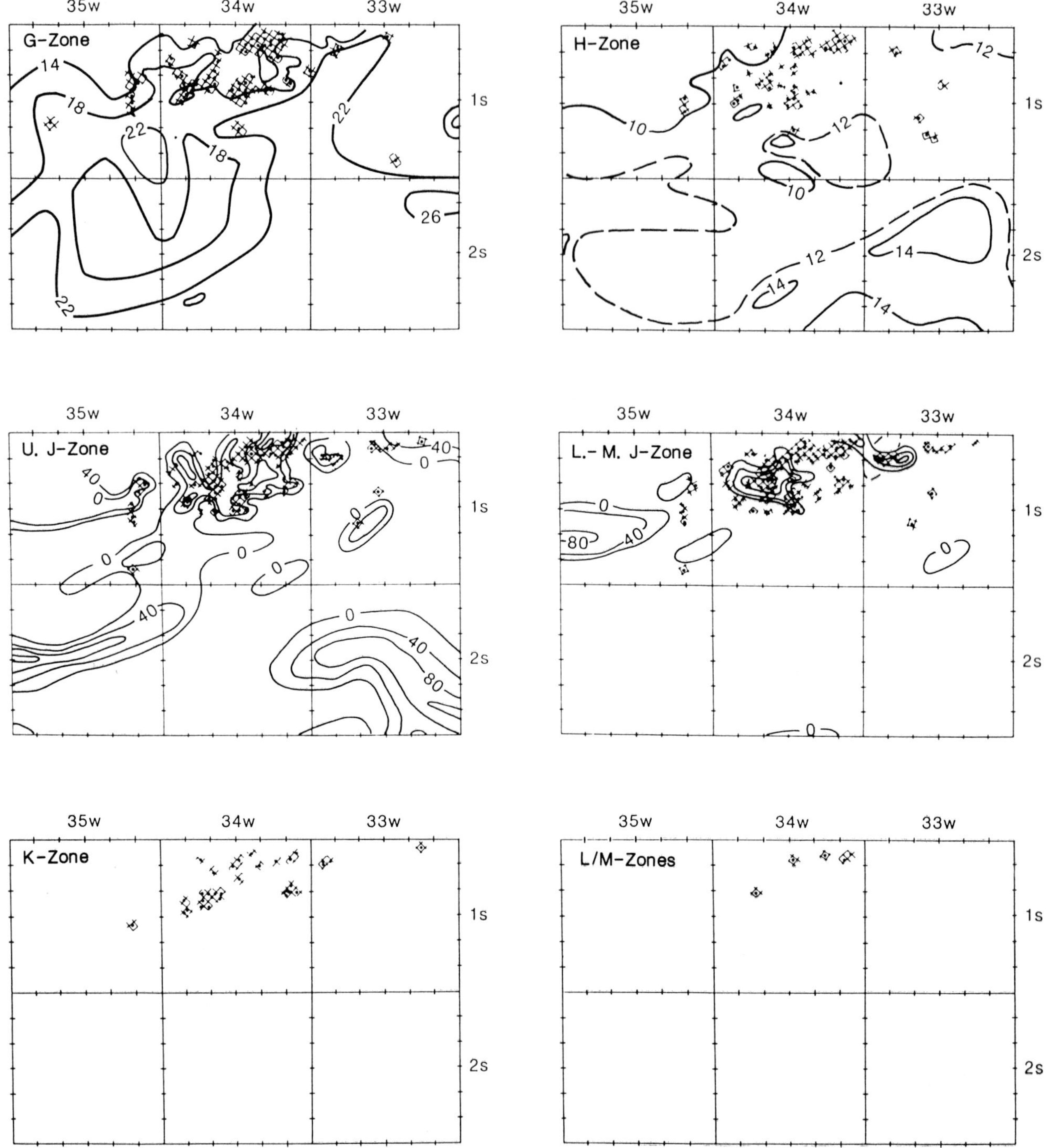

Figure 22. (Continued)

needs to be examined to define better the regional distribution of oil production.

EXPLORATION CONCEPTS—REGIONAL PLAY

The Lansing-Kansas City produces from a substantial area in the subsurface of western Kansas (Figure 3). The target reservoirs are porous carbonate rocks in which porosity distribution is highly variable and involves both primary and early- and late-stage diagenetic porosity. In general, productive carbonate facies are nonargillaceous carbonates, which may include grain-supported textures deposited in shallow water where fines (mud and silt particles) are winnowed out. Reservoirs can also include mud-supported carbonates, including broad, relatively thin bioherms of phylloid algae, which required

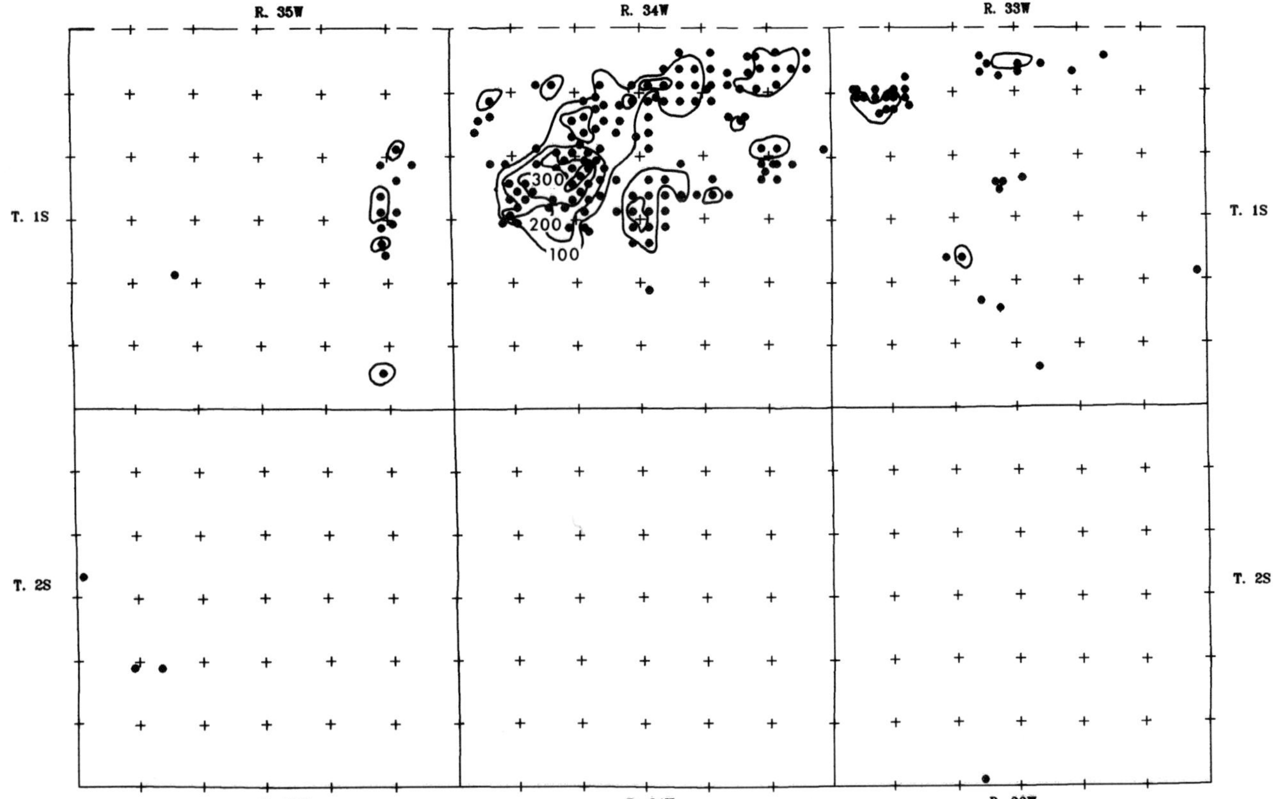

Figure 23. Map of initial oil potential. Six-township area defined in Figure 10. Plus symbols represent section corners, spaced 1 mi (1.6 km) apart. Contours correspond to initial oil potential at time wells were completed. Contour interval is 100 barrels of oil. Higher locations have higher productivity. Production is commingled in zones, particularly over crest of structure.

relatively clear-water conditions for their accumulation (Ebanks and Watney, 1985). Diagenesis in mud-supported carbonates plays a crucial role in porosity development. Early, freshwater diagenesis and subaerial exposure creates porosity through recrystallization and dissolution, but it also occludes existing porosity through cementation and clay/silt infiltration. Precipitation of calcite and baroque dolomite during burial commonly occludes porosity, but dissolution that post-dates all cementation has been observed to provide minor to substantial contributions to existing porosity and, in particular, it may enhance permeability (Anderson, 1989). Moreover, the preservation of porosity is a function of the abundance of sparry calcite cements, which precipitated during burial, that dominate the infilling of pores in northwestern Kansas. Patterns of cement distribution are suggested but are yet poorly understood. Understanding the preservation of porosity is as important as understanding its development (Anderson, 1989; Watney et al., 1989).

Most of the depositional and diagenetic factors that contribute to there being favorable reservoir rocks in the Lansing–Kansas City units seem related to the presence of paleohighs during and following deposition (Watney, 1980; DuBois, 1985). Thus a certain level of predictability of porosity exists via detection of paleohighs. While this strategy applies in northwest Kansas, carbonate grainstones in west-central and southwestern Kansas commonly accumulated on the flanks of paleohighs, creating reservoir rock that rims the highs (Watney and French, 1988).

Identification of paleohighs requires mapping critical lithofacies within each depositional sequence. Along northern Kansas and southern Nebraska, the upper shale of each sequence is thick enough to be easily mapped with wireline logs. Thinning of this shale will often indicate the position in the subsurface of a paleohigh that was due either to structural development or carbonate buildup (Watney, 1980; DuBois, 1985). This may or may not coincide with a present structure and, accordingly, suggests a method for locating potential stratigraphic traps.

The underlying cyclical carbonates and shales of the Marmaton and Cherokee groups (Middle Pennsylvanian) are present over most of western Kansas, and these offer similar opportunities for petroleum accumulation. Many fields have been found recently in these units, e.g., Celia and Celia South fields in southwestern Rawlins County.

The details of source, timing, and migration pathway of hydrocarbons that charge these reser-

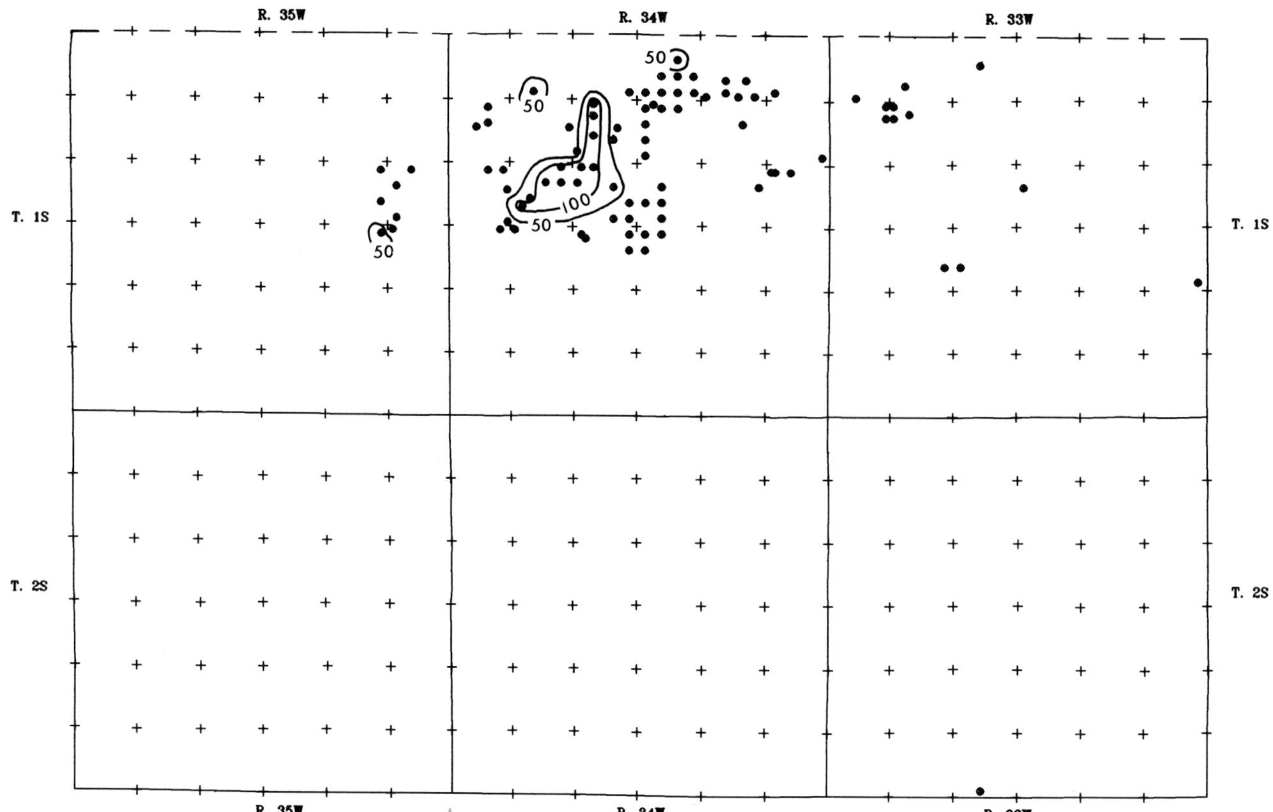

Figure 24. Map of initial water production. Six-township area defined in Figure 10. Contours correspond to initial water production when wells were completed. Contour interval is 50 barrels of water. Highest locations have highest water production.

voirs are being addressed but are still elusive. The fact that some reservoirs produce at economic rates only in certain areas, in spite of seemingly adequate porosity development, may be related to poorly understood relationships of timing and pathways of oil migration. More attention also needs to be given to defining and understanding basement geology and determining its role in influencing Phanerozoic reactivation of structures. Potential fields geophysics may be very useful in elucidating basement-related structures in areas of sparse control (Yarger, 1989).

ACKNOWLEDGMENTS

Thanks is extended to Robert Sampson for assistance in using and adapting new mapping software, Surface III. Thanks is also given to Edward Ackman, Delbert Costa, Shirley Paul, and Don Strong for background information on the initial exploration of the area. Roger Philpott was very helpful in providing information about the present activities and contacts relating to Cahoj. Thanks is given to Bruce Ruffin with Texaco, Inc., the major operator of Cahoj, for information about the field. Jim Anderson, who completed his M.S. thesis research on porosity development, cement, and fluid history, is acknowledged for his independent assessment of the formation of Lansing–Kansas City reservoirs. Jerry Clayton provided very useful data on the organic geochemistry of the hydrocarbon system affecting Cahoj. Appreciation is extended to Lea Ann Davidson for typing the manuscript, to Pat Acker and Renate Hensiek for assistance with graphic artwork, and John French for constructing Figure 7.

REFERENCES CITED

Anderson, J. E., 1989, Diagenesis of the Lansing and Kansas City groups (Upper Pennsylvanian), northwestern Kansas and southwestern Nebraska: Geological Society of America Abstracts with Programs, v. 21, n. 1, p. 1-2.

Beene, D. L., 1989, 1988 Oil and gas production in Kansas: Kansas Geological Survey Oil and Gas Production Data Set 88, 253 p.

Boardman, D. R., II, and P. H. Heckel, 1989, Glacial-eustatic sea-level curve for early Late Pennsylvanian sequence in north-central Texas and biostratigraphic correlation with curve for midcontinent North America: Geology, v. 17, p. 802-805.

Brewer, J. A., R. Good, J. E. Oliver, L. D. Brown, and S. Haufman, 1983, COCORP profiling across the southern Oklahoma aulacogen—overthrusting of the Wichita Mountains and compression within the Anadarko Basin: Geology, v. 11, p. 109-114.

Department of Energy, 1982, Abandoned oil fields in Kansas: DOE/BETC/IC-82/5, Bartlesville, Oklahoma, 280 p.

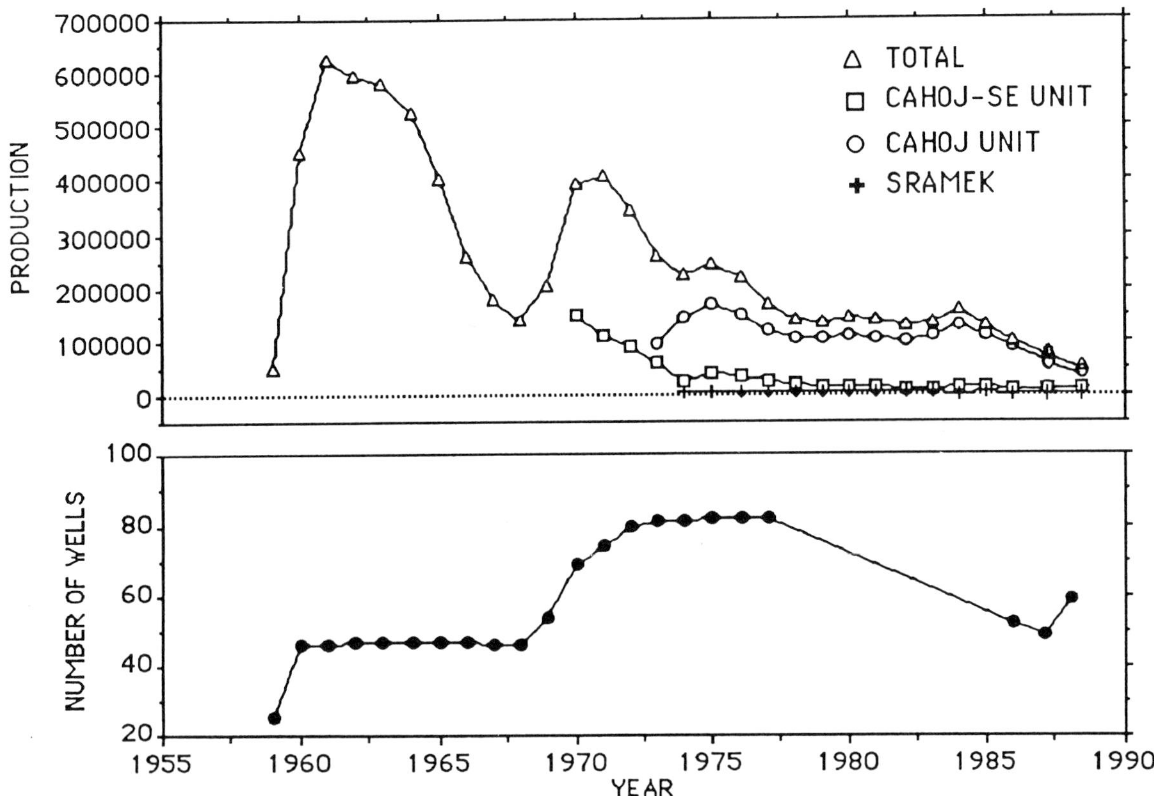

Figure 25. Annual production and well history of Cahoj field through 1988. Although maps include Cahoj field complex, these charts include only Cahoj field. The upper chart is the annual oil production for the field total and for major leases that underwent waterflooding. The waterflooded Cahoj lease accounts for over 90% of the oil production from the field. The lower diagram is an annual history of the number of wells. A gap in actual numbers of producing wells exists during the period of 1980 through 1985.

Dickinson, W. R., and H. Yarborough, 1979, Plate tectonics and hydrocarbon accumulation: American Association of Petroleum Geologists, Continuing Education Case Notes Series 1, 150 p.

DuBois, M. K., 1985, Application of cores in development of an exploration strategy for the Lansing-Kansas City "E" Zone, Hitchcock County, Nebraska: Kansas Geological Survey Subsurface Geology Series 6, Core Studies in Kansas, p. 120–132.

Dunham, R. J., 1962, Classification of carbonate rocks according to depositional texture, *in* W. E. Ham, ed., Classification of carbonate rocks: American Association of Petroleum Geologists Memoir 1, p. 108–121.

Ebanks, W. J., Jr., and W. L. Watney, 1985, Geology of Upper Pennsylvanian carbonate oil reservoirs, Happy and Seeberger fields, northwestern Kansas, *in* P. O. Roehl and P. W. Choquette, eds., Carbonate Petroleum Reservoirs: New York, Springer-Verlag, p. 241–250.

Galloway, W. E., M. S. Yancey, and A. P. Whipple, 1977, Seismic stratigraphic model of depositional platform margin, eastern Anadarko Basin, Oklahoma: American Association of Petroleum Geologists Bulletin, v. 61, p. 1,437–1,447.

Goebel, K. A., E. A. Bettis, III, and P. H. Heckel, 1989, Upper Pennsylvanian paleosol in Stranger Shale and underlying Iatan Limestone, southwestern Iowa: Journal of Sedimentary Petrology, v. 59, n. 2, p. 224–232.

Ham, W. E., and J. L. Wilson, 1967, Epeirogeny and orogeny in the central United States: American Journal of Science, v. 265, p. 332–407.

Harbaugh, J. W., and M. Ducastaing, 1981, Historical changes in oil field populations as a method of forecasting field sizes of undiscovered populations—a comparison of Kansas, Wyoming, and California: Kansas Geological Survey Subsurface Geology Series #5, 56 p.

Jenden, P. D., K. D. Newell, I. R. Kaplan, and W. L. Watney, 1988, Composition and stable-isotope geochemistry of natural gases from Kansas, Midcontinent, U.S.A.: Chemical Geology, v. 71, p. 117–147.

Kluth, 1986, Plate tectonics of the Ancestral Rocky Mountains, *in* J. A. Peterson, ed., Paleotectonic and sedimentation in the Rocky Mountain region, United States: American Association of Petroleum Geologists Memoir 41, p. 353–369.

Kumar, N., and R. M. Slatt, 1984, Submarine-fan and slope facies of Tonkawa (Missourian-Virgilian) sandstone in deep Anadarko Basin: American Association of Petroleum Geologists Bulletin, v. 689, p. 1,839–1,856.

Merriam, D. F., 1963, The geologic history of Kansas: Kansas Geological Survey Bulletin 162, 317 p.

Morgan, J. V., 1952, Correlation of radioactive logs of the Lansing and Kansas City Groups in Central Kansas: Journal of Petroleum Technology, v. 4, p. 111–118.

Newell, K. D., W. L. Watney, and J. R. Hatch, 1985, Time-temperature estimates of thermal maturation in north-central and northeast Kansas: Geological Society of America Abstracts with Programs, v. 17, p. 185–186.

Newell, K. D., W. L. Watney, S. W. L. Cheng, and R. L. Brownrigg, 1987, Stratigraphic and spatial distribution of oil and gas production in Kansas: Kansas Geological Survey Subsurface Geology Series #9, 86 p.

Newell, K. D., W. L. Watney, D. W. Steeples, R. W. Knapp, and S. W. L. Cheng, 1989, Suitability of high-resolution seismic method to identifying petroleum reservoirs in Kansas; a geological perspective, *in* D. W. Steeples, ed., Geophysics in Kansas: Kansas Geologic Survey. Report No. 226, p. 9–29.

Petroleum Information Corporation, Kansas Secondary Recovery Operations.

Prather, B. E., 1985, Depositional facies and diagenetic fabrics of the D-Zone cyclothem, Lansing-Kansas City groups,

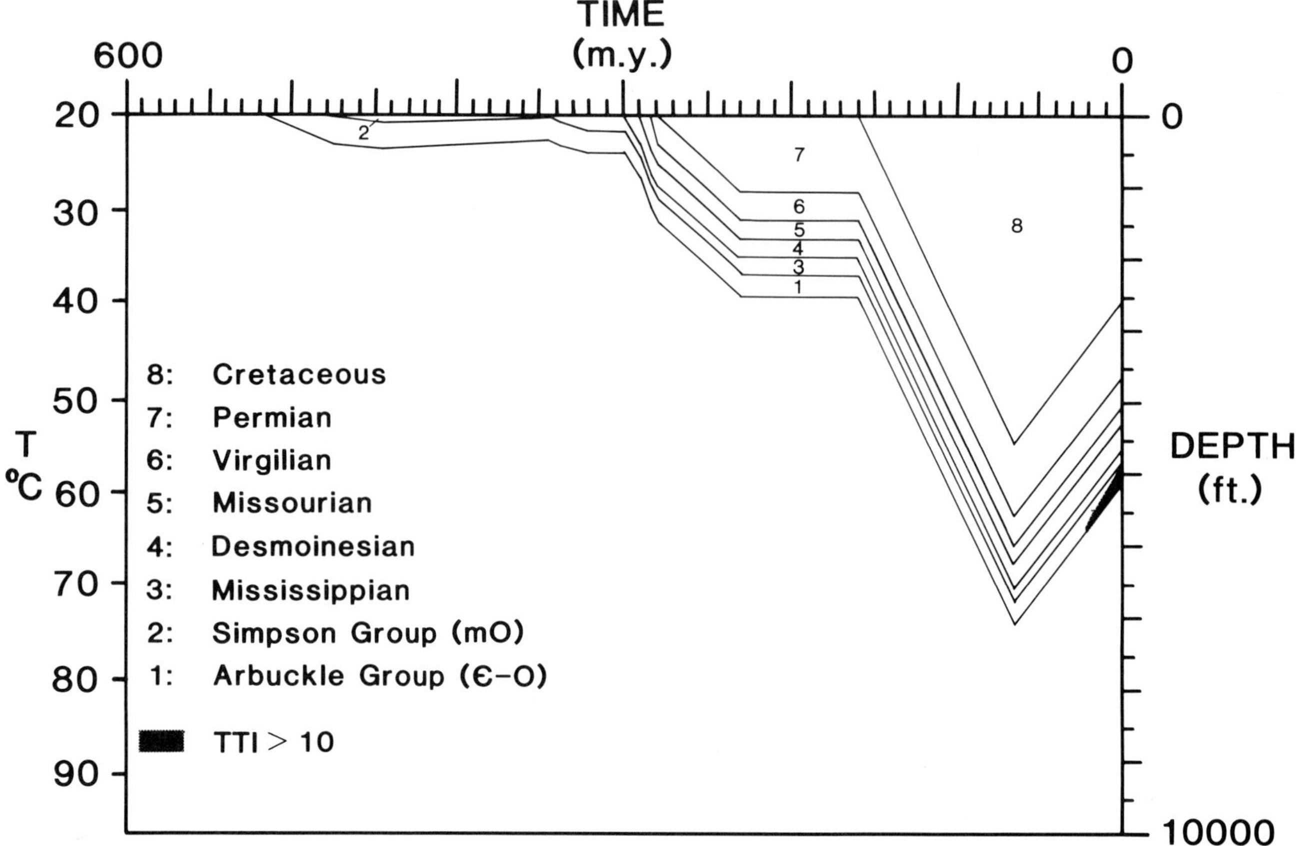

Figure 26. Time-depth burial history plot of various stratigraphic intervals, without consideration for compaction. Lopatin TTI calculation suggests that thermal maturation has been sufficient for initial oil generation (TTI > 10) in the Cambrian-Ordovician-age Arbuckle Group. Remainder of stratigraphic section including Lansing-Kansas City strata are immature according to these calculations.

Hitchcock County, Nebraska: Kansas Geological Survey, Subsurface Geology Series 6, Core Studies in Kansas, p. 133–144.

Rascoe, B., Jr., and F. J. Adler, 1983, Permo-Carboniferous hydrocarbon accumulations, Mid-continent, U.S.A.: American Association of Petroleum Geologists Bulletin, v. 67, p. 979–1,001.

Rogers, J. P., B. R. Stinson, and G. E. Morgan, 1983, Dry Creek Field, Nebraska: subsurface methods case history: American Association of Petroleum Geologists Bulletin, v. 67, p. 2208–2224.

Schmoker, J. W., 1986, Oil generation in the Anadarko basin, Oklahoma and Texas: modeling using Lopatin's method: Oklahoma Geological Survey Special Publication 86-3, 40 p.

Schutter, S. R., and P. H. Heckel, 1985, Missourian (early Late Pennsylvanian) climate in Midcontinent North America: International Journal Coal Geology, v. 5, p. 111–140.

Sloss, L. L., 1963, Sequences in the cratonic interior of North America: Geological Society of America Bulletin, v. 74, p. 93–114.

Walters, R. F., 1958, Differential entrapment of oil and gas in Arbuckle dolomite of central Kansas: American Association of Petroleum Geologists Bulletin, v. 42, p. 2133–2173.

Watney, W. L., 1979, Gamma ray-neutron cross-plots as an aid in sedimentological analysis, *in* D. Gill and D. F. Merriam, eds., Geomathematical and petrophysical studies in sedimentology: New York, Pergamon Press, p. 81–100.

Watney, W. L., 1980, Cyclic sedimentation of Lansing-Kansas City groups in northwestern Kansas and southwestern Nebraska: Kansas Geological Survey Bulletin 220, 72 p.

Watney, W. L., 1984, Recognition of favorable reservoir trends in Upper Pennsylvanian cyclic carbonates in western Kansas, N. J. Hyne, ed., *in* Limestones of the Mid-continent: Tulsa Geological Society Special Publication No. 2, p. 201–246.

Watney, W. L., 1985a, Resolving controls on epeiric sedimentation using trend surface analysis: Mathematical Geology, v. 17, p. 427–454.

Watney, W. L., 1985b, Evaluation of significance of tectonic, sedimentary, control versus eustatic control of Upper Pennsylvanian cyclothems in the western Midcontinent, *in* Recent interpretations of late Paleozoic cyclothems: Proceedings of 3rd Annual Meeting and Field Conference, Midcontinent Section of Society of Economic Paleontologists and Mineralogists, p. 105–140.

Watney, W. L., and W. J. Ebanks, Jr., 1978, Early subaerial exposure and freshwater diagenesis of the Upper Pennsylvanian cyclic sediments in northern Kansas and southern Nebraska: American Association of Petroleum Geologists Bulletin (Abs.), v. 62, p. 570–571.

Watney, W. L., and J. French, 1988, Characterization of carbonate reservoirs in the Lansing-Kansas City groups (Upper Pennsylvanian) in Victory Field, Haskell County, Kansas, *in* S. M. Goolsby and M. W. Longman, eds., Occurrence and petrophysical properties of carbonate reservoirs in the Rocky Mountain region: 1988 Carbonate Symposium, Rocky Mountain Association of Geologists, p. 27–46.

Watney, W. L., and S. E. Paul, 1983, Oil exploration and production in Kansas—present activity and future potential: Oil and Gas Journal, July 25, p. 193–198.

Watney, W. L., R. W. Knapp, J. A. French, Jr., and J. H. Doveton, 1988, Application of sequence stratigraphic analysis to thin cratonic carbonate-dominated shelf cycles (Upper Pennsylvanian) in the Midcontinent (abstract): American Association of Petroleum Geologists Bulletin, v. 72, p. 257.

Watney, W. L., J. C. Wong, and J. French, 1989, Computer simulation of coastal marine sedimentary sequences influenced by glacial-eustatic sea-level fluctuations: Abstracts, v. 3, p. (3-336)–(3-337), 28th International Geological Congress, Washington, D.C.

Yarger, H. L., 1989, Major magnetic features in Kansas and their possible geologic significance: Kansas Geological Survey Bulletin 226, p. 197–213.

SUGGESTED READING

Rascoe, B., Jr., and Adler, F. J., 1983, Permo-Carboniferous hydrocarbon accumulations, Mid-continent, U.S.A.: American Association of Petroleum Geologists Bulletin, v. 67, p. 979–1,001.

Watney, W. L., 1980, Cyclic sedimentation of Lansing-Kansas City groups in northwestern Kansas and southwestern Nebraska: Kansas Geological Survey Bulletin 220, 72 p.

Appendix 1. Field Description

Field name .... *Cahoj field*

Ultimate recoverable reserves .... *NA*

Field location:

Country .... *U.S.A.*

State .... *Kansas*

Basin/Province .... *Hugoton embayment*

Field discovery:

Year first pay discovered .... *Pennsylvanian Oread Limestone and Lansing-Kansas City groups zones 1959*

Discovery well name and general location:

First pay .... *Cahoj No. 1 SE SE NE Sec. 17, T1S, R34W*

Discovery well operator .... *Skelly Oil Company*

IP in barrels per day and/or cubic feet or cubic meters per day:

First pay .... *528 BOPD*

All other zones with shows of oil and gas in the field:

Age	Formation	Type of Show
None		

Geologic concept leading to discovery and method or methods used to delineate prospect, e.g., surface geology, subsurface geology, seeps, magnetic data, gravity data, seismic data, seismic refraction, nontechnical:

The geologic concept employed in search of this field was structural closure and proximity to shows of hydrocarbon. Earlier drilling was very limited. Seismic reflection profiling was used to delimit the structure.

Structure:

Province/basin type .... *Bally 221; Klemme IIA*

Tectonic history

Many of the major basin-uplift configurations observed today in northern Mid-Continent, including Central Kansas uplift, were result of major tectonic activity during Early Pennsylvanian. Hugoton embayment developed as broad area of subsidence north of the Anadarko basin in Permian-Pennsylvanian. Cahoj structure appears to have maintained relief by subsiding less rapidly than surrounding area. Reservoirs developed during times of regional sea-level oscillation affecting sedimentation and diagenesis in the vicinity of the field.

Regional structure

Cahoj field is located on a small, separate, structural closure adjoining the west flank of the Cambridge arch.

Local structure

Northeast-southwest-trending elongate dome.

Trap:

Trap type(s)

One anticlinal trap with multiple pays that show variable porosity development across structure due to depositional and diagenetic facies variation.

Basin stratigraphy (major stratigraphic intervals from surface to deepest penetration in field):

Chronostratigraphy	Formation	Depth to Top in ft (m)
Upper Cretaceous	*Niobrara Chalk*	*1120 (342)*
Jurassic	*Morrison Formation*	*2555 (779)*
Top Permian Cimarronian	*Blain Gypsum*	*2721 (830)*
Cimarronian	*Stone Corral Formation*	*2978 (908)*
Gearyan	*Chase Group*	*3122 (952)*
Top Pennsylvanian Virgilian	*Waubaunsee Group*	*3665 (1118)*
Virgilian	*Heebner Shale*	*3972 (1211)*
Missourian	*Lansing Group*	*4030 (1229)*
Des Moinesian	*Marmaton Group*	*4330 (1321)*
	Cherokee Group	*4466 (1362)*
Mississippian	*Osage Series*	*4664 (1422)*
Precambrian	*"Quartzite"*	*4712 (1437)*

Reservoir characteristics:

Number of reservoirs *11*

Formations .... *Shawnee Group Oread Limestone and Lansing-Kansas City groups*

Ages .... *Pennsylvanian (Virgilian and Missourian)*

Depths to tops of reservoirs .... *3950+ ft (1205 m)*

Gross thickness (top to bottom of producing interval) .... *350 ft (107 m)*

Net thickness—total thickness of producing zones

Average .... *14 ft (4.3 m)(total perforated interval)*

Maximum .... *60 ft (18 m)*

Lithology

Range of shelf carbonates affected by meteoric and burial diagenesis vary by zone and position in field; micritized; bioclastic grainstones and leached packstones; leached crinoid-brachiopod and phylloid algal wackestones; leaching create vugs, molds, and microcrystalline calcite spar replacement of micrite; minor sandstone/siltstone.

Porosity type .... *Moldic, vuggy, microcrystalline calcite porosity and intergranular porosity*

Average porosity *15%*

Average permeability .... *10 md*

Seals:

Upper

Formation, fault, or other feature .... *Lansing-Kansas City groups shales*

Lithology .... *Shale*

Lateral

Formation, fault, or other feature .... *Lansing-Kansas City groups shales*

Lithology .... *Carbonate (loss of porosity)*

Source:

Formation and age .... *Unconfirmed in field, but most common oil type in area is Mississippian-Devonian Chattanooga/Woodford*

Lithology .... *Shale*

Average total organic carbon (TOC) .... *NA*

Maximum TOC .... *Wide range, unsure of nature of source*

Kerogen type (I, II, or III) .... *NA*

Vitrinite reflectance (maturation) .... *NA*

Time of hydrocarbon expulsion .... *Hottest but not sufficient for area in late Tertiary*

Present depth to top of source .... *Source not present at location of field*

Thickness .... *NA*

Potential yield .... *NA*

Appendix 2. Production Data

Field name .... *Cahoj field*

Field size (complex: Cahoj, Cahoj West, Cahoj East, Cahoj Northeast, Cahoj Northwest, Cahoj South):

- **Proved acres** .... *2980 (complex)*
- **Number of wells all years** *184*
- **Current number of wells** .... *96 (producing)*
- **Well spacing** .... *40 ac reduced to 20 in selected areas under waterflood*
- **Ultimate recoverable** .... *NA*
- **Cumulative production (1988)** .... *9.36 million bbl*
- **Annual production (1988)** .... *144,000 bbl*
- **Present decline rate** *19%*
- **Annual water production** .... *NA*
- **In place, total reserves** .... *NA*
- **In place, per acre-foot** .... *NA*
- **Primary recovery** .... *NA*
- **Secondary recovery** .... *NA*
- **Enhanced recovery** .... *None*
- **Cumulative water production** .... *NA*

Drilling and casing practices:

- **Amount of surface casing set** .... *To 450 ft (137 m)*
- **Casing program** .... *5½-in. to TD*
- **Drilling mud** .... *Freshwater*

Completion practices:

- **Interval(s) perforated** .... *Multiple intervals in Lansing-Kansas City, variable between wells*
- **Well treatment** .... *15% HCl acid*

Formation evaluation:

- **Logging suites** .... *Older wells: IES, GR-sonic, microlaterolog followed later by GR-neutron-guard; modern wells: GR-neutron-density-sonic-resistivity*
- **Testing practices** .... *Drill-stem tests*
- **Mud logging techniques** .... *Not typically utilized*

Oil characteristics:

- **Type** .... *Paraffinic-naphthenic oils*
- **API gravity** .... *30–38°*
- **Base** .... *API*
- **Initial GOR** *0*
- **Sulfur, wt%** .... *NA*
- **Viscosity, SUS** .... *NA*
- **Pour point** .... *NA*
- **Gas-oil distillate** .... *NA*

Field characteristics:

- **Average elevation** .... *3150 ft (960 m)*
- **Initial pressure** .... *NA*
- **Present pressure** .... *NA*
- **Pressure gradient** .... *NA*
- **Temperature** .... *112°F (44.4°C)*
- **Geothermal gradient** .... *1.37°F/100 ft (25°C/km)*
- **Drive** .... *Solution gas/waterflood*

Oil column thickness .... *Multiple thin pay zones*
Oil-water contact .... *None*
Connate water .... *NA*
Water salinity, TDS .... *65,000 mg/L but variable*
Resistivity of water .... *0.07 ohm-m at 110°F*
Bulk volume water (%)

Transportation method and market for oil and gas:
Truck oil.

Prinos Field—Greece
Aegean Basin

P. PROEDROU
T. SIDIROPOULOS
Public Petroleum Corporation of Greece
Athens, Greece

FIELD CLASSIFICATION

BASIN: Aegean
BASIN TYPE: Backarc
RESERVOIR ROCK TYPE: Sandstone
RESERVOIR AGE: Miocene
PETROLEUM TYPE: Oil
TRAP TYPE: Faulted Anticline
RESERVOIR ENVIRONMENT OF DEPOSITION: Turbidite Channel Fill
TRAP DESCRIPTION: Roll-over faulted anticline with minor lateral stratigraphic closure

LOCATION

The Prinos oil field is located in the North Aegean sea, 6 km northwest of the Greek island of Thassos, in water depths of 31 m (Figure 1). The only adjacent field is the small gas reservoir of South Kavala, 11 km to the south. Both fields are traps in anticlines within the young taphrogenic Prinos basin. An ultimate oil production of 90 million bbl has been estimated.

HISTORY

Exploration in the area began with extensive seismic campaigns east and west of the Island of Thassos in the early 1970s. It was a time when major oil companies held concessions in different parts of the Aegean Sea.

Exploration was initiated by Oceanic Exploration Company of Greece in consortium with Hellenic Oil Company Inc., White Shield Greece Oil Corporation, and Colorado Greece Oil Corporation. A contract was signed with the Greek state in 1969 for the northern part of the Aegean Sea (Thrace Sea) from the Peninsula of Athos on the west to the Greek-Turkish border on the east. Later relinquishments have reduced the concession area to its present size of 1700 km^2 (Figure 1). The concession is presently operated by the North Aegean Petroleum Company (NAPC), which is owned by Wintershall Actien Gesellschaft, Denison Mines, Hellenic Oil Company Inc., and White Shield Greece Oil Corporation.

The first well, East Thassos 1, was drilled 20 km east of Thassos on a large anticline in 1971. It encountered an oil accumulation but with very low gravity oil.

The next two wells were drilled west of the island in 1972–1973 on a fault-bounded anticline combined with a stratigraphic pinch-out. This led to the discovery of the South Kavala gas field. Subsequent testing proved this field to be very small and for the time being not economical.

The existence of hydrocarbons in the area encouraged further drilling. The fourth well, Prinos 1, was drilled close to the top of the Prinos structure at the end of 1973 and was completed in February 1974. This was the discovery well for Prinos field. The Prinos 2 well, completed in April 1974, confirmed the discovery.

The structure itself consists of two anticlines separated by a small graben (Figure 2). The caprock is a thick salt section intercalated with clastics. A generalized stratigraphic column is shown by Figure 3. The producing zones are located not within the salt section, as at South Kavala, but immediately below it in a submarine fan of upper Miocene age (Figure 4). There are three reservoirs, designated A, B, and C, which in the central part of the field are found at depths of 2488.5 m, 2642 m, and 2743 m, respectively (Figure 5). The initial pressures for the three zones were 5735 psi (403.6 kg/cm^2), 5840 psi (411.0 kg/cm^2), and 6000 psi (422.0 kg/cm^2), respectively.

During 1975–1977, four delineation wells were drilled. The second delineation well (Prinos 4) proved that the northeast high, later named North Prinos, is completely separated from the rest of the structure by a fault, downthrown to the southwest (Figure 2). This reduced the presumed original area of Prinos field by nearly half. Additionally, the North Prinos reservoir also proved to be of limited extent, with a thinner pay zone and with a different oil quality. Another well drilled on this structural high, close to Prinos 4, the Prinos North 1, was nonproductive.

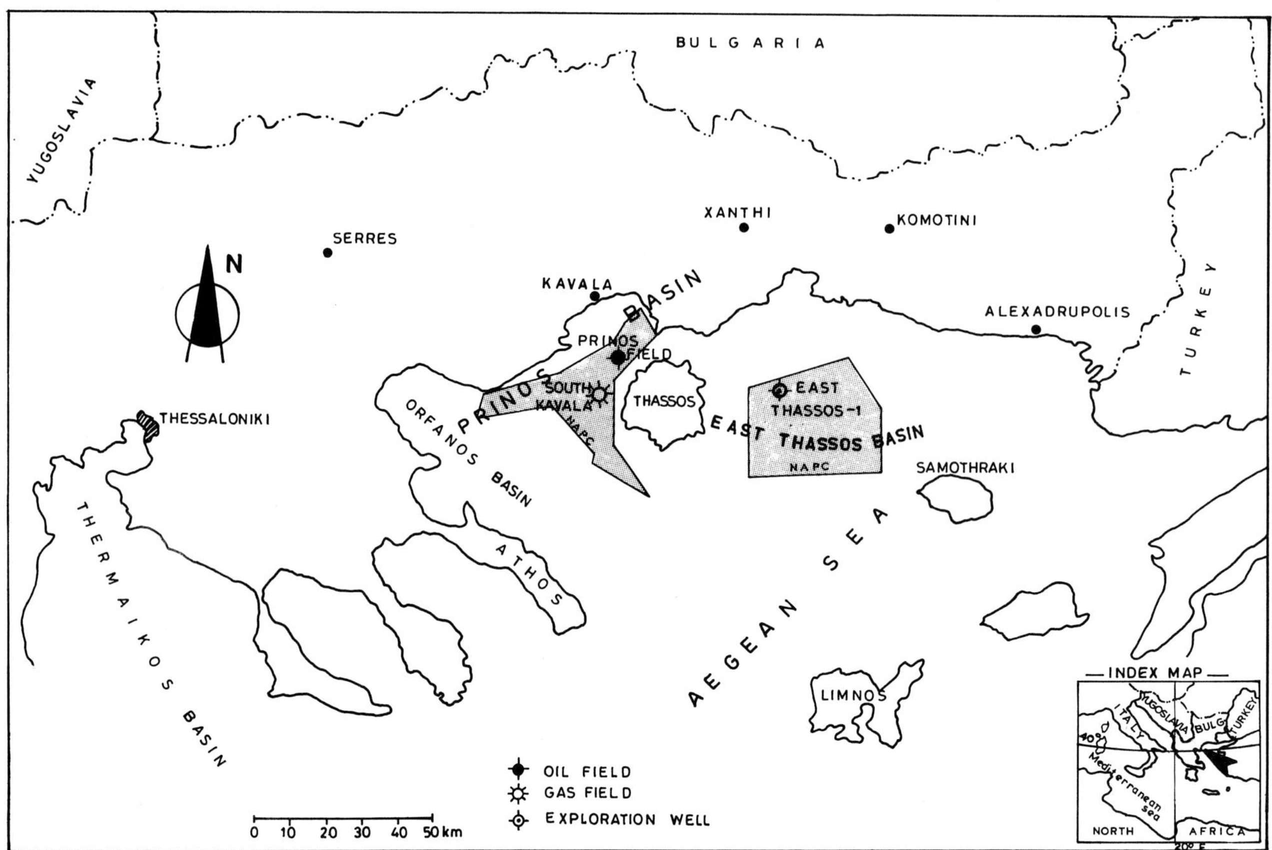

Figure 1. Location of NAPC's concession areas and Prinos and South Kavala fields.

Tests of reservoirs below the main Prinos zones were not successful.

DISCOVERY METHOD

The breakup and rapid subsidence of the Aegean Sea due to the extensional tectonism during the Tertiary resulted in the accumulation of thick sedimentary deposits in the newly formed grabens. Previous drilling in the nearby Thermaikos basin had indicated a sediment thickness of several thousand meters with encouraging gas shows. The backarc conditions of the area resulted in intensive tectonism accompanied by high thermal flow to the newly formed basins. Recent tectonism, rapid and thick sediment accumulations, and high heat flow were the key elements that made the North Aegean basins good prospective areas.

The first seismic campaign in 1970 showed the existence of two grabens east and west of the Thassos Island horst block, and a third, the Orphanos graben, to the southwest. Sediment thicknesses were estimated to be more than 5000 m, with a large number of fault-bounded structures.

Good seismic reflectors in the uppermost Miocene section were proved by drilling to be evaporites. The marine and lagoonal nature of the basin reflects the rapid subsidence and subsequent separation into taphrogenic basins, which at times were completely landlocked.

STRUCTURE

The Prinos basin began its development in early to middle Miocene time along northeast- and northwest-trending faults at the southern margin of the crystalline Rhodope Massif. The first deposits were continental with abundant coal occurrences. With continuing subsidence in upper Miocene time, the sea transgressed onto the older sediments. Shale was deposited in the central and deep parts of the basin, and submarine and slump fan deposits were deposited at the periphery. Evaporites a few meters thick intercalated in the marine deposits indicate the temporary restriction of the basin ("sub-evaporitic formation," Figure 4). Restriction became more pronounced during Messinian time, resulting in the deposition of about 800 m of salt alternating with clastics ("evaporites," Figure 3). By Messinian time the South Kavala ridge, the southern limit of the basin, had been uplifted and had isolated the basin from the rest of the sea except for periodic marine

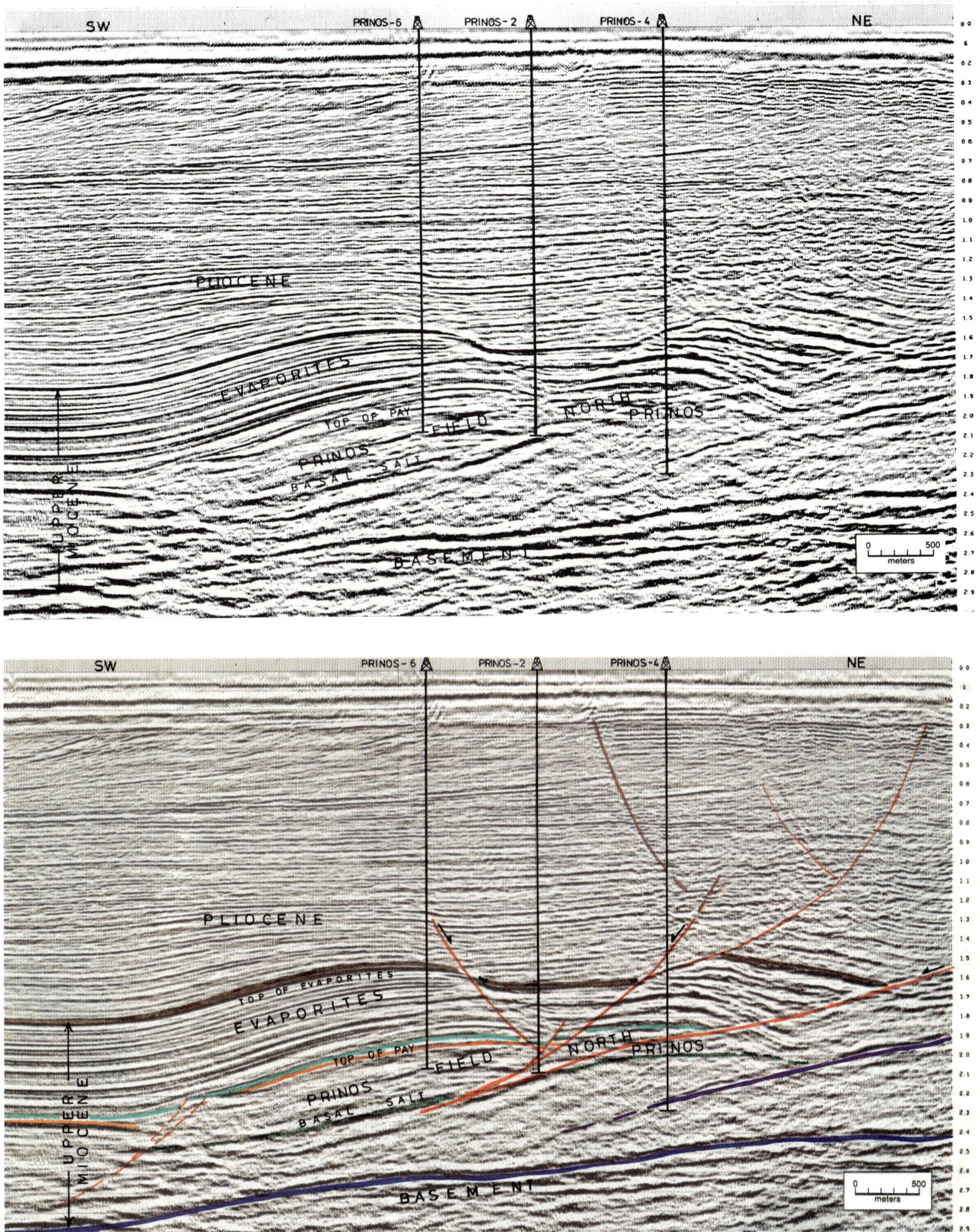

Figure 2. Northeast-southwest seismic section across the Prinos field (see Figure 6 for location). (A) Uninterpreted. (B) Interpreted. A southwest-dipping fault that formed the Prinos field roll-over and separates it from the North Prinos structure merges with a sole fault that underlies the evaporitic section in the North Prinos area. The graben between Prinos and North Prinos is the result of a northeast-dipping antithetic fault. Pre-Tertiary basement dips to the southwest. Note the poor reflection quality of the Prinos reservoir interval indicative of its complex depositional history.

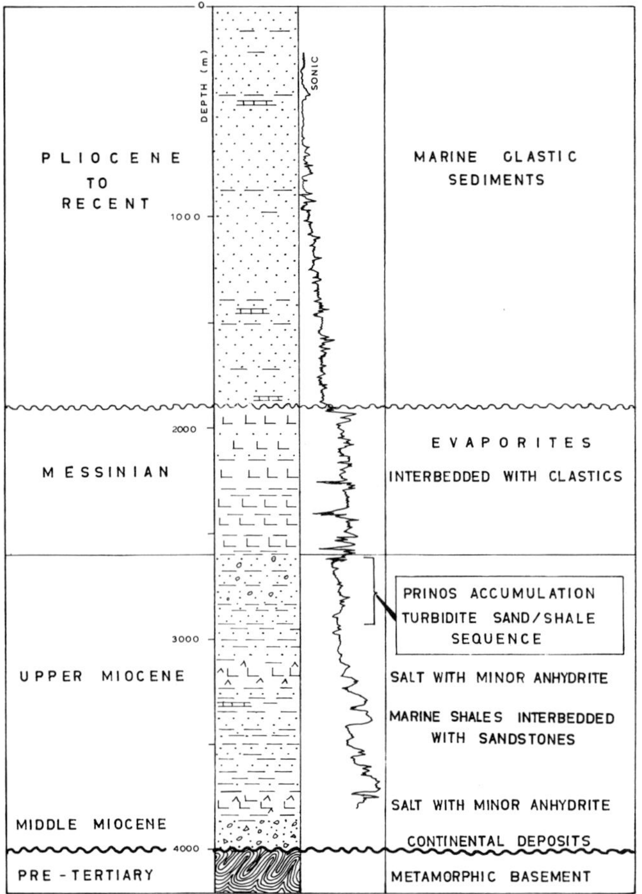

Figure 3. Generalized stratigraphic column of Prinos field area.

incursions. At the beginning of the Pliocene, marine sedimentation was reestablished over the entire basin ("post evaporitic formation"). Pliocene transgression resulted in the erosion of parts of these latest evaporite deposits (Figure 5). Temporary uplifting at the periphery of the basin in late Pliocene time resulted in erosion of parts of the margins of the older sediments.

All the structures in the Prinos basin are fault-bounded anticlines or stratigraphic pinch-outs, or a combination of both. Northwest- and northeast-trending faults dominate throughout the basin, and some of these existed from the beginning of basin development. Activation of the faults mostly occurred during sedimentation (syngenetic faults), resulting in roll-over folds along the downthrown side.

The Prinos structure is a broad dome-like anticline in front of a northwest-striking and southwest-dipping fault (Figures 6, 7, and 8). The fault began to form just above where the basement deepens to the southwest, creating a paleoslope for submarine fan deposition (Figure 9). Equivalent fan deposits do not exist on the upthrown side of this fault. Its activation continued into early Pliocene, dying out gradually. A secondary fault, dipping to the northeast and intersecting the first one below the reservoir, modifies the crest of the anticline, forming a local graben in the evaporitic section. Additional faulting complicates the structure. The North Prinos structure is even more complex, having a sole fault that underlies the evaporitic section and truncates some if not all of the reservoir beds (Kellogg, 1988).

The fan deposits dip steeply toward the south, west, and southwest, where they pinch out into marine shales.

STRATIGRAPHY—DEPOSITIONAL MODEL

Miocene sedimentation began with continental deposits and was followed by deposits of marine shales with interbedded sandstones (generalized stratigraphic column, Figure 3). Restricted environments, during which salt and anhydrite beds were formed, occurred at least three times during the Miocene. The first interval, 40–50 m thick, is found very close to basement. The second, a few tens of meters thick, accumulated just before deposition of the Prinos fan (main reservoir sandstones) and includes some dolomite. The third episode of restricted environment came at the end of the Miocene (Messinian) and resulted in the main evaporitic sequence within the Prinos basin; it consists of several thick salt layers interbedded with sandstone and shale and has a maximum thickness of 800 m. Deposition of marine clastics resumed in the Pliocene and has continued to the present.

The Prinos submarine fan, referred to as the *Prinos envelope*, was deposited along the downthrown side of a fault escarpment (Figure 9). Recognition of sedimentary structures and the relationship of different facies has led to the conclusion that these turbidites were deposited in channels, interchannels, and suprafan lobe environments during repeated prograding cycles. Four individual facies, A4/B2, C, D, and E, representative of the turbidite facies classification of Walker and Mutti (1973) that modifies the classical Bouma (1962) sequence, have been recognized by Robertson International Limited (unpublished report, 1981) and are adopted for this study. The most abundant and generally widespread facies, A4/B2, is considered to represent channel deposits of the upper and middle fan (Figure 9). Toward the center of the basin, but still within the perimeter of the Prinos field, they grade to suprafan lobe sediments of facies C that usually are overlain by facies A4/B2. Facies E represents channel margin deposits, whereas facies D is characterized by distal and lateral turbidites of the lower fan.

The sandstones of facies A4/B2, which form the coarsest-grained basal part of each cycle, are characteristically massive, with abrupt planar or scoured bases. They grade toward the top into

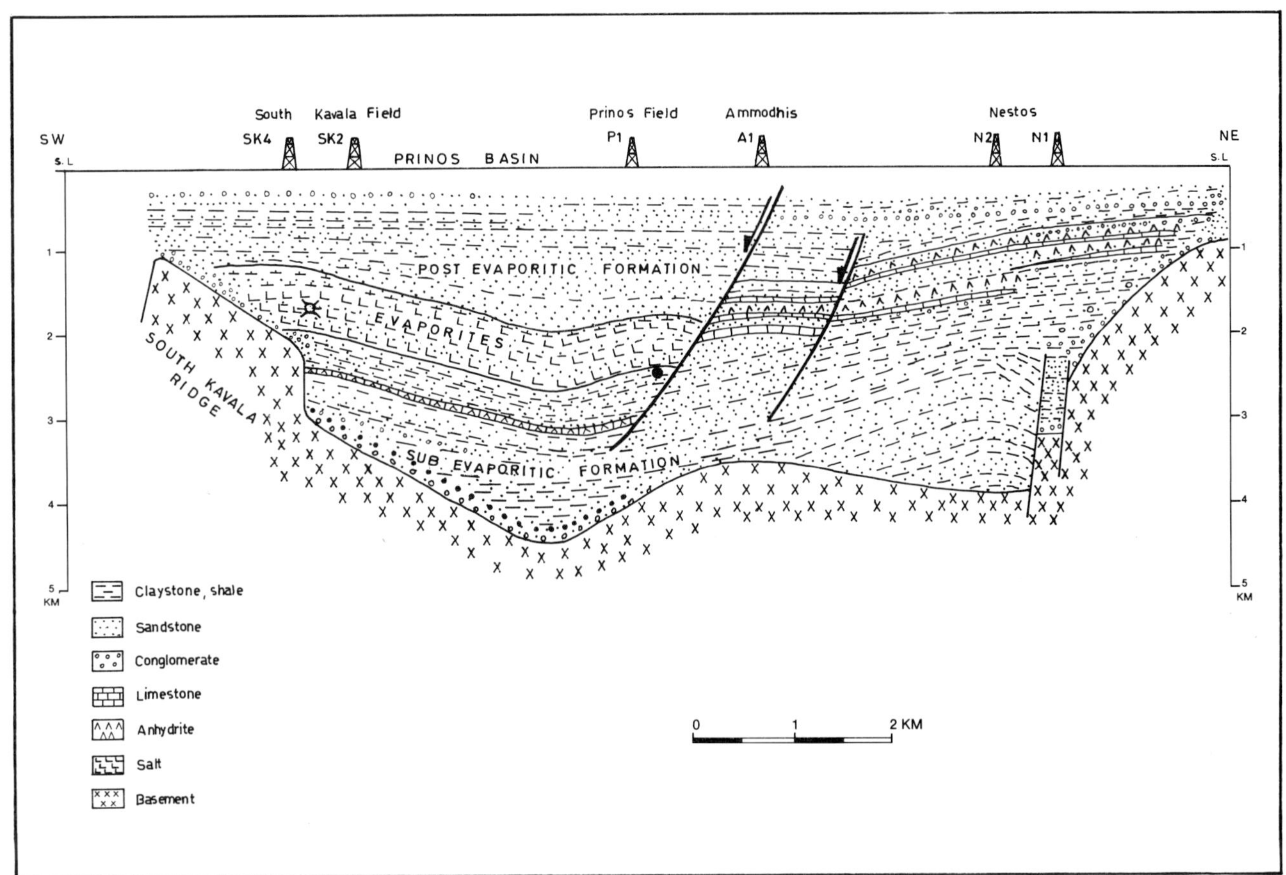

Figure 4. Simplified cross section of the Prinos basin parallel to the long axis. Structures are formed by roll-over on the downthrown sides of transverse synsedimentary faults. The Prinos field reservoirs are located immediately below the main evaporitic section, whereas at South Kavala field the reservoir is within the evaporitic section. All depths in kilometers below mean sea level.

medium to fine planar-bedded or rippled horizons. At the uppermost level, shale units of facies D and E were deposited (Figure 10). The sand units of facies C are regularly bedded and grade upward into shale.

The poor sorting and clay-poor matrix suggest rapid deposition. However, the large percentage of igneous and metamorphic rock fragments and of feldspars and the angular to subangular shape of the sand components indicate that the sediments were not transported over great distances nor subjected to much reworking.

The Prinos envelope consists entirely of turbiditic fan deposits overlying incised marine shales and capped by Messinian evaporites. It is subdivided into three separate reservoirs, A, B, and C, which are separated from each other by widespread shale units (Figure 5). Reservoir A is subdivided into two zones, A1 and A2. They are separated by a thick shale interval and are characterized mainly by thick channel and interchannel deposits with lobe accumulations at the channel front edges. The two lower reservoirs, B and C, are represented predominantly by the turbidite facies C, D, and E and less by thin channel deposits.

Definition of the various facies types in the Prinos field is mainly based on wireline log analysis, including gamma ray (GR), formation density compensated (FDC), compensated neutron log (CNL), caliper, high resolution dipmeter (HDT), and the dipmeter arrow plot. Petrographic and sedimentologic analysis of cored intervals and correlation of these to the wireline logs was the key for recognition of the different facies types in the other wells.

Fining- or coarsening-upward sequences indicated by the GR curve provide useful information for the recognition of cycle types and depositional environments. Density and neutron-porosity curves were also used for facies definition. The caliper log and the GR log were used mainly as supplementary criteria for the distinction between the E and D turbidite facies units.

HDT log analysis has provided useful information for distinguishing between A4/B2 and C facies. Furthermore, the evaluation of dipmeter patterns helped in the definition of the channel axes and sand-thickening directions.

Facies distribution maps indicate a north-northeast and northeast channel orientation (Fig-

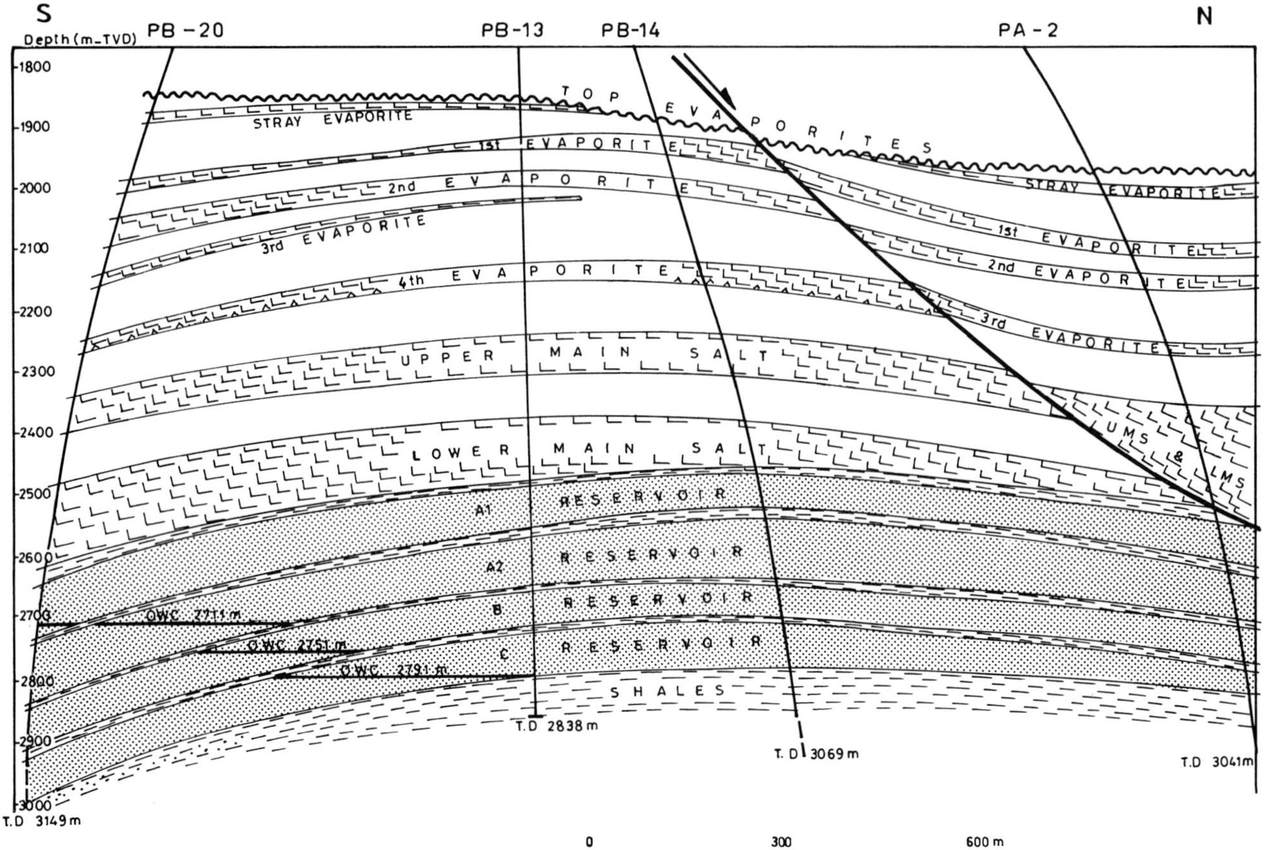

Figure 5. North-south cross section of the "Main Evaporite" and the underlying reservoir section, Prinos field (see Figure 6 for location). The three main reservoirs, A, B, and C, are each capped by shale beds and have separate oil-water contacts. The A reservoir is divided into upper and lower sands (A1 and A2) where a shale interbed is locally present; elsewhere the two units are in contact resulting in a common oil-water contact. UMS, "Upper Main Salt." LMS, "Lower Main Salt."

ure 9). The source area was of metamorphic and igneous rocks to the north and northeast. Isopach maps reveal a gradual thinning of the submarine fan deposits toward the southwest, southeast, and northwest, confirming the direction of sediment derivation (Figures 11 and 12).

To the south of the Prinos field, the reservoir of the South Kavala field is a turbidite fan deposit between two salt layers in the evaporitic section. Hydrocarbon shows also occur in these turbidite sandstones in other parts of the basin.

Marine shales below the Prinos fan and in laterally equivalent beds toward the central part of the basin are thought to be the oil source beds for the Prinos field. The maximum thickness of the subevaporitic shale section is believed to be over 1000 m. The total basin sediment thickness is about 5500 m. Shales within the evaporites also may contribute to oil generation in the deepest part of the basin.

TRAP

Prinos is a roll-over anticline bounded on the northeast by a sealing fault dipping toward the central deep part of the basin. Vertical closure is at least 300 m. The top is sealed by thick evaporites and by a persistent shale layer of about 20 m thickness just above the reservoir and below the evaporites (Figure 5).

The fault that formed the roll-over anticline is believed to be syndepositional and of late Miocene age. Fault movement continued during deposition of the lower part of the lower Pliocene sediments, but the fault dies out rapidly upward. All of the main faults of the trap are recognizable in the seismic data and confirmed by evaluation of dipmeter surveys. Local faults were identified by log correlations. The large faults that formed the basin are the only ones that can be observed in outcrops onshore (on Thassos Island).

Oil-water contacts for the three main reservoirs, A, B, and C, occur at depths of 2711 m, 2751 m, and 2791 m, respectively (Figures 5, 6, 7, and 8). The reservoirs are separated from each other by shale units that can be correlated in most wells. No internal unconformities exist except for scouring at the bottom of the channels.

The main northwest-trending fault that separates the Prinos field from North Prinos anticline is a

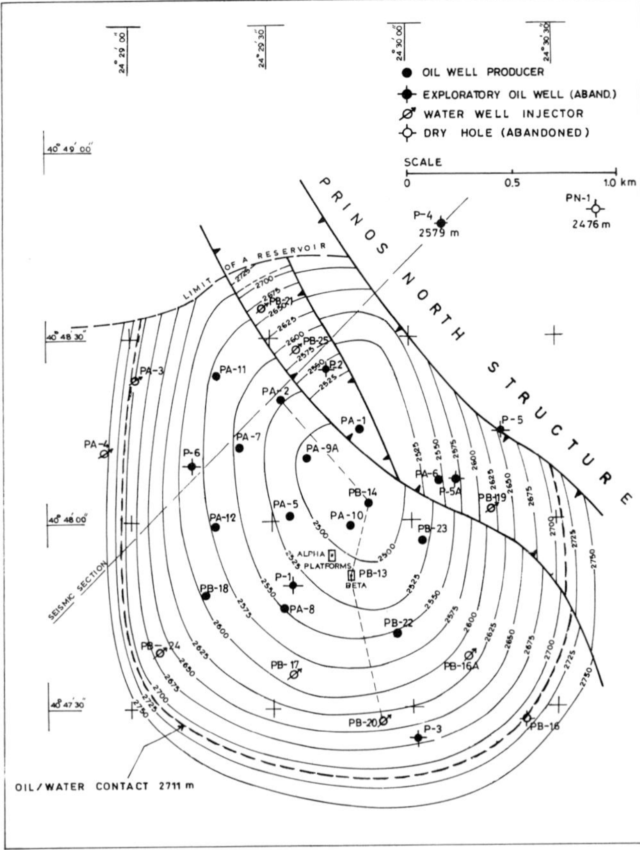

Figure 6. Structure map of Prinos field contoured on the top of the A reservoir. The structure is a dome-like roll-over on the downthrown side of a northwest-trending normal fault. A parallel northeast-dipping secondary fault modifies the crest of the structure. Note oil-water contact at 2711 m. All depths below mean sea level (MSL). Contour interval, 25 m.

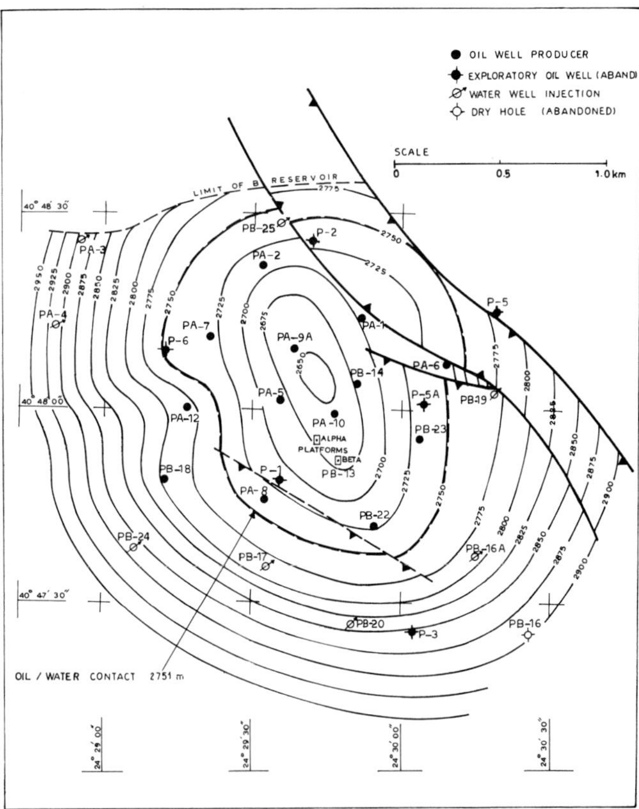

Figure 7. Structure map of Prinos field contoured on the top of the B reservoir. Note separate oil-water contact at 2571 m. All depths below MSL. Contour interval, 25 m.

sealing fault. The two structures contain quite different types of oil. Smaller faults, all parallel to the main one, cut through the main reservoirs or are restricted in the lower two reservoirs. They appear to act as barriers, although it is difficult to determine the role of faults versus facies deterioration.

Reservoirs

The Prinos reservoirs are the result of local facies development within the upper part of the upper Miocene marine shale section. The three sandstone units are intercalated with thin shale layers and are separated from each other by two unique shale horizons having fieldwide occurrence. These form three independent reservoirs, each with separate oil-water contacts.

The subsea depths at the highest parts of the reservoirs are as following:

	Top	Bottom
Reservoir A	2488.5 m	2636 m
Reservoir B	2642 m	2737 m
Reservoir C	2743 m	2785 m

Petrographically the sandstones are lithic-subgraywackes with acidic igneous, volcanic, metamorphic, and rare carbonate rock fragments.

The porosity is primary and governed by the depositional facies. Channel sandstones show higher porosities than suprafan lobe sandstones. Porosities in the thin turbidite sandstones are very low. Porosity decreases from top to bottom, as does permeability (Figure 13). The predominant A4/B2 turbidite facies unit consists of coarse and pebbly sandstone, which is poorly sorted, angular to subangular at the base, to medium, fine, and well sorted at the top (Figure 14). This facies has the best reservoir qualities, with maximum porosity and permeability values varying from 15 to 24% and from a few millidarcys to more than 2 darcys, respectively. Individual sandstone beds are continuous in the direction parallel to the

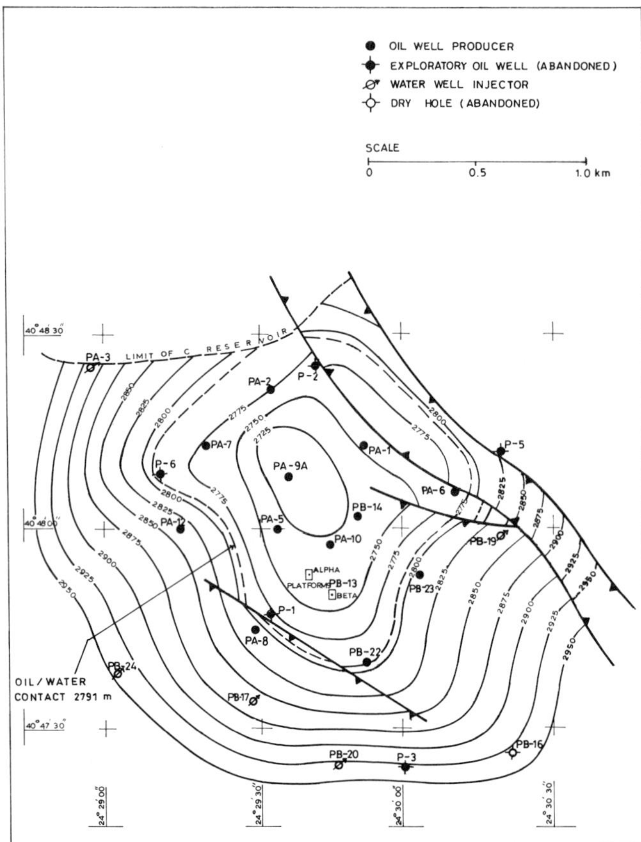

Figure 8. Structure map of Prinos field contoured on the top of the C reservoir. Note separate oil-water contact at 2791 m and presence of additional small faults. All depths below MSL. Contour interval, 25 m.

Figure 9. Facies distribution map of the A2 reservoir, which is representative of the main Prinos reservoirs. Note the northeast-trending channels characterized by deposits of turbidite facies A4/B2. These grade laterally to suprafan lobe deposits (facies C) and to lower fan classical turbidites (facies D). Overbank deposits (facies E) are observed along channel edges.

channel axis and may communicate with other channels, thus providing great lateral extension. Shale beds may also be quite continuous laterally and distally.

Facies C sandstone is generally medium to coarse grained, occasionally fine grained, thin-bedded and more argillaceous, grading into shale. Sorting is poor to good. Average porosities and permeabilities are less than in the A4/B2 facies, varying from 12 to 20% and to a maximum of a few hundred millidarcys, respectively. The detrital components of both facies consist primarily of quartz, acidic igneous rock fragments, and feldspar.

Facies D is shale with thin turbidite sandstone intercalations. The sandstones are mainly fine to medium grained, moderately to well sorted, with minor feldspar fragments. Detrital clay has been replaced by cementation. Facies E consists of alternating shales and fine-grained argillaceous sandstones with minor feldspar and detrital mica. Sandstones of both facies are generally thin bedded and tightly cemented and have very low porosities (around 10%) and permeabilities (few millidarcies).

Early oil migration appears to have inhibited loss of reservoir quality by diagenesis. Siderite and dolomite are the most common interstitial cements. Siderite generally replaces detrital clay, mudstone intraclasts, and locally interbedded shales, reducing the permeability but not the porosity. Silica cementation is very limited, and authigenic kaolinite is present in small amounts. Feldspar dissolution has enhanced secondary porosity to a small degree.

Most dolomite cementation has occurred in the vicinity of shale beds and increases in amount at the oil-water contact. The impact of diagenesis is quite strong below the oil-water contact, where porosity is greatly reduced. This suggests that dolomite cementation took place mainly after oil migration.

The average net pay for the A reservoir is 97 m; for the B reservoir, 22 m; and for the C reservoir, 23 m. The equivalent average porosities and permeabilities are 18% and 115 md for the A reservoir, 14.5% and 17 md for the B reservoir, and 15% and 25 md for the C reservoir. Horizontal permeabilities are generally higher than vertical ones by a few

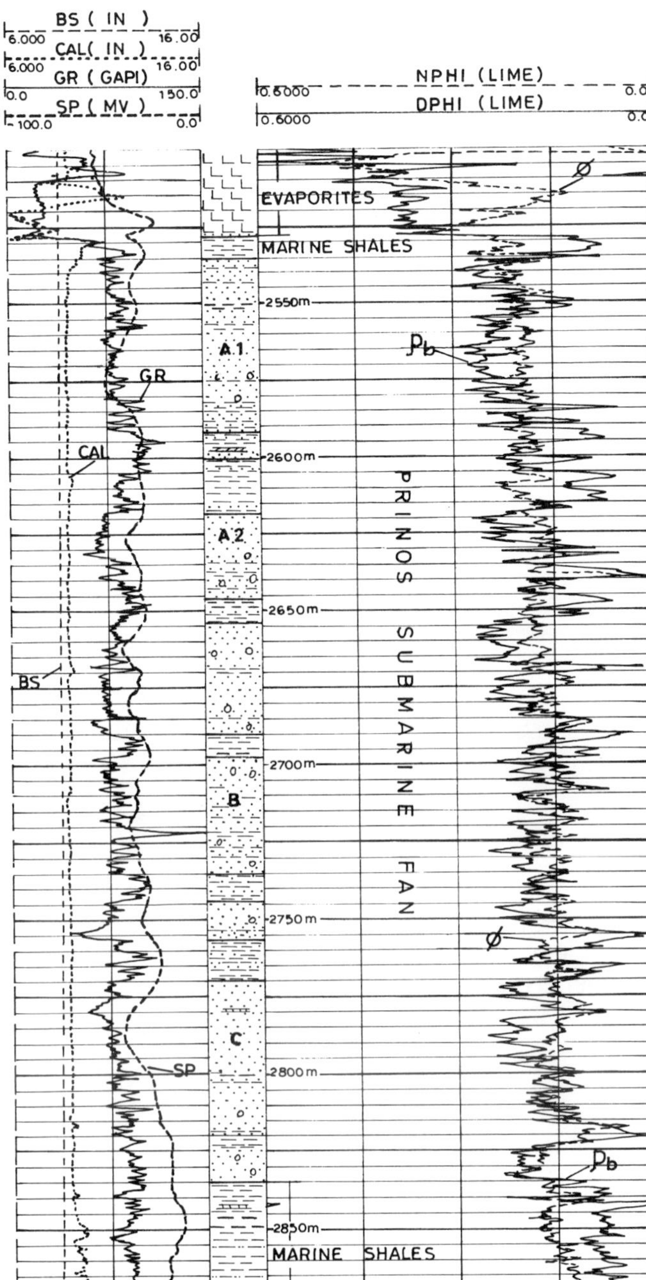

Figure 10. Lithological and wireline log of the reservoir interval in the PB-13 well. Gamma ray and density neutron response indicates fining upward of the channel sediments (e.g., 2735 m to 2756 m, etc.) and coarsening upward of overlying suprafan lobe deposits (e.g., 2775 m to 2800 m, etc.). Depths in meters below rotary table (R.T. elevation, 31.4 m above MSL).

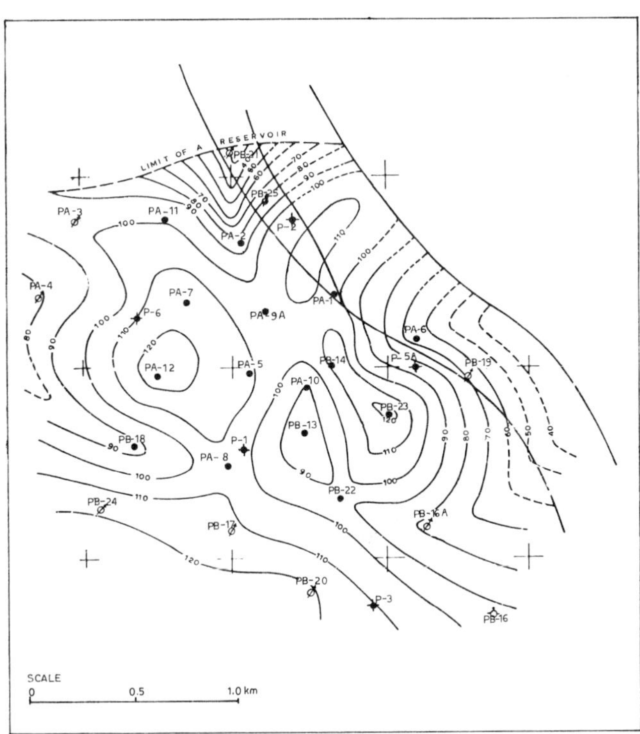

Figure 11. Isopach map of the A2 reservoir. The greatest thicknesses represent channel deposits. Contour interval, 10 m.

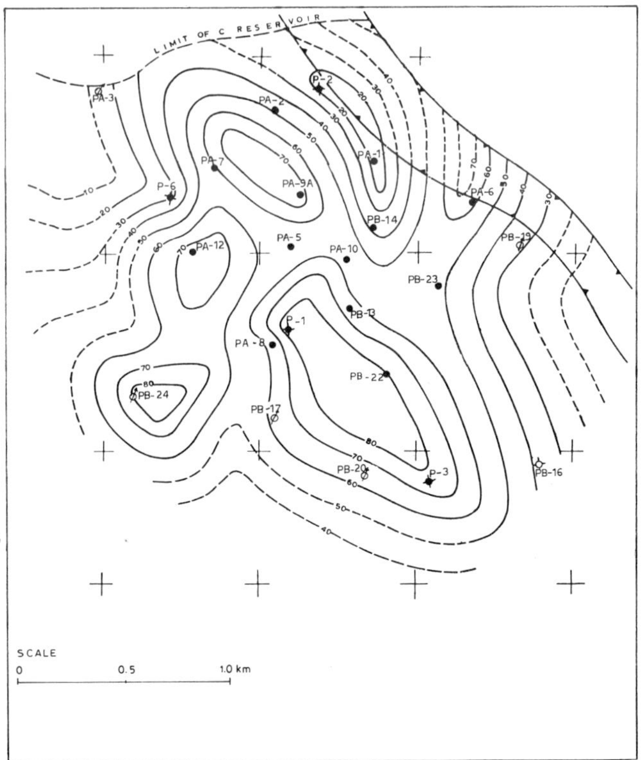

Figure 12. Isopach map of the C reservoir. Contour interval, 10 m. Evaluation of dipmeter patterns helped to define channel axes and reservoir facies developments.

hundred millidarcys in the more permeable zones. The differences are very small where permeabilities are less than 50 md. Good correlation exists between porosity and permeability values measured from cores. Both values decrease from the top to the bottom of the reservoir beds. The values decrease very rapidly below the oil-water contact (Figure 13).

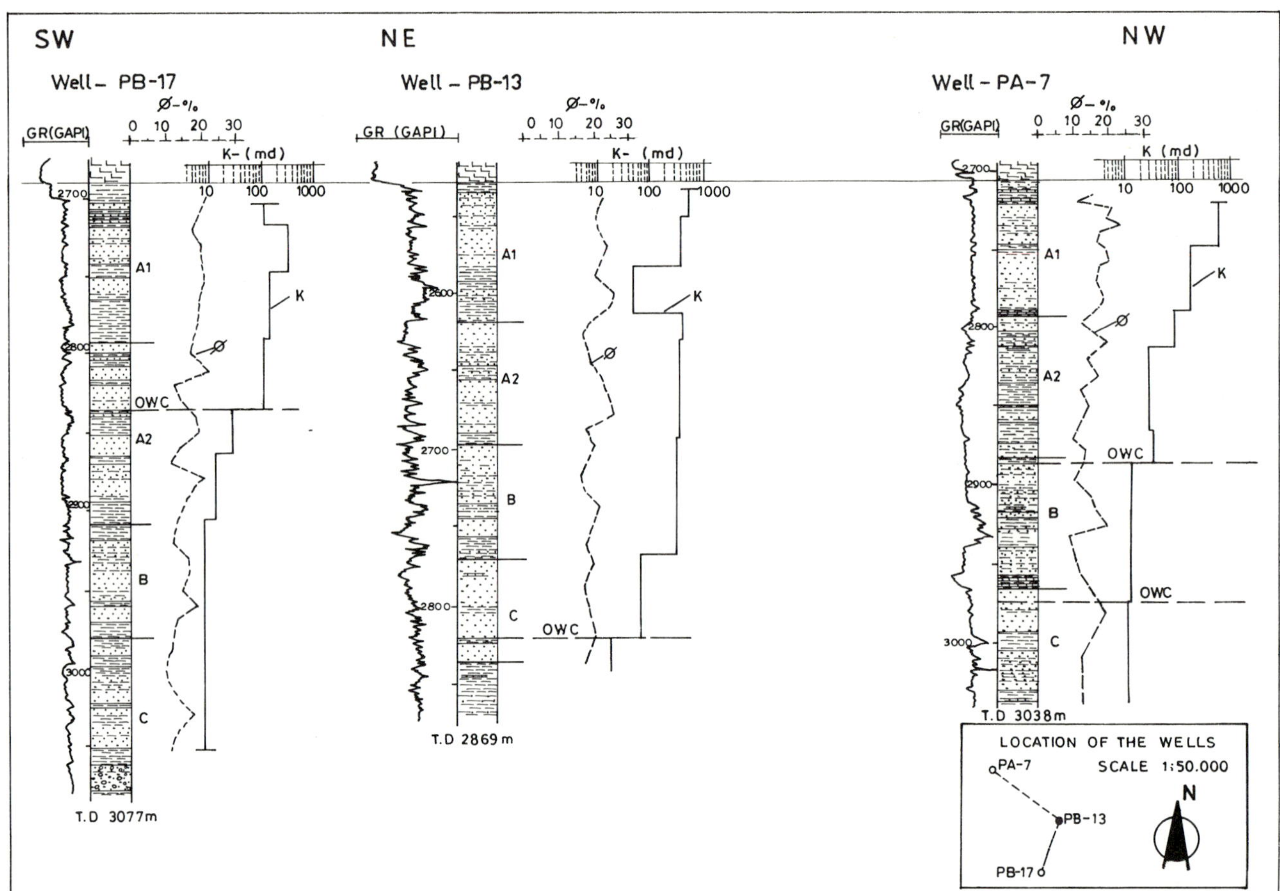

Figure 13. Porosity and permeability plots of three wells in Prinos field. Porosity values are calculated from log suites, and permeability values are averages for gross reservoir intervals. Note the general reduction in reservoir quality downwards and the sharper reduction below the oil-water contacts. All depths (measured depths) in meters below R.T.

Figure 14. Photomicrograph of facies A4/B2 in the Prinos field. Coarse sandstone with angular to subangular, poorly sorted quartz components. Intergranular porosity is filled with authigenic clay minerals and bituminous dead oil. Plane polarized light. Scale bar equals 0.16 mm.

Subdivision of the reservoirs into separate sandstone layers is based mostly on log characteristics and correlation of the intercalated shale beds. Pressure measurements taken during the production testing helped in correlating reservoir zones. The A reservoir has been subdivided into seven layers, and the B and C reservoirs into two each (Figure 15). Only in a few cases was it possible to correlate individual shale units with confidence across the field. Often the shales pinch out and the overlying and underlying sandstone layers merge into one; occasionally a sandstone layer grades laterally into a shaly to fine sandy section. Some individual sandstone bodies are completely isolated.

Thicknesses of the interbedded shale units vary from 1 to 30 m. Shale interbeds are thicker and reservoir quality is poorer in the southeast part of the field. Permeability deterioration of layer 3 of reservoir A is pronounced in the southeastern half of the field where the deposits are characterized exclusively by facies D and E.

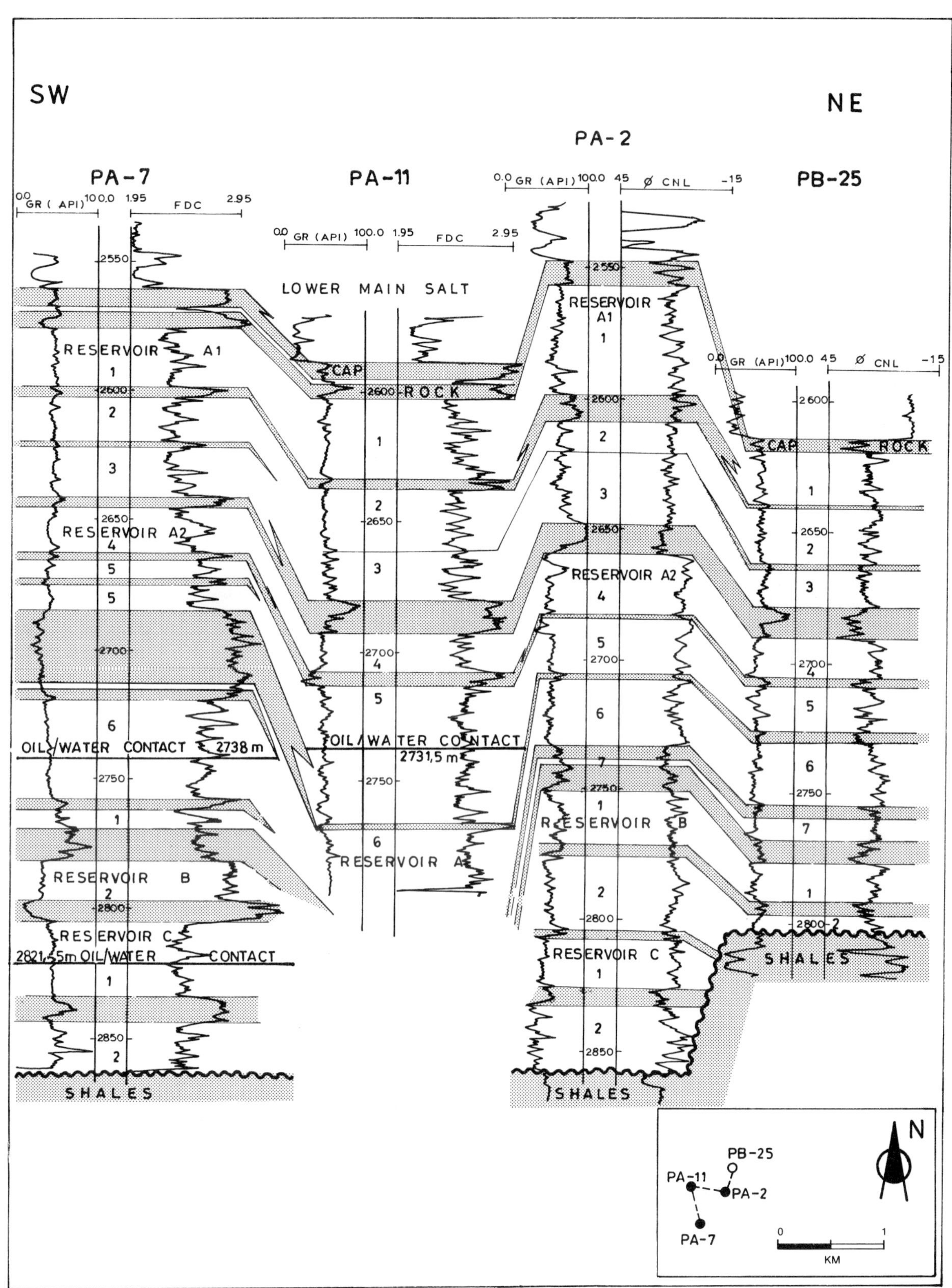

Figure 15. Detailed log correlations of sandstone layers of four wells in the Prinos field. Note the multiplicity of depositional units comprising the main reservoir intervals. Seven units separated by shale layers (shaded intervals) are correlated in the A zone and two each in the B and C zones. Correlation of units over great distances is very questionable. Note downcutting into underlying shale by the channels. True vertical depths (TVD) in meters below R.T.

Relative permeability curves show that a reservoir with an initial oil saturation of 70% will produce until oil saturation is reduced to 25%. Below this value the oil will not move (Figure 16). This establishes the effective limit of water flooding. Original water saturation for A, B, and C reservoirs has been calculated at 20%, 30%, and 28%, respectively.

Fluid flow is controlled by permeability, and permeability, in turn, is controlled by facies development. Channel deposits provide the highest flow rates and most persistent flow, followed by the suprafan lobe deposits. Conversely, interchannel and classical turbidite layers of local or broad extent either contribute less to the flow or act as barriers.

Faults cutting through the field act mostly as barriers to the distribution of injected water and result in separate compartments within the reservoir (Figure 6). For example, very limited communication exists between wells PB25 and PA2 in reservoir A in the northern part of the field.

Initially a uniform well spacing pattern was intended to cover the whole field. However, after the deposition model of the field was understood and channels and fan lobes were delineated, an attempt was then made to position development wells in these features. It was not always possible to penetrate permeable horizons of the three reservoirs with the same well. Often completely shaly horizons below permeable ones were encountered.

A U-shaped pattern of injection wells was drilled at the periphery of the field. The wells were located above the oil-water contact because of the reduced porosities and permeabilities below. It was hoped that this method would enhance the injected water transmissibility through the field. However, the result has been that an enormous amount of oil has been bypassed.

Secondary recovery started at the same time as initial production. The efficiency was estimated at 26 to 27%. Enhanced recovery in the later years, employing the gas lift method, is expected to increase the recovery efficiency in excess of 30%.

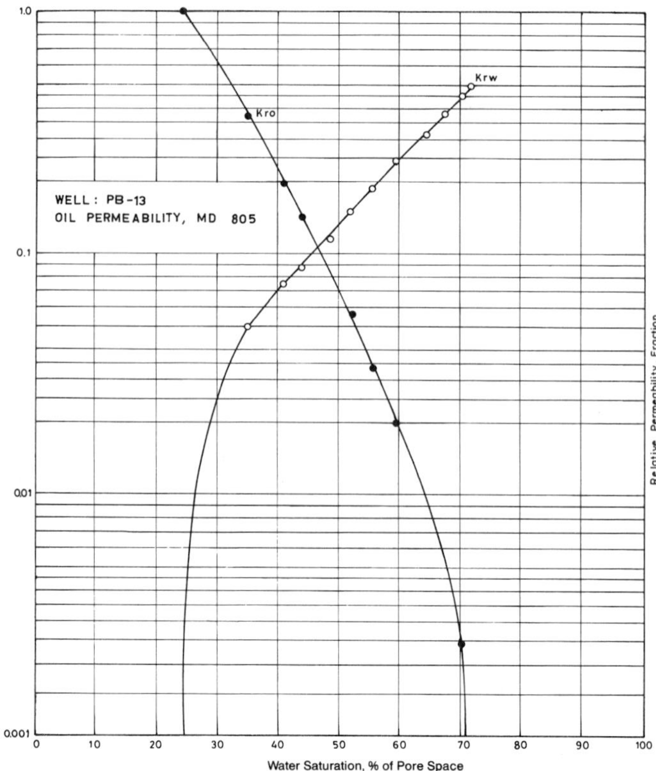

Figure 16. Relative permeability plots of a core from well PB-13 (2543.8 m depth, A reservoir). Initial oil saturation is about 70%. The diagram shows that oil flow from this reservoir will cease when oil saturation is reduced to about 25% (Core Laboratories UK Ltd., 1980.)

Source Rock and Oil Characteristics

Marine shales of upper Miocene and Messinian age were deposited under highly reducing conditions interrupted by hypersaline episodes. These beds are considered to be the source of the oil in the basin. Coals below the subevaporitic formation may have a good potential for gas generation. The oil source rock is characterized by a waxy sapropelic oil-prone kerogen with terrestrially derived organic matter frequently dominant (type III and I).

TOC values vary from 1.20 to 3.90% and HI from 100 to 500 mg HC/g to TOC. The potential yield ranges up to 13.500 ppm HC/g. Maturity measurements show vitrinite reflectance up to 1.10 and spore color index up to 9 (on a scale of 1 to 10) in the deepest parts of the basin penetrated (3500 m). The central portion of the basin, which is approximately 5500 m deep, has not been reached by drilling.

Prinos oil is aromatic-asphaltic type with 30° API at 18°C, with a sulfur content of 4% (Figure 17). The alkane to aromatic ratio is 1.5 and the asphaltene content is 4.4%. The initial GOR is 135 standard m^3/stock tank m^3 (767 ft^3/STB), with the gas consisting of 60% H_2S, 4% CO_2, and the rest wet gases from methane to butane.

According to the oil maturation plots, oil reached the threshold of maturity during latest Miocene time in the deepest part of the basin. Generation continued throughout the Pliocene and Recent (Figure 18). The present top of the maturation window starts at depths of 2500 m, corresponding to a vitrinite reflectance value of 0.55 and a spore color index of 5 (Figure 19). The top of the gas window is at a depth of 4000 m.

The main migration paths apparently were along high- to low-angle gravity faults that cross the basin.

EXPLORATION CONCEPTS

The Prinos field was found by drilling the crest of an anticlinal closure defined by marine seismic

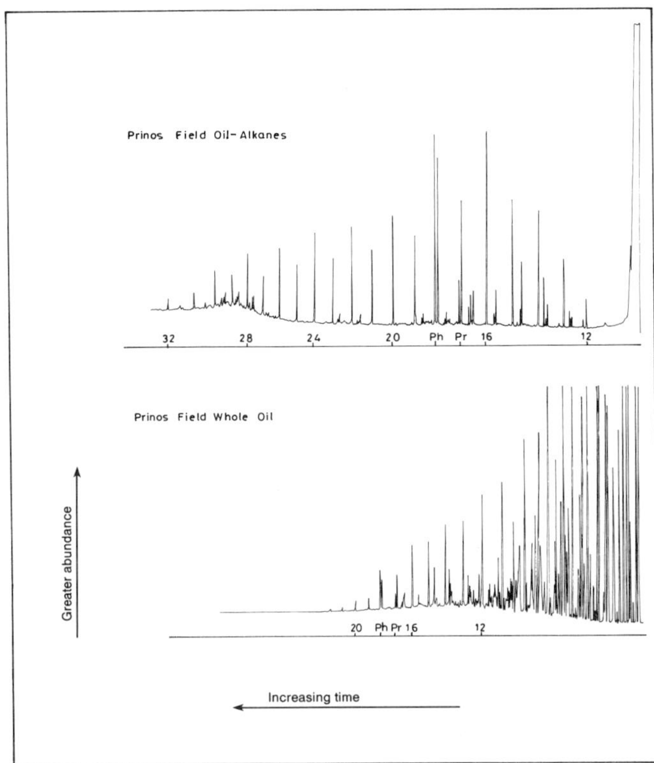

Figure 17. Gas chromatograms of Prinos field produced oil. (From Robertson Research International, Ltd. 1983.)

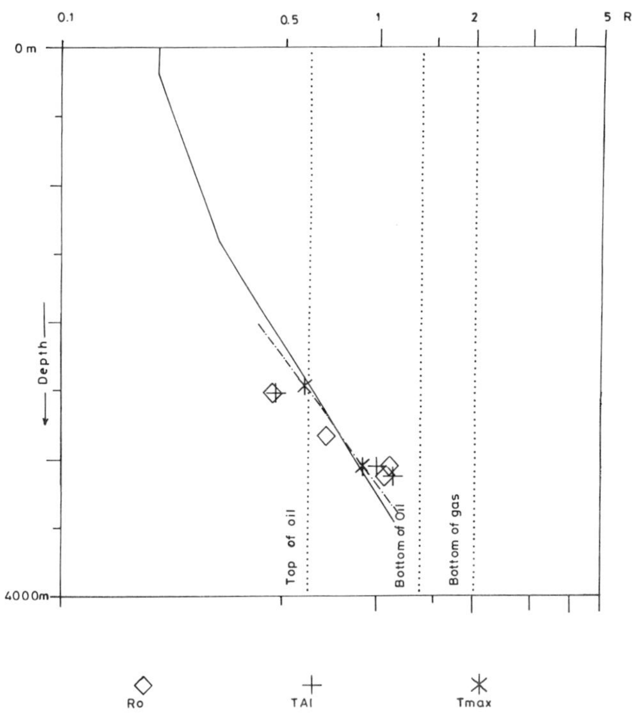

Figure 19. Plots of measured values of R_0, TAI, and T_{max} against depth from wells in the Prinos basin. The top of the oil generation window is at about 2500 m. (By Public Petroleum Corporation of Greece - EKY SA, 1988.)

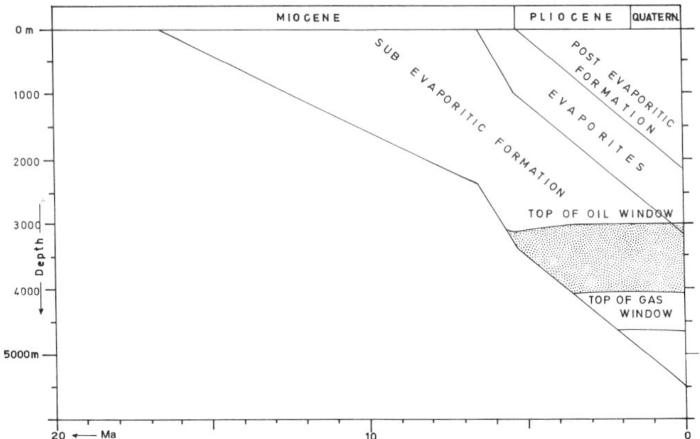

Figure 18. Oil maturation plot for the deepest part of the Prinos basin. A constant geothermal gradient of 4°C/100 m is assumed. Oil maturation began near the end of the Miocene and continued to Recent. (Diagram by Public Petroleum Corporation of Greece - EKY SA, 1988.)

surveys and mapped at the top of the Messinian evaporitic sequence. The main reservoirs were encountered beneath the evaporitic sequence. No wells in the field have been drilled to basement.

Other anticlinal prospects drilled in the basin have failed because of the absence of reservoir beds equivalent to those at Prinos. The reservoir in the small South Kavala gas field in the southern part of the basin is located between two evaporite beds within the evaporitic formation (Figure 4). Beds equivalent to the main Prinos pay zones are missing there.

The main lesson of the Prinos discovery for exploration in the North Aegean Sea is that small basins like Prinos, only 40 by 20 km in size, and of such young age can generate large amounts of hydrocarbons. Intensive graben formation isolated the different basins and led to the extensive evaporite deposition. This, in turn, resulted in favorable conditions for the generation and preservation of hydrocarbons. Contemporaneous growth faulting formed a large number of roll-over anticlines, which, together with stratigraphic pinch-outs, formed the trapping mechanisms.

Submarine fans may be very good reservoirs; however, maximum thickness and better petrophysical parameters are found in channel deposits. When exploring basins of Neogene age, the first objective should be a search for evaporite intervals and the location of associated structures. Unfortunately, the seismic reflection quality of events below evaporites may be of poor quality despite modern processing methods. High resolution seismic surveys have given slightly better results, but the velocity contrast between sandstone and shale is too low for distinguishing porous layers. Good knowledge of the depositional model of a submarine fan is a prerequisite for designing an efficient recovery system, which will require a large number of injection wells.

ACKNOWLEDGMENTS

The authors thank Dr. P. Sivenas, Managing Director of the Public Petroleum Corporation of Greece - EKY SA, for making available the field data and permitting the publication of this study.

REFERENCES CITED

Bouma, A. H., 1962, Sedimentology of some flysch deposits: Amsterdam, Elsevier, 168 p.

Kellogg, H. F., 1988, Seismic-structural evaluation, Prinos basin, northeastern Greece: Internal report, Public Petroleum Corporation of Greece, Athens.

Robertson Research International Ltd., 1981, Geological studies of five wells from the Prinos field, north Aegean Sea: Report No. 4531p/c, Volume I, internal report, North Aegean Petroleum Company, Athens.

Walker, R. G., and E. Mutti, 1973, Turbidite facies and facies associations, *in* G. V. Middleton and A. H. Bouma, eds., Turbidites and deep water sedimentation: SEPM Pacific Section Short Course, Anaheim, California, p. 119-157.

SUGGESTED READING

Hsu, K. J., 1972, Origin of saline giants: a critical review after the discovery of the Mediterranean evaporite: Earth-Science Review, v. 8, p. 371-396.

Krumbein, W. G., and L. L. Schloss, 1963, Stratigraphy and sedimentation, evaporite association: London, p. 581-585.

Lalechos, N., and E. Savoyat, 1979, La sedimentation Neogene dans le Fosse Nord Egean: VI Colloquium on the Geology of the Aegean Region, Vol. II, p. 591-603.

Pollak, W. H., 1979, Structural and lithological development of the Prinos-Kavala basin, Sea of Thrace, Greece: Annex Geologique Pays Hellenic tome hors serie, fash II, p. 1003-1011, VIInth International Congress on Mediterranean Neogene, Athens.

Proedrou, P., 1979, The evaporite formation in the Nestos-Prinos graben in the Northern Aegean Sea: Annex Geologique Pays Hellenic tome hors serie, fash II, P. 1013-1020, Athens.

Proedrou, P., 1986, New age determination of the Prinos basin: Bull. Geol. Soc. Greece, v. 20, n. 2, p. 141-147.

Speel, L., 1982, Development of the oil and gas fields Prinos and South Kavala: Oil Gas European Magazin, International Edition of Erdoel-Ergas Zeitschrift, v. 8, n. 2, p. 6-12.

Walker, R. G., 1978, Deep water sandstone facies and ancient submarine fans: models for exploration for stratigraphic traps: American Association of Petroleum Geologists Bulletin, v. 62, n. 6, p. 932-966.

Appendix 1. Field Description

Field name .... *Prinos field*

Ultimate recoverable reserves .... *90 million bbl*

Field location:

- **Country** .... *Greece*
- **Basin/Province** .... *Prinos basin/North Aegean Sea*

Field discovery:

- **Year first pay discovered** .... *Upper Miocene A, B, and C sandstone units 1974*

Discovery well name and general location:

- **First pay** .... *Prinos 1, zones A, B, and C, 6 km west of Thassos Island*

Discovery well operator .... *Oceanic Exploration Co. of Greece*

IP:

- **A sandstone** .... *25,000 bbl/day*
- **B sandstone** .... *810 bbl/day*
- **C sandstone** .... *1640 bbl/day*

All other zones with shows of oil and gas in the field:

Age	Formation	Type of Show
Miocene	*Sub-evaporitic series*	*Oil and gas*

Geologic concept leading to discovery and method or methods used to delineate prospect

The existence of grabens of young age in the Aegean Sea was well known. Seismic data indicated basins with enormous sediment accumulation and well-formed anticlines with fault-dependent closures.

Structure:

Province/basin type .... *North Aegean Sea, taphrogenic basin*

Tectonic history

The breakdown of the crystalline Rhodope massif led to the formation of the Prinos-Nestos graben during early Miocene. Northeast-southwest and northwest-southeast faults delineate the graben. A land barrier isolated the basin in uppermost Miocene resulting in the evaporite deposition.

Regional structure

Northwest-southeast synsedimentary fault, cutting the basin fill laterally, with a roll-over of the sediments along the downthrown side formed the dome-like Prinos anticline. Another structure genetically similar to Prinos exists close to the north of it.

Trap:

Trap type(s)

There is only one fault trap with three pays bound against this northwest-southeast-trending feature.

Basin stratigraphy (major stratigraphic intervals from surface to deepest penetration in field):

Chronostratigraphy	Formation	Depth to Top in m
Pliocene (Quaternary)	*Post-evaporitic*	*0–1810*
Messinian (upper Miocene)	*Main-evaporitic*	*1810–2470*
Upper Miocene	*Sub-evaporitic*	*2470–3030*

Reservoir characteristics:

- **Number of reservoirs** *3*
- **Formations** .... *Sub-evaporitic formation, the Prinos envelope*
- **Ages** .... *Upper Miocene*
- **Depths to tops of reservoirs** .... *A, 2488.5 m; B, 2642 m; C, 2743 m*

PRINOS

Gross thickness (top to bottom of producing interval) ... A, 158 m; B, 50 m; C, 56 m (including the shales)

Net thickness—total thickness of producing zones

Average ... A, 97 m; B, 22 m; C, 23 m

Maximum ... A, 140 m; B, 45 m; C, 50 m

Lithology

Turbiditic-fill channel sandstones, generally poorly sorted with intercalated dark claystones

Porosity type ... *Primary depositional porosity*

Average porosity ... *A, 18.1%; B, 14.5%; C, 15%*

Average permeability ... *A, 115 md; B, 17 md; C, 25 md*

Seals:

Upper

Formation, fault, or other feature ... *Formation*

Lithology ... *Evaporites*

Lateral

Formation, fault, or other feature ... *Fault*

Lithology ... *Claystone*

Source:

Formation and age ... *Sub-evaporitic, upper Miocene*

Lithology ... *Claystone*

Average total organic carbon (TOC) ... *2.0%*

Maximum TOC ... *3.90%*

Kerogen type (I, II, or III) ... *I and III (waxy sapropelic oil-prone)*

Vitrinite reflectance (maturation) ... $R_o = 0.40\text{–}1.10$

Time of hydrocarbon expulsion ... *End of upper Miocene to Pliocene*

Present depth to top of source ... *2500 m*

Thickness ... *Unknown (assumed few hundred meters)*

Potential yield ... *Unknown*

Appendix 2. Production Data

Field name ... *Prinos field*

Field size:

Proved acres ... *1000 (4 km²)*

Number of wells all years ... *24 (16 production + 8 injection)*

Current number of wells ... *22*

Well spacing ... *250–400 m*

Ultimate recoverable ... *90 million bbl oil*

Cumulative production (to end 1987) ... *52 million bbl oil and 1.018 (10^9) Sm^3 gas*

Annual production ... *8.4 million bbl*

Present decline rate ... *11% (1987)*

Initial decline rate ... *NA*

Overall decline rate ... *NA*

Annual water production ... *2.5–3 million bbl*

In place, total reserves ... *256 million bbl*

Primary and secondary recovery ... *51.6665 million bbl*

Enhanced recovery ... *333.5 thousand bbl (1987)*

Cumulative water production ... *9.619 million bbl*

Drilling and casing practices:

Amount of surface casing set .... *at 300–350 m*

Casing program

24-in. conductor pipes, 18⅝-in., 13⅜-in., 9⅝-in., 7-in. liner, 5-in. liner (in special cases)

Drilling mud .... *Sea water bentonite-CMC and supersaturated salt polymer*

Bit program .... *22-in., 17½-in., 12¼-in., 8½-in. (6-in. in special cases)*

High pressure zones .... *In shale lenses at the bottom of the evaporites*

Completion practices:

Interval(s) perforated .... *4 spf-scallop gun*

Well treatment .... *HCl + additives (included SCA 130 H_2S Scanvenger) xylene, diesel, diverter*

Formation evaluation:

Logging suites .... *LDT-CNL-GR, ISF-SLS-GR-SP, DLL-MSFL-GR, HDT, CBL-VDL, PLT, TDT*

Testing practices .... *Normal DST or RFT*

Mud logging techniques ... *Continuous gas recording with chromatographic analysis and H_2S detection*

Oil characteristics:

API gravity .... *30° at 60°F (15.6° C)*

Base .... *NA*

Initial GOR .... *135 standard m^3/stock tank m^3 (767 ft^3/STB)*

Sulfur, wt% *4%*

Viscosity, SUS .... *1.32 Engler at 50°F (10°C)*

Pour point .... *–23°C*

Field characteristics:

Average elevation .... *Sea depth 31 m*

Initial pressure .... *A, 5735 psi (403.6 kg/cm^2); B, 5840 psi (411.0 kg/cm^2); C, 6000 psi (422.2 kg/cm^2)*

Present pressure ..... *A, 4500 psi (316.7 kg/cm^2); B, 4000 psi (281.5 kg/cm^2); C, 4000 psi (281.5 kg/cm^2)*

Temperature .... *110°C (max 117°C)*

Geothermal gradient .... *3.5°C/100 m*

Drive .... *Water and gas injection*

Oil column thickness .... *190 m*

Oil-water contact .... *A, 2711 m; B, 2751 m; C, 2791 m*

Connate water .... *NA*

Water salinity, TDS .... *200,000 ppm*

Resistivity of water .... *0.015 ohm*

Bulk volume water (%) .... *A, 20%; B, 30%; C, 28%*

Transportation method and market for oil and gas:

Submarine pipe to desulfurization plant.

Rosario Field—Venezuela
Maracaibo Basin, Zulia State

ANGEL MOLINA
Maraven S. A.
Caracas, Venezuela

FIELD CLASSIFICATION

BASIN: Maracaibo
BASIN TYPE: Foredeep
RESERVOIR ROCK TYPE: Sandstone/Limestone
RESERVOIR ENVIRONMENT OF DEPOSITION: Fluvial/Deltaic
RESERVOIR AGE: Eocene/Cretaceous
PETROLEUM TYPE: Oil
TRAP TYPE: Faulted Anticline

LOCATION

The Rosario field is located in the western part of the Maracaibo basin of Venezuela, Zulia State, Catatumbo District (Figure 1), 40 km (25 mi) east of the Rio de Oro field and about 75 km (45 mi) north of the Tarra field. Physiographically, the area includes a vast plain bounded on the west by the Perija Cordillera foothills and on the east by Lake Maracaibo.

The estimated ultimate recovery for the field is about 50 million barrels: 32 million bbl of oil and 107 bcf of gas in Cretaceous reservoirs and about 18 million bbl of oil in the Eocene reservoir.

HISTORY

Exploration activities in the area began in 1913 and resulted in the discovery of the Tarra and Rio de Oro fields, located on anticlinal structures. These structures showed good surface expression and were associated with many oil seeps. The Rosario area to the north drew attention owing to the existence of a similar structure, but with a more subdued surface expression. Seismic surveys were carried out to confirm and define the structure, and two wildcat wells drilled on structure proved the existence of entrapped hydrocarbons.

The field was discovered in 1954 by the Compañia Shell de Venezuela, Ltd., through the drilling of the CR-3 well (Figure 2), located at about 175 km (110 mi) southwest of the city of Maracaibo. This well penetrated only 566 ft (173 m) into the Upper Cretaceous limestones, not penetrating the middle Cretaceous Aguardiente Formation. No oil or gas was obtained from the Cretaceous interval drilled, so drill-stem tests were made of the Eocene sandstones of the Mirador Formation; hydrocarbons had been indicated by ditch samples and high log resistivities. The Mirador interval was completed between 7690 and 7700 ft (2345 and 2348 m), with an initial production of 800 bbl/day of 28° API oil.

Later, in 1957, well CR-4 was drilled, which penetrated most of the Aguardiente Formation. Based on log resistivities and ditch samples that suggested the existence of hydrocarbons, these strata were tested and later completed between 14,143 and 14,238 ft (4314 and 4342 m) in very silty limestones and marls. Initial production was 4400 bbl/day of 40° API oil. Gas/oil ratio (GOR) was 1,740 ft^3/bbl.

The field has been developed through the drilling of eight additional wells. Of these, wells CR-6, CR-7, and CR-8 reached the Eocene reservoir, and CR-5 and CR-9 reached the Cretaceous. A seismic survey carried out in 1985 helped to obtain a more detailed definition of the field structure, and in 1986, wells CR-10, CR-11, and CR-12 were drilled and also completed in the Cretaceous Aguardiente Formation reservoir.

DISCOVERY METHOD

During the first phase of surface geology, only the surface features of a smooth anticlinal structure were observed. The second phase consisted of a seismic survey of the area to verify and define the subsurface structure and to establish the depths of prospective zones that might be correlatives of reservoirs found previously in the Tarra and Rio de Oro fields. The third phase included actual drilling and completion.

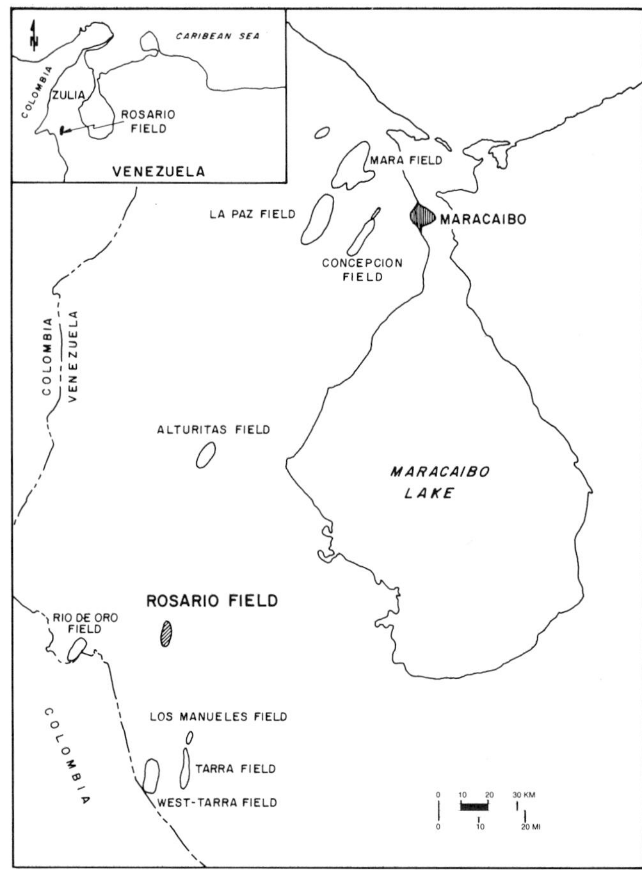

Figure 1. Index map of the western part of the Maracaibo basin showing the location of the Rosario and other fields.

Figure 2. Structural map of the top of the Lower Cretaceous Aguardiente Formation. Depths are below sea level and contour interval is 100 ft. Seismic section 51-U-18 is shown by Figure 3. A–A′ is the location of the cross section shown by Figure 4.

STRUCTURE

The Rosario field is located within the Maracaibo basin, which is a large triangular block bounded by the Oca, Santa Marta, and Bocono transcurrent fault systems. Movement along the Oca fault system was blocked by the Santa Marta system, which resulted in compressional forces from the east that formed a series of almost parallel, arcuate anticlines throughout the area. These fault systems are related to right-lateral strike-slip movement along the Caribbean–South American plate boundary.

During the Cretaceous, the Maracaibo basin was a vast platform belonging to the South American plate (González de Juana et al., 1980; Stobie, 1982). During this time, the deposition of 4000 ft (1220 m) of a transgressive marine sequence took place. The sediments consist of carbonates, sandstones, and shales. Later, during the Paleocene, compressional tectonic movements began to take place, and a regressive sequence was deposited. During the upper Eocene, strong tectonic movements associated with the Oca fault system resulted in a series of north-northeast-trending sinistral transcurrent faults that affected the whole basin. These faults were further developed owing to compressional forces during the Miocene and Pliocene, resulting in anticlinal folds along the fault trends. Sedimentary environments were mainly fluvial and coastal (Stobie, 1982).

The Rosario field was formed by one such fold. The structure map (Figure 2), seismic section 51-U-18 (Figure 3), and structural cross section A–A′ (Figure 4) show its major features. It consists of a north-northeast elongated dome. The eastern side has been affected by a reverse fault, with a throw of about 2000 ft (610 m) at the Cretaceous level (Albarracin, 1982). This fault extends up into the Eocene and Oligocene.

STRATIGRAPHY

The stratigraphy of the Rosario field is illustrated by Figure 5, which exhibits a type log of the field showing both of the existing reservoirs and the main lithology of each unit. Following is a brief description of the formations.

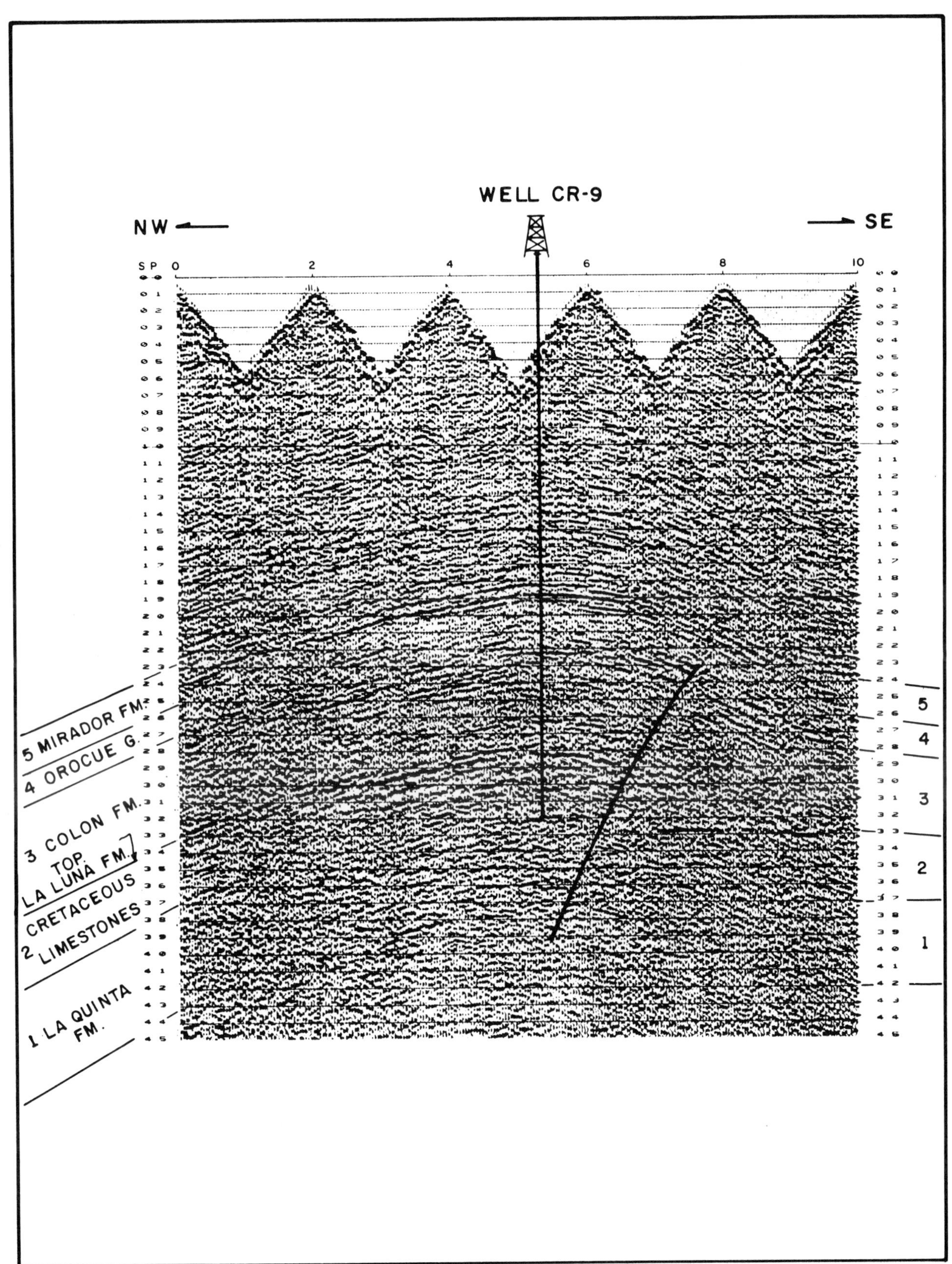

Figure 3. Seismic cross section 51-U-18 showing interpreted reflectors and the interpreted reverse fault on the east side of the structure. The location of the seismic line and of well CR-9 near the anticlinal axis is shown by Figure 2. Length of section shown is 15.5 km (9.6 mi).

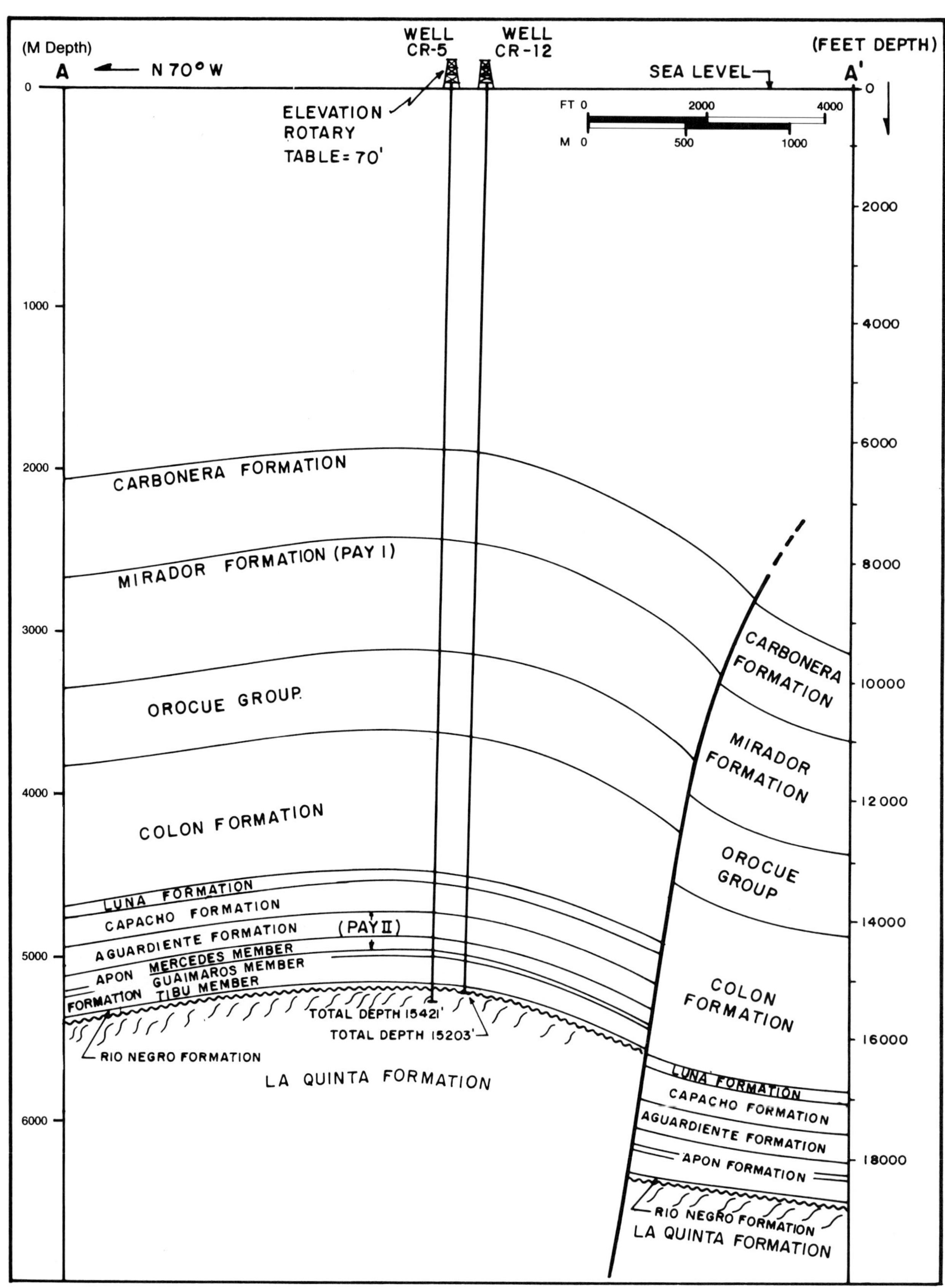

Figure 4. Structural cross section across the center of the structure perpendicular to the fold axis. Location is shown on Figure 2.

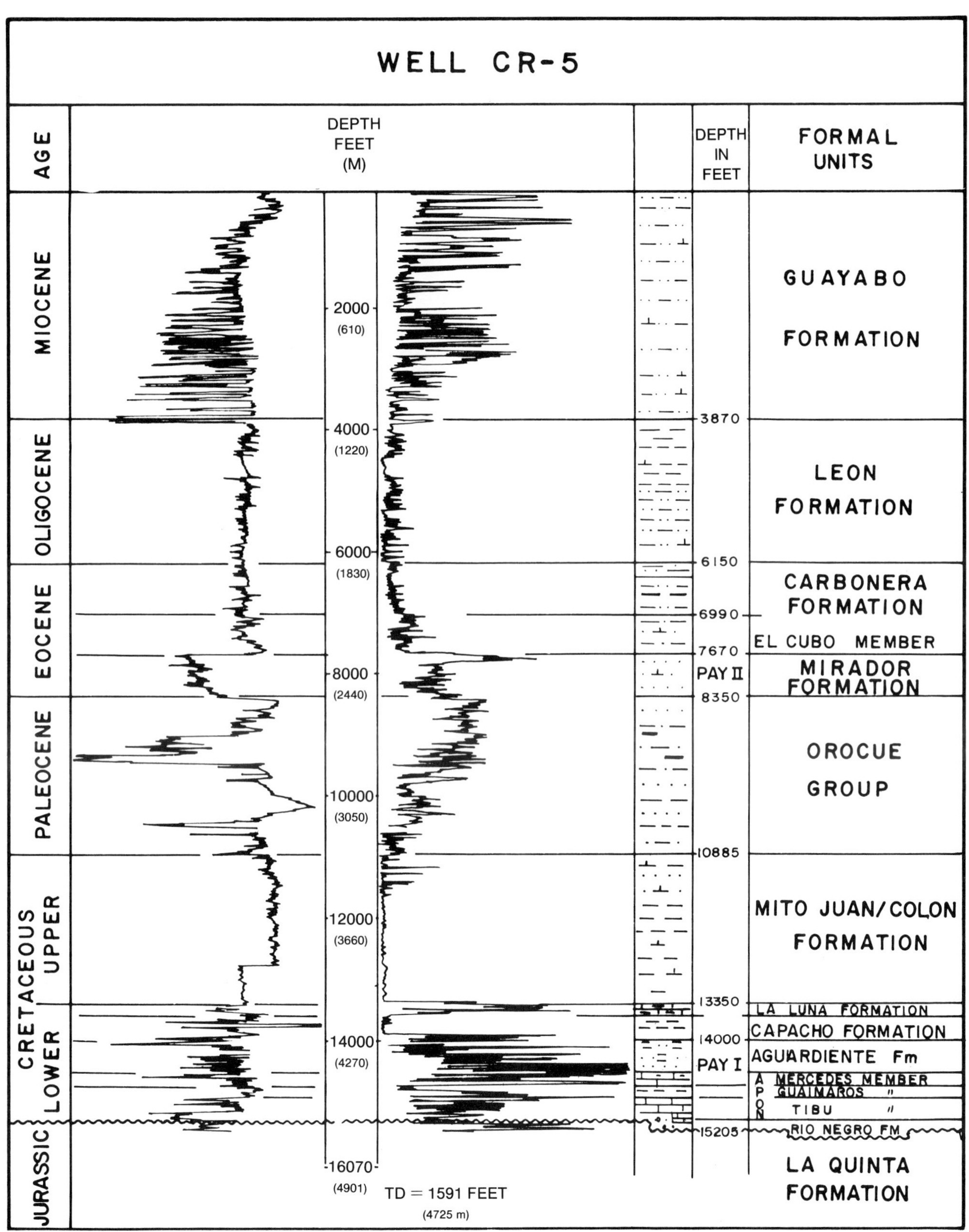

Figure 5. Stratigraphy and type log. Unquantified SP (left) and resistivity (right) curves are characteristic log-response signatures only.

Jurassic

The Jurassic La Quinta Formation consists of siltstones and sandstones with colors ranging from red to green. Some show a high percentage of pyroclasts.

Cretaceous

The Cretaceous Rio Negro Formation lies unconformably over the La Quinta Formation and consists of medium and coarse-grained sandstones, partly feldspathic, and is cemented with silica.

The lower part of the Apon Formation (Tibu Member) is characterized by recrystallized, dense, and impermeable limestones (originally calcareous muds), containing minor quantities of shales throughout the sequence. Its middle part (Guaimaros Member) consists of a sequence of shales, sometimes dolomitized, and is 100 ft (30.5 m) thick. Its upper part (Las Mercedes Member) consists mainly of limestones that are very sandy toward the top; they are interbedded with shales and sandstones.

The Aguardiente Formation consists mainly of fine- to coarse-grained sandstones cemented with silica and calcite. The lower part of this formation shows intercalations of sandy limestones and shales and consists only of shales near the top. This formation and the Mercedes Member make up the Cretaceous reservoir.

The Capacho Formation is mostly a shaly one. The lower, middle, and upper parts of this sequence consist of limestone strata. La Grita, Saboruco, and Guayacán members are part of this formation.

The La Luna Formation is considered to be the most important oil-generating source rock in the Maracaibo basin owing to its high organic matter content and high degree of maturity (Blaser, 1979; Lew, 1984; Talukdar et al., 1986). It consists of limestones interbedded with minor quantities of calcareous shales.

The Colon-Mito Juan Formation consists of a sequence of dark gray-colored shales that become progressively more silty upwards.

Paleocene

The Paleocene Orocue Group includes three formations: Catatumbo Formation, which consists of a sequence of silty shales intercalated with minor quantities of silty sandstones, siltstones, and coals; Barco Formation, consisting mainly of sandstones interbedded with siltstones, shales, and occasionally coal beds; and Los Cuervos Formation, mainly a silty one, its lower part consisting of siltstones with a high detrital coal content and its upper part more sandy.

Eocene

The base of the Eocene Mirador Formation consists mainly of medium-grained sandstones intercalated with siltstones and shales; its middle part consists of shales; and its upper section consists of sandstones interbedded with shales. It contains the second reservoir of the Rosario field.

Eocene–Oligocene

The Eocene–Oligocene Carbonera Formation consists of a lower interval of sandstones, silty sandstones, and minor quantities of siltstones and shales and an upper sequence of alternating siltstones, shales, and sandstones. It contains a thick regional stratum of coal 100 ft (30.5 m) above its base (Stobie, 1982).

Oligocene

The lower part of the Oligocene Leon Formation consists of siltstones and shales and its upper part is mainly shaly.

Miocene

The Miocene Guayabo Formation consists of siltstones with a conglomeratic sequence toward the top.

TRAPS

The field is associated with an elongated dome, faulted on its eastern side (Figure 4). Closure exists at the level of both the Eocene and Cretaceous reservoirs. The Cretaceous Aguardiente Formation has a closure of 1500 ft. The vertical seal of this reservoir consists of shales of the Capacho Formation. The vertical seal of the Eocene Mirador Formation can be attributed to the low permeability of the shaley sediments of the overlying unconformable Carbonera Formation. Laterally, the sandstones of the Mirador Formation lose their permeability in a gradual change to shaly facies.

Reservoirs

The Cretaceous reservoir is made up of the Mercedes Member of the Apon Formation and of the Aguardiente Formation. It is located at a depth of 14,000 ft (4270 m) and consists of a regressive sequence ranging from neritic marine sediments, with limestones interbedded with minor quantities of shales and calcareous sandstones, to shallow-water marine sediments related to a deltaic environment. The latter show a higher content of calcareous sandstones, and in the Aguardiente Formation less limestone and shale. The average porosity measured for this formation was 4.5%. Values of up to 20% were obtained toward the top, but with low permeability, between 0.1 and 4 md. The high porosity values corresponding to the calcareous sandstones may be the result of the dissolution of the carbonate cement during vadose diagenesis (Murray, 1960; Mazzulo, 1981). The higher permeability may be related to a better fluid transmission between the pores owing to fractures toward the crest of the structure and close to the main reverse fault on the east.

The second reservoir is located in the Mirador Formation at a depth of 7600 ft (2318 m). It consists of medium-grained sandstones intercalated with siltstones and shales, showing a higher content of the latter toward the top where these sediments truncate against the Carbonera Formation. It was deposited in a fluvial braided stream environment. It shows intergranular porosity, with an average value of 19%. Average permeability is about 150 md. Total pay-zone thickness is 1300 ft (396 m) and net thickness of oil-bearing sandstones averages 765 ft (233 m).

The Cretaceous reservoir contains 40° API oil. Initial GOR was 1800 ft^3/bbl. The Eocene reservoir contains 30° API oil. It is a naphthene-base crude with a sulfur content of 2%. Oil viscosity is 0.94 cp. Average daily production in 1989 was 2800 bbl per day and the cumulative production to this date has been 15 million bbl.

Faults

The Rosario anticline is bounded on the east by the main reverse fault, which is easily discernible on seismic sections (Figure 3).

Source and Migration

The most likely source rock all over the Maracaibo basin is the La Luna Formation. It has an average TOC value of 3.8%, the maximum being 9.6%. Kerogen is type II. Potential yield is about 10 kg of hydrocarbons per ton of rock. The La Luna Formation began to generate oil during the Eocene, and the process is still taking place today in different areas of the basin (Blaser, 1979; Talukdar et al., 1986). The initial migration path for hydrocarbons is considered to have begun in the east, where greater depths were reached sooner by the La Luna Formation than to the westward. Furthermore, overall dip during migration time has been toward the east.

EXPLORATION CONCEPTS

On the basis of vitrinite reflectance and subsidence-related data pertaining to the La Luna Formation (Blaser, 1979; Waples, 1980; Lew, 1984; Talukdar et al., 1986), it has been determined that hydrocarbon generation in and migration from this formation began during the Eocene. The structures currently found in the Maracaibo basin were formed during this time and were further developed during the Miocene and Pliocene. Hydrocarbons generated by the La Luna Formation during the Eocene thus migrated into the Aguardiente Formation stratigraphic traps and traps formed as a combination of these with recently formed structures. An important vertical seal already existed at this time, consisting of 2000 ft (610 m) of shales from the Colón Formation. Hydrocarbons generated after the Eocene period could have reached the spillpoint of the Rosario Formation trap. Part of these hydrocarbons may have migrated vertically from the Aguardiente Formation to the Mirador Formation along the main reverse fault.

The Rosario Field is similar to the Rio de Oro, Tarra, Mara, La Paz, and La Concepción fields located on the western coast of Maracaibo Lake. They consist of faulted anticlinal structures with similar reservoirs dating from the Cretaceous and Eocene.

Our current knowledge of this and other fields in the area suggest that exploration techniques used (detection, definition, and drilling of large structures) were suitable. As the outcrops of the Rosario structure are essentially eroded away and obscured by meander patterns of the Catatumbo river deltaic plain and by some farming activities, exploration methods such as SLAR and aerial photos do not show the anticline. However, airborne gravity *does* indicate the structural high of the Rosario field as well as other nearby structures. Hydrocarbon accumulations may, however, exist not within structures but within stratigraphic traps or fracture zones. In these areas, it has been observed that hydrocarbon production pertaining to Cretaceous carbonated rocks is associated with major fracture zones, which are thus more permeable. Drilling for these fractured reservoirs is considered to be highly risky, since the fracture zones within the carbonate rocks are difficult to locate, even though they generally tend to be adjacent to the main faults.

ACKNOWLEDGMENTS

The author thanks the management of Maraven S.A. and Petroleos de Venezuela for permission to publish this paper. I also thank Gordon Young and Donald Goddard for their suggestions in writing the paper and to the reviewers for their comments and suggestions.

REFERENCES CITED

Albarracin, J., 1982, Evaluación exploratoria en los distritos Perijá y Coón del Estado zulia durante 1977–1982: Unpublished report.

Blaser, R., 1979, Source rock and hydrocarbon generation in the Maracaibo Basin, Western Venezuela: Unpublished report.

González de Juana, C., J. Iturralde de Arozena, and X. Picard, 1980, Geologia de Venezuela y sus cuencias petroliferas: Ediciones Foninves, Caracas, 1031P, v. I and II.

Lew, M., 1984, Origen de los gases en el Eoceno del area de Maracaibo: Unpublished report.

Mazzulo, S., 1981, Facies and burial diagenesis of a carbonate reservoir: Chapman deep (Atoka) field, Delaware Basin,

Texas: American Association of Petroleum Geologists Bulletin, v. 65, p. 850.
Murray, R., 1960, Origin of porosity in carbonate rocks: Journal of Sedimentary Petrology, v. 30, p. 59.
Stobie, R., 1982, Geological evaluation and summary of Cretaceous exploration within the area of the Tarra anticline system. South Colon, Zulia State: Unpublished report.
Talukdar, S., O. Gallango, and M. Chin-A-Lien, 1986, Generation and migration of hydrocarbons in the Maracaibo Basin, Venezuela: an integrated basin study, *in* Advances in organic geochemistry: Organic Geochemistry, v. 10, p. 261-279.

Appendix 1. Field Description

Field name .... *Rosario field*

Ultimate recoverable reserves .... *32 million bbl oil, 107 bcf gas, Cretaceous; 18 million bbl oil, Eocene*

Field location:

- **Country** .... *Venezuela*
- **State** .... *Zulia*
- **Basin/Province** .... *Maracaibo basin*

Field discovery:

- **Year first pay discovered** .... *Mirador Formation (lower to middle Eocene) 1954*
- **Year second pay discovered** .... *Aguardiente Formation 1957*

Discovery well name and general location:

- **First pay** .... *CR-3, 175 km (109 mi) SW of Maracaibo City*
- **Second pay** .... *CR-4, 177 km (110 mi) SW of Maracaibo City*

Discovery well operator .... *Cia Shell de Venezuela, Ltd.*

- **Second pay** .... *Cia Shell de Venezuela, Ltd.*

IP:

- **First pay** .... *800 bbl/day (Eocene)*
- **Second pay** .... *4400 bbl/day (Cretaceous)*

All other zones with shows of oil and gas in the field:

Age	Formation	Type of Show
Jurassic	*La Quinta*	*Oil 42°API*
Upper Cretaceous	*Colon*	*Gas*

Geologic concept leading to discovery and method or methods used to delineate prospect

Surface geology showed the presence of an anticline, later confirmed by seismic data. Well CR-3 discovered the Eocene reservoir.

Structure:

Province/basin type .... *Continental wrench*

Tectonic history

Rosario field is located within the unstable zone bounded by the Oca, Perija, and Bocono transcurrent fault systems in northwestern Venezuela. These systems are related to right-lateral strike-slip movement along the Caribbean-South American plate boundary. Major period of tectonic activity occurred in the Tertiary, including the late Miocene Andean uplift.

Regional structure

Consists of a north-northeast-trending wrench fault system associated with several thrust faulted anticlines.

Local structure

A low-relief, north-northeast-trending elongated anticline bordered on its eastern flank by a reverse fault parallel to the main axis.

Trap:

Trap type(s)

The Rosario field has (1) anticlinal trap combined with reverse fault; (2) truncated Eocene sandstones against an unconformity

Basin stratigraphy (major stratigraphic intervals from surface to deepest penetration in field):

Chronostratigraphy	Formation	Depth to Top in ft (m)
Eocene-Oligocene	*Carbonera Formation*	*6100 (1860)*
Eocene	*Mirador Formation*	*7600 (2318)*
Paleocene	*Orocue Group*	*8400 (2562)*
Cretaceous	*Colon Formation*	*10,000 (3050)*
	La Luna Formation	*13,345 (4070)*
	Capacho Formation	*13,500 (4118)*
	Aguardiente Formation	*14,000 (4270)*
	Apon Formation	*14,550 (4438)*
	Rio Negro Formation	*15,200 (4636)*

ROSARIO

Reservoir characteristics:

Number of reservoirs *3*

Formations .... *Mirador, Aguardiente, Apon (Mercedes and Tibu Members)*

Ages .... *Eocene; Cretaceous*

Depths to tops of reservoirs .... *Mirador, 7600 ft (2318 m); Aguardiente, 14,000 ft (4270 m); Apon, 14,800 ft (4514 m)*

Gross thickness (top to bottom of producing interval) .... *Mirador, 750 ft (229 m); Aguardiente, 480 ft (146 m); Apon, 60 ft (18 m)*

Net thickness—total thickness of producing zones

Average .... *765 ft (233 m)*

Maximum .... *900 ft (274 m)*

Lithology

Mirador Formation: light gray, fine- to medium-grained sandstone intercalated with siltstones and shales

Aguardiente Formation: gray sandstones, fine to coarse grained, subrounded to subangular, variably but well cemented with silica and calcite; moderately well sorted and relatively tight, interbedded limestone and shales

Apon (Tibu): dark gray to gray-blue, hard, detrital, limestone; these are shelly in nature, fine to coarsely crystalline; subsidiary shale interbeds occur, scattered throughout

Porosity type .... *Mirador Formation: primary porosity; Aguardiente and Apon formations: intercrystalline and intergranular porosity and fracture porosity*

Average porosity .... *Eocene, 19%; Cretaceous, 4.5%*

Average permeability .... *Eocene, 150 md; Cretaceous, 1 md*

Seals:

Upper

Formation, fault, or other feature .... *Carbonera Formation*

Lithology .... *Shales interbedded with siltstones and coals*

Lateral

Formation, fault, or other feature .... *Fault and pinch-out*

Lithology .... *Shales*

Source:

Formation and age .... *La Luna, Cretaceous*

Lithology .... *Carbonaceous limestone, calcareous shale and chert*

Average total organic carbon (TOC) *3.8%*

Maximum TOC *9.6%*

Kerogen type (I, II, or III) .... *II*

Vitrinite reflectance (maturation) .... $R_o = 1.2$

Time of hydrocarbon expulsion .... *Eocene-Miocene*

Present depth to top of source .... *13,400 ft (4087 m)*

Thickness .... *150 ft (45.8 m)*

Potential yield .... *10 kg hydrocarbon per ton rock*

Appendix 2. Production Data

Field name ... *Rosario field*

Field size:

- **Proved acres** ... *2062 ac (835.1 ha)*
- **Number of wells all years** ... *12*
- **Current number of wells** ... *8 (producing)*
- **Well spacing** ... *1400 m (4600 ft)*
- **Ultimate recoverable** ... *Cretaceous, 32 million bbl oil and 107 bcf gas; Eocene, 18 million bbl oil*
- **Cumulative production** ... *Cretaceous and Eocene, 11.2 million bbl; Cretaceous only, 12.4 bcf gas*
- **Annual production** ... *Cretaceous, 0.3 million bbl oil and 0.3 bcf gas*
- **Present decline rate** ... *Eocene, 2% annually; Cretaceous, 12% annually*
    - **Initial decline rate** ... *Eocene, 3% annually; Cretaceous, 15% annually*
- **Annual water production** ... *200,000 bbl*
- **In place, total reserves** ... *Cretaceous, 107 million bbl oil and 194.6 bcf gas*
- **In place, per acre-foot** ... *NA*
- **Primary recovery** ... *Cretaceous, 32 million bbl oil and 107 bcf gas*
- **Secondary recovery** ... *NA*
- **Enhanced recovery** ... *NA*
- **Cumulative water production** ... *NA*

Drilling and casing practices:

- **Amount of surface casing set** ... *20-in. to 550 ft (168 m)*
- **Casing program**
  20-in. to 550 ft (168 m); 13⅜-in. to 4400 ft (1342 m); 9⅝-in. to 10,500 ft (4728 m); 7-in. to 13,800 ft (4209 m); liner 4½-in., 12,800 to 15,500 ft (3904–4728 m)
- **Drilling mud** ... *Lignosulfonate and invermul*
- **Bit program** ... *NA*
- **High pressure zones** ... *3400 ft (1040 m) of shales and siltstone of the Colon*

Completion practices:

- **Interval(s) perforated** ... *Eocene and Cretaceous, 7400 ft (2257 m), 7900–8000 ft (2410–2440 m)*
- **Well treatment** ... *Acid frac and wash*

Formation evaluation:

- **Logging suites** ... *Electrical, density, neutron (FDC/CNL) sonic and deviation logs*
- **Testing practices** ... *Standard production test, open hole, and some drill-stem test*
- **Mud logging techniques** ... *Cutting analyses; gas chromatograph*

Oil characteristics:

- **Type** ... *NA*
- **API gravity** ... *Eocene, 30° API; Cretaceous, 40° API*
- **Base** ... *Naphthenic*
- **Initial GOR** ... *1800 ft^3 gas/bbl oil*
- **Sulfur, wt%** ... *2%*
- **Viscosity, SUS** ... *0.94 cps*
- **Pour point** ... *<2°F (−16.7°C)*
- **Gas-oil distillate** ... *NA*

Field characteristics:

- **Average elevation** ... *80 ft (24.4 m)*
- **Initial pressure** ... *Eocene, 4000 psi (27,580 kPa); Cretaceous, 10,500 psi (73,298 kPa)*
- **Present pressure** ... *Cretaceous, 5800 psi (39,990 kPa)*
- **Pressure gradient** ... *0.74 psi/ft (16.7 kPa/m)*

Temperature .... *300°F (148.9°C)*
Geothermal gradient .... *0.03°F/ft (0.055°C/m)*
Drive .... *Water*
Oil column thickness .... *1500 ft (458 m)*
Oil-water contact .... *Unknown*
Connate water .... *NA*
Water salinity, TDS .... *NA*
Resistivity of water .... *0.16 ohm/m²m*
Bulk volume water (%) .... *NA*

Transportation method and market for oil and gas:
From the field via pipeline, gas is sent to La Fria for industrial use; oil is sent via pipeline to oil dock on Lake Maracaibo for export